SELECTED WORKS
OF
YAKOV BORISOVICH ZELDOVICH

Ya. B. Zeldovich, ca. 1946

SELECTED WORKS
OF
YAKOV BORISOVICH ZELDOVICH

Editor of English Edition
J. P. Ostriker

Coeditors of English Edition
G. I. Barenblatt
R. A. Sunyaev

Technical Supervisor of English Edition
E. Jackson

Translators of English Edition
A. Granik
E. Jackson

VOLUME I
CHEMICAL PHYSICS AND HYDRODYNAMICS

PRINCETON UNIVERSITY PRESS
PRINCETON, NEW JERSEY

QD
450.2
.Z44213
1992

Library of Congress Cataloging–in–Publication Data

Zel´dovich, ÎA. B. (Îakov Borisovich)
  [Khimicheskaîa fizika i gidrodinamika. English]
  Chemical physics and hydrodynamics / editor of English edition
J. P. Ostriker; coeditors of English edition G. I. Barenblatt, R. A.
Sunyaev; technical supervisor of English edition E. Jackson;
translators of English edition A. Granik, E. Jackson.
      p. cm. -- (Selected works of Yakov Borisovich Zeldovich;
v. 1)
    Translation of: Khimicheskaîa fizika i gidrodinamika.
    Includes bibliographical references.
    ISBN 0-691-08594-3
    1. Chemistry, Physical and theoretical. 2. Hydrodynamics.
I. Ostriker, J. P. II. Barenblatt, G. I. III. Sunyaev, R. A.
IV. Title. V. Series: Zel´dovich, ÎA. B. (Îakov Borisovich).
Selections. English. 1992 ; v. 1.
QC3.Z44213   1992 vol. 1
[QD450.2]
530 s--dc20
[539]   91-14813

24066246

Contents

Part One

I. Adsorption and Catalysis

II. Hydrodynamics. Magnetohydrodynamics. Heat Transfer. Self-Similarity

III. Phase Transitions. Molecular Physics

IV. Theory of Shock Waves

Part Two
Theory of Combustion and Detonation

I. Ignition and Thermal Explosion

II. Flame Propagation

III. Combustion of Powders. Oxidation of Nitrogen

IV. Detonation

Preface to the English Edition of the Selected Works of Ya. B. Zeldovich

There has been no physical scientist in the second half of the twentieth century whose work shows the scope and depth of the late Yakov Borisovich Zeldovich. Born in Minsk in 1914, he was the author of over 20 books and over 500 scientific articles on subjects ranging from chemical catalysis to large-scale cosmic structure, with major contributions to the theory of combustion and hydrodynamics of explosive phenomena. His passing in Moscow in December of 1987 was mourned by scientists everywhere. To quote Professor John Bahcall of the Institute for Advanced Study: "We were enriched in Princeton as in the rest of the world by his insightful mastery of physical phenomena on all scales. All of us were his students, even those of us who never met him." In his range and productivity, Zeldovich was the modern equivalent of the English physicist Raleigh (1842-1919) whose name is associated with phenomena ranging from optics to engineering.

The breadth of Zeldovich's genius (characterized as "probably unique" by the Soviet physicist Andrei Sakharov) was alternately intimidating or enthralling to other scientists. A letter sent to Zeldovich by the Cambridge physicist Steven Hawking, after a first meeting in Moscow, compares Zeldovich to a famous school of pre-war mathematicians who wrote under a single fictitious pseudonym: "Now I know that you are a real person, and not a group of scientists like Bourbaki."

No selection from an opus of such scope can capture its full range and vigor. While basing ourselves primarily on the Russian edition, published by the Soviet Academy of Sciences in 1984–1985, we were delayed repeatedly as important and hitherto untranslated (but frequently cited) papers were brought to our attention as clearly warranting inclusion. Zeldovich played a major role in re-editing the Russian edition before translation and in choosing additional material for the present work. All told, this edition is approximately 15% longer than the Russian edition and the second volume contains one largely new section: *Physics, Personalia*, including impressions of Einstein and Landau, and ending with *An Autobiographical Afterword*.

Because he wrote in Russian during a period when relations between that culture and the western world were at an historically low ebb, international recognition for Zeldovich's achievements were slower to arrive than merited. Within the Soviet Union his accomplishments were very well recognized, in part, due to his major contributions to secret wartime work. As the world's leading expert on combustion and detonation, he had naturally been drafted early into the effort for national survival. He had written entirely prescient papers in 1939 and 1940 (included in this volume) on the theoretical possibility of chain reactions among certain isotopes of uranium. The physicist Andrei Sakharov wrote that "from the very beginning of Soviet work on the atomic (and later the thermonuclear) problem, Zeldovich was at the very epicenter of events. His role there was completely exceptional." Zeldovich was intensely proud of his contributions to the wartime Soviet scientific effort and was the most decorated Soviet scientist. His awards include the Lenin Prize, four State Prizes, and three Gold Stars.

As a corollary to internal recognition, of course, Zeldovich's scientific work was burdened by the enormous handicaps of isolation, secrecy and bureaucracy in a closed society, made more extreme for him by restrictions due to defense work. He was not permitted to attend conferences outside of the Soviet Block until August 1982 at age 68, when he delivered an invited discourse "Remarks on the Structure of the Universe" to the International Astronomical Union in Patras, Greece. When asked then by this Editor when he was last out of the Soviet Union, he answered without hesitation "sixty eight years ago," i.e., only in a prior life. Previous to that meeting, his access to preprints, normal correspondence, all of the human interchange of normal scientific life, were severely circumscribed with contacts increasing as he moved out of defense work. Then, as international relations improved, international acclaim followed. Elected in 1979 as a foreign associate of the U. S. National Academy of Sciences, he had already been made a member of the Royal Society of London and other national scientific academies. Despite having turned relatively late in his scientific career to astrophysics, his accumulated achievements in that area, rewarded with the Robertson Prize of the U. S. National Academy of Sciences for advances in cosmology, put him among the world's leading theoretical astrophysicists.

The science is of course more interesting than the honors it wins for the scientist. Let me note just two items from astrophysics, my own specialty, where Zeldovich showed extraordinary vision and imagination. He argued shortly after their discovery that quasars were accreting black holes, and that the universe was likely to have a large-scale porous structure, anticipating in both cases the standard paradigms for interpreting these cosmic phenomena. In addition, he was among the first to realize that the early universe could be used as our laboratory for very high energy physics, leaving as fossils strange particles and cosmic microwave background fluctuations.

If the matter is more important than the recognition, it was also true, for Zeldovich in particular, that the manner was as significant as the matter. He always proceeded by a direct intuitive *physical* approach to problems. Even in areas where his ultimate accomplishment was a mathematical formulation adopted by others such as the "Zeldovich number" in combustion theory or the "Zeldovich spectrum" and the "Zeldovich approximation" to linear perturbations in cosmology, the reasoning and approach are initially and ultimately physical and intuitive. His view was that if you cannot explain an idea to a bright high school student, then you do not understand it. He backed up this conviction, and his interest in the education of young scientists, with the book *Higher Mathematics for Beginners*, which presented in a clear and intuitive way the elementary mathematical tools needed for modern science. Here again Zeldovich was in good company; from Einstein to Feynman, the greatest physicists have felt that they could and *should* make clear to anyone who cared to listen, the excitement of modern science.

The value of Zeldovich's papers, unlike those of most scientists, has outlived the novelty of his results. But, inevitably, one must question the logic of republishing scientific papers. Is not all valid scientific work included in and superseded by later work. Of course there is a value in collecting, for the record, in one place the major works of a truly great scientist. The fact that we include with each paper, commentaries (often revised from the Soviet edition) by the author on the significance of these papers will further enhance their value to historians and philosophers of science. But Zeldovich was almost above all else the teacher, the founder of a school of today's world famous scientists and author of widely read texts at all levels. He had strong views on *how* science should be done and how it should be taught. To him, the "how" of the scientific method, of his own scientific method, was central; it was what he most wanted to communicate in making his work available to a broader audience.

We are happy to be able to provide a complete enlarged edition of the works of this great scientist for the English-speaking world. We would like to thank the Academy of Sciences of the USSR for permission to utilize (a) *Izbrannye Trudy: I. Khimicheskaĭa Fizika i Gidrodinamika* and (b) *II. Chastitsy, Iadra, Vselennaĭa*, but especially offer our thanks to Professors G. I. Barenblatt and R. A. Sunyaev for their dedication and expertise in closely reading (Volumes One and Two, respectively) the entire manuscript in its English edition.

J. P. Ostriker
19 January 1990

Я.Б.ЗЕЛЬДОВИЧ

Избранные труды

ХИМИЧЕСКАЯ ФИЗИКА И ГИДРОДИНАМИКА

Под редакцией
академика Ю. Б. ХАРИТОНА

ИЗДАТЕЛЬСТВО «НАУКА»
МОСКВА
1984

The Scientific and Creative Career of Yakov Borisovich Zeldovich

1. Introduction

The Editors feel that this volume of selected works by Ya.B. Zeldovich, *Chemical Physics and Hydrodynamics*, as well as a second volume, *Particles, Nuclei, and the Universe*, will be of great scientific interest to the reader, be he a chemist, physicist, or astronomer.

The present edition, undertaken in connection with the seventieth birthday of Academician Yakov Borisovich Zeldovich, is an original exposition of the now-classical results of a scientist, results which remain actual even today. Some were obtained by Ya.B.[1] as early as the thirties, before World War II, while many papers were published between 1947–1986. Some have already become bibliographical rarities, as, for example, "The Theory of Combustion and Detonation of Gases," published in 1944 and included in this volume. In spite of the fact that a significant part of the results presented here is widely known, having been incorporated into textbooks and monographs, we can heartily recommend these collected works to scientists, graduate and undergraduate students; the material offered here has not become obsolete not only because we have included some of Ya.B.'s most recent papers. In this introductory article and in the commentaries to individual papers the present state of each problem is briefly described and references are made to the most important later works.

However, most important for the reader, especially the young reader, is the style of presentation in these original articles by Ya.B. By studying these papers one may not only obtain specific information about a particular question, but also learn about the formulation of scientific problems and the creative scientific process. In the original papers the "scaffolding" used by the author during the "construction of the building" is still there. One may see just which questions were left unclear or solved erroneously by his predecessors, and what obstacles, psychological and scientific, the author, Ya.B., had to overcome to produce new results. One can even see his emotions as he wrote a paper or a monograph. In textbooks and modern monographs, as a rule, there predominates a natural tendency to "straighten up" the actual course of development of science, to show rather how, in retrospect, it should have proceeded. The original publications, and in particular the papers by Ya.B. collected here, show how science is actually built.

In terms of their subject matter the two volumes are basically independent, and so one may envision a physical chemist interested only in this first part or an astronomer interested only in *Particles, Nuclei, and the Universe*. A certain connection between the two volumes does, however, exist. The present volume

[1] In this book we will use the abbreviation Ya.B., as he is often referred to in conversation by his friends and colleagues.

contains a biographical outline describing all of Ya.B.'s creative life, including general characterizations of papers which, in terms of content, are related to the next volume. The second book also includes several articles by Ya.B. on general subjects and his scientific autobiography.

This introductory article has turned out much longer than those in many other anniversary editions. Even if this is not justified, then at least it is explained by the unusual breadth of Ya.B.'s scientific interests. Indeed, where else can one find a scientist who, having begun with problems of adsorption, catalysis, and chemical kinetics, moves on to ignition and combustion of gases and powders and detonation, then to nuclear chain reactions, critical mass, and nuclear power, then immerses himself in questions of elementary particle physics, and, finally, plunges into cosmological structure and the first moments of the creation of the universe ... and in each area brings about a fundamental breakthrough and leaves several generations of thankful students and colleagues to continue to develop the results from the heights already achieved ... participates in experiments to verify predictions he has made and often waits many years for someone to discover the predicted phenomenon ... and takes an active part in the technological realization of possibilities revealed in his theoretical calculations, participating in the direction of a large collective of theorists and experimentalists ... and generalizes the results of his work in numerous monographs.

The Editors of this book felt obliged to do everything possible to help the reader, especially the younger reader, to extract a maximum of useful information from Ya.B.'s works. We felt we could best do this by describing in detail the creative course of Ya.B.'s life and work.

Ya.B. was born on March 8, 1914, at the home of his grandfather in Minsk, but after mid-1914 the family lived in Petrograd. Ya.B. began his scientific career when he was seventeen years old. After graduating from high school at fifteen, he entered a school for laboratory assistants associated with the Leningrad Institute, "Mekhanobr."[†] The students were paid stipends, and to repay them they had to spend at least three years on assigned jobs. It so happened that the students were taken on an excursion to the Physico-Technical Institute. Ya.B. liked it there, and he in turn was liked.

The head of one of the Institute departments, S. Z. Roginskiĭ (later a Corresponding Member[‡] of the USSR Academy of Sciences), became interested in the serious questions asked by the youth. They agreed that Ya.B. would work at the Institute for a couple of hours after classes. The date that this work began is recorded as March 15, 1931. The head of the chemical physics section, N. N. Semenov, assigned Ya.B. a report to present at a seminar: "Transformations of Ortho–Para-Hydrogen in the works of the German physical chemist Bonhöffer." Ya.B. showed both enthusiasm and understanding. First impressions had been confirmed and soon the influ-

[†]An institute of mechanical, as opposed to thermal, processing of solids.

[‡]Members of the USSR Academy of Sciences (AS USSR) fall into two categories: full members (academicians) and corresponding members.

ential director of the Physico-Technical Institute, Academician A. F. Ioffe, signed a letter to "Mekhanobr" requesting that they "release Ya.B. to science." The transfer (*via* the unemployment office, which still existed in the USSR!) was effected, and on May 15, 1931, Ya.B. began his work in the chemical physics section which, by that time, had become an independent institution, the AS USSR Institute of Chemical Physics.

Ya.B. was assigned to S. Z. Roginskiĭ's department of heterogeneous reactions. However, he was immediately spotted by the theorists of the chemical physics section (the theoretical department in this section was headed by AS USSR Corresponding Member Ya. I. Frenkel). They helped Ya.B. to grasp the foundations of theoretical physics. They taught him constantly, persistently, and patiently, and thus Ya.B. received a higher education without entering a university or institute. He had little choice: studies at a university would have required him to abandon his own work for 4–5 years. Such a loss of time did not seem reasonable, and this was understood by all his colleagues at the Institute. The absence of a university degree did not bother anyone.

Ya.B. himself considers his teachers to be L. D. Landau, S. Z. Roginskiĭ, N. N. Semenov, and Yu. B. Khariton. Nonetheless, Ya.B. is an exceptionally original person in science. He does not repeat anyone, yet at the same time, from his first steps in science, he was accustomed to learning from everyone with whom life brought him together. And even now, famous and renowned as few are, he continues to learn from anyone who can offer anything new.

The variety of problems in chemistry, classical physics, nuclear physics, and cosmology to which Ya.B. has made fundamental contributions is so great that it is hard to believe that it has all been done by a single person.

After meeting and talking to Ya.B., the famous English physicist and mathematician S. Hawking wrote to him: "Now I know that you are a real person and not a group of scientists like the Bourbaki."

In a first approximation the 70 years of Ya.B.'s life can be divided into four periods: 1914–1930 — childhood and high school; 1931–1947 — the Institute of Chemical Physics, the study of adsorption, catalysis, phase transitions, hydrodynamics, and, most importantly, the theory of combustion and detonation with application to rocket ballistics, and the first papers on nuclear chain reactions; 1947–1963 — work on the creation of a new technology, nuclear physics and elementary particle physics, and a textbook, *Higher Mathematics for Beginners*; 1964–1987 — astronomy, including application of the general theory of relativity, and cosmology.

We move now to a description of basic directions in the scientific work of Academician Ya.B. Zeldovich.

2. Adsorption and Catalysis[2]

Ya.B. began his work at the Institute of Chemical Physics as a laboratory assistant in the catalysis laboratory. He stayed in this laboratory for four years. While there he began postgraduate work and defended his thesis for the Candidate of Sciences (Ph.D) degree, "Problems of Adsorption," in February, 1936.

Together with his colleagues in the laboratory he studied problems of the crystallization of nitroglycerin, oxidation of hydrogen on a platinum catalyst, and oxidation of CO on manganese catalysts. The first and the third problems were of practical significance.

The most fundamental results, presented in three articles (**1, 2, 3**),[†] are related to the *theory* of adsorption and catalysis. However, the simultaneous experimental work left its imprint on these theoretical papers by aiding in the selection of the most relevant problems and providing concreteness in the approach.

The interconnection between theory and experiment was realized in this research in the most direct fashion. In subsequent work as well his own experimental work helped Ya.B. to better understand a problem and influenced the entire style of his work as a theorist.

Let us turn to the theory of the adsorption isotherm, i.e., the dependence of the quantity q of a substance adsorbed on the surface of an adsorbent on the gas pressure p or on the concentration of the substance being adsorbed (the "adsorbate") in a solution. This dependence is studied at constant temperature, whence the name "isotherm."

At the beginning of the thirties there was a deep, puzzling contradiction between numerous experiments and theory. The simple theory developed by Irving Langmuir (USA) led to the expression $q = ap/(p + b)$, which at low pressure gives $q = ap/b = kp$, where $k = a/b$. Thus, for $p \ll b$, $q \ll a$, a linear dependence (direct proportionality) between q and p at low pressure and sparse surface coverage is a general feature of the theory. Meanwhile, in a very large number of cases and over a broad range of variation of p and q, a fractional dependence was observed—the so-called Freundlich isotherm,

$$q = Cp^{1/n},$$

where $n > 1$, i.e., the exponent is less than one, which corresponds to stronger binding of the first portions of the adsorbate.

Ya.B. advanced an essentially new idea: the adsorbent surface is composed of sectors with differing levels of adsorption activity, and each surface may be characterized by a specific distribution function of the sectors according to their activity level. He found an effective method of determining

---

[2]Professor O. V. Krylov took part in the writing of this section.

[†]Here and below, numbers in boldface type refer to papers in the present volume.

this distribution function using the known (measured) adsorption isotherm. It may be noted here that I. Langmuir had earlier considered adsorption on a surface with two or three different types of sectors, but he did not provide an explanation of the Freundlich isotherm.

In essence, I. Langmuir had in mind an aggregate of monocrystals with several kinds of faces. The linear law of adsorption on each face (at low pressures) could yield nothing more than a linear law for the whole aggregate.

Ya.B. considered that the most important adsorbents—porous coal, silica gel, and the powdered manganese dioxide which he had studied experimentally—are amorphous substances, i.e., they do not have clearly articulated crystalline structure. Only thus is it possible to obtain a large developed surface—the most important feature of an adsorbent. In this case, it is natural to consider all the possible values of adsorption activity and a smooth distribution function of surface sectors according to their level of activity.

How may we check this conception? The adsorption activity reflects the adsorption energy, i.e., the bonding energy between an adsorbed atom or molecule and the corresponding surface sector. More precisely, this activity depends on the ratio of the bonding energy to the energy of thermal motion, i.e., to the temperature at which the adsorption is measured. Knowing the activity distribution function of the sectors, we may proceed to a function of distribution according to the heat of adsorption. But this means that, from the adsorption isotherm measured at one temperature, we may calculate the isotherm for any other temperature.

When applied to the Freundlich isotherm, $q = Cp^{1/n}$, Ya.B.'s conception leads to an unusually simple conclusion: the exponent is directly proportional to the absolute temperature,

$$1/n = T/T_1,$$

where T_1 has a particular value for each adsorbent–adsorbate pair.

Experiments (including those of Ya.B. himself) provided excellent confirmation of this conclusion.

Let us note that this dependence of n or $1/n$ on T is valid only for $T < T_1$, when $1/n < 1$, so that the isotherm has the form of an upwardly convex curve.

Ya.B. elegantly showed that an integral of a certain form yields different kinds of expressions (exponential with a fractional power or linear) for different parameter values.

Ya.B.'s paper (1) remains a model of precise macroscopic analysis of a complicated situation.

However, Ya.B. goes further. Analysis of the Freundlich isotherm leads to exponential dependence of the number of sectors with a given value, α, of the adsorption heat on Q, namely, $\rho(Q) = \text{const} \cdot \exp(-\alpha Q)$. A natural explanation of this dependence lies in the fact that the adsorption heat, Q, is linearly related to the energy, U, necessary to create an area of a given type, $Q = \beta U$. In this case, the distribution law found for Q is transformed into an analogous law, $\rho(U) = \text{const} \cdot \exp(-\alpha \beta U)$. Now it acquires the clear physical meaning of the Boltzmann distribution.

The quantity $(\alpha \beta)^{-1}$ acquires the meaning of effective temperature, which characterizes the energy properties of the surface. It is assumed that, when the catalyst is prepared, fluctuations occur in the distribution of atoms on the surface which in fact lead to its inhomogeneity; later the fluctuations "freeze," and the experimenter works with a well-defined, inhomogeneous, but immutable surface.

Significantly later, foreign scientists reached a similar conclusion regarding the Freundlich isotherm. In the USSR, a theory of adsorption on an inhomogeneous surface was developed independently by M. I. Temkin of the Karpov Physico-Chemical Institute in connection with electrochemical research by Academician A. N. Frumkin. M. I. Temkin's work on a logarithmic isotherm was cited in [74] and published in [75]. The theory of adsorption and catalysis on an inhomogeneous surface was especially extensively developed by S. Z. Roginskiĭ.

In his next paper (**2**), Ya.B. posed a theoretical question which did not arise from experiments known at that time.

If fluctuations and restructuring of the surface are possible during the preparation of an adsorbent, then might not this process occur *during* the process of adsorption and desorption as well? There arises a new concept of a "homogeneous-in-the-mean" surface: at each moment this surface consists of different sectors, but over the course of the fluctuations every sector passes through all possible states. For each sector the probability of a given state (or the amount of time that the sector is in a given state) is the same as for any other sector.

This is a new, more complicated, yet also more realistic and fruitful concept of statistical homogeneity, as opposed to the simple "static" homogeneity of a fixed monocrystal face. Together with statistical homogeneity, the concepts of fluctuations and of the characteristic time of these fluctuations are introduced.

The next point of fundamental significance is that the very presence of the adsorbate on the surface must change the equilibrium distribution of the surface sectors according to their structure and energy of formation U, since the adsorption energy Q partially compensates for the energy used to form a sector of a given type.

Now the kinetic concept of surface fluctuations and relaxation enters into the theory of adsorption. A nontrivial dependence is predicted between the adsorbed quantity and the rate of pressure or concentration change of the adsorbent. Also predicted is the phenomenon of hysteresis during adsorption and desorption over a time comparable to the relaxation time of the surface.

It is shown that for very rapid relaxation the surface once again behaves as a homogeneous surface. This explains why Langmuir's isotherm is rigorously applicable, not only to an ideal homogeneous fixed monocrystal face, but to a rapidly fluctuating fluid surface as well. The significance of this work goes beyond the boundaries of adsorption problems.

Adsorption of a reagent is a necessary stage of catalysis, and of enzymatic catalysis in particular, i.e., the assembly of protein or transport RNA on a DNA matrix.

Enzyme or DNA mutations may be likened to the fluctuations of a surface. In principle, therefore, in developing upon this work of Ya.B., one might consider a natural, physical explanation of directed mutations and of the influence of the environment (an adsorbate entering a reaction) on the mutations.

Ya.B. expressed these thoughts in a presentation in 1960 on the occasion of S. Z. Roginskiĭ's sixtieth birthday, but he never published them, realizing that they were too raw and that their scientific realization would require a deep understanding of biological processes.

Since then some progress has been made in this direction (completely independently of Ya.B.!): in violation of the "central dogma" of molecular genetics, influence of transport RNA on DNA was discovered. It is possible that the ideas of how reacting substances influence enzymes will prove significant in the study of the prebiological stage of evolution.

All these considerations are given here simply to emphasize the depth and potential significance of what Ya.B. actually did in this paper.

We should note that this article by Ya.B. apparently remained little noticed in its time. In any case, we are unaware of any reference to it in the works of other authors. This is explained by the fact that its ideas were far ahead of their time. Only in recent years, due to the wide application of physical methods in studies of adsorption and catalysis, have the changes in the surface (and volume) structure of a solid body during adsorption and catalysis been proved. Critical phenomena have been discovered, phenomena of hysteresis and auto-oscillation related to the slowness of restructuring processes in a solid body compared to processes on its surface. Relaxation times of processes in adsorbents and catalysts and comparison with chemical process times on a surface were considered in papers by O. V. Krylov in 1981 and 1982 [1] (see references at end of Introduction).

Finally, paper **3** is devoted to the specifics of chemical reactions on porous catalysts. Catalysts in the form of porous granules with highly developed inner surfaces are precisely those most frequently used in industry.

Ya.B. analyzes the problem of penetration of reacting substances into the granules. He shows that, together with the well-known extreme cases of reaction throughout the entire volume of the granule and reaction restricted to the granule's surface, there is an important intermediate region of parameters where the reaction occurs inside the granule, but only in a layer adjacent to its surface. The thickness of this layer depends in turn on the concentration of the reacting substances and on the temperature. Therefore the observed effective activation heat and order of the reaction are variable with respect to the actual characteristics of elementary events on the surface.

For the first time a graph of the logarithm of the reaction rate as a function of the inverse temperature is given in which the slope of the curve decreases by a factor of two as the temperature increases in the transition to the internal diffusion region, and falls almost to zero in the external diffusion region. This graph is now reproduced in almost all textbooks.

Let us mention Ya.B.'s contribution to the theory of multi-stage adsorption, reflected in his Candidate of Sciences thesis.

Let us also touch on some experimental papers by Ya.B. on adsorption and catalysis which are not included in the present book.

Together with S. Z. Roginskiĭ, he studied adsorption and catalytic oxidation of CO on MnO_2 [2]. He obtained the equation of the adsorption isotherm. He showed that the kinetics of activated adsorption of CO on MnO_2 obey an exponential law of decrease of the adsorption rate as a function of the quantity of adsorbed substance. This fact was interpreted as "self-poisoning" of the surface by the bicarbonate acid which forms. It was later explained by the inhomogeneity of the surface with a uniform distribution of adsorption centers according to their activation energies. Subsequent studies [3] showed the wide applicability of this equation. In the literature it was called the Roginskiĭ-Zeldovich equation, and sometimes the Elovich equation (after one of S. Z. Roginskiĭ's collaborators who was also studying problems of adsorption).

It was further found that strong chemosorption of CO on MnO_2 and oxidation of CO to CO_2 have a common intermediate stage—the so-called weak adsorption of CO:

$$MnO_2 + CO \rightarrow MnO_2(CO) \begin{array}{l} \longrightarrow MnO(CO_2) \\ \overset{O_2}{\longrightarrow} MnO_2 + CO_2. \end{array}$$

A system in the weak adsorption stage may possess an increased level of activity. The fruitfulness of this concept was recently confirmed in studies of the initial stages of adsorption (see, for example, [4, 5]). Particles in a presorption state can have an increased energy level; they may, for example, be in an oscillatory- or electron-excited state. Ya.B. also studied the catalytic oxidation of hydrogen [6].

3.   Hydrodynamics. Magnetohydrodynamics. Heat Transfer. Self-Similarity

Problems of how chemical reactions (including catalytic reactions and combustion) run under real conditions naturally led Ya.B. to hydrodynamics, heat transfer and problems of turbulence. Another significant factor was his contact with the prominent scientist D. A. Frank-Kamenetskiĭ, who joined the Institute of Chemical Physics in 1935 with broad interests in these areas and in similarity theory.

Ya.B.'s early works (**4, 5**) were strikingly deep and far ahead of their time. It is difficult to imagine that their author was a young man of 23. In the first of these papers (**4**) he established the extremum property of heat transfer in a fluid at rest, analogous to the well-known extremum property of dissipation for viscous fluid flows. But what is remarkable here is not only the classically simple result. In this paper a very important physical quantity appeared for the first time—"the rate of decay of temperature inhomogeneities"—which plays for the temperature field exactly the same role as does the rate of energy dissipation for the velocity field in a viscous fluid.

A few years later, A. N. Kolmogorov and A. M. Obukhov constructed a remarkable theory of local structure of developed turbulent flows which became one of the greatest achievements of classical physics in the twentieth century. Here the dissipation rate turned out to be the basic governing parameter of this theory in the inertial interval of turbulence scales, and this then yielded the famous Kolmogorov-Obukhov "two thirds law". In 1949, A. M. Obukhov developed a corresponding theory for temperature fields which proved to be of exceptional practical importance since temperature fluctuations determine the dispersion of light in the atmosphere. The governing parameter in this theory turned out to be the same decay rate of temperature inhomogeneities first introduced by Ya.B.

In the second paper (**5**) Ya.B. obtained the now classical similarity laws of ascending convective flow development (for both laminar and turbulent flows). These laws are now widely used by geophysicists in studies of atmospheric and oceanic convection. In 1953–1954 A. S. Monin and A. M. Obukhov independently used the same fruitful ideas for other conditions and obtained similarity laws for shear flow in a density stratified medium in a gravity field.

We note that later too—even in recent years—Ya.B. has turned his attention to this sphere of problems: we have here a paper by Ya.B. devoted to diffusion in a one-dimensional fluid flow (**6**), papers on hydrodynamics and thermal processes in shock waves which are reviewed in the next section, and on the hydrodynamics of the Universe in the next volume in the section devoted to astrophysics and cosmology. Here we consider only Ya.B.'s papers on magnetohydrodynamics or, more precisely, on the problem of magnetic field generation in the motion of a conducting fluid.

The paper "A Magnetic Field in the Two-Dimensional Motion of a Conducting Turbulent Fluid" (**7**) began a series of studies on magnetohydrodynamics. The reader can see that this is not a long paper, however, very important results are obtained in it. To correctly appraise it we must recall that at the time of its publication it seemed almost self-evident that turbulent motion of a sufficiently well-conducting fluid is unstable with respect to spontaneous generation of a magnetic field. In other words, any initial value of the magnetic field, however small, must grow with time. Indeed, a magnetic field is frozen into an ideally conducting fluid. Chaotic turbulent motion of such a fluid therefore "entangles" the magnetic field lines and stretches them, which leads, it would seem, to growth of the field. Ya.B. was the first to find that such reasoning is, in any case, not conclusive.

In the special case of two-dimensional flow considered in the paper, the initial field does in fact increase, but only by a finite factor which increases with the magnetic Reynolds number. Subsequently this increase is replaced by exponential decay. The proof is based on the fact that the vector potential equation for a magnetic field in the two-dimensional case has the character of the diffusion equation in a moving medium which naturally describes perturbation decay. The error in the "naive" argument above is that, during entanglement of the magnetic field lines, there occurs not only an increase of the field, but also a decrease in its characteristic spatial scale, which in the two-dimensional case yields energy flux into the small-scale region of the field where dissipation related to finite conductivity necessarily occurs.

Another, no less important result obtained in this paper is the calculation of the diamagnetic susceptibility of a turbulent conducting fluid. Because magnetic field lines are frozen into a chaotically moving, ideally conducting fluid, the average of the field (not the mean square of the field!) in such a fluid in steady state after a sufficiently long time is zero, so that the ideally conducting turbulent region "expels" the external magnetic field. For finite conductivity, because the field lines slip by the fluid, this expulsion is incomplete, and the turbulent fluid acts like a diamagnetic with finite, but small, magnetic permeability. In the paper an estimate is made of this permeability. The result is not specific to the two-dimensional case—in the three-dimensional case as well the magnetic permeability decreases without limit as the so-called magnetic Reynolds number increases.

These results were further generalized in a joint paper by Ya.B. and A. A. Ruzmaikin, "The Magnetic Field in a Conducting Fluid Moving in Two Dimensions" (**8**). It is shown that the proof of the statement regarding decay of the magnetic field can be carried out even without assuming that the velocity field of the fluid is a function of only two coordinates. It is essential only that the motion itself be two-dimensional, i.e., that the velocity component normal to the surface be zero everywhere. In the simplest case the velocity lies in the x,y-plane, with $v_x = v_x(x, y, z, t)$, $v_y = v_y(x, y, z, t)$, and the only condition is that $v_z \equiv 0$.

It is quite interesting that exponential growth of the field is impossible in the case of motion along plane surfaces and along the surfaces of spheres enclosed within one another. In motion along other surface systems, for example, cylinders or ellipsoids, exponential growth of the field is possible, however, it is specific and slow for low magnetic viscosity. Thus, the concept of a "slow dynamo" was introduced as an intermediate case between absence of field generation and rapid generation independent of the magnetic viscosity. This question, as well as Ya.B.'s own concrete realization of a fast dynamo in three-dimensional non-steady motion, is considered in more detail in the section "Mathematics in the Works of Ya.B. Zeldovich."

Subsequently Ya.B.'s colleagues and students substantially advanced both the general theory of magnetic field generation and its practical applications to the Sun, galaxies, and other astrophysical objects.

Along with the methods of similarity theory, Ya.B. extensively used and enriched the important concept of self-similarity. Ya.B. discovered the property of self-similarity in many problems which he studied, beginning with his hydrodynamic papers in 1937 and his first papers on nitrogen oxidation (**25, 26**). Let us mention his joint work with A. S. Kompaneets [7] on self-similar solutions of nonlinear thermal conduction problems. A remarkable property of strong thermal waves before whose front the thermal conduction is zero was discovered here for the first time: their finite propagation velocity. Independently, but somewhat later, similar results were obtained by G. I. Barenblatt in another physical problem, the filtration of gas and underground water. But these were classical self-similarities; the exponents in the self-similar variables were obtained in these problems from dimensional analysis and the conservation laws.

In 1956, Ya.B. encountered a different problem, the problem of a short-duration impulse (**9**), in which the self-similarity proved to be completely different. This problem was studied independently by a West German physicist, K. von Weizsäcker, who published very similar results somewhat earlier. The problem, which has various technological and astrophysical applications, may be outlined as follows: a half-space filled with gas is adjacent to a vacuum. The boundary of the half-space is impacted so that the boundary for some time penetrates into the gas and is then withdrawn. The motion then proceeds as follows: a strong shock wave propagates in the gas, and at the same time the gas on the opposite side expands into the vacuum. It turns out that the motion quickly enters a self-similar regime, just as in the problem of strong thermal waves. This problem also involves conservation laws—those of energy and momentum. However, the exponents in the similarity variables, and in particular in the law of the shock wave propagation, cannot be determined from the conservation laws. The reason for this is deep and non-trivial: the transition to the self-similar regime occurs non-uniformly in space. The exponents of the self-similar variables, as it works out, are determined not from the conservation laws and dimensional analysis, but rather from the global condition of existence of the self-similar solution, in full analogy with the propagation velocity of a travelling wave of flame in the combustion problem; this analogy, as subsequently became clear, is of a very deep nature. Problems of this kind were also studied before Ya.B. in the works of the German physicist K. Guderley, and of the Soviet scientists L. D. Landau and K. P. Staniukovich (1942–1944) on convergent strong shock waves. However, it was precisely in the aforementioned paper by Ya.B. in 1956 that the essential difference between these problems and the classical self-similarities was emphasized. Such self-similarities were separated into a special class and called self-similarities of the second kind.

4.   Phase Transitions. Molecular Physics

The problems of phase transition always deeply interested Ya.B. The first work carried out by him consisted in experimentally determining the nature of "memory" in nitroglycerin crystallization [8]. In the course of this work, questions of the sharpness of phase transition, the possibility of existence of monocrystals in a fluid at temperatures above the melting point, and the kinetics of phase transition were discussed. It is no accident, therefore, that 10 years later a fundamental theoretical study was published by Ya.B. (**10**) which played an enormous role in the development of physical and chemical kinetics. The paper is devoted to calculation of the rate of formation of embryos—vapor bubbles—in a fluid which is in a metastable (superheated or even stretched, $p < 0$) state. Ya.B. assumed the fluid to be far from the boundary of absolute instability, so that only embryos of sufficiently large (macroscopic) size were thermodynamically efficient, and calculated the probability of their formation. The paper generated extensive literature even though the problem to this day cannot be considered solved with accuracy satisfying the needs of experimentalists. Particular difficulties arise when one attempts to calculate the preexponential coefficient.

The significance of this research, however, is by no means exhausted by this particular, though important, problem. It turned out that the technique developed here may be transferred almost without change to a large number of kinetics problems in which the slow decay of non-equilibrium systems of widely varying physical nature is considered. We may speak here not only of the kinetics of other phase transitions of the first kind (the generality of the work in this respect was indicated by Ya.B. himself), but also of such outwardly dissimilar phenomena as, say, formation of vortex rings in the supercritical motion of liquid helium, "three-particle" recombination of electrons in a gas, and formation of a stream of "escaping" electrons in a plasma. In all cases the governing factor is precisely the slowness of the process, which allows application of the Fokker-Planck equation or the diffusion equation for the appropriate variable (the meaning of these variables, of course, is different in different problems). Further, in all cases it turns out, just as in Ya.B.'s basic paper, that the process rate is determined by the "bottleneck," i.e., by a comparatively small region of parameter values where the process occurs most slowly. The diffusion coefficient varies little in this region and may be determined outside of the region, where it has a direct macroscopic meaning. Further, the boundary condition imposed on the distribution function downstream from the "bottleneck" turns out to be insignificant, which allows one to obtain a closed-form solution even when the form of this condition cannot be established exactly.

In the problem of bubble production the size of the embryo is the independent variable in the diffusion equation. The value of the diffusion coefficient itself is determined by solving the hydrodynamic equations describing the growth of a bubble in a viscous fluid.

By way of illustration, we note that in the recombination problem mentioned above the energies of the electrons are the variables in the diffusion equation, the bottleneck is the region of energies near the boundary of the continuous spectrum, and the slowness of the process is related to the small amount of energy transfer from an electron to a heavy particle in one collision. In the problem of "escaping" electrons in a plasma, slowness is ensured by the weakness of the electric field, the independent variable in the diffusion equation is the momentum component along the field, and the bottleneck is determined, as in the kinetics of new phase formation, by the saddle point of the integral.

Turning to molecular physics, we note first papers by Ya.B. which are close to the problem of phase transition. We begin with the theory of interaction of an atom with a metal (**11**). By applying quantum-mechanical perturbation theory to the interaction of the virtual dipole moment of an atom with conducting electrons of the metal, the dependence on distance of the attractive force of the atom to the surface is obtained. The calculation led to a slow, $\sim l^{-2}$, law for the potential energy decay with distance. This paper was published in 1935, and for many years remained essentially the only one devoted to the subject.

Later, considering the problem from a macroscopic point of view, H. Casimir and D. Polder (Netherlands, 1948), and E. M. Lifshitz (1954), obtained a different, more rapid law of interaction decay. Only recently L. P. Pitaevskiĭ showed that the contradiction does not indicate an error: Ya.B. studied an extreme case of large Debye radius, and this case is realizable in principle.

Of particular value, especially in stimulating related experimental research, was a joint paper by Ya.B. and L. D. Landau, "On the Relation Between Liquid and Gaseous States of Metals" (**13**), in which the authors discuss the form of the phase equilibrium curve between the dielectric and conducting phases of liquid metals. Because transition between the metallic and dielectric phases at absolute zero may occur only as a transition of the first kind, the P,T-plane must have a line of transition of the first kind between the metal and dielectric, terminated by a critical point. There are then two fundamentally different possibilities. Either this critical point coincides with the critical point of fluid–vapor transition (in this case the entire transition curve coincides), or the fluid–vapor and metal–dielectric transition curves branch at some triple point, and each transition has its own critical point. The experimental situation, due to difficulties related to the necessity of working at high temperatures and pressures, long remained unclear. Only recently has it become apparent that, in fact, both variants are feasible: the first in cesium and the second in mercury.

In a paper in 1938 (**12**), "Proof of the Uniqueness of the Solution of the Equations of the Law of Mass Action," Ya.B. showed by means of an elegant mathematical investigation that the equations of thermodynamic equilibrium in a mixture of chemically reacting ideal gases (or in ideal solutions) always have a solution and that this solution is unique. The significance of this result has grown in recent years as systems with several stable states or nondecaying oscillations have begun to attract a great deal of attention, and a new field has emerged—synergetics—which studies such situations. Thanks to Ya.B.'s work, we are certain that the non-trivial character of synergetic systems is due either to the non-ideal nature of the system (in the sense of molecular interactions and phase transitions), or to the fact that the system is open. The steady state equations describing an ideal system with chemical reactions, influx of reagents, and outflow of reaction products, at first glance resemble the equilibrium equation. Therefore, it is especially interesting to establish a very general property of the equilibrium equations which distinguishes them from the equations for an open steady state system.

A number of Ya.B.'s papers were devoted to the properties of states close to the fluid–vapor critical point, to periodic crystallization, and to the limiting chemical kinetic laws of bimolecular and chain reactions. We will not, however, attempt to take the place of the bibliography at the end of the book which provides a complete list of Ya.B.'s papers on the subjects of this volume.

Let us mention several papers by Ya.B. on various problems of molecular physics and quantum mechanics which have not been included in this volume. Among the problems considered are the peculiar distribution of molecules according to their oscillatory modes when the overall number of oscillatory quanta does not correspond to the temperature of translation [9], the influence of the nuclear magnetic moment on the diffusion coefficient [10] and on absorption of light by prohibited spectral lines [11].

The study of electron interaction with a chaotic radiation field is essentially similar to problems in molecular physics. However, because of its significance for cosmology, the papers on this subject have been placed in the next volume (see also Ya.B.'s review [12]).

The study of quantum systems in a periodic field and the introduction of the concept of quasi-energy are also of much interest; a brief article on this subject is placed in the next volume. Here we restrict ourselves to a reference to the review paper [13].

Ya.B.'s contribution to the theory of unstable states is systematized in his monograph *Scattering, Reactions and Fission in Non-Relativistic Quantum Mechanics* [14], and one of his papers will be included in the book *Particles, Nuclei, and the Universe*.

5. Theory of Shock Waves*

Ya.B.'s work on shock waves was of immense practical and theoretical importance. The monographs he wrote received worldwide recognition. The publication in 1946 of a small book, *Theory of Shock Waves and Introduction to Gasdynamics* [15] was a major event in the literature on fluid mechanics. The problem was that gasdynamics was usually presented as a mathematical science, one aimed at mathematicians and theoreticians. In Ya.B.'s book, for the first time, the same basic material was presented by a physicist with the deep penetration into the physical essence of phenomena characteristic of Ya.B., with his own, original vision of the physical world, and with simplicity and accessibility. One is astounded by the insight of individual remarks, findings, and associations which allow one to see a phenomenon from an unusual angle, to understand something which went unnoticed before.

Generations of physicists entering the field of gasdynamics and using it for the solution of both practical and theoretical problems learned it from the book *Theory of Shock Waves and Introduction to Gasdynamics.*

In 1963 an extensive monograph written by Ya.B. and Yu. P. Raizer, "The Physics of Shock Waves and High Temperature Hydrodynamic Phenomena," was published. A second, expanded edition of the monograph came out in 1966 [16]. Simultaneously, in 1966, an English translation of the second edition was published in the USA. In this book, besides the authors' papers on shock waves, explosions, high temperature phenomena and related problems, all the experience gained in the use of physical concepts in gasdynamics was generalized and systematized. The book has become a reference for anyone who works in the fields of high temperatures, plasma, explosions, and high-speed flows. According to the Science Citation Index, for many years the book has been one of the most frequently cited.

Let us mention several specific papers by Ya.B. on shock waves. In a

*Professor Yu. P. Raizer took part in the writing of this section.

paper written in 1946 (**15**) the problem of the structure of a shock wave in a gas with retarded excitation of some degrees of freedom was considered for the first time. Earlier, A. Einstein [18] and others had studied the influence of processes with slow relaxation (dissociation and excitation of the internal degrees of freedom) on sound propagation. In Ya.B.'s paper these ideas are carried over to the nonlinear process of shock wave propagation. Different compression regimes are feasible—smooth or with a pressure discontinuity— depending on the wave amplitude.

Several years later the process analyzed in Ya.B.'s article became the basis for the most powerful method for experimental study of physico-chemical kinetics in gases at high temperatures—shock tubes. For two decades practically all measurements of the probabilities of excitation of molecular oscillations and of dissociation of molecules, i.e., everything that was needed for calculations of the motion of space vehicles through the atmosphere, were performed in shock tubes by recording current parameters in the relaxation layer. We may note that even in the experimental study of ignition in a shock tube, the first work was done by Ya.B. with Ya. T. Gershanik and A. I. Rozlovskiĭ [19].

Since as far back as the time of the classical works on gasdynamics only compression shock waves were known, while rarefaction occurred without discontinuities. Rarefaction waves are continuous in space. This is stated by Zemplen's theorem and is related to the fact that in a rarefaction discontinuity the entropy would decrease, which is impossible. But this is so only if the adiabate in the pressure-volume diagram is convex down. This fact was also known. The thermodynamic properties of practically all substances satisfy this condition.

In 1946 Ya.B. pointed out (**14**) a possible case where the opposite situation occurs. This happens near the critical point where the differences between a vapor and a fluid are obliterated. In a substance under near-critical conditions rarefaction should propagate as a discontinuity, and compression— as a continuous process. Many years later, at the end of the seventies, this prediction of Ya.B. was confirmed experimentally in Novosibirsk by a group working under Academician S. S. Kutateladze. At present, only two cases are known when rarefaction shocks occur: in solid bodies in the region of polymorphous transformations (this had been observed long ago), and near the critical point, as Ya.B. predicted.

At the end of the fifties, Ya.B. gave a qualitative picture of the structure of shock waves with radiation transfer taken into account [20]. In front of a compression shock there is a layer heated by radiation from the compressed gas. Behind the discontinuity there is a temperature peak. The simultaneously developed quantitative theory of these effects allowed detailed explanation of the experimentally observed patterns of luminescence of the front in strong shock waves and of the radiation in the early stage of a fire ball in a strong explosion [21]. These and a few other works on shock waves were presented in the first review paper on the physics of strong shock waves by Ya.B. and Yu. P. Raizer [22].

In 1963 the remarkable phenomenon of gas breakdown by laser radiation was discovered in experiments, which laid the ground for broad new directions in plasma physics and the physics of the interaction of laser radiation with a substance. Soon Ya.B. and Yu. P. Raizer developed the cascade theory of laser breakdown [23]. Hardly a single article of the great number devoted to optical breakdown manages without a reference to this work.

Ya.B. is a theoretical physicist. However, a characteristic peculiarity of his scientific style is his interest primarily in those problems in theoretical physics which allow immediate experimental verification by one method or another, and his interest in the methodology and feasibility of real physical experiments. Therefore the reader should not be surprised at the presence of a number of experimental works in his bibliography. The most interesting of these from a methodological standpoint is his new method for studying substances at high pressures by reflection and refraction of light on the surface of a shock wave propagating in a transparent substance [24–26]. Study of the dependence of the reflection and polarization coefficients of reflected light on the incident angle allows, in principle, determination of the complex refraction coefficient of a medium at pressures up to hundreds of thousands of atmospheres. It turns out, in particular, that water stays transparent up to a pressure of 144 thousand atm, with significant deviations from the Lorenz-Lorentz formula observed.

6.   Theory of Combustion and Detonation

Ya.B.'s papers on combustion and detonation initiated a new stage in the development of this science in which the ideas and methods of gasdynamics, gas-kinetic theory, and the effects of molecular transport and the actual kinetics of high-temperature chemical reactions were brought together logically and consistently. A Soviet school of specialists on combustion was formed and received worldwide recognition; one of its recognized founders was Ya.B.[*] It was no accident that this school emerged at the end of the thirties at the Institute of Chemical Physics. It was here that Academician N. N. Semenov developed the chain theory of chemical reactions and the theory of thermal explosion.

The combustion theory could not have been created without a clear understanding of the kinetics of chemical reactions and without the creative atmosphere, initiated by A. F. Ioffe, which N. N. Semenov fostered and expanded. Under these conditions, Ya.B. introduced clarity into the deep

---

[*]The formation of this stage is excellently characterized by collections of papers from the period 1920–1950 [27–30].

understanding of processes of heat transfer and hydrodynamics and developed mathematical methods adequate to the problems of combustion theory (see section 7 below regarding these methods).

Ya.B.'s studies of combustion and detonation are diverse and multidirectional. They include the chemical thermodynamics of combustion, propagation of exothermic chemical transformation fronts, deflagration and detonation theory, thermo-diffusion and chemo-kinetic processes in combustion and at high temperatures in general, and gasdynamics of flows in the propagation of non-uniform flame fronts and in detonation.

Among Ya.B.'s interests were: the combustion of gases and solid rocket fuels, of condensed liquid explosives and powders, the combustion of premixed fuel compounds, and diffusive combustion. In every one of his lines of inquiry he obtained fundamental results which served as starting points for numerous theoretical and experimental studies in the USSR and worldwide.

The scope of the modern science of combustion is significantly broader than it was a few decades ago. Together with the traditional application of combustion in energy installations—to obtain mechanical work, heat, electrical energy, and to maintain transportation systems, etc.—new applications have been developed such as the production of new materials through combustion, use as a source of information about chemical kinetics at high temperatures and pressures, and the production of a high-temperature, laser-active medium.

In the present collection of Ya.B.'s papers we have included primarily his first articles which formulate the basic ideas and concepts that were then worked out mathematically and thoroughly verified both in later papers by Ya.B. and in the papers of his many students. Moreover, the modern science of combustion all over the world is developing now along the paths outlined in Ya.B.'s papers.

Indicative in this respect is the monograph, "Theory of Combustion and Detonation of Gases," published in 1944 (**16**). In it the basic concepts of flames and detonation waves in gases, their characteristic properties and possibilities, and the interaction within them of gasdynamic, molecular and kinetic processes, are presented in concise form;[3] explanations are given for previously inadequately understood phenomena of propagation limits, thermodiffusive flame instability, and the peculiar influence of small chemical admixtures. The style of presentation is appealing for its deep argumentation; theoretical reasoning is supported by numerous experimental studies both in the Soviet Union and abroad, with many of them performed in the combustion laboratory of the AS USSR Institute of Chemical

---

[3]Detailed and rigorous study of the most important problems is done mainly in subsequent articles in the second part of this book, and in a series of monographs and papers (see the bibliography at the end of this volume). However, this does not diminish the significance of the monograph included here, which presents the main aspects in the clearest and most concise form.

Physics, which Ya.B. headed until 1947. Close interaction between theorists and experimentalists ensured extraordinarily rapid progress in the science of combustion during this period. Among those involved in collaboration and in all the discussions with Ya.B. were N. N. Semenov, Yu. B. Khariton, A. F. Belyaev, D. A. Frank-Kamenetskiĭ, K. I. Schelkin, O. I. Leipunskiĭ, S. M. Kogarko, P. Ya. Sadovnikov, G. A. Barskiĭ, V. V. Voevodskiĭ and other well-known representatives of the Soviet school of combustion.

Despite the diversity of the studies being carried out, they had a single ideological and methodological platform: at their foundation was the strong dependence of the chemical reaction rate on temperature, and various related threshold phenomena. To obtain the basic laws of combustion, asymptotic methods were used, complemented by an explicitly physical interpretation.

Ya.B. showed that exothermic chemical reactions in a flow which arise due to the strong dependence of their rates on the temperature lead to a discontinuous, jump-like transition from one reaction regime to another, despite the fact that the chemical reaction rate itself is a smooth, continuous function of the temperature, pressure, reagent composition, and other parameters. The reason lies in the nonlinearity of the basic equations of combustion theory, which generally have several solutions, some stable and some not. For an exothermic reaction in a jet with intensive mixing (an ideal mixing reactor) there are stable low- and high-temperature regimes, and an unstable transient regime. The transitions between the first two correspond to ignition and extinction (experimentally they are recorded, for example, by a sudden change in luminosity).

An important feature of combustion phenomena is their hysteretic character: for example, extinguishing a high-temperature chemical reaction turns out to be possible only when conditions are provided such that the cold system is far from the threshold of ignition. Realization of a particular combustion regime depends on the history of the process, i.e., whether the initial gas temperature was high or low.

The present volume contains a paper by Ya.B. (**17**) and a joint article with Yu. A. Zysin (**17a**) which are devoted to the study of a reactor with complete mixing. In the section where ignition is considered Ya.B.'s results essentially reproduce Academician N. N. Semenov's theory of thermal explosion (1928), which relates to a single reaction event involving a given portion of a substance in a closed vessel. It is interesting, however, that when one takes reagent consumption into account, the transition from slow reaction to explosion in a closed vessel is not, rigorously speaking, discontinuous. In a reactor with ideal mixing, as the time approaches infinity, we obtain an absolutely sharp discontinuity. The theory of extinction in a reactor has no precursors, and for energy science this extreme case is no less interesting than that of thermal explosion.

In the second paper (**17a**), by taking into account the heat transfer to the vessel walls, a stable reaction regime is discovered which cannot be obtained by continuous variation of the external conditions. These features of the equations of combustion theory and the basic patterns of exothermic reaction in a jet studied by Ya.B. have recently been used widely in, for example, the modern theory of chemical reactors.

However, most important and complicated are those processes in which the chemical combustion reaction occurs in space and time. Let us note a brief paper by Ya.B. on the theory of gas ignition by a heated surface (**18**). This work may be considered a generalization of D. A. Frank-Kamenetskiĭ's theory of thermal explosion. However, when part of the surface has a high temperature, Ya.B. was able to formulate a general principle of ignition which is applicable under the broadest variety of geometric and gasdynamic conditions.

Nevertheless, the most typical general feature of a reaction is the existence of fronts of chemical transformation which are able to propagate, without being extinguished, in a hot mixture with a constant velocity: at subsonic speed for a laminar flame (or deflagration front), at supersonic speed for a detonation wave (see below for a more detailed discussion of this paper).

The strong dependence of the chemical reaction rate on the temperature allowed Ya.B., in his work with D. A. Frank-Kamenetskiĭ (**19**), to find the structure of a laminar flame: he isolated in the flame a narrow zone of chemical transformation adjacent to the region of maximum combustion temperature, and a wider zone of heating in which the chemical reaction can be neglected. In each of these zones simplifications of the basic equations of the theory are possible which allow them to be integrated, i.e., allow one to find the temperature and concentration distributions; the integration results to a simple analytical formula for the velocity of flame propagation, known in world literature as the Zeldovich–Frank-Kamenetskiĭ formula (ZFK-model of thermal flame propagation).

The Zeldovich–Frank-Kamenetskiĭ formula related the velocity of flame propagation to real, Arrhenius-type chemical kinetics, and thus raised flame experiments to the rank of kinetic experiments which allow one to obtain important information (activation energy, reaction order, pressure dependence, etc.) about the process of chemical reaction at high temperatures and pressures in a broad range of variation of the composition of the combustible mixture. In the combustion of energy fuels more than half of the energy is supplied by the reaction $CO + \frac{1}{2}O_2 = CO_2$. On the initiative of N. N. Semenov a detailed kinetic study of this reaction was carried out at high temperatures by measurement of the flame velocity [31, 32]. In the absence of hydrogen admixtures, water vapors, or other hydrogenous compounds, the flame propagates quite slowly. The rate of the reaction between CO and oxygen turned out to be proportional to the concentration of hydrogen, which plays the role of a necessary catalyst. We should also mention experiments on the influence of various flegmatizers on the flame velocity in near-threshold mixtures. These experiments form the basis of modern methods of accident prevention in work with combustible mixtures. Thanks to the work of Ya.B., combustion theory has become a part of chemical physics.

Initially developed for simple, single-stage schemes of chemical transformation and for the case of similar temperature and concentration distributions, the theory of normal flame propagation was generalized by Ya.B. and his followers to complex chemical transformations with branching and non-branching chain reactions, with sequential and parallel stages and several separate reaction zones, and with a large concentration of intermediate active centers.

At the basis of analytic solutions with complex chemical transformation mechanisms in flames there remained the fundamental assumption of the narrowness of the chemical reaction zones compared to the heating and diffusion zones; this assumption is valid for large reaction activation energies, which correspond to realistic situations. Considering the importance of this basic assumption, at the Ninth International Colloquium on the dynamics of explosions and reacting systems (France, 1983) specialists from different countries working in the field of combustion decided to introduce the use of the dimensionless Zeldovich number, $Ze = E(T_B - T_0)/RT_B^2$ (E is the activation energy, T_B and T_0 are the maximum temperature in the reaction zone and the initial temperature, respectively, and R is the universal gas constant). In the asymptotic method developed by Ya.B., Ze is a substantially large quantity.

In 1980 Ya.B., together with G. I. Barenblatt, V.B. Librovich, and G. M. Makhviladze, published a fundamental monograph, "The Mathematical Theory of Combustion and Explosion" [33].

The ideas and methods of combustion theory have found wide application in various areas of physics, biology, and mechanics. We note such phenomena as the propagation of an impulse along a nerve, the formation of a neck (a characteristic wavelike narrowing) in polymer extension, the propagation of laser breakdown and discharge, the zones of ionization of a gas by ultra-high-frequency radiation, and the structure and propagation of cracks in elastic materials.

Let us turn to detonation theory. By the turn of the century the rule for calculating the velocity of a detonation wave and other parameters using only thermodynamic data and the conservation laws was known. It appeared that the chemical kinetics of the transformation of the original explosive substance or mixture into the final reaction products did not play any role, and that it was enough to consider only the initial and final states.

Despite good agreement between this rule and experiment, a certain inconsistency remained in its justification. The laws of conservation and thermodynamics allowed a solution with increased detonation velocity and pressure (compared with those calculated according to the Chapman-Jouguet "rule"), but this solution proved incompatible with the condition of expansion of the explosion products after detonation. One half of the rule was thus justified: the pressure may not be higher than that allowed by the Chapman-Jouguet rule. However, no mechanical or thermodynamic considerations prohibited detonation with an increased velocity, but decreased pressure.

It was necessary to consider the kinetics of the chemical reaction, and this was first done by Ya.B. Previously, the kinetics had been considered in connection with the detonation capacity of combustible materials. Yu. B. Khariton in 1939 had formulated a general principle: a substance reacts in a detonation wave, but at the same time it also flies apart under the influence of high pressure. The ability of a charge to detonate depends on the relation between these two processes. In a paper by Ya.B. published in May, 1940 (27), an idealized process without any losses is considered. The conservation equations are valid; however, it turns out that not all states satisfying the conservation laws occur during the reaction. Ya.B. succeeded in providing a logically flawless justification of the Chapman-Jouguet rule. The problem had been ripe. Apparently independently, in 1942 W. Doering in Germany arrived at an analogous result, and in 1943 J. von Neumann in the USA did as well. We should also note that in 1940 a Soviet scientist, A. A. Grib (a student of Academician S. A. Khristianovich), had been working on detonation theory. Because of the war he was not able to publish his results (which provided a less complete analysis of the chemical reaction, but a more detailed picture of the hydrodynamics of the expansion of the explosion products) until 1944.

Ya.B.'s research on detonation waves received worldwide recognition. In the literature outside the USSR the term "ZND-theory," after Zeldovich, von Neumann, and Doering, is widely accepted.

The most important new conclusion of the theory turned out to be the fact that in front of the zone of reaction products, whose state is determined by the Chapman-Jouguet rule, there is a certain amount of initial combustible material compressed by the shock wave. In this compressed material the pressure is approximately twice as high as the final pressure. The significance of the existence of a zone of such increased pressure is obvious not only for the theory, but also for accident prevention. Many studies are devoted to experimental proof of the existence of this zone. Perhaps still the most convincing and practically useful is a paper written by Ya.B. in collaboration with S. M. Kogarko which confirmed the conception of an increased pressure zone (28). Concrete perceptions of the conditions of the chemical reaction have changed substantially (see the commentary to 28). However, Ya.B.'s basic conclusions—the principle of velocity selection and the existence of this increased pressure zone—remain valid even today.

Ya.B. performed interesting research on the limits of propagation of deflagration and detonation waves in channels, and on the concentration limits of combustion in gases (**20**). The strong sensitivity of the chemical reaction rate to temperature variation in the wave caused by heat transfer to the channel walls produces an avalanche-like process: the heat transfer decreases the reaction rate and, thus, the wave velocity, and this in turn promotes even greater heat transfer. At the limit the wave propagation velocity turns out to be only one and a half to two times less than the adiabatic velocity. Modelling of this phenomenon for different combustible mixtures and for channels of different sizes is determined by the Peclet number, which is constructed from the adiabatic propagation velocity. This number was introduced into practice in papers by Ya.B. Modern calculations for various fire-retarding devices are based on concepts developed by him.

The concentration limits of flame propagation in large vessels, when thermal losses due to heat transfer are small, are explained by Ya.B. from the standpoint of thermal losses by radiation. The flame velocity at the concentration limits also turned out to differ significantly from zero.

Combustion is extraordinarily rich with a variety of instabilities. Bending of the flame front causes redistribution of the thermal and diffusion fluxes within it: flame sectors which are convex with respect to the combustible mixture end up in a different temperature-concentration regime than concave sectors, and therefore propagate through the combustible mixture with a different velocity. Ya.B.'s paper with N. P. Drozdov (**21**) shows that in combustible mixtures in which the diffusion coefficient of the deficient reagent exceeds the coefficient of thermal conductivity the plane flame front ceases to exist—it disintegrates into individual islands enriched by the easily diffused component and possessing an increased temperature. These islands are able to move independently in the combustible mixture with a velocity which exceeds the planar flame velocity. Thermodiffusion phenomena are the reason for the formation of flames with complex structures (cellular, polyhedral, pulsating, etc.) which presently attract the attention of numerous researchers. (Regarding this and, in particular, subsequent papers by Ya.B. see the commentary to **21**, cited above.)

Thermal expansion of a gas in a curved flame front leads to the formation of gasdynamic vorticity in the combustion products and is the cause of a flame instability discovered by L. D. Landau, and also by G. Darriet (France), in 1944. It turned out, however, that this instability was very reluctant to exhibit itself in experiments! The first explanation of such a phenomenon—using the example of a spherical flame—was given by A. G. Istratov and V. B. Librovich. Ya.B. and his coauthors [34] proposed a method for calculating rapid combustion in a tube containing an elongated flame front (stabilized combustion in a flow), and also spontaneous flame propagation in a tube. The evolution of the surface of a curved flame front in time, as well as the presence of a tangential component of the flow velocity along the flame, makes the flame front more stable with respect to hydrodynamic perturbations and explains the observed deviation of experimental facts from the results of the hydrodynamic instability theory proposed by L. D. Landau.

In diffusion combustion of unmixed gases the combustion intensity is limited by the supply of fuel and oxidizer to the reaction zone. The basic task of a theory of diffusion combustion is the determination of the location of the reaction zone and of the flow of fuel and oxidizer into it for a given gas flow field. Following V. A. Schvab, Ya.B. considered (**22**) the diffusion equation for an appropriately selected linear combination of fuel and oxidizer concentrations such that the chemical reaction rate is excluded from the equation, so that it may be solved throughout the desired region. The location of the reaction zone and the combustion intensity are determined using simple algebraic relations. This convenient method, which is universally used for calculations of diffusion flames, has been named the Schvab-Zeldovich method.

However, in this paper Ya.B. went further and considered the chemical kinetics. He determined the limit of intensification of diffusion combustion, which is related to the finite chemical reaction rate and the cooling of the reaction zone, for an excessive increase of the supply of fuel and oxidizer. If the temperature in the reaction zone decreases in comparison with the maximum possible value by an amount approximately equal to the characteristic temperature interval (calculated from the activation energy of the reaction), then the diffusion flame is extinguished. The maximum intensity of diffusion combustion, as Ya.B. showed, corresponds to the combustion intensity in a laminar flame of a premixed stoichiometric combustible mixture.

With Ya.B.'s active participation Soviet science achieved great success in the theory of combustion and in the practical use of solid rocket fuels (powders). This volume contains Ya.B.'s basic ground-laying paper (**24**).

As in the theory of detonation, specialists on internal ballistics were working on thermodynamic and hydrodynamic theories. Conceptions of the mechanism of combustion in the thirties were quite primitive.

Meanwhile, in 1938 at the AS USSR Institute of Chemical Physics A. F. Belyaev showed that the combustion of liquid explosives occurs in the gas phase after their evaporation. In analogy with this, Ya.B. proposed a theory of combustion of a solid powder (**24**) according to which the powder is heated in the solid phase and then decomposes, transforming into a gas; it is only in the gas phase, at some distance from the surface, that the bulk of the chemical energy is released. Ya.B. also pointed out the peculiar effects of non-steady combustion of a powder.

The heated layer in a condensed powder substance plays the role of an inertial heat accumulator: the heat stored in the layer during slow (low-pressure) combustion goes for additional heating of the combustion products at increased pressure and increases the combustion rate compared to the rate of the steady process. In contrast, as the pressure is decreased part of the heat from the reaction zone is spent on the creation of a wider heated layer; the reaction zone is cooled and combustion may cease. Ya.B. developed a theory for the combustion rate of a powder with small pressure variations, and determined the conditions under which undesirable high frequency oscillations occur in a powder combustion chamber. Discovered at the AS USSR Institute of Chemical Physics by O. I. Leipunskiĭ, the increase in the combustion rate of a powder when a gas flow is blown over its surface (the blowing or erosion effect), was given a clear gasdynamic interpretation in a paper by Ya.B. [35] (written in 1943, published in 1971). Of extraordinary importance for the internal ballistics of solid propellant rockets is the possibility of extinguishing the powder by a rapid decrease in pressure; a theory of powder extinction by a sharp, jump-like drop in pressure, or by a smooth, but sufficiently deep change in pressure, was developed by Ya.B. in the 1940s. It has found extensive application in modern industrial equipment which uses a powder as a working agent.

A special place among kinetic studies in combustion is occupied by work on nitrogen oxidation. Begun at the AS USSR Institute of Chemical Physics in the mid-thirties on the initiative of N. N. Semenov, research to determine the feasibility of fixation of atmospheric nitrogen for the production of mineral fertilizers has today found application in the development of environmental protection measures for toxic components of combustion products, including nitrogen oxide. In December, 1939, Ya.B. defended his doctoral dissertation on "The Oxidation of Nitrogen in Combustion and Explosions." It was precisely these studies, in which D. A. Frank-Kamenetskiĭ, P. Ya. Sadovnikov, A. A. Rudoy, A. A. Kovalskiĭ, and others actively participated, that led Ya.B. to the problems of combustion and detonation.

The publication of papers on nitrogen oxidation was delayed (**26** and a monograph, "The Oxidation of Nitrogen in Combustion," written together with P. Ya. Sadovnikov and D. A. Frank-Kamenetskiĭ). In these works Ya.B. studied in detail the kinetics of nitrogen oxidation; having proved that oxidation occurs *via* a non-branching chain reaction with an equilibrium concentration of active centers, he calculated the formation of nitrogen oxide in a closed vessel and during the sudden "hardening" of a reacting, rapidly expanding gas. The kinetic scheme of nitrogen oxidation which Ya.B. demonstrated is a central one in the internationally accepted practice of calculating environmental pollution by the exhaust products of internal combustion engines, by coal power stations, and by chemical factories, and rightfully bears Zeldovich's name.

The restricted space of this introductory article does not allow us to give further attention to Ya.B.'s papers on the physical chemistry of combustion and detonation.[4] In Ya.B.'s own words, the science of combustion and detonation is dear to him because it provided him, at the beginning of his life and career, with a wide-open field of activity both as a theorist and as a bold experimenter. Important problems of a fundamental nature were being solved, problems which were related to important technological applications. The tremendous contribution which Ya.B. made to this science is the pride and showpiece of the Soviet school of combustion. Ya.B.'s work on the theory of detonation and combustion was recognized by awarding him the State Prize of the USSR in 1943.

The science of detonation and combustion developed actively after World War II. A significant role in this development was played by the work of scientists at the AS USSR Institute of Chemical Physics, particularly that of A. G. Merzhanov and his colleagues in the field of combustion, including condensed systems, the work of AS USSR Corresponding Member K. I. Schelkin, Y. K. Troshin, A. N. Driemin, B. V. Novoshilov and G. B. Manelis, and also of scientists at the AS USSR Siberian Branch—AS USSR Corresponding Member R. I. Soloukhin and his colleagues—in the field of detonation, and the work of scientists at the AS USSR Institute of Problems of Mechanics—Y. B. Librovich and his colleagues. New concepts appeared, especially in problems of the stability of solutions found earlier. Let us emphasize, however, that the work of Ya.B. was not refuted; rather it was built upon, and it served as a starting point for the future development of combustion science. In 1984, the International Institute of Combustion awarded Ya.B. the Lewis gold medal for his brilliant accomplishments in research on combustion processes.

7.   *Mathematical Aspects of the Combustion Theory*

The phenomena of ignition and extinction of a flame are typical examples of discontinuous change in a system under smooth variation of parameters. It is natural that they have played a substantial role in the formation of one of the branches of modern mathematics—catastrophe theory. In Ya.B.'s work it is clearly shown that steady, time-independent solutions which arise asymptotically from non-steady solutions as the time goes to infinity are discontinuous. It is further shown that transition from one type of solution to the other occurs when the first ceases to exist. The interest which this set of problems stirred among mathematicians is illustrated by I. M. Gel'fand's

---

[4]The next section, which deals with the mathematical aspects of this group of papers, is something of a supplement to this one.

well-known programmatic paper [36].

Even greater interest among mathematicians was stimulated by problems of state propagation. As is known, in 1937 the fundamental paper by A. N. Kolmogorov, I. G. Petrovskiĭ, and N. S. Piskunov (KPP) appeared, and, independently, a paper by R. A. Fisher (USA), regarding the propagation of a biologically predominant species. In 1938, the aforementioned paper by Ya.B. and D. A. Frank-Kamenetskiĭ (ZFK) on flame propagation (**19**) was published.

At issue in both cases were solutions describing propagation with constant velocity. The velocity itself here should be defined as the eigenvalue of the parameter for which the equation has a solution. Such a statement of the problem was unusual for mathematicians. Evidence of the interest it caused may be found, in addition to I. M. Gel'fand's article, in review articles by the American scientists D. Aronson and H. Weynberger (1975 and 1978 [37,38]), and many others.

The difference in the results of KPP and ZFK is characteristic. In the KPP problem the velocity has a continuous spectrum, bounded from below by the condition of a non-negative solution. The ZFK method, meanwhile, yields in the typical case one definite value for the propagation velocity. The difference is completely explained by the fact that, unlike the biological problem of KPP, the chemical reaction rate at and near the initial temperature is typically negligible. In practice, the problem reduced to construction of a travelling-wave-type solution for a set of coupled diffusion and heat transfer equations (of a combustible gas or for the combustion products). It was rigorously proved that, allowing for a shift in the direction of propagation, the solution of the problem exists and is unique.

This work had a happy fate; it was continued and developed by scientists around the world. It is enough to say that the first generalization of Ya.B.'s work on flame propagation to multi-component mixtures was done by the famous American hydrodynamicist T. von Kármán. Later, many hundreds of papers were devoted to problems of the mathematical theory of combustion, considering both theoretical problems of flame propagation, even questions of existence and uniqueness, and purely practical problems, including calculations of the flame propagation velocity in specific gas mixtures. A review of these papers, albeit far from complete, may be found in monograph [33].

We note also research by Ya.B. performed together with G. I. Barenblatt in 1957 [39], which has proved to be a key to understanding not only the problems of stability and flame propagation to which it was directed, but a far wider range of phenomena as well. This is the problem of the stability of invariant solutions to problems in mathematical physics. The question was posed thus: what is flame stability? Let us perturb the temperature and concentration distribution in a flame. Which flame shall we call stable? We recall that the solution of the problem of a flame as a travelling wave was determined with accuracy to within translation. In the later paper, it was noted that the definition of flame stability should also be translation-invariant: a situation in which the perturbed distribution of temperature and concentration tends to the translated, rather than the original, distribution should not be considered unstable. As it turned out, gas flames are stable in this sense under fairly general assumptions. One may also speak of the zeroth (time-independent) perturbation mode, which has now been found in many problems of theoretical physics.

Finally, comparatively recently, Ya.B., A. P. Aldushin, and S. I. Khudyaev (**23**) completed a theory of flame propagation which considers the most general case of a mixture in which the chemical reaction occurs at a finite rate at the initial temperature as well. In this work the basic idea is followed through with extraordinary clarity: flame propagation represents an intermediate asymptote of the general problem of a chemical reaction occurring in space and time. At the same time, the relation between the two types of solutions (KPP and ZFK) is completely clarified.

Let us turn to Ya.B.'s paper on ignition of a combustible gas mixture by a heated wall (**18**). By the time this paper was written, N. N. Semenov and D. A. Frank-Kamenetskiĭ had already done work on gas ignition in closed vessels under a variety of assumptions. In N. N. Semenov's work the gas was assumed to be ideally mixed and the temperature of the gas in the vessel to be constant. D. A. Frank-Kamenetskiĭ considered a gas at rest, so that heat transfer occurred only through molecular thermal conduction.

Ya.B.'s paper was the first to consider the external problem: ignition of a reacting gas mixture by a heated body. It was clear that he had to begin with the simplest case of a heated wall; problems of ignition by a wire or ball were solved later. Here again a basic, classically simple result was obtained: at the limit of ignition of the mixture the thermal flux from the heated wall vanishes. An expression for the minimum rate of heat transfer far from the body sufficient to prevent ignition was also obtained. However something else was significant as well: just as in the problem of calculating the flame propagation velocity, this problem had all the distinctive features of asymptotic analysis. Certain terms which arose only a decade and a half later were not used in the paper; otherwise the paper (**18**) consistently applied a technique which took final shape in the mid-fifties in the works of S. Kaplun, P. Lagerstrom, M. Van-Dyke, J. Cole, and others—the method of matched asymptotic expansions—which now plays a very important role in fluid and gas mechanics. Furthermore, group-theoretical methods, a great rarity at the time, also played a significant role in this paper and were decisive in the solution of a difficult nonlinear problem of mathematical physics.

8.   Chain Fission of Uranium[5]

The discovery in 1938–1939 of nuclear fission of uranium, which led ultimately to the discovery of nuclear power, heralded a new, extraordinarily fruitful stage in Ya.B.'s scientific activity. His interests were concentrated on the study of the mechanism of fission of heavy nuclei and, what proved particularly important, on the development of a theory of the chain fission reaction of uranium. During 1939–1943 Ya.B. wrote several papers which laid the foundation for this subject and were of fundamental value. We note that four of these papers, written in collaboration with Yu. B. Khariton, were done practically in two years before the war. The papers of this series form the foundation of modern physics of reactors and nuclear power; they are widely known and do not require special commentary—a short review of the basic physical results is eloquent enough.

The paper of 1939 [1*], "On the Chain Decay of the Main Uranium Isotope," studies the effects of elastic and non-elastic neutron moderation and concludes that chain fission reactions by fast neutrons in pure metallic natural uranium are impossible. The 1940 paper, "On the Chain Decay of Uranium under the Influence of Slow Neutrons" [2*], is classic in the best sense of this word; its value is difficult to overestimate. The theoretical study performed showed clearly that the effect of resonance absorption of neutrons by nuclei of ^{238}U is a governing factor in the calculation of the coefficient of neutron breeding in an unbounded medium; it was concluded that a self-sustained chain reaction in a homogeneous "natural uranium–light water" system is impossible.

The second paper of 1940 [3*], entitled "Kinetics of Uranium Chain Decay," is no less significant than the first. This pioneering work yielded a whole series of brilliant results: for the first time, the need to take into account the role of delayed neutrons in the kinetics of chain nuclear reactions was shown (it is precisely the delayed neutrons which ensure easy control of nuclear reactors), the influence of heating on the kinetics of a chain process was considered in detail, and a number of conclusions were reached which are of much importance for the theory of reactor control. This same paper predicted the formation in the process of chain fission of new, previously unknown, nuclei which strongly absorb neutrons, a prediction which was later fully confirmed.

In the 1941 paper with Yu. B. Khariton [40], the problem of the critical size of a sample of ^{235}U in the fission of nuclei by fast neutrons was considered. The calculations showed that, in order to sustain a chain fission reaction by fast neutrons in a sample of ^{235}U surrounded by a heavy neutron reflector, it is sufficient to have only ten kilograms of pure ^{235}U isotope. Here also a theory is given which allows calculation of the critical mass of

[5]Here and below we denote by * references to articles by Ya. B. Zeldovich which may be found in the second volume of his selected works, *Particles, Nuclei, and the Universe.*

^{235}U dissolved in light water.

Ya.B.'s unpublished 1943 paper, "The Age Theory of Neutron Moderation," is closely related to his studies of the war period. The age theory, developed independently of E. Fermi, forms the basis for calculation of a reactor by slow neutrons. It was in this paper that the famous "age equation" was obtained.

Ya.B.'s work on the problem of the chain fission of nuclei, together with his work on detonation and shock waves, were the scientific foundation of Ya.B.'s practical activity in a collective which was carrying out a very important state assignment. This activity was acknowledged by awarding Ya.B. the highest decorations of the USSR.

At the same time, the study of uranium fission guided Ya.B. toward problems of microphysics, the theory of elementary particles and the nucleus. Thus, the work of 1939–1941 became a decisive turning point in Ya.B.'s life and career.

9.   The Theory of Elementary Particles

Ya.B.'s contribution to the theory of elementary particles came primarily during 1950–1960. Therefore it does not seem superfluous to provide a brief description of the situation during that time.

The 1950s saw the beginning and flowering of research on elementary particles in accelerators built specially for this purpose. Beginning with the study of pions, discovered not long before in cosmic radiation, the decade was marked by the discoveries of most of the strange particles, and was crowned by the discovery and study of hadron resonances in accelerators. In the theory of strong interactions the decade began with construction of an isotopic classification of the hadrons and ended with the discovery of $SU(3)$-symmetry. At the same time, the dispersion approach in the theory of strong interactions was formed.

In the mid-fifties the violation of parity was discovered, and a universal theory of weak interactions—the $(V\text{-}A)$–theory—was created. Construction of composite hadron models was begun. The first non-abelian gauge theory was developed.

During these years Ya.B. made important contributions to a whole series of the above trends. In 1952 he formulated the law of nuclear (baryon) charge conservation [4*], extending it to the unstable particles recently discovered in cosmic radiation and subsequently called strange particles.

In 1953 Ya.B. introduced the law of conservation of lepton (neutrino) charge [5*] as one of the basic rigorous laws of nature. Both of these laws are important for the classification of elementary particles and the processes in which they participate.

Thorough testing of these conservation laws has so far revealed no violations. Thus, recent (1983) experiments demonstrated that the lifetime of a proton exceeds 10^{31} years (these experiments sought the decay of a proton into a positron and neutral pion predicted by the Grand Unification Theory). Modern theoretical predictions regarding baryon instability and nonconservation of baryon charge do not change the practical applicability of this conservation law throughout laboratory physics.

In 1954 Ya.B. was the first [6*] to direct attention to the importance of measuring the β-decay of a charged π-meson. He proved that for the scalar variant of the weak interaction this decay is prohibited if the pion is a composite one, while for the vector variant the decay is allowed and the square of its matrix element is twice that of the vector matrix element in neutron decay. Continuing the theoretical investigation of this decay in a 1955 paper [7*], Ya.B. and S. S. Gershtein concluded that the constant of the vector interaction of nucleons is not modified by strong interactions, just as the electric charge of strongly interacting particles is not modified by virtual particles. This observation by Ya.B. and S. S. Gershtein, made at a time when it was generally accepted that a scalar, rather than vector interaction occurs in β-decay, played an important role in the creation of a universal theory of weak interactions.

In 1957, formulating within the framework of this theory the idea of conserved vector current, M. Gell-Mann and R. Feynman (USA) revived the hypothesis of Ya.B. and S. S. Gershtein. In a report at the Rockberk conference in 1960, R. Feynman said, "The idea that if there is a vector current in β-decay then it can be made a conserved one was first proposed by S. S. Gershtein and Ya. B. Zeldovich. M. Gell-Mann and I were not aware of this when working on the idea." Experimental discovery of the decay, $\pi^+ = \pi^0 + e^+ + v$, was made in 1962. Subsequent measurements of the probability of this decay in laboratories at the Joint Institute for Nuclear Research, CERN, and at research centers in the USA, showed that it satisfies the relation predicted by Ya.B. with a high degree of accuracy.

In 1955 Ya.B. and G. M. Gandelman [8*] indicated that exact measurement of the magnetic moment of an electron is a promising method of determining the limits of applicability of quantum electrodynamics at small distances. In 1956 V. B. Berestetskiĭ, O. N. Krokhin, and A. K. Khlebnikov noticed that measurement of the magnetic moment of a muon allowed even smaller distances to be probed. Since then, study of the magnetic moments of the muon and electron has become a classical method for finding the limits of applicability of quantum electrodynamics. Today, theoretical and experimental values of the magnetic moment agree with an accuracy of order 10^{-9} for an electron and 10^{-8} for a muon. Recently it has become customary to interpret this agreement as an indication of the fact that the dimensions of the internal structure of leptons do not exceed 10^{-16} cm.

In 1958 Ya.B. proposed a technique for finding neutral resonances using the method of deficient mass [9*]. The search for resonances by this method, when a peak is observed in the spectrum of deficient masses, was later used successfully in the discovery of a series of unstable particles. The authors of these experimental papers arrived at this method apparently without knowing of Ya.B.'s work, but this does not diminish his priority.

Far ahead of the physics of its time, Ya.B.'s paper of 1959 [10*] presents arguments in favor of the existence of a weak interaction of neutral currents which does not preserve parity (electron–electron and proton–proton) and proposes experiments to seek this interaction. It is remarkable that Ya.B.'s considerations were based on the hypothesis that a neutrino and electron form an isotopic doublet, and that the weak interaction must be isotopically invariant. This idea, proposed independently by Ya.B. and not long before by S. A. Bloodman (USA), was very bold. After all, at the time it was assumed that the concept of isotopic spin did not apply to leptons. The experiments proposed by Ya.B. were performed two decades later when, in 1978 in Novosibirsk, the rotation of a polarization plane of a laser beam passing through bismuth vapor was found, and at Stanford the difference between cross-sections of nucleon interactions with left- and right-polarized electrons was demonstrated. This latter effect was predicted by Ya.B. for muons as well, and was recently observed in an experiment at CERN. It should be emphasized that the scheme proposed by Ya.B. in 1959 contained only diagonal neutral currents, in complete agreement with the later theory of electro-weak interaction and with experiment. Ya.B. turned to the question of the possible role of weak interactions in explaining left-right asymmetry in biology later, following other researchers (see his paper with D. B. Saakyan [41]).

Also in 1959, Ya.B. proposed an experiment on the transformation, $K_2^0 \rightarrow K_1^0$ on electrons and offered a theory for this phenomenon [42]. The experiment was later carried out at the Serpukhov accelerator.

When we speak of the papers of the 1950s, we must not fail to mention articles published in 1954–1960 on muon catalysis (for a review with complete bibliography, see [43]). Various aspects of cold muon catalysis in nuclear reactions of hydrogen, deuterium, and tritium were considered. In recent years the ideas which lay at the foundation of these papers have been further developed and verified. In particular, it was Ya.B. [40] who noted the decisive role of muon adhesion to the helium nucleus which is formed in the reaction: this process restricts the number of reaction steps, and the future for applying this process in energy engineering depends on it.

In giving an overall evaluation of the series of papers on elementary particle theory published by Ya.B. in the 1950s, it must be noted that it was performed during a period when Ya.B.'s primary and very important activity was the creation of a new technology. In the almost complete absence of personal interaction and contacts with leading physicists during this period, his choice of timely problems and his ability to obtain important results appears even more surprising.

Let us turn to papers on the theory of elementary particles published by Ya.B. in the 1960s and 1970s. The 1960s brought into the physics of elementary particles the quark hypothesis. Theorists were on the verge of creating a quantum chromodynamics, a theory of quark–gluon interaction.

During these years Ya.B.'s unremitting attention was directed to quarks. He organized a true offensive all along the front, studying various quark models, analyzing possible physico-chemical methods for seeking free quarks, and elucidating the possible behavior of quarks in the early stages of the evolution of the Universe. He not only initiated, but even personally participated in an experiment seeking free quarks [44]. In 1965 he published a brief review, "Quarks and the Classification of Elementary Particles for Pedestrians" [45], remarkable for its simplicity and enthusiasm, and a detailed study, "Quarks: Astrophysical and Physico-Chemical Aspects" [11*], [46] (in collaboration with L. B. Okun and S. B. Pikelner).

Taken together, these works gave rise to a contradiction which led theorists to the idea of quarks confinement. At the same time, these papers gave birth to a new direction which Ya.B. began to develop in the 1960s, a bridge between cosmology and elementary particle physics.

In 1961 the first paper on this subject came out—"On the Upper Density Limit of Neutrinos, Gravitons and Baryons in the Universe" written with Ya. A. Smorodinskiĭ [47]. In 1966 Ya.B. and S. S. Gershtein [12*] established an upper cosmological limit for the mass of a muonic neutrino. This limit, about $100\,eV$, is four orders more accurate than that given by the best laboratory measurements. The point is that for any larger neutrino mass their combined mass in the Universe would be so large that it would be incompatible with the limits established by the known age of the Universe. This paper had great influence both on the development of elementary particle physics and on the development of cosmology; it has entered into textbooks of both disciplines. In connection with this paper, we should also mention the research of one of Ya.B.'s students, V. F. Schvartsman [48], in which a theoretical limit is established for the number of different types of neutrinos ($n < 10$) by their influence on the relation between the rates of occurrence of hydrogen and helium in the Universe. All of modern "cosmological elementary particle physics" is very much indebted to these two works.

Beginning in the mid-sixties, Ya.B. repeatedly tried to evaluate the energy density of a vacuum, to find the magnitude of the so-called cosmological constant [13*]. His interest in the physics of vacuum grew in the 1970s when, while exploring an idea of D. A. Kirzhnitz and A. D. Linde on phase transition, Ya.B., I. Yu. Kobzarev and L. B. Okun [49] constructed the theory of the domain structure of a vacuum and its possible influence on the cosmological expansion of the Universe.

Vacuum domains should appear during the cooling of the Universe if spontaneous violation of SP-symmetry occurs in nature. In addition, contiguous domains should differ from one another by the sign of the SP violation. The domains should be separated by very thin and very heavy walls, the theory of whose motion was developed by Ya.B. and his coauthors.

This work initiated a whole new line of research at the intersection of elementary particle theory and cosmology (the theory of unstable vacuum, vacuum bubbles, vacuum strings, etc.).

An interim summary of much of the research of Ya.B. and his students was given in a review by Ya.B. and A. D. Dolgov [50] which gained worldwide recognition and has become a standard reference in practically all the papers on this subject both in the USSR and abroad.

Let us also note Ya.B.'s analysis of how well the conservation of electric charge has been verified (his paper together with L. B. Okun on the stability of an electron [51]), and his paper investigating the possibility of gravitational annihilation of the baryon charge [52].

We note several very general formulations of the problem. A striking example of this is Ya.B.'s 1967 paper [14*], in which he considers the possibility of a theory in which the bare photon field is absent, while the observed electromagnetic field is created entirely by quantum fluctuations of a vacuum. This bold idea, which extends to electrodynamics an earlier idea about gravitational interaction (in part, under the influence of Ya.B.'s papers on the cosmological constant), has not yet been either proved or disproved. However, both ideas have elicited lively discussion in the scientific literature.

As an example of Ya.B.'s role in the development of particle physics we quote from an article by G. I. Budker on colliding beams, "The idea of antiparallel particles is not new; it is a trivial consequence of the theory of relativity. As far as I know, it was first expressed by Academician Zeldovich, though in a very pessimistic context. His pessimism is quite understandable" [53]. As is well known, G. I. Budker and his collaborators and their foreign colleagues were able to overcome the enormous difficulties which caused this pessimism. The role of colliding beams in particle physics is impossible to overestimate.

10. Nuclear Physics

Forming a natural group with Ya.B.'s papers on uranium fission and elementary particles are a small, but fundamentally important group of papers on nuclear physics which have had major repercussions.

The first was a brief note in 1956 on the possibility of cold neutron storage [15*]. It is known that a neutron moving with sufficiently small velocity experiences complete internal reflection in its fall from a vacuum to a wall made, for example, of graphite. According to Ya.B., this phenomenon may be used for long-term "storage" of neutrons. The idea was experimentally realized, first in the USSR by AS USSR Corresponding Member L. F. Shapiro and his colleagues, and later abroad as well. Thanks to this idea it became possible to measure with high accuracy the electric dipole moment and life span of a neutron.

The storage of cold neutrons (the idea and the experiment) was registered officially* as a discovery in 1958. Ya.B. was awarded the Kurchatov gold medal for this work.

The second paper, in 1960, discusses the existence of nuclei at the threshold of stability having excessive numbers of neutrons [16*]. Its most remarkable result was the prediction of a new helium isotope, Helium-8. This isotope was soon discovered experimentally and the decay chain was traced (due to the weak interaction $^8He \rightarrow {}^8Li \rightarrow {}^8Be$ with subsequent $^8Be \rightarrow 2^4Be$).

After this article there followed a series of other publications, reviews and monographs by Ya.B. in collaboration with A. I. Baz, V. I. Goldanskiĭ and V. Z. Goldberg [54], listed in the commentary. Overall, the study of nuclei at the threshold of stability has now become a broad field of inquiry in nuclear physics.

Let us note, finally, two ideas of Ya.B. which have not yet found experimental confirmation. We refer to the possibility of the existence of excited states of nuclei with an anomalously long life span—with a large angular momentum [17*] or with an anomalously large value of the isotopic spin [18*]. Perhaps their publication in this edition will help to experimentally verify the nontrivial ideas of Ya.B.

11. Astrophysics and Cosmology

Ya.B. actively entered astrophysics and cosmology at the beginning of the sixties. Today, twenty years later, few people in the world can rival his influence on the development of astrophysics and cosmology. His ideas inspire new work not only among theoreticians. The largest radiotelescopes, optical devices, orbital x-ray observatories list among their experimental and observational accomplishments the discovery of effects predicted by Ya.B. These include gigantic voids in the Universe, surrounded by clusters of galaxies, and x-ray sources which gain energy by accretion to black holes and neutron stars, and the decrease in relic radiation intensity in the directions of clusters of galaxies surrounded by hot intergalactic gas.

The chief problem (and one suited to the scale of his talent), which Ya.B. has set himself and toward whose solution he has persistently moved the last twenty years, is the question of the properties and origin of the large-scale

---

*There exists in the USSR a practice of official registration of scientific discoveries. *Translator's note.*

structure of the Universe, of the reasons for the appearance of the original density perturbations, of the law of their growth in the course of the cosmological expansion, of the peculiarities of compression of matter in the nonlinear stage of growth, and, finally, of observable manifestations of all these stages. The resulting picture is often called the theory of galaxy formation, although it would be more correct to call it the theory of formation of the large-scale structure of the Universe.

Ya.B.'s entry into astrophysics coincided with the second revolution in astronomy, which was marked by rapid development of experimental research methods in radioastronomy and the launching into space of instruments which are sensitive in the x-ray and ultraviolet ranges of the spectrum. During the period from 1964 to 1972, discoveries were made of quasars, relic radiation, radiopulsars, and compact x-ray sources emitting radiation due to accretion, i.e., the collapse of matter onto neutron stars and black holes. Ya.B. actively began work in this, for him, new field several years before the most important new observations appeared. His papers during these years were devoted mainly to the application of general relativity theory to astrophysical objects and to the Universe as a whole. He was among the founders of a new field of science—relativistic astrophysics—which addresses such problems as the last catastrophic stages of the evolution of stars, the discovery of black holes, the physics of the early stages of the expansion of the Universe, and the theory of supermassive stars, with masses ranging from hundreds of thousands up to billions of solar masses. He studied in detail the properties of black holes and the processes which occur in their vicinity. Earlier, black holes had been considered only as a possible product of the evolution of sufficiently massive stars. Ya.B. showed in 1962 [19*] that even a small mass can collapse if its density is sufficiently high. From this followed a paradoxical conclusion, that the equilibrium of any mass is always metastable with respect to gravitational collapse.

Cosmology is based on the assumption that matter in the early stage of evolution of the Universe was of extraordinarily high density. From this, in 1966, Ya.B. and I. D. Novikov came to the conclusion that the generation of small black holes was possible in the early stages of evolution. Finally, it was shown in a very general form that the collapse of any nonsymmetrical object leads to the creation of an external observable metric which is wholly determined by conserved quantities [55] (with A. G. Doroshkevich and I. D. Novikov).

In 1964 Ya.B. [20*], and independently of him E. Salpiter (USA), showed that a black hole may be found by its influence on the surrounding gas. The heating of the gas produces radiation which may then be detected. It was only after the publication of these papers that astronomers realized that black holes could be observed. It was, in fact, the papers of Ya.B. and his students that pointed out the very important role of accretion of matter which makes "dead stars"—neutron stars and black holes—visible. It was also his idea, remarkable for its elegance, to seek black holes in double star systems by means of their influence on the motion of a "bright companion" which is easily observed in optical radiation (with O. Kh. Guseinov, 1966) [21*].

Under Ya.B.'s guidance the theory of disc accretion was developed and received recognition and experimental verification. We note that all this work was basically performed before the experimental discoveries. Still awaiting experimental confirmation is the burst of neutrino radiation accompanying the collapse of a star, which Ya.B. examined together with O. Kh. Guseinov in 1965 [22*].[†]

Ya.B.'s gift for foreseeing an experimental breakthrough and preparing well in advance the theoretical foundation for its subsequent study is surprising and worthy of great admiration. It was precisely this quality of Ya.B. which in large measure helped to transform Moscow in the 1960s and early 1970s into the world capital of relativistic astrophysics. Many of today's leading American and European theoretical astrophysicists, directly or indirectly, are his disciples.

These papers also in large measure initiated the construction of the theory of quasars. To this day, three quasar models are discussed: the supermassive star, the packed star cluster, and accretion on supermassive black holes. Ya.B. has made a major contribution to the investigation of all three models. As early as 1965, he and M.A. Podurets [23*] discovered that, like a single massive star, star clusters evolve, transforming into a black hole.

The launch of the American satellite "Uhuru" brought about a revolution in X-ray astronomy. The discovery of double X-ray sources, neutron stars, and black holes (the three candidates at present) fully confirmed the correctness of the accretion course chosen long before by Ya.B. [24*], and made him a leading theoretical authority in international X-ray astronomy.

Ya.B. has worked extensively on the problem of the origin of magnetic fields, stars, and galaxies, actively developing the theory of dynamos. These studies (in particular, that of 1956, paper **7**) and one written with A. A. Ruzmaikin (1980, paper **8**) are related to his research in hydrodynamics and are reviewed in Sections 3 and 12 of this Introduction. But his major field of endeavor is cosmology.

In the early sixties the problem of the beginning of cosmological expansion was deeply theoretical and far from verification by observation. Ya.B. persistently supported the search for methods of experimental verification of the question of whether the Universe was "cold" or "hot" in the early stages of its evolution. At first, Ya.B. sought an alternative to the hot model (1962) [25*].

[†]This prediction was confirmed in 1987 when scientists in the United States and Japan detected prompt neutrinos from SN1987a. This event was the first naked eye supernova to be seen since the 1604 event witnessed by Tycho Brahe.

On his initiative, A. G. Doroshkevich and I. D. Novikov [56] constructed a global spectrum of the electromagnetic radiation in the Universe and showed that relic radiation in thermodynamic equilibrium can be found in the centimeter region. The discovery of relic radiation answered the question of what model to choose for the Universe. Ya.B. became an ardent proponent of the theory of a hot Universe (see the 1966 review [26*]). He was one of the first in the world to understand what a powerful tool relic radiation represented for discovery of the Universe's past. His reviews of 1962–1966, which became the basis for excellent books written later with I. D. Novikov [57–59], contain practically all the ideas which have now become the methods for studying the large-scale structure of the Universe. These include the question of dipole and quadrupole anisotropy, and of angular fluctuations of relic radiation, the problem of nuclear synthesis reactions in the hot Universe, and the quark problem, first raised by Ya.B. together with L. B. Okun and S. B. Pikelner (1965) [11*].

Ya.B. was among the first to realize that the quantum properties of gravitation should exhibit themselves near a singularity. He showed that, if the beginning of the expansion is anisotropic, quantum effects near the singularity lead to intensive generation of particle–antiparticle pairs, thus decreasing anisotropy, i.e., making the Universe more isotropic. This will be discussed in more detail below.

Ya.B. (with V. G. Kurt and R. A. Sunyaev, 1968) [27*] solved the very important problem of the process of hydrogen recombination in the Universe. Scattering of photons by free electrons turns out to be a very important process over the course of a long stage in the evolution of the Universe. Following A. S. Kompaneets, Ya.B. studied this process in detail and found its interesting features (with R. A. Sunyaev, 1972) [28*]. The cosmological consequences are presented in three papers (in 1969 with R. A. Sunyaev [29*], and in 1970 [30, 31*]). These studies found the types of spectrum distortion typical for various stages of the Universe, and found for each of these stages restrictions on the energy yield which are related to the dissipation of turbulent and potential motions, the decay of unstable particles and the first black holes, and the annihilation of matter and antimatter.

At the same time, a paper written with R. A. Sunyaev (1972) [32*] was published in which a decrease in the intensity temperature of relic radioradiation directed toward galaxy clusters with hot intergalactic gas was predicted. This phenomenon was later named the Zeldovich-Sunyaev effect, and has today become part of the observational programs of the largest radiotelescopes in the world, and has served to justify projects for specialized cosmic submillimeter observatories. This effect, which transforms a cluster of galaxies into "negative extended radiosources," simultaneously makes them extremely bright sources in the submillimeter range. Together with the X-ray observations this effect creates an opportunity for measurement of the absolute size of a galaxy cluster, i.e., it provides astronomers with a "standard rod" which allows determination of the distance to distant objects, Hubble's constant, and so on (with R. A. Sunyaev, 1980) [33*]. The effect also enables one to find the peculiar velocity of clusters with respect to relic radiation and to determine the mean density of the Universe.

Calculations of weak radiosources show that they have undergone strong cosmological evolution: their number has decreased relative to galaxies by thousands of times between the moment corresponding to a red shift of $Z \sim 2-3$ and the present. In his only paper on this subject, Ya.B. (with A. G. Doroshkevich and M. S. Longair, 1970) [60] introduces a now commonly accepted function of the evolution of radio sources.

Ya.B. called his most important achievement in cosmology—the theory of the formation of the large-scale structure of the Universe—the "pancake theory."[†] The evolution of international public opinion on this subject is both interesting and instructive. In 1970 Ya.B. published a paper [34*] in which he showed that the commonly accepted picture up to then of spherical compression of protoclusters of galaxies during the nonlinear stage of their evolution was highly improbable. Compression along one of the axes must dominate, and as a result flat structures—"pancakes"—must form. It was then indicated that one-dimensional compression must lead to the appearance of shock waves, and in the cooling zone the gas density must increase, which makes possible the formation and separation of galaxies (written jointly with R. A. Sunyaev and A. G. Doroshkevich [35*]; see also the review [36*]). Among the conclusions of this theory were the prediction of the existence of gigantic voids in the Universe (with S. F. Shandarin) [37*], and a characteristic lattice structure [38*]. For almost a decade work on this theory was performed only in the USSR. Then came a sensational event—the discovery of the gigantic voids. Observations of galactic superclusters confirmed some details of the theory. Today it is in the limelight, and further research and specification are taking place in major universities throughout the world. Everyone has become interested in the theory and Ya.B.'s approach has attracted the interest of prominent mathematicians. Among the scientists who, together with Ya.B., made major contributions to the theory is V. I. Arnold. The joint work of V. I. Arnold, S. F. Shandarin, and Ya.B. has convinced even sticklers for rigour of the flawlessness of the mathematical approach.

Ya.B.'s love for and unrelenting interest in the theory of elementary particles have maintained his interest in the problem of the earliest stages of expansion of the Universe, in the search for the reasons for the creation of our world, and in the nature of the first metric and density fluctuations which later led to the formation of the observed structure of the Universe.

Let us recall the problem of neutrino mass. Immediately after receiving

[†]It is also sometimes referred to by its Russian equivalent, *bliny*.

information on the results of V. A. Liubimov's group, Ya.B. threw his colossal creative potential into the development of a variant of the theory of galaxy formation which assumes that the dominant contribution to the mean mass density in the Universe is that of massive neutrinos. Over a brief period five papers appeared (in particular, one with A. G. Doroshkevich, R. A. Sunyaev, and M. Yu. Khlopov, 1980) [39*–41*] in which all the most important aspects of the theory are captured, the basic directions of its future development are understood, and the validity of the pancake theory in this variant as well is shown.

It is remarkable that the problem of perturbation growth or damping in a gas of collisionless gravitating particles had been earlier solved by Ya.B. (with G. S. Bisnovatyĭ-Kogan) [61] applied to stellar gas. This solution turned out to be very important in the problem of the cosmological role of massive neutrinos; it determines the damping scale of the first density perturbations and gives the mass of galaxy clusters.

The most profound, difficult, and yet fundamentally important results were obtained over the last few years in the study of quantum phenomena in gravitational fields. It is sufficient to cite two examples—the quantum "evaporation" of black holes and the "inflationary" period of expansion in the very young Universe. Neither of these results belong to Ya.B.: the evaporation in its final form was calculated by S. Hawking (England), the inflationary stage was studied by L. E. Gurevich (Leningrad), D. A. Kirznits, A. D. Linde, and A. A. Starobinskiĭ (Moscow), and its astronomical implications were clearly formulated by A. Guth (USA). Nonetheless, even these examples convincingly demonstrate the major role which Ya.B. played in the overall development of science which requires the collective efforts of scientists.

Let us follow Ya.B.'s role in solving the two questions above. Evaporation of black holes is meaningful only for very small black holes whose masses are very small compared to stellar masses. The immediate significance of the possibility of black minihole formation for cosmology has already been noted. Here we note only that without the concept of a small black hole, there can be no question of quantum phenomena in its vicinity or, in particular, of its evaporation.

Is the gravitational field able, in principle, to produce real, tangible particles in a void, or in the terms of general relativity theory, in curved space-time?

Today there is no doubt of this. However, it is appropriate to recall that it was S. Hawking who in 1970 published a paper proving the impossibility of particle generation. The proof was based on the principle of energy dominance: roughly, the gist of the proof is that the energy density is always greater than the pressure. Since the energy density includes the mass density multiplied by the square of the speed of light, the principle of energy dominance is always correct by a large margin for conventional or compressed matter. However, Ya.B. (with L. P. Pitaevskiĭ, 1971) [42*] immediately showed that in curved space the creation of particles is preceded by polarization of the vacuum, i.e., by a change of the ground state of fields in space. The principle of energy dominance is inapplicable to the energy or pressure of vacuum polarization. Thus S. Hawking's objection was removed and the path opened to the study of particle generation.

R. Penrose (England) noticed that a rotating black hole is able to give up its rotational energy in a fairly complicated process of radioactive decay of a particle in the vicinity of such a black hole.

Ya.B. noted (1970) [43*] that any rotating body, including a rotating black hole, reflects with amplification certain waves which fall upon that body or hole. Hence he concluded that a rotating hole can also spontaneously give off corresponding waves, for example, electromagnetic ones. Ya.B. carried out the electrodynamic calculations for a cylinder, and his students, A. A. Starobinskiĭ and S. M. Churilov, performed analogous calculations for a black hole in the framework of general relativity. Therefore, although they did not reach the final result, Ya.B. and his students laid the ground for S. Hawking's discovery.

It is interesting that, in an interview with the *New York Times*, S. Hawking said that while in Moscow he learned of the work of Ya.B. and his colleagues on particle generation by a rotating black hole, but did not like the mathematical method used in the work. Solving the problem differently, using a more thorough method, he arrived at the most important result—the evaporation of any black hole and the transformation of the entire mass of the black hole into radiation.

Clear formulation of the requirements of the theory which follow from an analysis of modern observations [62] played a major role in the development of the model of the earliest Universe. A very promising direction was worked out here by A. A. Starobinskiĭ, who inherited from Ya.B. both the formulation of the quantum effect problem (1971) [44*] and a profound creative interest in cosmology as a whole.

Recently, Ya.B. has been working on a "complete" cosmological theory which would incorporate the creation of the Universe (1982) [45*]. Let us mention finally that it was Ya.B. who recently gave a profound formulation of the question of the cosmological constant, i.e., the energy density in Minkovsky space (see Section 9). More precisely, the question is formulated thus: is the Minkovsky space a self-consistent solution of the equations for all possible fields and the equations of general relativity?

In classical field and particle theory the answer is trivial. In quantum theory (without which description of nature is impossible!), the answer is not only nontrivial, but has not even been found to this day. In the development of science, not only the solution to a problem, but even its very statement is a significant step. Ya.B. did not solve the problem (nor has it yet been solved by anyone else), but he has essentially formulated a very important requirement of a future theory which will unify fields, particles, and gravitation, namely the condition that the plane vacuum energy be extremely small (or equal to zero). The examples given are instructive because they exhibit Ya.B.'s social-scientific, catalyzing role, a scientist who is genuinely glad for the successes of other scientists and who aids these successes with his own enormous scientific potential, vitality, friendly nature and optimism.

International recognition of Ya.B.'s achievements in the development of cosmology has been been expressed, in particular, by his election in 1970 at the 14th General Assembly of the International Astronomical Union as the first President of the Cosmological Commission of the International Astronomical Union, created at this Assembly. Ya.B.'s election to the Royal Society of London (Britain's Academy of Sciences) and to the US National Academy of Sciences in 1979 was in significant measure related to his work in astronomy. And in 1983 the Pacific Astronomical Society (USA) awarded Ya.B. the Katharine Bruce gold medal, its highest award, "given for a lifetime of distinguished contributions to astronomy." In 1984, Ya.B. was rewarded with the highest award of the Royal Astronomical Society—a gold medal—for his great contribution to cosmology and astrophysics.[†]

12. Mathematics in the Work of Ya. B. Zeldovich[*]

Mathematics in Ya.B.'s work is not restricted to the standard arsenal of well-known methods; some of his achievements are essentially mathematical discoveries and rank with the most modern research by mathematicians.

Remaining above all a physicist, Ya.B. always departs from a concrete physical problem. But in those cases where the results have a general mathematical character, they are also applicable to physical situations which go far beyond the boundaries of the original problem. Thus, for example, Ya.B.'s analysis (1970) [34*] of particle cluster formation in a dust medium in his theory of the formation of the large-scale structure of the Universe simultaneously describes the appearance of optical caustics, as was shown, in particular, in his article with A. V. Mamaev and S. F. Shandarin [63].

The original physical problem is posed as follows: in accordance with modern conceptions of the early stages in the development of the Universe, at a time when it was 1000 times "less" than it is now, matter was distributed almost uniformly, and particle velocities formed a smooth vector field. But now we observe extreme inhomogeneity in the distribution of

---

[†]Note that Ya.B. also received the Robertson award of the US National Academy of Sciences in 1987.

[*]The Editors felt it appropriate to enlist the aid of an eminent mathematician, V. I. Arnold, to write the present section.

matter: there are galaxies, stars, planets, etc. The question arises, how did these inhomogeneities form?

Ya.B.'s theory explains the appearance of the largest-scale inhomogeneities from initially small fluctuations in the original velocity and density field.

The mathematical technique necessary here—the singularity theory of Lagrange mapping—was created quite recently, and Ya.B.'s paper in 1970 on "Hydrodynamics of the Universe" [34*] was one of the first in this rapidly developing field of mathematics.

The theory of mapping singularities (a branch of mathematics related to differential topology which represents a multidimensional generalization of the study of minima and maxima of functions) is not a part of the standard mathematical baggage of the physicist. Therefore we shall explain this subject using a very simple model—a system of noninteracting particles freely moving in Euclidean space.[6]

Let us assume that initially a particle at point x has the velocity $v(x)$. After a time t it arrives at the point $x + tv(x)$. We thus obtain a mapping of three-dimensional space onto itself (a point x corresponds to a point $x + tv(x)$) which depends on the time t as a parameter.

For small t this mapping is one-to-one, but as t increases faster particles begin to overtake slower ones and the one-to-one property is lost. To understand what is happening it is convenient to consider the phase space (x, p). The initial velocity field is represented in this space by a submanifold of the intermediate dimension $p = v(x)$ (Fig. 1).

The particle motion is represented in the phase space by a phase flow. The phase flow drags the manifold L_0 to a new location. The manifold L_0 remains smooth during this process. However, the new manifold L_t, beginning from a certain t, will no longer map uniquely onto the configuration x-space. In Fig. 1, we see that a three-flow configuration is formed: at a single point x there are particles with three different velocities.

The mapping of the manifold L_t onto the configuration x-space has critical points (A and B in Fig. 1). The corresponding points of the configuration space, a and b, are called caustic points.

At the caustic points the density of the medium is infinite (here particles which initially occupied points infinitesimally close to one another collide). Figure 1 describes a one-dimensional medium. In a three-dimensional space the caustic points form surfaces. Ya.B.

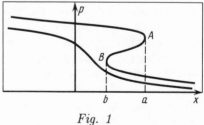

Fig. 1

[6]In the theory of formation of the large-scale structure of the Universe which Ya.B. constructed the expansion of the Universe and mutual attraction of particles are also taken into account, but the qualitative results are close to the results of the model problem.

called these surfaces pancakes, and the entire theory is known as the pancake theory.

If the initial velocity field is a potential one, $v = \partial S/\partial x$, then it will remain potential subsequently as well, even when it becomes multivalued. In mathematics, multivalued potential fields are called Lagrange manifolds. More precisely, a submanifold of an intermediate dimension in phase space is called Lagrange if $\int p\,dx$ on it depends only on the end points, and not on the path of integration.

The phase flow of the Hamilton equations preserves the Lagrange property. Thus, the manifold L_t at any t is a Lagrange manifold (even if the particles move not by inertia, but in any potential field). And so, Ya.B.'s pancake theory is a theory of the caustics which form in the mapping of Lagrange manifolds from phase space onto configuration space.

The caustics change with time. At certain moments they experience qualitative reconstructions (metamorphoses). For example, for small t there is no caustic at all, but at some later moment it appears for the first time.

The metamorphosis of caustic creation was studied by Ya.B. in detail. He demonstrated that after a small time τ following its creation, the newborn caustic has the form of an almost exactly elliptical saucer, where the axes of the ellipse are of order $\tau^{1/2}$, the depth is of order τ, and the thickness is of order $\tau^{3/2}$; along the edge of the saucer there is a semi-cubic upwardly turned edge. In the vicinity of this edge the particle density is especially high.

In optical terms the Lagrange manifolds are the same as what Hamilton called ray systems (a system of normals to a smooth surface may serve as an example). In this case, the caustic is the point of the concentration of light. Thus, Ya.B.'s pancake theory in particular predicts that the first caustics in a weakly inhomogeneous system of rays have the form of saucers (in a dispersive medium the saucers may become visible).

In the modern mathematical theory of Lagrange singularities the metamorphosis of saucer formation is the first in a long list (related to the classification of Lie groups, catastrophe theory, etc.). But Ya.B.'s pancake theory was constructed two years prior to these mathematical theories and, thus, Ya.B.'s work anticipated a series of results in catastrophe theory and the theory of singularities. Many later mathematical studies in the theory of singularities and metamorphoses of caustics and wave fronts were performed under the influence of Ya.B.'s pioneering work in 1970 on the pancake theory [34*].

In the years following Ya.B. returned several times to this topic. With the emergence of the hypothesis that a significant portion of the Universe's mass is concentrated in massive neutrinos, studies of higher-degree singularities which form in the process of pancake growth became very timely [64]. Analysis of the collisionless model, both theoretical and by means of computer simulation, led Ya.B. to conclude that the structure was cellular, with matter clustered predominantly on the cell walls, with especially high density along certain lines and even greater density at the exception points.

Current observational data seem to confirm this picture: when the red shifts of a large number of galaxies were measured, it turned out that their three-dimensional distributions have large voids, and this confirms a cellular structure of matter distribution. However, the question of the objectivity of a cellular or mesh structure is a subtle one since the eye tends to organize patterns even in a random distribution of points.

Today, convincing proofs have been obtained that galaxy clusters on the largest scale are grouped along threads or surfaces, rather than in clusters.

Objective criteria which clearly distinguish cluster structures from cellular and mesh structures have been developed by S. F. Shandarin (1983) [65] on the basis of percolation theory, whose usefulness for differentiating structures was conjectured by Ya.B. The percolation of cellular- or mesh-structure objects on a union of r-neighborhoods begins to occur at smaller r than in the case of a system of independent, randomly located points or clusters with the same (average) density.

Other recent achievements of Ya.B. in the pancake theory are related to the topology of the system as a whole (1982) [38*]: if we assume that colliding particles stick together, then over time a cellular structure should form.

The pancake theory today is perceived by mathematicians as a chapter contributed by Ya.B. to the general mathematical theory of singularities, bifurcations and catastrophes which may be applied not only to the theory of large-scale structure formation of the Universe, but also to optics, the general theory of wave propagation, variational calculus, the theory of partial differential equations, differential geometry, topology, and other areas of mathematics.

Ya.B. again extracted a precisely posed mathematical problem in the magnetic dynamo problem. From the standpoint of physics, the problem is one of the mechanism of magnetic field generation of the Sun or a star.

The physical mechanism is this: inhomogeneity of the velocity field leads to a flow which exponentially stretches the force lines of the initial small magnetic field. The growth continues until the forces generated by the magnetic field become equal to the hydrodynamic forces and begin to influence the flow.

In the theory of a kinematic dynamo the magnetic field is assumed not to have reached these values and not to influence the flow. The question is, will the initially random magnetic field grow exponentially?

The mathematical task extracted here by Ya.B. is formulated as follows.

We consider a vector field v in three-dimensional Euclidean space whose divergence is zero, with boundary conditions periodic in all three coordinates (the velocity field of an incompressible fluid). We assume that this field is given and does not change with time.

Let us consider a second field, B, also with zero divergence and the same periodicity (a magnetic field). The advection of the magnetic field B by the velocity field v is governed by the "frozenness" condition: an infinitesimally small vector proportional to B is advected by the fluid flow as if it were drawn onto the particles (and stretched together with them).

Besides this advection, there is also a "diffusion" mechanism with a small coefficient of magnetic viscosity D. Thus, the final equation of the evolution of the magnetic field B has the form

$$\frac{\partial B}{\partial t} = -(v, \nabla)B + (B, \nabla)v + D\Delta B$$

The velocity field v generates a kinematic dynamo (for a given D) if this equation has an exponentially increasing solution, $B = \exp(\gamma t)B_0$, $\mathrm{Re}\,\gamma > 0$.

The increment γ depends on the magnetic viscosity D. The dynamo is called fast (this concept was introduced by Ya.B.) if the increment remains positive (and greater than a certain positive number) for arbitrarily small values of the magnetic viscosity D.

The problem of a fast kinematic dynamo consists in the following: does there exist a field v which generates a fast dynamo? To date, this mathematical problem has not been solved for a steady flow v, and it is difficult to say what the answer will be.

Ya.B. proved (1956, **7**) that a two-dimensional flow, $v_x(x,y)$, $v_y(x,y)$, is not capable of generating a kinematic dynamo (for the special case of axisymmetric flow and magnetic field geometry this had been noted by the English scientist T. G. Cowling in 1934).

Later, in a joint paper with A. A. Ruzmaikin (**8**) this theorem was generalized to the case where the two-dimensional flow depends on all three coordinates, x, y, z. This same paper introduced the concept of a *slow* dynamo, whose increment tends to zero or becomes negative as the magnetic viscosity approaches zero. It is interesting that in the case of axisymmetric flow considered by Cowling, the antidynamo theorem is, generally speaking, incorrect. It is sufficient to abandon the condition of axial symmetry of the magnetic field to obtain a slow dynamo.

The general "antidynamo theorem" of Zeldovich is related to the fact that in the two-dimensional, singly-connected[7] case, a field of divergence 0 is given by a *scalar* which is invariantly related to it (a "streamline function" or "Hamilton function"). If the field is frozen into the fluid then the corresponding scalar is carried with the flow and, in particular, the integral of its square is conserved at $D = 0$ and decreases for $D > 0$, which is in fact why a dynamo is impossible.

---

[7]The influence of the flow region topology was studied by Ya.B. together with V. I. Arnold, A. A. Ruzmaikin, and A. D. Sokolov [66].

In the three-dimensional case this reasoning does not work. Moreover, velocity fields with zero divergence are known which exponentially stretch a frozen magnetic field. For example, a computer experiment (M. M. Henon, France, 1966) demonstrated that a field with, say, the components ($\cos y + \sin z, \cos z + \sin x, \cos x + \sin y$) has this property. In agreement with the ideas of the modern theory of dynamical systems, any generic three-dimensional field has the property of leading to exponential divergence of particle paths (the only exceptions are singular velocity fields of a special type, for example, such that all their flow lines are closed or fit onto two-dimensional closed surfaces).

Thus, at $D = 0$ in a steady flow the magnetic field will evidently be strengthened by almost any three-dimensional velocity field, e.g., the one written out above (this has not been proved with mathematical rigor, but it has been established with practical reliability in computer experiments).

However, it is not at all clear how a magnetic field will behave for small finite values of the magnetic viscosity D. The problem is that a mode of the field B which is increasing at $D = 0$ is, generally speaking, strongly broken up (saturated with high-order harmonics). Diffusion quickly extinguishes high-order harmonics; the problem is to find which will win out: particle expansion by the flow, which increases the magnetic field increase, or diffusion, which decreases it.

It is interesting to note that a fast steady kinematic dynamo is built on appropriate, though unusual, three-dimensional manifolds. In these examples the most rapidly growing mode at $D = 0$ of the field B is not broken up. It would appear that in regions of normal three-dimensional space or in a periodic problem this situation is impossible.

The fact that a mechanism of exponential expansion of fluid particles is possible at all, and that it should lead to the dynamo effect, was explained by Ya.B. and S. I. Vainshtein (see [67]) with the help of a graphic model (Fig. 2). A ring (*1* in Fig. 2) expands and contracts like a rubber band (*2*). The ring's volume does not change, but its length (and hence the magnetic field directed along the ring) increases.

Ya.B. assumes that 1) a fast dynamo based on this mechanism can be generated by a time-periodic three-dimensional fluid flow, but that 2) a fast kinematic dynamo (in three-dimensional space with steady flow and periodic boundary conditions) is impossible.

To date neither the one nor the other has been formally proved with complete rigor.

The special role of three-dimensionality in this field amplification mechanism becomes especially clear when it is compared with an externally similar cycle proposed earlier by the Swedish scientist H. Alfven (1950) [68]. In the Alfven cycle (Fig. 3, borrowed from [68]), an initial tube (*1*) is stretched, and opposite sides of the ring (*2*), which is filled with magnetic field lines,

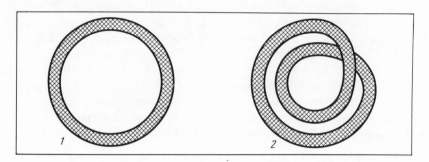

Fig. 2

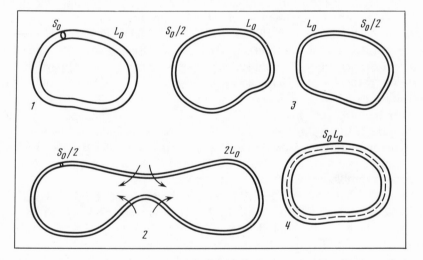

Fig. 3

are brought together as the fluid moves. Then a break occurs, forming two independent rings (*3*) which are superimposed upon one another; this eventually leads to doubling of the field (*4*), as in Zeldovich's example. Unlike the latter, however, this cycle may be realized by plane flows as well, but one of its stages—the breaking of the ring—requires violation of the frozenness of the field. In this sense Alfven's example serves as an illustration of a "slow dynamo," and not of a fast one as in Zeldovich's example.

The velocity field, $v_x = \cos y + \sin z \ldots$, mentioned earlier generates a kinematic dynamo only at certain restricted values of Re^{-1} (of order 0.01–0.05 according to calculations by E. I. Korkina, see [69]).

The problem of a kinematic dynamo in a steady velocity field can be treated mathematically as a problem of the effect of a small diffusion or round-off error on the Kolmogorov-Sinai entropy (or Lyapunov exponent) of a dynamical system which is specified by the velocity field v. This problem, on which Ya.B. worked actively, therefore has a general mathematical nature as well, and each step toward its solution is simultaneously a step forward in several seemingly distant areas of modern mathematics.

Ya.B. is the author of a textbook highly rated by many mathematicians, *Higher Mathematics for Beginners and its Applications to Physics* [70] (the latest edition, written together with I. M. Yaglom, is *Higher Mathematics for Beginning Physicists and Engineers* [71]).

The need for writing this book was dictated by the peculiar situation which had become established in mathematics in the middle of the twentieth century. The deductive-axiomatic style, which emerged at the end of the nineteenth century in works on the foundations of mathematics, found many adherents. (As Bertrand Russell said in promoting the axiomatic method, the latter has many advantages similar to those of stealing versus doing honest work.) By the mid-twentieth century use of this method had become widespread in the teaching of mathematics as well.

In practical terms, this basically means that an enormous number of concepts developed in the process of a long historical evolution are given to students in the form of logically flawless, but absolutely incomprehensible, definitions (whose meaning, like the prophecies, is revealed only to a chosen few).

Important meaningful facts, like the Pythagorean theorem, are then transformed into easily overlooked definitions. Of course, the substitution of definitions for theorems decreases the number of pages in textbooks, but then it does not leave the student any hope of understanding why it is necessary to consider these particular definitions. Moreover, it creates the impression that mathematics is the study of corollaries of arbitrary axioms.

Unfortunately, so many mathematical texts have been written in this language that, by the middle of the twentieth century, the threat of a complete break between mathematics and the rest of the natural sciences had become a reality.

The burden of "rigorous" justification forced authors to throw out of textbooks on analysis any nontrivial and truly interesting applications. A comparison of typical purged modern textbooks with the brilliant courses written fifty years before (for example, by the German mathematicians R. Courant and D. Hilbert, and others) clearly shows how much of importance was lost with the introduction of axiomatic and set-theoretical methods in the teaching of mathematics.

Ya.B. was among the first to raise the alarm over this trend and to begin to actively correct it. His *Higher Mathematics for Beginners and its Applications to Physics* became an important milestone on the path toward revival of meaningful courses in mathematics. Today, even some of the most fervent opponents of the book have taken positions close to it. And, indeed, it is difficult not to agree with Newton that the ability to quickly calculate the area of a figure is more important for a student than the ability to prove the existence of a midpoint of any segment or the existence of an integral of a continuous function.

And, while Ya.B.'s textbook allowed many "beginning physicists and technicians" to master meaningful mathematics without losing time on justifications, the book also turned some mathematicians toward interesting real problems in physics and technology.[8]

Comparing the approaches of physicists (R. Hooke) and mathematicians (himself), Newton wrote that "Mathematicians that find out, settle & do all the business must content themselves with being nothing but dry calculators & drudges & another that does nothing but pretend & grasp at all things must carry away all the invention as well of those that were to follow him as of those that went before."

In the textbook (as in his papers), Ya.B. sets himself the task of uniting the advantages of both approaches: Hooke's freedom in "grasping at all things" when constructing mathematical models of reality, and Newton's rigorousness in studying these models:

> In the investigation of Nature everything is necessary: mastery of mathematical technique and overcoming mathematical difficulties, boldness of ideas and physical intuition, skillfully staged experiment and mathematical modelling—all these very different approaches are equally necessary, and only their union can lead to progress [71, p. 469].

Let us simply mention two other textbooks: *Elements of Applied Mathematics* (with A. D. Myshkis) [72] and *Elements of Mathematical Physics. Vol. 1. Non-Interacting Particle Medium* (with A. D. Myshkis) [73] which are a continuation of *Higher Mathematics for Beginners and its Applications to Physics*, published in 1960, and are written with the same pedagogical goals.

---

[8]The Editors feel that they should note other positive responses by mathematicians to Ya.B.'s book as well. In 1968 the RSFSR Ministry of Education, on the recommendation of the Scientific Council of the AS USSR Institute of Applied Mathematics chaired by M. V. Keldysh, recommended the book "Higher Mathematics for Beginners and its Applications to Physics" as a textbook for physico-mathematical schools and elective study. A very prominent American mathematician, Lippman Bers, wrote, "Finally, I would like to indicate how I was influenced by two very nonstandard books on mathematical analysis: a historical survey written by the outstanding mathematician O. Toeplitz, and a textbook written by a well-known physicist, Ya. Zeldovich (unfortunately not yet available to the British or American reader)."

* * *

The years have no power over Ya. B. Zeldovich. He continues to work with youthful ardor. Lately he pays more attention to cosmology, but because of old attachments he directs the Scientific Council of the AS USSR on the problem of "Theoretical Foundations of Combustion Processes."

Ya. B. Zeldovich's papers, as always, attract the close attention of Soviet and foreign scientists.

The next book of Ya.B.'s selected works (*Particles, Nuclei, and the Universe*) will contain biographies and essays on the works of A. Einstein [46*], the outstanding Soviet physicist, Academician B. Konstantinov [47*], and the eminent Soviet scientist, Professor D. A. Frank-Kamenetskiĭ. These articles characterize not only the people they describe, but they also reveal the aspirations and world view of Ya.B. himself, just as does his self-critical scientific autobiography [49*].

The USSR values highly the scientific work of Ya. B. Zeldovich. He was elected a full member of the Academy of Sciences of the USSR, three times awarded the title of Hero of Socialist Labor (the highest Soviet award), and rewarded with many orders. Ya. B. Zeldovich is a Lenin Prize winner, and four times he was a winner of the State Prize of the USSR. He has been elected a foreign member of many academies in other countries: The German Academy Leopoldina, the American Academy of Arts and Sciences, the U. S. National Academy of Sciences (Cambridge, USA), the Royal Society (London), and the Hungarian Academy of Sciences.

As before, Ya. B. Zeldovich publishes papers on his own and with his students and colleagues. As always, he is full of life and wit. The years and international fame, just as for the majority of truly great scientists, have not burdened him with self-importance.

We warmly wish Yakov Borisovich good health and many more years of hard, intense work.

The Editors
Moscow, 1984

References

1. *Krylov O. V.*—Kinetika i kataliz **22**, 15–29 (1981); in: Sbornik trudov vsesoĭuznoĭ konferentsii po mekhanizmam kataliticheskikh reaktsiĭ [Proceedings of the all-Union conference on mechanisms of catalytic reactions]. Novosibirsk: Nauka, 35 p. (1982).
2. *Roginskiĭ S. Z., Zeldovich Ya. B.*—Acta physicochim. URSS **1**, 554–594 (1934); Zeldovich Ya. B.—Zh. fiz. khimii **6**, 234–242 (1935); Acta physicochim. URSS **1**, 449–469 (1934).

3. *Aharoni C., Tomkins F. C.*—In: Advances in Catalysis. N.Y., London: **21**, 1-12 (1970).

4. *Gomer R.*—Discuss. Faraday Soc. **28**, 29-36 (1959).

5. *Palmer R. L.*—Chem. Technol. **80**, 702-709 (1978).

6. *Zeldovich Ya. B., Roginskĭ S. Z.*—Acta physicochim. URSS **8**, 361-363 (1932).

7. *Zeldovich Ya. B., Kompaneets A. S.*—In: Sbornik, posvĭashchenny 70-letiĭu akademika A. F. Ioffe [Seventieth Birthday Festschrift for A. F. Ioffe]. Moscow: Izd-vo AN SSSR, 61-71 (1950).

8. *Roginskĭ S. Z., Sena L. A., Zeldovich Ya. B.*—Phys. Ztschr. Sowietunion **1**, 630-639 (1932).

9. *Zeldovich Ya. B., Ovchinnikov A. A.*—Pisma v ZhETF **13**, 636-638 (1971).

10. *Zeldovich Ya. B., Maksimov L. A.*—ZhETF **70**, 76-80 (1975).

11. *Zeldovich Ya. B., Sobelman I. I.*—Pisma v ZhETF **21**, 368-373 (1975).

12. *Zeldovich Ya. B.*—Uspekhi fiz. nauk **115**, 161-197 (1975).

13. *Zeldovich Ya. B.*—Uspekhi fiz. nauk **110**, 139-151 (1973).

14. *Baz A. I., Zeldovich Ya. B., Perelomov A. M.* Rasseĭanie, reaktsii i raspady v nerelĭativistskoĭ kvantovoĭ mekhanike [Scattering, Reactions and Fission in Non-Relativistic Quantum Mechanics]. 2nd Ed., rev. and exp. Moscow: Nauka, 480 p. (1971).

15. *Zeldovich Ya. B.* Teoriya udarnykh voln i vvedenie v gazodinamiku [Theory of shock waves and introduction to gasdynamics]. Moscow: Izd-vo AN SSSR, 186 p. (1946).

16. *Zeldovich Ya. B., Raizer Yu. P.* Fizika udarnykh voln i vysokotemperaturnykh gidrodinamicheskikh yavleniĭ [see next reference]. 2nd Ed., rev. and exp. Moscow: Nauka, 686 p. (1966).

17. *Zeldovich Ya. B., Raizer Yu. P.* Physics of shock waves and high temperature hydrodynamics phenomena. N.Y.: Acad. press, Vol. 1 464 p., Vol. 2 451 p. (1966).

18. *Einstein A.* Sobr. nauchnykh trudov [Collected Scientific Works] **3**, 423-429. Moscow: Nauka (1966).

19. *Zeldovich Ya. B., Gershanik Ya. T., Rozlovskĭ A. I.*—Zh. fiz. khimii **24**, 85-95 (1950).

20. *Zeldovich Ya. B.*—ZhETF **32**, 1577-1578 (1957).

21. *Raizer Yu. P.*—ZhETF **33**, 101-104 (1957).

22. *Zeldovich Ya. B., Raizer Yu. P.*—Uspekhi fiz. nauk **63**, 613-641 (1957).

23. *Zeldovich Ya. B., Raizer Yu. P.*—ZhETF **47**, 1150-1161 (1964).

24. *Zeldovich Ya. B., Kormer S. B., Sinitsyn M. V., Kuryapin A. I.*—Dokl. AN SSSR **122**, 48-50 (1958).

25. *Zeldovich Ya. B., Kormer S. B., Krishkevich G. V., Yushko K. B.*—Dokl. AN SSSR **158**, 1051-1063 (1964).

26. *Zeldovich Ya. B., Kormer S. B., Krishkevich G. V., Yushko K. B.*—Fizika goreniĭa i vzryva **5**, 312-315 (1969).

27. Sbornik Stateĭ po Teorii Goreniĭa [Collected Articles on the Theory of Combustion]. Moscow, Leningrad: Oborongiz (1940).

28. Khimicheskaĭa kinetika i tsepnye reaktsii: K semidesĭatiletiĭu N. N. Semenova [Chemical Kinetics and Chain Reactions: On the Seventieth Birthday of N. N. Semenov]. Moscow: Nauka (1966).

29. Teoriya goreniĭa i vzryva [Theory of Combustion and Explosion]. Moscow: Nauka (1981).

30. Teoriĭa goreniĭa porokhov i vzryvchatykh veshchestv [Theory of Combustion of Powders and Explosives]. Moscow: Nauka (1982).

31. *Zeldovich Ya. B., Semenov N. N.*—ZhETF **10**, 1116–1136, 1427–1440 (1940).

32. *Barskiĭ G. A., Zeldovich Ya. B.*—Zh. fiz. khimii **25**, 523–537 (1951).

33. *Zeldovich Ya. B., Barenblatt G. I., Librovich V. B., Makhviladze G. M.* Matematicheskaĭa teoriĭa goreniĭa i vzryva [Mathematical Theory of Combustion and Explosion]. Moscow: Nauka, 478 p. (1980).

34. *Zeldovich Ya. B., Istratov A. G., Kidin N. I., Librovich V. B.*—Preprint of the AS USSR Institute of Problems of Mechanics. Moscow, No. 143 (1980); Combust. Sci. and Technol. **24**, 1–13 (1980); Arch. Combust. **1**, 181–202 (1981).

35. *Zeldovich Ya. B.*—Fizika goreniĭa i vzryva **7**, 463–476 (1971).

36. *Gelfand I. M.*—Uspekhi mat. nauk **14**, 87–94 (1959).

37. *Aronson D. G., Weynberger H. F.*—In: Lecture Notes in Mathematics. N.Y.: Springer, Vol. 44, 5–49 (1975).

38. *Aronson D. G., Weynberger H. F.*—Adv. Math. **30**, 33–76 (1978).

39. *Barenblatt G. I., Zeldovich Ya. B.*—PMM **12**, 856–859 (1957).

40. *Zeldovich Ya. B., Khariton Yu. B.*—Uspekhi fiz. nauk **23**, 329–348 (1941).

41. *Zeldovich Ya. B., Saakyan D. B.*—ZhETF **78**, 1233–1236 (1980).

42. *Zeldovich Ya. B.*—ZhETF **36**, 1952–1953 (1959).

43. *Zeldovich Ya. B., Gershtein S. S.*—Uspekhi fiz. nauk **71**, 581–630 (1960).

44. *Braginskiĭ V. B., Zeldovich Ya. B., Martynov V. K., Migulin V. V.*—ZhETF **52**, 29–39 (1967).

45. *Zeldovich Ya. B.*—Uspekhi fiz. nauk **86**, 303–314 (1965).

46. *Zeldovich Ya. B., Okun' L. B., Pikelner S. B.*—Uspekhi fiz. nauk **87**, 113–124 (1965).

47. *Zeldovich Ya. B., Smorodinskiĭ Ya. A.*—ZhETF **41**, 907–911 (1961).

48. *Shvartsman V. F.*—Pisma v ZhETF **9**, 315–317 (1969).

49. *Zeldovich Ya. B., Kobzarev I. Yu., Okun L. B*—ZhETF **67**, 3–11 (1974).

50. *Zeldovich Ya. B., Dolgov A. D.*—Uspekhi fiz. nauk **130**, 559–614 (1980).

51. *Zeldovich Ya. B., Okun' L. B.*—Phys. Lett. B **78**, 597–598 (1978).

52. *Zeldovich Ya. B.*—ZhETF **72**, 18–21 (1977).

53. *Budker G. I.*—Uspekhi fiz. nauk **89**, 553–575 (1966).

54. *Baz' A. I., Zeldovich Ya. B., Goldanskiĭ V. I., Goldberg V. Z.*—Lëgkie i promezhutochnye ĭadra vblizi granits nuklonnoĭ stabilnosti [Light and Intermediate Nuclei Near the Boundaries of Nucleon Stability]. Moscow: Nauka, 172 p. (1972).

55. *Doroshkevich A. G., Zeldovich Ya. B., Novikov I. D.*—ZhETF **49**, 170–181 (1965).

56. *Doroshkevich A. G., Novikov I. D.*—Dokl. AN SSSR **154**, 809–811 (1964).

57. *Zeldovich Ya. B., Novikov I. D.* Relĭativistskaĭa astrofizika [Relativistic Astrophysics]. Moscow: Nauka, 654 p. (1967).

58. *Zeldovich Ya. B., Novikov I. D.* Teoriĭa tĭagoteniĭa i evolĭutsiĭa zvëzd [Theory of Attraction and Evolution of Stars]. Moscow: Nauka, 484 p. (1971).

59. *Zeldovich Ya. B., Novikov I. D.* Stroenie i evolĭutsiĭa Vselennoĭ [Structure and Evolution of the Universe]. Moscow: Nauka, 735 p. (1975).

60. *Zeldovich Ya. B., Doroshkevich A. G., Longair M. S.*—Month. Notic. Roy. Astron. Soc. **147**, 139–144 (1970).
61. *Bisnovatyĭ-Kogan G. S., Zeldovich Ya. B.*—Astrofizika **5**, 425–432 (1969).
62. *Zeldovich Ya. B.*—Uspekhi fiz. nauk **133**, 479–503 (1981).
63. *Zeldovich Ya. B., Mamaev A. V., Shandarin S. F.*—Uspekhi fiz. nauk **139**, 153–163 (1983).
64. *Arnold V. I., Shandarin S. F., Zeldovich Ya. B.*—Journ. Geophys. Astrophys. Fluid Dynamics **20**, 111–125 (1982).
65. *Shandarin S. F.*—Pisma v Astron. Zh. **9**, 195–199 (1983).
66. *Arnold V. I., Zeldovich Ya. B., Ruzmaikin A. A., Sokolov D. D.*—Dokl. AN SSSR **266**, 1357–1361 (1982).
67. *Zeldovich Ya. B., Vainshtein S. I.*—Uspekhi fiz. nauk **106**, 431–457 (1972).
68. *Alfven H.*—Tellus **2**, 74–83 (1950).
69. *Arnold V. I.*—Uspekhi mat. nauk **38**, 226–235 (1983).
70. *Zeldovich Ya. B.* Vysshaĭa matematika dlĭa nachinaĭushchikh i ee prilozheniĭa k fizike [Higher Mathematics for Beginners and its Applications to Physics]. Moscow: Fizmatgiz, 460 p. (1960); 5th Ed. Moscow: Nauka, 560 p. (1970).
71. *Zeldovich Ya. B., Yaglom I. M.* Vysshaĭa matematika dlĭa nachinaĭushchikh fizikov i tekhnikov [Higher Mathematics for Beginning Physicists and Engineers]. Moscow: Nauka, 510 p. (1982).
72. *Zeldovich Ya. B., Myshkis A. D.* Elementy prikladnoĭ matematiki [Elements of Applied Mathematics]. 3rd Ed. Moscow: Nauka, 592 p. (1972).
73. *Zeldovich Ya. B., Myshkis A. D.* Elementy matematicheskoi fiziki. T. 1 Sreda iz nevzaimodeĭstvuĭushchikh chastits [Elements of Mathematical Physics. V. 1 Non-Interacting Particle Medium]. Moscow: Nauka, 351 p. (1973).
74. *Šlygin A., Frumkin A.*—Acta physicochim. URSS **3**, 791–808 (1935).
75. *Temkin M. I.*—Zh. fiz. khimii **15**, 266–332 (1941).

PART ONE

I

Adsorption and Catalysis

1

On the Theory of the
Freundlich Adsorption Isotherm[*]

§**1.** By far the majority of available experimental data on the adsorption
of gases on the surface of a solid substance totally fails to agree with the
simple laws derived by Langmuir. Moreover, even the linearity of the adsorption isotherm at very low surface coverage, which follows for adsorption
on a uniform surface from the general laws of statistical mechanics, independently of any assumptions regarding the interaction between adsorbed
particles, the dependence of the potential energy on distance, etc.[1]—even
this linearity of the initial portion of the isotherm represents not so much
an unambiguous result of experiments as the assumptions of theorists [1].

Of the various empirical equations proposed for the adsorption isotherm,
the most significant is undoubtedly the isotherm

$$q = Cp^{1/n}. \tag{1}$$

Its value for the description of experimental data is convincingly elucidated in
the well-known monograph by McBain [2, p. 5]. This formula was introduced
originally by analogy with the distribution of a substance between phases in
which it exists in various states of association. However, experimental values
very soon showed the impossibility of this explanation: it allows neither
fractional values of n which vary smoothly with temperature, nor such values
as $n = 10$ for the adsorption of iodine on starch [3].

---

[*]Acta Physicochimica URSS 1 6, 961–974 (1934).
[1]We exclude only the possibility of dissociation in the absorbed layer.

Isotherm (1) is often obtained [1] under conditions which physically exclude the dissociation of adsorbed molecules, for example, in the adsorption of O_2 and Ar on SiO_2 at low temperatures [4].

Let us note that since complete adsorption is limited by the size of the surface, at least under conditions which are far from condensation, for greater surface coverage deviations of experimental data from equation (1), where $\lim q = \infty$, unavoidably occur. It would be incorrect, however, to consider that this equation satisfies the experimental data only insofar as it coincides in practice with the Langmuir isotherm in some middle portion [5]. In many cases an isotherm which satisfies the Langmuir equation at high surface coverage (correctly describing the saturation of the surface) definitely diverges from it at low coverage, following the nonlinear law (1) (cf. 4 below).

§2. We shall consider several proposed explanations of isotherm (1). Chakravarti and Dhar [6] derived isotherm (1) from the conception of multiple adsorption. Here, if one adsorbed particle occupies n empty locations on a surface, Chakravarti and Dhar[2] assume that the probability of adsorption is proportional to $(1 - \vartheta^n)^n$, while the probability of desorption is ϑ^n.

However, such expressions may be obtained only by assuming that n empty sectors occupied by a particle being adsorbed may be located arbitrarily far from one another, and that any n occupied locations on the surface, if freed simultaneously, will release from the surface a molecule of the adsorbed substance.

In other words, the formulas which Chakravarti and Dhar write correspond to the assumption of dissociation of an adsorbed particle into n separate parts. The authors themselves reject this assumption at the beginning of the article.

Zeise [7], apparently independently of them, considered a special case of multiple adsorption: adsorption of a molecule at two empty locations of the surface. In addition, besides the proportionality of the probability of adsorption to $(1 - \theta^2)$ (in our notation), he also makes quite strange assumptions about the mechanism of adsorption in his paper. Regardless of the fact that he is considering an equilibrium at which the temperature of the adsorbent does not differ from the temperature of the surrounding gas, Zeise writes: "Escape (of adsorbed molecules) from a unit of the surface must consist of those molecules which are knocked out by molecules (from the gas phase) which collide with them, since we cannot consider that molecules of gas, once adsorbed, may be freed solely by the effect of the *insignificant* impulses of the *thermal motion* of *molecules* of the underlying solid body."[3]

But most extraordinary is the fact that, if we accept Zeise's assertion, it turns out that both adsorption and desorption are proportional to the pressure in the gas phase. The equilibrium adsorbed amount turns out not to

---

[2]We denote by ϑ the occupied portion of the surface; $\vartheta = 1 - \theta$ in the notation of Chakravarti and Dhar.

[3]Words in parentheses and emphasis mine.—Ya. Z.

depend on the pressure. Zeise manages to produce an isotherm only because, several lines later, he forgets about the proportionality of the probability of desorption, $\vartheta = x\mu$, to the number of collisions with the surface, μ, and subsequently treats ν as a constant.

We mentioned above the general agreement on the linearity of the initial sector of the adsorption isotherm on a uniform surface. The isotherms derived by Chakravarti and Dhar, $q = Cp^{1/n}/(b + p^{1/n})$, and by Zeise, $q = [ap/(p + b)]^{1/2}$, are interesting in that, together with the nonlinearity of the isotherm at the beginning ($q \sim p^{1/n}$; $q \sim \sqrt{p}$), they also reflect the tendency of adsorption towards saturation for $p \to \infty$.

A number of derivations of isotherm (1) are based on the well-known formula of Gibbs,

$$q = \frac{1}{RT}\frac{d\sigma}{d\log p} = \frac{p}{RT}\frac{d\sigma}{dp},$$

where equation (1) is obtained with the help of some incomplete assumption as to the dependence of the surface tension σ on q: $\pi\Omega = nRT$ where $\pi = \sigma_0 - \sigma$ is the surface pressure, Ω is the surface occupied by one gram molecule of adsorbed substance, the reciprocal of q (Rideal [8]); or $\sigma = \sigma_0(1 - q/q_\infty)$ (Chakravarti and Dhar [6]); or, finally, the *ad hoc* $(\sigma_0 - \sigma)/\sigma = Cp^{1/n}$ (Freundlich [9]).

Meanwhile, it is evident that with sufficient dilution any adsorbed substance behaves, independently of its two-dimensional mobility, as a two-dimensional ideal gas with the equation of state $\pi\Omega = RT$. Together with the Gibbs equation, this is one proof (if such is in fact necessary) of the linearity of the initial part of the isotherm on a uniform surface.

§3. For some number of elementary sectors with given properties (heat of adsorption, vibrational volume) arbitrarily distributed on a surface, the adsorption isotherm of Langmuir,

$$q = \frac{ap}{p + b}, \tag{2}$$

should strictly obtain (under the usual assumptions that not more than one molecule can be adsorbed in each sector and that mutual interactions may be neglected). In this formula, b depends upon the properties of the sectors, while a is proportional to their number.

The isotherm is linear for $p \ll b$; for p of the same order as b the increase of q begins to lag strongly behind that of p. Therefore, the nonlinearity of the experimental adsorption isotherm in some region of pressures should be considered (again, in the absence of dissociation or strong interaction between adsorbed particles) as proof of the presence on the surface of points with values of b from the Langmuir formula which are (for a given gas) of the order of magnitude of the pressures under consideration.

A rational explanation of equation (1) will be obtained if we find a distribution of points of the surface according to their adsorptive capacity such that the superposition of the Langmuir isotherm on each "kind" of points leads to $q = Cp^{1/n}$. Mathematically, the problem is formulated as an integral equation of the first kind,

$$q(p) = p \int_0^\infty \frac{a(b)}{p+b} \, db, \tag{3}$$

where $q(p)$ is the equation of the observed adsorption isotherm, and $a(b)$ is the function sought. Convergence of the integral $\int_0^\infty a(b) \, db$ to q_∞ corresponds to the finite limit of adsorption for $p \to \infty$. In Langmuir's [5] original interpretation of adsorption on a non-uniform surface, we encounter two kinds of formulas: for adsorption on a crystal surface, Langmuir writes

$$q = \sum_i \frac{a_i p}{p + b_i}, \tag{4}$$

while for adsorption on an amorphous surface, where continuous variation of b is possible, he gives the following formula for q:

$$q = \int \frac{ap}{p+b} \, dS. \tag{4a}$$

The integral is taken over the entire geometric surface, without separating out or collecting together points with equal values of b or introducing the function $a(b)$.

We shall now proceed to the solution of integral equation (3). Mathematics gives no general method for the solution of equations of this type (the equation is singular). This situation more than justifies the method we apply, one of guessing rather than finding the solution, based on the special properties of the "nucleus" of the equation of the function, $p/(p+b)$. Therefore the emphasis shifts to verification of the solution obtained. We shall check the correctness of the "guessed" solution by the irreproachable method of substituting it into the exact integral equation. In this same way we shall also find the numerical coefficients.

We schematize the Langmuir isotherm with b given by replacing it with two straight lines,

$$q = ap/b, \quad 0 < p \le b; \qquad q = a, \quad b \le p, \tag{5}$$

which qualitatively describe the initial and saturation portions of the isotherm. After this substitution, the solution is obtained very simply:

$$q(p) = \int_0^p a(b) \, db + \int_p^\infty a(b) p \, \frac{db}{b}. \tag{6}$$

Differentiating equation (6) twice with respect to p, we easily find

$$q'(p) = \int_p^\infty a(b) \, \frac{db}{b}$$

and

$$q''(p) = \frac{-a(p)}{p}, \qquad a(p) = -pq''(p), \tag{7}$$

whereby equation (3) is (approximately) solved. It is easy to see that this solution correctly reflects the basic properties of the true $a(p)$; the requirement that $a(p) > 0$ gives, for a superposition of the Langmuir isotherms, the obvious $q''(p) < 0$.

It is also easy to verify that, automatically,

$$\int_0^\infty a(b)\, db = q_\infty. \tag{8}$$

Formula (7) is a mathematical expression of the fact that the nonlinearity of the isotherm for some p_1 is related to the presence of sectors of the surface for which $b = p_1$.

§**4.** Let us return to equation (3); from it we easily find the corresponding expression

$$a(b) = Ab^{1/n-1}. \tag{9}$$

In order to make $\int_0^\infty a(b)\, db$ convergent, it is sufficient to break off $a(b)$ at some b_0. The infinite value of $b^{1/n-1}$ at $b = 0$ is of no consequence since $\int_0^{b_0} b^{1/n-1}\, db$ converges.

After substitution into formula (3), the expression found for $a(b)$ gives

$$q(p) = Ap \int_0^{b_0} \frac{b^{1/n-1}}{p+b}\, db. \tag{10}$$

In finite form the integral cannot be taken for an arbitrary value of n. Let us separately consider its asymptotic behavior at $p \ll b_0$ and $p \gg b_0$. At low pressure

$$q(p) = Ap^{1/n} \int_0^{b_0/p} \frac{x^{1/n-1}}{1+x}\, dx, \qquad x = \frac{b}{p}, \tag{11}$$

and, taking the limit,

$$\lim_{p \to 0} q(p) = \frac{A\pi p^{1/n}}{\sin(\pi/n)} \tag{12}$$

since, as we know,

$$\int_0^\infty \frac{x^{m-1}}{1+x}\, dx = \frac{\pi}{\sin m\pi}, \qquad 0 < m < 1. \tag{13}$$

Expressing A in terms of the saturation, which we determine by formula (8), we obtain

$$\lim_{p \to 0} q(p) = \frac{q_\infty \pi b_0^{-1/n} p^{1/n}}{n \sin(\pi n)} \cong q_\infty (p/b_0)^{1/n}. \tag{14}$$

Formulas (12) and (14) prove that the distribution (9) which we have found actually leads to isotherm (3) at low pressure.

For $p > b_0$, we expand $p/(p+b) = 1/(1+b/p) = 1 - b/p + \ldots$, and integrate by terms, after which we obtain

$$q(p) = q_\infty \left(1 - \frac{b_0}{p(n+1)} + \ldots\right). \tag{15}$$

With accuracy to within higher order terms, expression (15) coincides with the expansion of separate Langmuir isotherms in $1/p$ at high pressure:

$$\frac{ap}{p+b'} = a(1 - \frac{b'}{p} + \ldots); \quad b' = \frac{b_0}{n+1}.$$

Thus, distribution (9) gives us an adsorption isotherm which not only coincides with the given Freundlich equation (1) for small p, but also describes the tendency of q to saturation for large p which occurs in experiment (cf. §1). Precisely this character, incidentally, is exhibited by the experimental data of Zeise [7], Chakravarti and Dhar [6], and Roginskiĭ and the author [10].

§5. The quantities b with which we characterized the adsorptive properties of points on the surface can be written in the form $g\exp(-Q/RT)$, where g varies only slightly with the temperature compared to the exponent. Assuming that the value of g is the same for all points on the surface, we may find a distribution function of points on the surface according to the heat of adsorption of the gas, $A(Q)$:

$$A(Q)\,dQ = a(b)\,db = D'e^{-Q/nRT}\,dQ, \quad Q > Q_0, \tag{16}$$

where D' is independent of Q; Q_0 is determined from $b_0 = g\exp(-Q_0/RT)$.

Denoting nRT by γ, and normalizing for q_∞ by a formula analogous to (8), we easily obtain

$$A(Q) = q_\infty \gamma e^{Q_0/\gamma} e^{-Q/\gamma} = De^{(Q_0-Q)/\gamma}, \quad Q > Q_0. \tag{17}$$

The form of distribution (17) recalls a Boltzmann expression with modulus of distribution γ. Attempts at a direct physical explanation of this result are thwarted by the obvious dependence of γ, not only on the state of the surface, but also on the nature of the gas whose adsorption proceeds according to equation (1). Nevertheless, formula (17) makes very plausible the experimentally observed constancy of the functional dependence $A(Q)$ itself which leads to equation (1). It seems natural that with training or sintering of the surface, the liberation or destruction of points with different heats of adsorption may proceed in such a way as to preserve the exponential relation between A and Q, changing only the constants D, Q_0, and especially γ.

The growth in n, and therefore in γ, for the adsorption of CO_2 on charcoal in proportion to the increase of the training temperature of the charcoal for a given adsorption temperature was observed by Magnus and Cahn [11]. The same picture was very clearly obtained in the work cited above for CO_2 on MnO_2 [10].

In both cases the isotherms followed equation (1) down to the very lowest pressures.

§6. $A(Q)$ in formula (17) does not explicitly contain either the temperature of the adsorption experiment or the obtained values of n. Let us now proceed backwards in order to find, for a given $A(Q)$, the dependence of the constants n and C of isotherm (1) on the temperature (considering that the surface, and hence $A(Q)$, is constant).

We ignore the temperature dependence of g. Recalling how we introduced γ, we easily find

$$n = \frac{\gamma}{RT}, \quad \frac{1}{n} = \frac{RT}{\gamma}. \tag{18}$$

The exponent $1/n$ of isotherm (1) is linearly dependent on the temperature. At sufficiently high temperatures when, by (18), $1/n > 1$, the theory gives simply a linear course of the isotherm at the beginning; an isotherm of the form $q = Cp^{1/n}$, $1/n > 1$, cannot be obtained by superposition of the Langmuir isotherm. Indeed, in this case $a(b)$ still has the form of equation (9) with $1/n > 1$, but we may no longer seek the asymptotic form of the beginning of the isotherm by formulas (11) and (12) since

$$\int_0^\infty \frac{x^{m-1}}{1+x}\,dx, \quad m > 1$$

diverges. Instead, we represent

$$q(p) = Ap \int_0^{b_0} \frac{b^{1/n-1}}{p+b}\,db$$

and take advantage of the existence at $1/n - 1 > 0$ of

$$\lim_{p \to 0} \int_0^{b_0} \frac{b^{1/n-1}}{p+b}\,db = \int_0^{b_0} b^{1/n-2}\,db = \frac{b_0^{1/n-1}}{1/n-1}.$$

Expressing A in terms of q_∞, we obtain in place of formula (14)

$$\lim_{p \to 0} q(p) = \frac{q_\infty b_0^{-1} p}{1/n - 1} \cong \frac{q_\infty p}{b_0'} = Fp. \tag{19}$$

The smooth rise of $1/n$ with temperature to unity and the sudden stop at this value is very clearly observed in the work of Urry [4]. We borrow from him (Fig. 1) the curve $1/n - T$ for O_2 on SiO_2. The explanation of this regularity given by Urry is essentially that isotherm (1) (below 195°K) is caused by capillary condensation, and that the linear ones (above 195°K) are due to ordinary molecular adsorption. Even if we ignore the fact that the isotherms (1) are often observed under conditions when capillary condensation is absolutely excluded, Urry's own data force him to take T_K for oxygen in capillaries as 195° instead of the usual 155°, and to apply the theory of capillary condensation to capillaries with radii down to 10^{-8}. All this makes his explanation little convincing.

Let us finally consider the temperature dependence of the coefficients of $p^{1/n}$ and p in the formulas

$$q = Cp^{1/n}, \qquad q = Fp$$

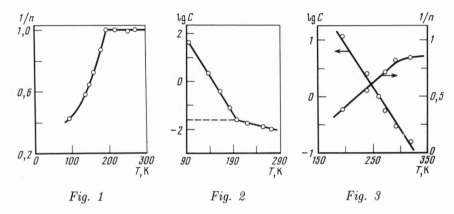

Fig. 1   Fig. 2   Fig. 3

Discarding the factors $n \sin(\pi/n)$ and $1/n - 1$ of order unity in formulas (14) and (19), we find

$$\log C = \log q_{\infty} + \frac{Q_0}{nRT} - \frac{1}{n}\log g$$
$$= \log q_{\infty} + \frac{Q_0}{\gamma} - \frac{RT \log g}{\gamma} = K_1 - \frac{RT \log g}{\gamma}, \qquad (20)$$

where C varies linearly with T (and not with $1/T$ as one might expect) up to $T = \gamma/R$ (at which $1/n = 1$). This pattern is well corroborated by Urry's data (Fig. 2), and also by the data of Homfray [12] (Fig. 3) on the adsorption of CO on charcoal at temperatures of 200–300°K, where one may no longer speak of condensation. From the slope of the curve we may calculate g even further—to the probability of desorption of an individual molecule. For this we have[4]

$$w = \alpha e^{-Q/RT},$$

where α in both cases has the very plausible value of $5 \cdot (10^{11}$–$10^{12})$ sec^{-1}. The dependence of F ($q = Fp$, $T > \gamma/R$) on T is trivial:

$$\log F = K_2 + \frac{Q_0}{RT}. \qquad (21)$$

Some difficulties arise for n close to 1; here $1/[n\sin(\pi/n)]$, corresponding to formula (14), and $1/(1/n - 1)$, corresponding to formula (19), tend to infinity. This is because, as we approach $n = 1$, each of the integrals (11), as well as the corresponding integrals for formula (19), shows ever-diminishing convergence and, therefore, the difference between the asymptotic course of the isotherm for $p \to 0$ and its actual course with nonzero p and q increases. Substituting the distribution (17) into the initial formula (3), written with the integration variable $Q(b) = g\exp(-Q/RT)$, it is easy to show that for any $p \neq 0$, $q(p, T)$ depends smoothly on the temperature.

[4]Under the simplest assumptions.

In this connection, it is interesting to consider questions of the heat of adsorption on a non-uniform surface, of the fraction of the specific heat of the adsorbed gas related to redistribution of the gas on the surface for a change in temperature, and so on—both for the general case and for a distribution of the form (17). The elaboration of these problems, however, at present would necessarily be nothing more than a mathematical exercise. We note only that for the distribution equation (17), the smaller the value of q, the sharper is the maximum of the specific heat near $T = \gamma$, while the differential heat of absorption at $T = 0$ obeys the equation

$$(Q)_q = Q_0 + \gamma \log(q_\infty/q). \tag{22}$$

In conclusion, I would like to emphasize that the main object of this part of the paper has not been a quantitative theory of adsorption on non-uniform surfaces, but a struggle for the correct qualitative interpretation of a specific set of experimental data.

In the theory of adsorption there are more examples than anywhere else of "ridiculously good quantitative coincidence of different theories applied to the same data" (McBain [2], p. 439).

Summary

1.   A critical examination is made of various proposed explanations for the deviation of the adsorption isotherm from linearity at low surface coverage.

2.   An integral equation for adsorption on non-uniform surfaces is derived and an approximate method for its solution T is given.

3.   A distribution of surface points according to the heat of adsorption corresponding to the isotherm $q = Cp^{1/n}$, is found.

4.   Under the simplest assumptions, the following relations are established for the temperature variation of the constants n and C:

$$\frac{1}{n} = \frac{T}{T_K}, \qquad \log C = AT + B,$$

for the temperature region $T < T_K$ and

$$\frac{1}{n} = 1, \qquad \log C = \frac{D}{T} + E$$

for $T > T_K$, where A, B, D, E and T_K are constants.

This work was carried out independently of and simultaneously with Temkin's work on adsorption on non-uniform surfaces (performed at the L. Ya. Karpov Physico-Chemical Institute). The difference in methods, objectives and, partly, results justifies the separate publication of our articles.

Catalysis Laboratory   *Received*
Institute of Chemical Physics   *November 17, 1934*
USSR Academy of Sciences, Leningrad

References

1. *Hückel E.* Adsorption und Kapillarkondensation. Leipzig, 52–61 (1928).
2. *McBain J. W.* Adsorption of Gases and Vapors on Solids. London (1932).
3. *Kuster H.*—Lieb. Ann. Chem. **283**, 360 (1894).
4. *Urry K.*—Phys. Chem. **36**, 1836 (1932).
5. *Langmuir I.*—J. Amer. Chem. Soc. **40**, 1361 (1918).
6. *Chakravarti H., Dhar J.*—Kolloid Ztschr. **43**, 377 (1907).
7. *Zeise H.*—Ztschr. Phys. Chem. **136**, 385 (1928).
8. *Rideal M.*—Surface Chem., 184 (1931); *Henry P.*—Philos. Mag. **44**, 689 (1922).
9. *Freundlich I.*—Kapillarchemie. 2. Aufl. B. **1**, 82 (1930).
10. *Roginskiĭ S. Z, Zeldovich Ya. B.*—Acta Physicochim. URSS **1**, 554 (1934).
11. *Magnus K., Cahn A.*—Ztschr. anorg. und allg. Chem. **155**, 205 (1926).
12. *Homfray Ida.*—Ztschr. Phys. Chem. **74**, 120 (1911).

Commentary

This paper by Ya.B. was translated and published, with a few changes, in the collection "Statistical phenomena in heterogeneous systems,"[1] which was devoted especially to the theory of non-uniform surfaces and to statistical phenomena in adsorption and catalysis. In the review article by V. I. Levin in this collection the priority of Ya.B.'s article in statistical research on the theory of adsorption and catalysis is emphasized. The article also cites articles by other authors who came to similar conclusions, but later than Ya.B. The significance of Ya.B.'s work for the theory of catalysis is elucidated in detail in S. Z. Roginskiĭ's book, "Adsorption and Catalysis on a Non-Uniform Surface."[2] After this a summary of this paper by Ya.B. has entered into the majority of monographs and textbooks on catalysis. Thus, in the course of Thomas and Thomas[3] the derivation of the adsorption isotherm on a non-uniform surface is given in full and referred to as classical.

[1] *Zeldovich Ya. B.*—In: Statisticheskie yavleniĭa v geterogennykh sistemakh [Statistical Phenomena in Heterogeneous Systems]. Moscow: Izd-vo AN SSSR, 238–247 (Problems of Kinetics and Catalysis, Vol. 7) (1949).

[2] *Roginskiĭ S. Z.* Adsorbtsiĭa i kataliz na neodnorodnoĭ poverkhnosti [Adsorption and Catalysis on a Non-Uniform Surface]. Moscow, Leningrad: Izd-vo AN SSSR, 643 p. (1948).

[3] *Thomas J. M., Thomas W. J.* Introduction to the Principles of Heterogeneous Catalysis. N.Y.: Acad. Press (1967), pp. 43–45.

2

Adsorption on a Uniform Surface*

The concept of a uniform surface is widely used in theoretical work on adsorption. A closer examination of this concept, and of the unexpected conclusions to which it may lead upon suitable selection of the various possibilities, may therefore be of some interest.

Let us start with an analogy. An ideal crystal, in which all the atoms are exactly located at the nodes of a geometrically perfect space lattice, can be conceived only on classical grounds and at absolute zero. However, it is impossible to accept this somewhat naive concept because of the uncertainty principle and thermal agitation at $T \neq 0°$K. This does not, however, mean that the idea of crystallinity loses all definiteness or that, for instance, a crystal can melt in a continuous process, as Frenkel [1] seems to suggest.

Nevertheless, the definition of crystallinity must be somewhat modified. Landau [2] introduces the function $\rho(x, y, z)$ as the probability of finding a particle (atom, electron, ion) in a particular position. In a crystal this function is periodic at any distance in a coordinate system x, y, z referred to some fixed group of particles; in any disorderly state (gas, liquid), it becomes constant at sufficiently great distances.

Analogously, one can imagine a uniform surface as a collection of identical potential holes only at $0°$K. As a result of thermal agitation at $T \neq 0°$, sectors with heats of adsorption differing from the mean will be continuously formed and destroyed on a surface which was originally ideally uniform. A uniform surface may be defined as one on which $\rho(E)$ is the same for all surface elements, where $\rho(E)\, dE$ is the probability that the heat of adsorption on a given element lies between E and $E + dE$.

In this way the uniformity of the surface, although violated instantaneously, is retained on the average, provided the time element is large compared with the time of relaxation of the thermal motion on the surface. We should immediately note that this time of relaxation is by no means necessarily of the order of 10^{-13} sec (the period of atomic oscillations in the lattice) since not only is simple displacement of atoms about their equilibrium positions possible, but also much more complicated and slower processes—for example, exchange of foreign dissolved atoms between the surface and the bulk of the crystal. From our point of view such a surface can be called uniform if each atom of the surface has the same probability of being replaced

*Acta Physicochimica URSS **8** 5, 527–530 (1938).

by an atom of the dissolved substance, irrespective of the state of all the other parts of the surface.

It is important to emphasize the fact that the process of adsorption itself changes the equilibrium function $\rho(E)$ by facilitating the formation of sectors with high heats of adsorption.

Thus, complete establishment of adsorption equilibrium involves two fundamentally different processes: the adsorption of gas on a surface in a given state with a given distribution, $\rho(E)$, and the change of the surface due to the presence of the adsorbed substance, which induces a tendency towards a new equilibrium distribution $\rho(E, q)$, where q is the quantity of adsorbed matter.

Assuming that the second process is rapid, we obtain the following standard picture of adsorption on a uniform surface: the equilibrium concentration q, which depends on the pressure of the gas, is determined by the Langmuir isotherm. The only difference from the standard picture is that the statistical sum for all states of the adsorbed molecule in a potential hole must be replaced by a combination of two statistical sums for all states of the adsorbed molecule and for all possible states of the surface element. This, of course, has no effect on the form of the Langmuir equation. Under very simple assumptions the kinetics of establishment of equilibrium will also not differ from those on a uniform surface. Thus, the initial velocity is proportional to the pressure and approaches equilibrium exponentially.

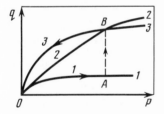

Adsorption on a uniform surface

1 — Rapid adsorption on a uniform surface
2 — Dotted line is equilibrium isotherm
A–B — Slow adsorption over prolonged gas action
3 — Rapid adsorption after prolonged gas action

An unusual dependence for adsorption on a uniform surface arises when it is assumed that the rate of change of the surface is considerably slower than the rate of adsorption (see figure). If adsorption and desorption occur rapidly, the state of the surface remains practically unaltered and we then get an adsorption isotherm corresponding to a non-uniform surface with a distribution $\rho(E, 0)$ of the heat of adsorption (curve *1*) [3]. However, when the time interval is considerable, slow adsorption accompanies changes in the properties of the surface, and the amount of gas adsorbed approaches that given by the Langmuir isotherm (curve *2*, point *B*), which describes a state of complete equilibrium (see above).

The state of the surface at equilibrium, $\rho(E, q)$, differs from its equilibrium state *in vacuo* $\rho(E, 0)$.

If adsorption and desorption occur rapidly, then under prolonged gas action a new isotherm is obtained (curve *3*) which corresponds to a significantly activated surface.

The activation in this case is obtained as a result of work done in the hysteresis loop enclosed between the adsorption and desorption curves, which do not coincide when, after the attainment of equilibrium, rapid desorption is carried out.

The adsorption kinetics will also be characteristic: rapid adsorption of an amount corresponding to the instantaneous state of the surface (point A), followed by a slow change in the surface which is accompanied by further adsorption (A–B).

This hypothetical picture of the process of adsorption resembles in some respects activated adsorption. However, it is precisely on this point that our ideas are not new. Taylor and Pace [4], upon finding that the rates of activated adsorption of light and heavy hydrogen were identical, concluded that "... the spacing of the surface atoms is an important determining factor in the problem of activation energies of such slow processes of adsorption. This forces us to attribute the activation energy to a slow temperature-sensitive process of production of surface atoms favorably spaced for adsorption." It is also stated in their paper that Evans was examining the problem theoretically, but the author was unable to find any publication by Evans on this question.

In a somewhat peculiar and unconvincing form, similar ideas were advanced in a previous note [5], in which it was pointed out that during strong adsorption the surface tension of the adsorbent may become less than zero. In this case, spontaneous increase of the adsorbent surface becomes thermodynamically possible.

Summary

1. A definition is given of a uniform surface which remains valid when thermal agitation and surface fluctuations are present.

2. The adsorption process is examined on the basis of various assumptions regarding the relation between the time of relaxation of the surface and the time of adsorption.

Physico-Chemical Laboratory   *Received*
USSR Academy of Sciences, Leningrad   *January 11, 1938*

References

1. *Frenkel J.*—Acta Physicochim. URSS **3**, 633 (1935).
2. *Landau L.*—Sov. Phys. **11**, 26 (1937).
3. *Zeldovich Ja.*—Acta Physicochim. URSS **1**, 961 (1934); here art. **1**.
4. *Pace J., Taylor H.*—J. Chem. Phys. **2**, 578 (1934).
5. *Zeldovich Ya. B.*—Zhurn. Fiz. Khimii **5**, 924 (1934).

3

On the Theory of Reactions
on Powders and Porous Substances*

A chemical reaction frequently turns out to be related to the process of transport of the reacting substances (for instance, in the case of a reaction between a gas and a condensed phase or in heterogeneous catalysis).

Two limiting cases may be considered: 1) for a reaction rate which is considerably larger than the rate of transport, the observed macroscopic reaction kinetics is completely determined by the transport conditions and does not reflect in any way the actual reaction rate on the surface or its dependence on temperature, concentration of the reacting substances, surface activity, etc. ("diffusion region"); 2) for a small chemical reaction rate, the concentrations of the reacting substances in close vicinity to the surface where the reaction is taking place do not appreciably differ from the concentrations observed above the catalyst which are used to judge progress of the process, and the macroscopic kinetics coincide fully with the actual kinetics of the reaction ("kinetic region").

When all the points of the surface are identically accessible to the reacting substances, a detailed consideration of the intermediate region between the two above limiting cases will lead to simple results in the simplest case of a monomolecular reaction (addition of the chemical and diffusion inverse rates, or "resistances") [1]. Any other more complex reaction kinetics can be easily examined by means of the so-called quasi-stationary method [2]. By applying the limiting expressions of the reaction and transport rates, and by choosing in each case that which yields the lower value for the rate, we will never exaggerate the rate of the reaction by more than two times in the case of a monomolecular reaction, or by more than 2.6 times in the case of a bimolecular reaction. The condition of equality of the two rates determines the location of the intermediate region.

However, this kind of examination is inapplicable to reactions on porous substances. In this case, the accessibility of various sectors of the active surface, which lie more or less deeply inside the catalyst, varies, and we find ourselves in the kinetic region only when the rate of the chemical reaction is smaller than the rate of diffusion of the substance to the *least accessible* sectors of the surface.

*Zhurnal fizicheskoĭ khimii **13** 2, 163–168 (1939).

In the diffusion region, the rate of transport to the *most easily accessible* sectors of the active surface which are located outside the catalyst must be smaller than the reaction rate on these sectors.

The intermediate region becomes enormously spread out and encompasses an interval in which the reaction rate varies in proportion to the extent that the rate of diffusion or other transport of the substance to the most easily accessible parts of the active surface exceeds the rate of transport to the least accessible parts.[1]

Let us find the law of variation of the observed reaction rate in this intermediate region. In this region, the rate of transport to the surface of the catalyst is significantly greater than the observed reaction rate; consequently, the concentration of the reacting substances on the surface of the catalyst does not differ from the macroscopic concentration C_0. On the other hand, the deeply-lying sectors of the active surface do not participate in the process.

It is characteristic of the intermediate region that the depth of the actively working layer of catalyst varies with the reaction rate.

Since this depth is significantly smaller than the dimensions of the piece of catalyst, and at the same time is much greater than the diameter of individual pores, the phenomenon can be represented schematically by introducing the effective coefficient of diffusion through the porous substance (which depends on the number of the pores and their diameters), and by examining a layer of the catalyst of indefinite depth with a flat surface.

The distribution of the concentration is given by the equation:

$$D\frac{d^2c}{dx^2} = KSf(c),$$

where D is the above-mentioned effective diffusion coefficient; c is the current concentration of the reacting substance, depending on the distance x from the surface of the catalyst; $KSf(c)$ is the expression for the rate of consumption of the reacting substance in the chemical reaction per unit volume of the catalyst, where K is the constant of the reaction rate per unit surface, S is the specific surface of a unit volume of the porous substance, and $f(c)$ represents the rate as a function of concentration: $f(c) = C$ for a monomolecular reaction, $f(c) = C^2$ for a bimolecular one, and so on. The boundary conditions are $C = C_0$ at $x = 0$, $C = 0$ at $x = \infty$.

The integration can be performed in quadratures even for an arbitrary form of $f(c)$.

It is sufficient for us, however, to note that the boundary conditions do not contain quantities with the dimension of length, and the dependence of the width of the reaction zone on the rate constant and on the coefficient of diffusion are uniquely determined from dimensional considerations:

$$C = C_0\,\varphi(x\sqrt{KS/D}, C_0).$$

[1]Cf. the more precise formulation given at the end of the article.

The form of the function φ can be established only by integrating the equation. However, even without this we may conclude that for a given concentration C_0 the effective depth of the reaction zone is

$$\xi \sim \sqrt{D/KS}.$$

The macroscopically observed reaction rate, W, is proportional to the product of the rate constant and the magnitude of the active surface:

$$W \sim KS'\xi \sim \sqrt{DKS}.$$

Thus, in the intermediate region the observed reaction rate turns out to be proportional to the square root of the rate constant.

If, in accordance with Arrhenius' law,

$$K \sim e^{-A/RT},$$

where A is the heat of activation of the reaction, then in the intermediate region

$$W \sim \sqrt{K} \sim e^{-A/2RT},$$

i.e., the "activation heat" of the observed process is half the actual heat of activation of the heterogeneous chemical reaction.

For an n-th order reaction, $f(c) = C^n$, analogously to the change in the activation heat of the observed process the change in the order of the reaction can also be found. The equations

$$D\frac{d^2c}{dx^2} = KSc^n,$$

with boundary conditions $C = C_0$ at $x = 0$, $C = 0$ at $x = \infty$, may be reduced to the single equation

$$\frac{d^2\eta}{d\varepsilon^2} = \eta^n,$$

with boundary conditions $\eta = 1$ at $\varepsilon = 0$, $\eta = 0$ at $\varepsilon = \infty$, if we introduce the dimensionless variables

$$\eta = \frac{C}{C_0}; \qquad \varepsilon = \frac{x}{C_0^{(1-n)/2}\sqrt{D/KS}}.$$

In this way, having once found

$$\eta = \Psi(\varepsilon),$$

we obtain for any C_0, K, D, S the general solution

$$C = C_0\Psi\left[\frac{x}{C_0^{(1-n)/2}\sqrt{D/KS}}\right].$$

The observed reaction rate is

$$\int_0^\infty KSC^n\,dx = KSC_0^n C_0^{(1-n)/2}\sqrt{\frac{D}{KS}}\int_0^\infty \Psi^n(\varepsilon)\,d\varepsilon \simeq C_0^{(n+1)/2}\sqrt{DKS}$$

(this last is accurate to within a numerical factor).

The order of the reaction in the intermediate region, $(n + 1)/2$, is the mean between the first order in the diffusion region and the true n-th order of the reaction in the kinetic region.

It should be noted that all that has been said concerning the order of the reaction refers to the change in the partial pressure of the reacting substances at constant total pressure. When the total pressure changes, the change in the coefficient of diffusion in the gaseous phase, which is inversely proportional to the pressure, is superimposed on the above change in partial pressure. The diffusion coefficient in pores whose radius is less than the mean free path of a molecule at the given pressure is independent of the pressure. We cannot dwell here on these relations.

The relations which arise out of these considerations are represented in Fig. 1.

Curve I shows the observed reaction rate (on a logarithmic scale) as a function of the reciprocal of temperature for a smooth piece of catalyst with only the external surface active so that all the points are equally accessible, and the calculations given in [1, 2] are applicable. Curve II refers to a porous sample under the same conditions, so that the rates in the purely diffusion region coincide.

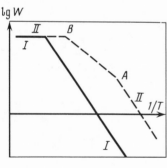

Fig. 1

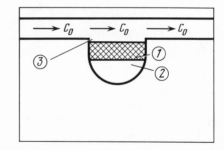

Fig. 2

1: Catalyst. 2: $C = C_0$ in the kinetic region; $C \ll C_0$ in the transition and diffusion regions. 3: $C = C_0$ in the kinetic and transition regions; $C \ll C_0$ in the diffusion region.

In the kinetic region the rate on a porous material is greater than that on a smooth one, in the same proportion as the ratio of the total surface area of the pores to the external surface of the smooth sample.

A characteristic break, A, appears on curve II at the transition from the kinetic region to the intermediate one, with the activation heat decreasing by half.

In cases when a decrease in the activation heat with increasing temperature was observed experimentally, it was usually explained by a change in the reaction mechanism. The present paper, perhaps, will allow us to attribute this decrease, at least in certain instances, to the elimination of part of the surface of the porous catalyst.

An experimental proof of our suggested explanation could be given for simplified conditions by measuring the concentration of the reacting substances below a layer of catalyst deposited on a screen (Fig. 2). For this it should be established that precisely in the region where the activation heat decreases, the concentration below the layer sharply decreases, so that the deepest layers of the catalyst are eliminated (the reacting gases are passed over the catalyst, but not through it).

The condition of transition from the intermediate region to the kinetic region (point A in Fig. 1) can be mathematically expressed as the penetration of the reaction to the middle of the catalyst:

$$\xi = C_0^{(1-n)/2}\sqrt{D/KS} = r_0, \qquad r_0^2 C_0^{n-1} KS/D = 1,$$

where r_0 is the radius (or, in general, the dimension) of the catalyst.[2]

This equation can be interpreted as the equality of the rate of diffusion to the depth $r_0 D(C_0/r_0)$ and the reaction rate in a layer whose depth is $KSC_0^n r_0$ (both values are with respect to a unit external surface of the layer).

The condition of transition to the diffusion region (point B in Fig. 1) is determined by the equality of the reaction rate in the pores and the rate of supply by diffusion:

$$\alpha C_0 = C_0^{(n+1)/2}\sqrt{DKS}, \qquad \frac{DKSC_0^{n-1}}{\alpha^2} = 1,$$

where α is the coefficient of diffusion exchange (analogous to the coefficient of heat transfer).

In the intermediate region the rate constant K varies within the limits

$$\frac{D}{r_0^2 C^{n-1}S} < K < \frac{\alpha^2}{DSC_0^{n-1}}.$$

The ratio of the limits of variation of K is $\alpha^2 : D^2/r_0^2$.

The observed reaction rate $W \sim \sqrt{K}$ varies in the intermediate region in the ratio $\alpha : D/r_0$, i.e., in precisely the ratio of the rate of supply to the most easily accessible parts of the surface and the rate of diffusion to the middle of the catalyst.

For a rate of supply to the surface, if $\alpha \leq D/r_0$, the intermediate region vanishes altogether; the entire process of compression of the reaction zone in the catalyst will occur when the reaction rate is already determined by the supply of the reacting substances to the surface, and will not be reflected in the observed rate of the process.[3]

---

[2]Damköhler [3] derives the condition of penetration of the reaction to the middle of the catalyst in this form.

In contrast, when the catalyst is placed in a strong current of the reacting gases, even before the diffusion region is reached (corresponding to a very large reaction rate), the reaction zone will contract to a depth of the order of the diameter of individual pores, and will not change thereafter (our equations are then no longer applicable). We obtain a second kinetic region in which the active surface differs little from the visible surface of the catalyst. It may be anticipated here that the dependence of the observed reaction rate on the temperature will assume the unusual form shown in Fig. 3.

In Fig. 3 we have the kinetic region I in which the entire internal surface of the catalyst is active. In the intermediate region II the activation heat drops by half; it again doubles to the original value of the actual activation heat in region III, where only the visible part of the surface is active.[4]

The rates obtained by extrapolation from I and III are in a constant

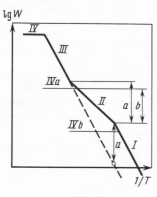

Fig. 3

ratio equal to the ratio of the total internal surface of the catalyst to its visible surface; the reaction rate over the intermediate region varies in the same proportion.

It is only at even higher temperatures that the diffusion region IV appears, a region in which the reaction rate depends very weakly on the temperature.

If the rate of supply to the surface were smaller (IVa), we would be unable to observe the second kinetic region. At still lower rates, the intermediate region (IVb) will also vanish.[5] So let us now summarize.

---

[3]In this case, for $x = 0$, the boundary condition $C = C_0$ is replaced by another, $-D\,dc/dx = \alpha C_0 = W_0$. Then we have

$$C = W_0 \xi f(x/\xi) \big/ L,$$

where f satisfies the equation $f' = f^n$ with boundary conditions $f' = -1$ at $x = 0$; $f = 0$ at $x = \infty$;

$$\xi^{n+1} = D^n \big/ (W_0^{n-1} KS),$$

so that

$$C(0) \sim W_0^{2/(n+1)} K^{-1/(n+1)}.$$

[4]The temperature at which the activation heat decreases by half grows as the dimensions of the piece of catalyst are decreased: $T_2 - T_1 = 2RT_0T_1 \log(r_1/r_2)\big/ A$.
Note made by the author in proof (1939—Ed. note).

[5]In Figs. 1 and 3, for the sake of graphical clarity, individual sections of the curves $\log W - 1/T$ are joined at corners. In reality, of course, the transitions are smooth and rounded. By analogy with the case mentioned at the beginning of a surface, all of whose points are equally accessible, we may expect that this rounding occupies an interval of no more than 0.7–1.5 on the ordinate axis ($\log W$).

When a chemical or catalytic reaction takes place on the surface of a porous substance or a powder, the depth to which the reaction penetrates is inversely proportional to the square root of the reaction rate constant over a wide range of varying conditions, for instance, of temperature. In this region, the observed rate of the process is directly proportional to the square root of the rate constant; this corresponds to an activation heat twice as small as the actual one.

The order of the reaction is the mean between the first and the true order. Conditions are given which define the region in which these relations hold.

Physico-Chemical Laboratory   *Received*
USSR Academy of Sciences, Leningrad   *June 24, 1938*

References

1. *Davis T., Hottel W.*—Industr. and Eng. Chem. **26**, 749 (1934).
2. *Frank-Kamenetskiĭ D. A.* Report at a meeting on topochemical kinetics. Leningrad, May 11 (1938).
3. Chem.-Ing. **3**, Th. 1, 2 (1937).

Commentary

The role of this article was described in sufficient detail in the Introduction. The paper is cited in a number of monographs on catalysis as the first work on the internal diffusion kinetics of processes in powders and porous substances (see, for example, the monograph by G. K. Boreskov).[1] Figure 3 from Ya.B.'s paper is reproduced in many textbooks and encyclopedic articles on catalysis.[2] In the introduction to this book, D. A. Frank-Kamenetskiĭ names Ya.B. as one of his teachers. The region of internal diffusion in kinetics on porous bodies is often called the Zeldovich region (see also [3]).

[1] *Boreskov G. K.* Kataliz v proizvodstve sernoi kisloty [Catalysis in the Production of Sulfuric Acid]. Moscow, Leningrad: G.Kh.I., p. 80 (1954).
[2] *Frank-Kamenetskiĭ D. A.* Diffuziĭa i teploperedacha v khimicheskoi kinetike [Diffusion and Heat Transfer in Chemical Kinetics], 2nd Ed. Moscow: Nauka, Chap. 2 (1967).
[3] *Thiele E. W.*—Ind. Eng. Chem. **31**, 916–929 (1939).

II

Hydrodynamics. Magnetohydrodynamics.

Heat Transfer. Self-Similarity

4

The Asymptotic Law of
Heat Transfer at Small Velocities
in the Finite Domain Problem*

As was shown by G. Helmholtz [1], in a purely viscous regime the energy dissipation, and hence the drag as well, are minimal. Similarly, it may be shown that heat transfer is minimal in a purely conductive regime in a fluid at rest.

Let us consider heat transfer in a closed vessel filled with fluid. All of the vessel's walls are impermeable to the fluid. Some of the walls are thermally insulated, the rest are maintained at different fixed temperatures. It is the heat transfer between these regions heated to different temperatures which interests us. We will assume that the fluid is incompressible. We multiply both sides of the thermal conduction equation

$$c\frac{\partial T}{\partial t} = \nabla(k\nabla T) - c\mathbf{v}\nabla T \tag{1}$$

by T, integrate the result over the whole volume and apply Gauss's theorem. Noting that $\mathbf{v}_n = 0$ at the walls, and div $\mathbf{v} = 0$, and considering $c = \text{const}$, we find:

$$\frac{1}{2}\frac{\partial}{\partial t}\int cT^2 \, dV = \int Tk(\nabla T)_n \, ds - \int k(\nabla T)^2 \, dV. \tag{2}$$

*Zhurnal eksperimentalnoĭ i teoreticheskoĭ fiziki **7** 12, 1466–1468 (1937).

The quantity $k(\nabla T)^2$ is the dissipation of heat, or of the temperature inhomogeneity, just as $\mu[2(\partial u/\partial x)^2 + \ldots + (\partial w/\partial x + \partial v/\partial z)^2 + \ldots]$ is the energy dissipation. Indeed, in a completely thermally insulated system, $(\nabla T)_n \equiv 0$; and the decrease in $\int cT^2\, dV$ (which measures the temperature inhomogeneity for a constant value of $\int cT\, dV$) eventually becomes equal to $2\int k(\nabla T)^2\, dV$.

In a steady or mean steady turbulent regime, $(\partial \int cT^2\, dV)/\partial t = 0$ exactly or approximately over time intervals which are large compared to the pulsation period. We will show that $\int Tk(\nabla T)_n\, dS$ may be considered a measure of the heat transfer.

At the thermally insulated walls $\nabla T_n = 0$; it remains only to find the integral in the regions with the given temperatures. We obtain

$$\int_s Tk(\nabla T)_n\, dS = \sum_i T_i \int_{s_i} k(\nabla T)_n\, dS = \sum_i T_i Q_i = \int k(\nabla T)^2\, dV. \quad (3)$$

Here the index i distinguishes the different regions, Q is the overall heat flux from outside into the system through the i-th region, which is maintained at the temperature T_i. If we have only two surfaces with temperatures T_I and T_{II}, then $Q_I = -Q_{II}$ by the conservation of energy. Referring the heat transfer to some surface S_0, we determine the heat transfer coefficient

$$\alpha = \frac{Q_I}{S_0(T_I - T_{II})} = \frac{T_I Q_I + T_{II} Q_{II}}{S_0(T_I - T_{II})^2}. \quad (4)$$

The quantity $\sum(T_i Q_i)/(T_n - T_m)^2 S_0$, with arbitrarily chosen values S_0, T_n, and T_m, may, for T_i given, be considered a reasonable generalization of the heat transfer coefficient to the case of several heat sources.

Then, by formula (3),

$$\alpha = \int \frac{k(\nabla T)^2}{(T_n - T_m)^2 S_0}\, dV. \quad (5)$$

If T_i is given and S_0, T_n, and T_m are properly chosen, then the quantity α is proportional to the dissipation, $\int k(\nabla T)^2\, dV$.

If T is given in some regions of the surface, and $(\nabla T)_n = 0$ in the others, then the condition of the minimum

$$\delta \int k(\nabla T)^2\, dV = 0 \quad (6)$$

yields the equation

$$\nabla(k\nabla T) = 0, \quad (7)$$

i.e., precisely the equation of the steady temperature field in a fluid at rest (cf. Eq. (1)).[1]

Thus, we have proved that in any given system the coefficient of heat transfer is at a minimum when the system is at rest. Any motion of the fluid can only increase it.

---

[1]The proof that the extremum sought is unique, and that it is a minimum, is well known.

It is also possible to find the limiting law for this increase in the steady regime.

The original equation is

$$\nabla(k\nabla T) - c\mathbf{v}\nabla T = 0. \tag{8}$$

We seek

$$T = T_0 + \epsilon T_1, \tag{9}$$

where T_0 satisfies the boundary conditions and equation (7), and ϵT_1 is a small correction caused by the motion of the fluid. So, neglecting the product of the small quantities $\epsilon\mathbf{v}$ we have

$$\epsilon\nabla(k\nabla T_1) - c\mathbf{v}\nabla T_0 = 0. \tag{10}$$

For a similar change in the velocity field we have

$$\epsilon T_1 \approx \frac{vdc}{kT_0}, \qquad \epsilon \approx \text{Pe}, \tag{11}$$

where Pe is the Peclet number, vdc/k.

Substituting into equation (5) and performing the calculations, we obtain ($\text{Nu} = \alpha d/k$, the Nusselt number)

$$\text{Nu} = \text{Nu}_0 + a^2\text{Pe}^2, \tag{12}$$

where the term with Pe to the first power, proportional to $\int k\nabla T_0\nabla T_1\, dV$, vanishes due to the extremal properties of T_0 and to the boundary conditions imposed on T_1.

In free convection at small Grashof numbers (Gr) we have laminar self-similarity of the Peclet number (Pe), which is no longer a governing parameter proportional to Gr, and once again in the limit of small Gr

$$\text{Nu} = \text{Nu}_0 + b^2\text{Gr}^2. \tag{13}$$

Similarly, developing G. Helmholtz's result mentioned at the beginning, it is easy to obtain for the drag coefficient at small values of the Reynolds number

$$\xi = \frac{c^2(1 + d^2\text{Re}^2)}{\text{Re}}. \tag{14}$$

The results (12)–(14) are valid for the finite domain only. In the external problem, because of the infinite extent of the fluid, for arbitrarily small Pe, Gr, and Re there may be regions in which the correction ϵT_1, or its analogue in the velocity field, is no longer small compared to T_0 (or v_0, as appropriate), and our analysis is inapplicable. After all, it was through just such reasoning that Oseen [2], for example, obtained the asymptotic formula for the drag coefficient of a sphere

$$\xi = \frac{24(1 + 3\text{Re}^1/16)}{\text{Re}},$$

which would be impossible (Re^1 instead of Re^2) in the finite domain.

Summary

In the finite domain problem, for small Pe, Gr, and Re, we have the following asymptotic formulas: $\mathrm{Nu} = \mathrm{Nu}_0 + a^2\mathrm{Pe}^2$; $\mathrm{Nu} = \mathrm{Nu}_0 + b^2\mathrm{Gr}^2$; $\xi = c^2(1 + d^2\mathrm{Re}^2)/\mathrm{Re}$.

Physico-Chemical Laboratory   *Received*
USSR Academy of Sciences, Leningrad   *July 5, 1937*

References

1. *Helmholtz G.*—Verh. Naturhist.-Med., 223 (1868).
2. *Oseen F.* Neuere Methoden und Ergebnise in der Hydrodynamik. Leipzig (1927).

Commentary

The author considered the principle result of his work, as we see from the summary, to be the asymptotic formulas for the heat transfer (the Nusselt number) at small values of the external parameters (the Peclet, Grashof and Reynolds numbers). However, the article also incidentally introduces for the first time an important characteristic of temperature fields—the rate of decay of the measure of temperature inhomogeneity, $k(\nabla T)^2$. The analogy between this quantity and the energy dissipation in a flow of viscous fluid is indicated. Twelve years later this quantity independently appeared in a paper by A. M. Obukhov[1] as a governing characteristic of the local temperature field in developed turbulent flow. It was precisely the rate of decay that entered into formulations of the basic hypothesis of similarity of the temperature fields in developed turbulent flows. It figures in the asymptotic formulas for structural functions of the temperature field in the inertial interval of turbulent flow scales, and for scales much smaller than the Kolmogorov scale. Of course, in turbulent flow this quantity is stochastic, and as a result of the intermittent character of turbulence the averaged characteristics are influenced by its fluctuations at scales significantly larger than the scales of the inertial interval. Therefore, just as for structural functions of the velocity field, deviations from the classical similarity laws occur here.

A detailed exposition of these questions may be found in the monograph by A. S. Monin and A. M. Yaglom.[2]

[1] *Obukhov A. M.*—Izv. AN SSSR, Ser. Geog. i Geofiz. **13** 21, 58–69 (1949).
[2] *Monin A. S., Yaglom A. M.* Statisticheskaĭa Gidrodinamika. [Statistical Hydrodynamics]. Moscow: Nauka, Part 2., 345 and following pages (1967).

5

The Asymptotic Laws
of Freely-Ascending Convective Flows[*]

In studying the motion of submerged jets it was natural to assume that at large distances from the nozzle the phenomenon would become self-similar, and that the nozzle's diameter would disappear as a governing parameter. Combining this self-similarity with the assumption of proportionality of the mixing path length l (Prandtl) to the submerged jet width b, Tollmien [1] easily derived the asymptotic laws of turbulent propagation of the jet:

$$u = \left(\frac{P_1}{\rho x}\right)^{1/2} f_1\left(\frac{y}{x}\right) \qquad \text{for a slit} \qquad (1)$$

and

$$u = \frac{(P_2/\rho)^{1/2}}{x} f_2\left(\frac{y}{x}\right) \qquad \text{for a round nozzle} \qquad (1a)$$

The jet is directed along the x-axis, u is the velocity, P_1 is the momentum of the jet per unit length of the slit, and P_2 is the full momentum of the round jet. The functions f_1 and f_2 are found by integration of the ordinary differential equations. The result is in satisfactory agreement with experiment.

Schlichting [2] used an analogous method to study the laminar diffusion of a jet where he obtained

$$u = \left(\frac{P_1}{\rho \nu x}\right)^{1/3} f_2(y\rho^{1/3}/\nu^{2/3}P_1^{1/3}x^{2/3}) \qquad \text{for a slit} \qquad (2)$$

and

$$u = \left(\frac{P_2}{\rho \nu x}\right) f_4(yP_2^{1/2}/\nu\rho^{1/2}x) \qquad \text{for a round nozzle} \qquad (2a)$$

As before, $P_1 = \rho u_0^2 h$ and $P_2 = \rho u_0^2 d^2$, where h is the slit width, d is the nozzle diameter, and u_0 is the velocity in the nozzle or slit.

In the present note we use the same methods to find similar expressions for convective flows freely-ascending under the action of gravity and density differences. We direct the x-axis upward in the direction of the flow. Instead of conservation of momentum (which will not exist because of the action of buoyancy forces), we write the conservation of heat flux

$$\int u\theta c\rho\, dt = Q_1 \qquad (3)$$

$$\int u\theta c\rho y\, dy = Q_2 \qquad (3a)$$

[*]Zhurnal eksperimentalnoĭ i teoreticheskoĭ fiziki **7** 12, 1463–1465 (1937).

The first formula relates to the plane case, analogous to jet flow from a slit, such as an ascending flow over a long horizontal cylinder; Q_1 is the quantity of heat given off by a unit length of the cylinder. The second formula is written for a radially-symmetric flow over an isolated heated body, and Q_2 is the total amount of heat given off by it. As usual, c is the specific heat and $\theta = T - T_0$, where T_0 is the temperature of the medium at infinity. We neglect molecular and turbulent diffusion of heat along the flow compared to the molar transfer. We assume that the phenomenon is self-similar at a sufficiently large distance from the heated body. Going to average quantities, accurate to within numerical factors, we obtain

$$u\theta c\rho b = Q_1, \tag{4}$$

$$u\theta c\rho b^2 = Q_2 \tag{4a}$$

(b is the width of the flow).

We first write the equations of motion for the turbulent case

$$\rho u \left(\frac{\partial u}{\partial x} \right) = \frac{\rho \left/ \partial u/\partial y \right.}{\partial^2 u/\partial y^2} - \frac{\partial p}{\partial x} - g\rho. \tag{5}$$

We assume that on every horizontal plane the pressure is the same everywhere; we determine the value $\partial p/\partial x$ from the value of $\partial p/\partial x = -g\rho_0$ at infinity (for $y \to \infty$).

Finally, substituting $\rho = \rho_0(1 - \beta\theta)$ [or $\rho_0 = \rho(1 + \beta\theta)$; $(\beta\theta \ll 1)$, where β is the expansion coefficient $(\beta = 1/T$ for an ideal gas)], we obtain the usual expression for the buoyancy force for equations of free convection

$$\rho u \left(\frac{\partial u}{\partial x} \right) = \frac{l^2\rho \left/ \partial u/\partial y \right.}{\partial^2 u/\partial y^2} + g\rho\beta\theta. \tag{6}$$

Consistently replacing $\partial/\partial x$ by $1/x$, $\partial/\partial y$ by $1/b$, and taking into account $l \approx b$, we obtain

$$\frac{\rho u^2}{x} = \frac{\rho u^2}{b} = g\rho\beta\theta. \tag{7}$$

The equation of heat transfer does not contribute anything new even though the temperature field is not similar to the velocity field due to the presence of the buoyancy force. (This situation holds identically in turbulent and laminar flows.)

Combining (7) with (4) and (4a) we obtain in the plane case

$$u = \left(\frac{g\beta Q_1}{c\rho} \right)^{1/3} f_5\left(\frac{y}{x}\right); \qquad \theta = \left(\frac{Q_1}{c\rho} \right)^{2/3} (g\beta)^{-1/3} x^{-1} \varphi_5\left(\frac{y}{x}\right) \tag{8}$$

and in the radially-symmetric case

$$u = \left(\frac{g\beta Q_2}{c\rho x} \right)^{1/3} f_6\left(\frac{y}{x}\right); \qquad \theta = \left(\frac{Q_2}{c\rho} \right)^{2/3} (g\beta)^{-1/3} x^{--5/3} \varphi_6\left(\frac{y}{x}\right). \tag{8a}$$

It is interesting here that in the plane case the velocity at the flow axis is constant.

In laminar flow the equation of motion,

$$u\left(\frac{\partial u}{\partial x}\right) = \nu\left(\frac{\partial^2 u}{\partial y^2}\right) + g\beta\theta, \tag{9}$$

under the assumption of self-similarity should be

$$\frac{u^2}{x} = \frac{\nu u}{b^2} = g\beta\theta. \tag{10}$$

Combining (10) with (4) and (4a), we find in the plane case

$$u = (g\beta Q_1/c\rho)^{2/5}\nu^{1/5}x^{1/5}f_7[y(g\beta Q_1/c\rho)^{1/5}/x^{2/5}\nu^{3/5}];$$

$$\theta = (Q_1/c\rho)^{4/5}(g\beta)^{-1/5}\nu^{-2/5}x^{-3/5}\varphi_7[y(g\beta Q_1/c\rho)^{1/5}/x^{2/5}\nu^{3/5}] \tag{11}$$

and in the radially-symmetric case

$$u = (g\beta Q_2/c\rho)^{1/2}\nu^{-1/2}f_8[y(g\beta Q_2/c\rho)^{1/4}/x^{1/2}\nu^{3/4}];$$

$$\theta = (Q_2/c\rho)\nu^{-1}x^{-1}\varphi_8[y(g\beta Q_2/c\rho)^{1/4}/x^{1/2}\nu^{3/4}]. \tag{11a}$$

The functions f_5–f_8 and φ_5–φ_8 may be determined from a system of two ordinary differential equations. In addition, the form of the functions f_7, f_8, φ_7, and φ_8 depends on the value of the Prandtl number. It is interesting that along a laminar ascending flow, in both the plane and radially-symmetric cases, the Reynolds number (defined as ub/v) increases as $x^{3/5}$ and $x^{1/2}$, respectively.[1] Consequently, at a sufficient height disruption of the laminar flow and transition to turbulent flow should take place.

The values Q_1 and Q_2 in our formulas may be expressed in terms of the dimensions and temperature of the heated body if the Nusselt number is known as a function of the Grashof number in the case under consideration. If Nu $= \psi(\text{Gr})$ then, within a numerical factor, $Q_1 = \lambda\theta_0\psi(\text{Gr})$, $Q_2 = \lambda d\theta_0\psi(\text{Gr})$, where λ is the heat conductivity of the medium, θ_0 is the temperature, and d is the size of the heated body.

The formulas are also applicable to ascending flows over flames. In this case Q is proportional to the fuel consumption. The temperature field is always similar to the concentration field of combustion products in a turbulent flow; in a laminar flow this is so only if $D = \lambda/c\rho$ (D is the diffusion coefficient, λ is the thermal conductivity). Here, in order for self-similarity to occur the condition $x \gg d$ is insufficient (d is the size of the burner nozzle); it is also necessary that the total momentum of the ascending flow be much greater than the momentum of the gases in the burner nozzle. Since the momentum of an ascending flow (u^2b in the plane case, u^2b^2 in the axially-symmetric case) continuously increases with x as x^1, $x^{4/3}$, $x^{4/5}$, and x^1 in the four cases considered, at sufficient height this condition is always fulfilled. In the laminar case the region of self-similarity is bounded from above by the regime's critical region. Backward extrapolation of these asymptotic laws to the motion and temperature of the fluid near the heated body itself is, obviously, inadmissible.

[1] $\text{Re} = (g\beta Q_1 x^3/c\rho\nu^3)^{1/5}$ and $\text{Re} = (g\beta Q_2 x^2/c\rho\nu^3)^{1/4}$.

Summary

Asymptotic laws are found for freely-ascending convective flows in the laminar and turbulent planar and radially-symmetric cases in the form $u = x^m f(y/x^n)$, $\theta = x^k \varphi(y/x^n)$.

Physico-Chemical Laboratory
USSR Academy of Sciences, Leningrad

Received
July 5, 1937

References

1. *Tollmien W.*—ZAMM **6**, 468 (1926).
2. *Schlichting H.*—ZAMM **13**, 260 (1933).

Commentary

The fundamental self-similarity laws of the evolution of freely-ascending laminar and turbulent convective flows have found numerous applications, above all in geophysics.[1]

It should be mentioned that, as is often the case in a first draft, the author based his treatment of the turbulent convection on certain assumptions which in fact were not necessary, in particular the semi-empirical concept of L. Prandtl. Moreover, even in the analysis of laminar convection, the author [cf., for example the transition from equation (9) to equation (10)], to derive the asymptotic laws, resorts to simplifications of the equations which are really not necessary. Actually, it is possible to manage without these assumptions so that Zeldovich's asymptotic laws (8), (8a), (11), and (11a) may be obtained by simple dimensional analysis under the most general assumptions.

It is clear, for example, that the velocity, u, and the temperature, θ, in planar developed turbulent convection (when the role of molecular transport is negligibly small) depend only on the following quantities: the longitudinal coordinate, x, the transverse coordinate y, the buoyancy parameter $g\beta$ (it is obvious that the quantities g and β, in the case of a strong gravitational field when accelerations of convective flows are small compared to the acceleration of gravity, appear only as a product, and not separately; in the theory of convection the approximation of a strong gravitational field is called the Boussinesq approximation), and also on the "temperature flux," $Q_1/c\rho$. The dimensions of the governing parameters have the form

$$[y] = [x] = L, \quad [g\beta] = L\theta^{-1}T^{-2}, \quad [Q_1/c\rho] = L^2\theta T^{-1}. \qquad (I)$$

Here L is the dimension of length, T—of time, and θ—of temperature. The scales of velocity, u_0, and temperature, θ_0, are uniquely determined by the governing parameters (I):

$$u_0 = \left(\frac{g\beta Q_1}{c\rho}\right)^{1/3}; \quad \theta_0 = \left(\frac{Q_1}{c\rho}\right)^{2/3} \frac{(g\beta)^{-1/3}}{x},$$

whence the law (8) is obtained directly, without resorting to the approximate equation (5).

[1] Monin A. S., Yaglom A. M. Statisticheskaĭa Gidrodinamika. Moscow, Nauka, Part 1, 639 p. (1965).

6

Exact Solution of the Diffusion Problem in a Periodic Velocity Field and Turbulent Diffusion*

We write the exact solution of the diffusion in a velocity field with a single Fourier-component. The expression for the effective diffusion contains the molecular diffusion coefficient as a factor; this ensures correct behavior of the result with respect to time reversal.

The diffusion in a velocity field with a wide velocity spectrum supposedly describing turbulence is considered in the spirit of cascade-renormalization ideas. For the latter case of isotropic turbulence, we construct an ordinary differential equation for the turbulent diffusion coefficient.

With certain restrictions on the parameters, we obtain an answer which agrees with elementary conceptions of the characteristic mixing length.

1. In Moffat's review [1] an excellent account is given of the relation between the elementary approach and modern statistical methods in turbulence theory. In particular, by analogy with kinetic theory we easily derive for the diffusion coefficient of a scalar admixture the expression

$$D_t = \frac{u'l}{3},\tag{1}$$

where u' is the turbulent velocity and l is the mixing length. In an effort to make (1) more exact, it is natural to take

$$u' = \langle u^2 \rangle^{1/2}; \quad l = \frac{\iint \mathbf{u}(\mathbf{r} + \mathbf{n}l)\,\mathbf{u}(\mathbf{r})\,|l|\,d^3r\,d^2n}{\int |\mathbf{u}(\mathbf{r})|^2\,d^3r},\tag{2}$$

where $\mathbf{n}$ is the unit vector, so that l is defined as the correlation length.

Analogously, we may write

$$D_t = \frac{\langle u^2 \rangle \tau}{3},\tag{3}$$

where τ is the characteristic correlation time.

In the spectral representation the velocity is

$$\mathbf{u}(\mathbf{x}, t) = \sum_{i=1,2} \int u_{ik\omega} e^{i\mathbf{n}\mathbf{x}}\,d^3\mathbf{k}\,dt,\tag{4}$$

*Doklady AN SSSR **266** 4, 821–826 (1982).

where $i = 1, 2$ corresponds to the two directions perpendicular to $\mathbf{k}$. In this representation it would be natural to write

$$D_t = \sum_{i=1,2} \int |u_{ik}|^2 \omega^{-1} \, d^3\mathbf{k} \, d\omega. \tag{5}$$

In practical terms, if the relation

$$\omega_{\max} \sim |\mathbf{k}||\mathbf{u}| \tag{6}$$

holds for the frequency at which $\mathbf{u}_{\mathbf{k}\omega}$ reaches a maximum, and for the spectrum which is bounded by the maximum energy-carrying scale, then expressions (1), (3) and (5) yield one and the same answer, in reasonable agreement with experiment. However, investigation of the transformation properties of the diffusion coefficient does not allow us to be satisfied with such an elementary theory. In fact, in the diffusion equation,

$$\frac{\partial n}{\partial t} = D \Delta n, \tag{7}$$

D is a scalar with respect to transformations of the three-dimensional space, but changes sign when t is replaced by $-t$. In molecular-kinetic theory this property of D is related to the well-known problems of irreversibility and entropy increase, the more so since by the dissipation theorem,

$$\frac{d}{dt} \int (n - \bar{n})^2 \, dV = -2D \int (\nabla n)^2 \, dV.$$

But in the statistical theory of turbulence the situation is more complicated.

Expression (1) has this property only until we define u' according to (2), since $\sqrt{\langle u^2 \rangle}$ does not change sign when $-t$ is substituted for t. In exactly the same way relations (3) and (5) are unsatisfactory.

Therefore we must take a new approach to the diffusion problem. This new approach is related to a certain exact solution for a simple velocity field (see § 2), which we will then attempt to generalize to a turbulent field with a set of harmonics and scales.

2. An Exact Solution. We define[1]

$$u_x = 2v \cos ky \cos \omega t, \quad u_y = u_z = 0. \tag{8}$$

We define a field n which has a gradient along the x-axis on the average over time and which is independent of t, y, and z:

$$\bar{n} = n_0 - ax. \tag{9}$$

The exact solution has the form

$$n(x, y, t) = n_0 + ax + b \cos ky \cos(\omega t + \varphi). \tag{10}$$

We easily find the parameters b and φ by substituting $n(x, y, t)$ into the equation. We simultaneously verify that (10) is an exact solution of the

---

[1]The factor 2 is convenient since then $\langle u^2 \rangle = v^2$; averaging is performed over t and y.

equation

$$\frac{\partial n}{\partial t} + u_x \frac{\partial n}{\partial x} = D\left(\frac{\partial^2 n}{\partial x^2} + \frac{\partial^2 n}{\partial y^2}\right); \tag{11}$$

here D is the molecular diffusion coefficient.

We obtain

$$-\omega b \sin(\omega t + \varphi)\cos ky + 2av\cos\omega t\cos ky = -Dk^2\cos(\omega t + \varphi)\cos ky \tag{12}$$

and, as a consequence,

$$b^2 = \frac{4a^2 v^2}{\omega^2 + D^2 k^4}, \quad \tan\varphi = \frac{-\omega}{Dk^2}. \tag{13}$$

In the unperturbed solution, obviously, (9) is the flux density,

$$J_x = -Da. \tag{14}$$

In the exact solution it is obvious that the time average of the flux is still directed along the x-axis. The dissipation theorem (Zeldovich, 1937) may be written in the form

$$\langle\mathbf{J}\rangle\langle\nabla n\rangle = -D\langle(\nabla\mathbf{n})^2\rangle = -D_{\text{eff}}\langle\nabla n\rangle^2.$$

Thus,

$$\langle\nabla\mathbf{n}\rangle = -a; \quad (\nabla n)^2 = a^2;$$
$$\langle\nabla\mathbf{n}\rangle^2 = a^2 + b^2\cos^2\omega t\cos^2 ky;$$
$$\langle(\nabla n)^2\rangle = a^2 + \frac{1}{4}b^2 k^2. \tag{15}$$

We carry out the averaging over both time and volume, in this case, over y. We find

$$D_{\text{eff}} = D\left(a^2 + \frac{1}{4}b^2\right) \Big/ a^2 = D[1 + v^2 k^2/(\omega^2 + D^2 k^4)]. \tag{16}$$

The heat flux, J_x, can be easily found by direct averaging:

$$J_x(x,y,t) = D\frac{\partial n}{\partial t} + u(x,y,t)\,n(x,y,t) \tag{17}$$

Here the product $\overline{\cos\omega t\cos(\omega t + \varphi)}$ will enter into the formula and, thus, so will the phase shift, φ.

It is significant that (16) yields D_{eff}, having D as a factor, so that for $D = 0$, $D_{\text{eff}} = 0$. Therefore, D_{eff} changes sign when t is replaced by $-t$, just as D does.

As applied to turbulence, despite the dissimilarity between field (8) and the stochastic field, it is natural to assume $\omega^2 \sim u^2 k^2 \beta$ (β is a coefficient of order unity) which gives

$$D_{\text{eff}} = D\frac{1 + v^2}{\beta v^2 + D^2 k^2}.$$

We introduce the Reynolds (Peclet) number

$$\text{Re} = \frac{vl}{D} = \frac{v}{Dk},$$

and obtain

$$D_{\text{eff}} = D\frac{(1+\beta)\text{Re} + 1}{\beta\text{Re} + 1} < D\frac{1+\beta}{\beta}$$

And so, even at large Reynolds number in a simple, exactly solvable model, the diffusion increase is quite limited! (cf. Howell's result at $\beta = 0$ [1]). The reason, obviously, lies in the regularity of the motion in which a particle moving along the x-axis unfailingly returns back after every half-period.

3. *Generalization to Turbulent Flow.*   Realistic application of the solution obtained in § 2 to a turbulent flow is evidently possible only when a cascade of motions is considered. As the independent variable it is convenient to choose the inverse of the wave vector modulus, $\bar{\lambda} = k^{-1}$, and then to integrate over this variable from zero to some maximum energy-carrying turbulent scale, $\lambda_{\max}$. It is not necessary here that large-scale motion be absent, however, this motion is considered to be regular and not turbulent. Returning to the region of interest, $\bar{\lambda} < \bar{\lambda}_{\max}$, we rewrite the equation, first considering the cascade λ_{n-1}, λ_n, λ_{n+1}, where the heat conductivity at each subsequent (greater) scale is calculated by substitution in formula (16) of the smaller-scale turbulent heat conductivity in place of the molecular heat conductivity:

$$D_{n+1} = D_n\left[1 + \frac{v_n^2}{\beta v_n^2 + \bar{\lambda}_n^2 D_n^2}\right].$$

We consider that the cascade occurs at equal intervals on the logarithmic scale of the quantity $\bar{\lambda}$. We replace the difference equation by a differential one. In doing so, we introduce an empirical coefficient, α which accounts for the fact that, in heat transfer in a given direction, only the component of the turbulent velocity in this direction takes part. Finally we propose the equation (omitting the overbar in $\bar{\lambda}$)

$$\frac{d\ln D}{d\ln\lambda} = \frac{\alpha v^2(\lambda)}{\beta v^2(\lambda)f + \lambda^{-2}D^2}$$

with the boundary condition $D = D_0$ at $\lambda = 0$. This equation can be solved exactly for $\alpha = \text{const}$, $\beta = \text{const}$, and the power dependence $v(\lambda) = v(\lambda/\lambda_{\max})^n$.

We recall that the Kolmogorov spectrum has exactly the same form with $n = 1/3$. Deviations from the power spectrum in the region of small λ can hardly change the result substantially.

Let us denote $\lambda v(\lambda) = r$, then

$$d\ln\lambda = (1-n)^{-1}d\ln r.$$

With respect to the variable r, the equation is homogeneous:

$$\frac{d \ln D}{d \ln r} = \frac{\alpha(1-n)r^2}{\beta r^2 + D^2}$$

Using general principles, we construct the equation for $p = D/r$:

$$\frac{d \ln D}{d \ln r} = 1 + \frac{1}{2}\frac{d \ln p^2}{d \ln r} = \alpha(1-n)\frac{1}{\beta + p^2},$$

$$\frac{1}{p^2}\frac{dp^2}{dr} = 2\left[\frac{\alpha(1-n)}{\beta + p^2} - 1\right] = 2\frac{p^2 + \beta - \alpha(1-n)}{\beta + p^2},$$

$$\ln r = -\int_\infty^p \frac{(\beta + p^2)\, dp^2}{2p^2[p^2 + \beta - \alpha(n-1)]} + \text{const.}$$

The boundary conditions are

$$r = 0, \quad \ln r = -\infty, \quad D = D_0, \quad p = \frac{D_0}{r} = \infty.$$

We satisfy them by substituting p_1 for ∞ and considering const $= \ln(D_0/p_1)$ since for $p^2 \gg \beta$, $p^2 \gg \alpha(n-1)$. Asymptotically,

$$\ln r = \int_p^{p_1} \frac{dp^2}{2p^2} + \ln \frac{D_0}{p_1} = \ln p_1 - \ln p + \ln D_0 - \ln p_1,$$

hence we have identically

$$\ln r = -\ln p + \ln D_0 = -\ln D + \ln r + \ln D_0, \qquad D \equiv D_0.$$

At large r we shall have to extend the integral into the region of smaller p,

$$\ln r = \ln D_0 - \ln p_1 + \int_p^{p_1} \frac{(\beta + p^2)\, dp^2}{2p^2[p^2 + \beta - \alpha(n-1)]}.$$

We assume that $\alpha(n-1) > \beta$, and denote

$$\alpha(n-1) - \beta = p_{\max}^2 > 0.$$

In the denominator, in the square brackets, the expression $p^2 - p_{\max}^2$ will appear. For large r it is obvious that $p \to p_{\max}$. In order to obtain the approximation law in explicit form, it is convenient to write

$$\ln p_1 = \int_p^{p_1} \frac{dp^2}{2p^2} + \ln p.$$

Subtracting one integral from the other, we remove the divergence at infinity, and again replace the artificially introduced p_1 by ∞:

$$\ln r = \ln D_0 - \ln p + \int_p^\infty \frac{1}{2p^2}\left[\frac{p^2 + \beta}{p^2 - p_{\max}} - 1\right] dp^2.$$

We substitute $p = D/r$ and obtain

$$D = D_0 \exp\left[\int_{D/r}^\infty f(p)\, dp\right].$$

We substitute $D/D_0 = z$ and arrive at the transcendental equation

$$z = \exp \int_{D_0 x/r}^0 f(p)\, dp.$$

The form of the function f is such that the integral diverges logarithmically when the lower limit approaches p_{max} from above:

$$J = A - B \ln \left(\frac{D_0 z}{r} - p_{max} \right).$$

Iterating the approximation, we find $z = p_{max} r/D_0$; the next approximation is

$$\frac{p_{max} r}{D_0} = e^A - \left(\frac{D_0 z}{r} - p_{max} \right)^{-B},$$

$$z = \frac{p_{max} r}{D_0} + \left(e^A - \frac{p_{max} r}{D_0} \right)^{-1/B} = f \left(\frac{r}{D_0} \right).$$

The numbers A and B may be expressed in terms of p_{max} and β, i.e., in terms of α, β, and n, but we are not going to do this. Thus, the answer is obtained in the form

$$D = D_0 f \left(\frac{r}{D_0} \right).$$

Here $r = v(\lambda)\lambda$, i.e., by order of magnitude it becomes the turbulent diffusion coefficient. The ratio r/D_0 is the Peclet number—the analogue of the Reynolds number.

In complete accord with a simple numerical evaluation, $f \to p_{max} r/D_0$ and the magnitude of D_0 decreases; the turbulent diffusion is independent of the molecular diffusion coefficient. Let us carefully consider the structure of the quantity $r = v\lambda$. It is obvious that in a turbulent flow we cannot directly determine a quantity which is linear in the fluctuation velocity. It is no accident that the square of the velocity figured in the original equation. It is precisely the mean square of the velocity and its spectral representation that may be determined in a turbulent flow. Therefore, consistently performing all the calculations, we obtain

$$D = D_0 \sqrt{ p_{max}^2 \frac{v^2 \lambda^2}{D_0^2} } + O(1).$$

It is fundamentally important here that the factor p_{max} is not obtained, let alone the corrections. The cascade method is not sufficiently accurate.

Important also (and apparently new) is the fact that the resulting expression has the correct transformation properties with respect to replacement of t by $-t$, just as the molecular diffusion coefficient. In the sense of the derivation, the square root here is an approximation of a function which always remains positive. If we had immediately written

$$D = \lambda \sqrt{v^2},$$

as is usually done, we would not have obtained the correct properties of D (for more details, see the beginning of the paper). The type of solution found here depends substantially on the assumption $\alpha(n-1) - \beta > 0$. For some other relation between parameters which violates this inequality we would have obtained a different asymptotic of D, one which grows as λ^m, where, however, $m < 1 - n$. But, apparently, in isotropic turbulence, which naturally occurs in the dissipative energy transfer from a larger scale to a smaller one, this situation does not arise.

Thus, in this paper we have obtained an exact solution of the diffusion equation for one-dimensional motion of an incompressible fluid, and determined the effective diffusion coefficient. We have constructed an approximate theory of turbulent diffusion as a cascade process of motion interaction on different scales. We have obtained an expression for the turbulent diffusion coefficient with the correct transformation properties under time reversal.

The M. V. Keldysh Institute of Applied Mathematics   *Received*
USSR Academy of Sciences, Moscow   *May 31, 1982*

References

1. *Moffatt H. K.*—J. Fluid Mech. **106**, 127 (1981).

Commentary

This paper poses for the first time the problem of the transformation properties of the diffusion coefficient under time reversal. The increase of the effective diffusion coefficient in the flow compared to the molecular coefficient has a very clear physical meaning: the flow increases the concentration gradients. Apparently, the first to perform similar calculations for flow in a cylindrical pipe was G. I. Taylor;[1] the effective diffusion coefficient understandably turned out to be proportional to the molecular diffusion coefficient, even though the coefficient of proportionality may be very large. The analysis in Ya.B.'s paper, just as that in G. I. Taylor's paper, is based on the derivation and investigation of an exact solution and is performed for a periodic velocity field. Ya.B.'s arguments relating to turbulent two- or three-dimensional diffusion rest, in addition to the solution obtained, on a number of rather intuitive assumptions. At the same time, the independence obtained for turbulent diffusion from the molecular diffusion coefficient in these approximations seems natural enough.

It is interesting to note that in a study by A. P. Mirabel and A. S. Monin,[2] written after Ya.B.'s paper, the expression for the turbulent diffusion coefficient in a two-dimensional turbulent field differs from the one obtained by Ya.B. by a factor which is equal to some logarithmic power of the ratio of the mixing scale to the energy-supply scale.

[1] *Taylor G. I.*—Proc. Roy. Soc. A **218**, 186–203 (1953); **225**, 473–477 (1954).
[2] *Mirabel A. P., Monin A. S.*—Dokl. AN SSSR **268**, 975–978 (1983)

7

A Magnetic Field in the Two-Dimensional Motion of a Conducting Turbulent Fluid*

The problem of magnetic fields arising spontaneously in the motion of a fluid was considered by Batchelor [1]. He came to the conclusion that the magnetic field increases without limit for sufficient conductivity in a given velocity field. His conclusion was based on nonrigorous considerations of the analogy between the magnetic field and a velocity vortex.

In the present work, the special case of two-dimensional motion is considered: $v_z = 0$, v_x and v_y depend only on x and y; the fluid is incompressible, div $\mathbf{v} = 0$. In this case, we have succeeded in treating the problem completely rigorously. The results differ substantially from the conclusions of Batchelor: in two-dimensional motion, in the absence of external fields, the initial magnetic field can increase no more than a certain number of times and then certainly decays. In the presence of external fields on the boundaries of the region of motion, the fields in the moving fluid in a steady state are proportional to the external fields. In the absence of mean, regular flow, the turbulently moving, conducting fluid behaves as a diamagnet with permeability μ, inversely proportional to the intensity of the turbulent mixing.

Following Batchelor, we set up the equation in the quasi-stationary approximation, neglecting the displacement current and the density of free charges. We employ $c = 1$ and the Heaviside system (without 4π), $\varphi = $ scalar potential, $\mathbf{A} = $ vector potential, div $\mathbf{A} = 0$, $\mathbf{J} = $ current, div $\mathbf{J} = 0$; the specific resistance of the fluid is r.

The equations have the form:

$$r\mathbf{j} = \mathbf{E} + \mathbf{v} \times \mathbf{H}; \quad \mathbf{H} = \operatorname{curl} \mathbf{A}; \quad \mathbf{E} = \frac{\partial \mathbf{A}}{\partial t} - \nabla\varphi; \quad \mathbf{J} = \operatorname{curl} \mathbf{H} = \nabla^2 \mathbf{A}. \quad (1)$$

It follows from this system that

$$\frac{\partial \mathbf{A}}{\partial t} + \mathbf{v} \times \operatorname{curl} \mathbf{A} = r\nabla^2 \mathbf{A} + \nabla\varphi. \quad (2)$$

Taking the curl of (2), we obtain the equation used by Batchelor,

$$\frac{\partial \mathbf{H}}{\partial t} + \operatorname{curl}(\mathbf{v} \times \mathbf{H}) = r\nabla^2 \mathbf{H}. \quad (3)$$

We now turn to the case of two-dimensional motion of an incompressible

*Zhurnal eksperimentalnoĭ i teoreticheskoĭ fiziki **31** 1, 154–155 (1956).

fluid. The equation for H_z is separated out; employing div $\mathbf{v} = 0$, we get

$$\frac{\partial H_z}{\partial t} + v_x \frac{\partial H_z}{\partial x} + v_y \frac{\partial H_z}{\partial y} = \frac{dH_z}{dt} = r\nabla^2 H_z. \tag{4}$$

Equation (4) is entirely analogous to the heat conduction equation in a moving fluid.

It is easy to verify that in the absence of external fields H_z only decreases. If in some particle H_z is maximal, then in the neighborhood $\nabla^2 H_z < 0$, so that $dH_z/dt = 0$, i.e., the maximum decays. In order to be certain of this, we set $H_z = 0$ at infinity or at the boundaries of the region and find, integrating by parts,

$$\frac{d}{dt} \int H_z^2 \, dV = \int H_z \frac{\partial H_z}{\partial t} \, dV = -r \int (\nabla H_z)^2 \, dV. \tag{5}$$

Let us consider the field which is obtained after the elimination of H_z. In this field $j = j_z$, $\varphi = 0$, $E = E_z$. $\mathbf{A}$ has only one component, A_z, which we denote by a in what follows. The two-dimensional vector of the magnetic field with components H_x and H_y we denote by $\mathbf{h}$. Expanding (3), we obtain the equation for $\mathbf{h}$:

$$\frac{\partial \mathbf{h}}{\partial t} + (\mathbf{v} \cdot \nabla)\mathbf{h} = \frac{d\mathbf{h}}{dt} = (\mathbf{h} \cdot \nabla)\mathbf{v} + r\nabla^2 \mathbf{h}. \tag{6}$$

Along with the dissipation term, $r\nabla^2 \mathbf{h}$, this equation contains the term $(\mathbf{h} \cdot \nabla)\mathbf{v}$ which describes the growth of the magnetic field with the stretching of the magnetic force lines noted by Batchelor. Thus, we cannot assert that $\mathbf{h}$ or $\mathbf{h}^2$ is eliminated and that $\int h^2 \, dV$ decreases monotonically.

For us it is basic that the equation for a has the form of the heat conduction equation and that a cannot increase (see above, the behavior of H_z). In the two-dimensional case and for $H_z = \varphi = 0$, we easily obtain from (2)

$$\frac{\partial a}{\partial t} + (\mathbf{v} \cdot \nabla)a = \frac{da}{dt} = r\nabla^2 a. \tag{7}$$

Hence, in the absence of external fields,

$$\frac{d}{dt} \int a^2 \, dV = -r \int (\nabla a)^2 \, dV = -r \int h^2 \, dV. \tag{8}$$

We imagine a distribution of a which is characterized by an amplitude a_0 and a length scale L which exceeds the maximum scale of the turbulent pulsation, l. We denote the pulsating velocity by u; the turbulent coefficient of diffusion, the coefficient of thermal conductivity and the effective turbulent kinematic viscosity are all expressed by the formula $\kappa = ul$. For an initial uniform distribution of a, obviously,

$$\left(\sqrt{\bar{h}^2}\right)_0 \approx \frac{a_0}{L}. \tag{9}$$

The macroscopic leveling of a in turbulent exchange is characterized by a time of order $\tau = L^2/\kappa = L^2/ul$ so that, taking (8) into account,

$$r\bar{h}^2 = \frac{d\bar{a}^2}{dt} = \frac{-\bar{a}^2}{\tau} = -\frac{ul}{L^2}\bar{a}^2. \tag{10}$$

Thus, after a time of order τ, the mean field increases to the quantity

$$\sqrt{\bar{h}^2} = \frac{a_0}{L}\sqrt{\frac{ul}{r}} = \sqrt{\frac{ul}{r}}(\sqrt{\bar{h}^2})_0, \tag{11}$$

but then both $\bar{a}^2$ and $\bar{h}^2$ fall exponentially with a period of order τ. The quantity ul/r is similar to the Reynolds number of turbulent motion, ul/ν (ν is the molecular kinematic viscosity). We denote it by $\mathrm{Re}_m = ul/r$ [2]. As is evident from (11), the increase in the field is limited to $\sqrt{\mathrm{Re}_m}\,h_0$; in the two-dimensional case, the possibility is excluded of increase of the field from arbitrarily small quantities (for example, depending on the fluctuations) to a finite value for finite r (even for smaller ν).

We consider turbulent motion in a closed region of size L at whose boundaries $v_n = 0$ (the index n is the normal to the surface S of the region) in the presence of external fields. It follows from (7) that the quantity $h_t = (\mathrm{grad}\,a)_n$ should be continuous at the boundary. From the analogy between (7) and turbulent heat conduction, noting that r plays the role of molecular thermal conductivity and considering the flow a as a heat flow, we find

$$r(\mathrm{grad}\,a)_n|_S \approx \frac{a_0\kappa}{L}, \tag{12}$$

where a_0 is of the order of magnitude of the difference in a at the different boundaries of the region. Consequently, the mean (not the mean square!) value of the field in the volume, $\bar{h} = a_0/L$, and the value of the tangential field at the boundary of the region where the fluid is at rest, are related by

$$\bar{h} = \frac{r}{\kappa}h_t|_S. \tag{13}$$

By analogy with macroscopic electrodynamics, we consider the currents in the turbulent fluid as molecular ones, the averaged actual field $\bar{h}$ we denote by $\mathbf{B}$, and we introduce $\mathcal{H}$; here $\mathrm{curl}\,\mathcal{H} = 0$ in the region where there are only unordered, turbulence-dependent currents. From (13) we obtain $\mathbf{B} = (r/\kappa)\mathcal{H} \cong \mathcal{H}/\mathrm{Re}_m$. Thus, macroscopically, the turbulent conducting fluid acts like a diamagnet[1] with very small permeability $\mu \sim 1/\mathrm{Re}_m$.

Apropos of the analogy noted by Batchelor between vortex velocity and a magnetic field, it should be noted that for the realization of truly steady turbulence a supply of mechanical energy is necessary. The supply of energy comes about either through nonpotential volume forces or through the motion of the surfaces bounding the fluid. With these factors taken into consideration, the set of equations and boundary conditions for a vortex does

---

[1]Czada [3] concluded that a conducting turbulent fluid is a paramagnet with large permeability. His analysis, based on a consideration of the damping of the field [3, p. 140] and of the energy of the field [3, p. 143], is not convincing since the energy of the pulsating components of the field can be regarded macroscopically as part of the turbulent energy, while the damping of the field may be related not only to the conductivity, but also to transformation of the energy into mechanical form.

not resemble the equation and boundary conditions for a magnetic field in the absence of external magnetic fields and outside electromotive forces.

Once again, we note that direct, step-by-step consideration of the three-dimensional case has not yet been possible, and the question of the character of growth of field in the three-dimensional case remains open.

Institute of Chemical Physics   *Received*
USSR Academy of Sciences, Moscow   *March 30, 1956*

References

1. *Batchelor G. K.*—Proc. Roy. Soc. London A **201**, 405 (1950).
2. *Elsasser W. M.*—Rev. Mod. Phys. **22**, 1 (1950).
3. *Czada I. K.*—Acta phys. hung. **1**, 235 (1952).

Commentary

In this brief work a number of important results are obtained on the evolution of initial perturbations of a magnetic field in a conducting turbulent flow. The paper studies plane motion in which the velocity vector lies in the x,y-plane and depends only on these coordinates. Under these conditions the initial magnetic fields decay; this, in particular, refutes for the two-dimensional case G. Batchelor's assertion that in a sufficiently well-conducting fluid the chaotic magnetic field infinitely increases. However, the decay occurs differently for different field components. The field component parallel to the z-axis decreases monotonically, while the component in the x,y-plane initially increases by order of magnitude to $\sqrt{\mathrm{Re}_m}$, where Re_m is the magnetic Reynolds number. This non-monotonic behavior is related to the competition of two processes: the field increase due to the expansion of the force lines frozen into the fluid, and the increase of dissipation due to the decrease of the characteristic dimension of field variation in the process of turbulent mixing.

It is important to emphasize that these magnetic fields may be absolutely arbitrary and, in particular, it is not necessary for them to possess the symmetry of the velocity field of the fluid. In this sense the theorem proved in this paper is much stronger than an analogous theorem considered by Cowling and Phil for the axisymmetric case.[1] They succeeded only in proving the decay of a magnetic field whose distribution, like the velocity field, is axisymmetric.

The paper also proves that the fluid region where turbulent motion occurs expels the external magnetic field like a diamagnetic body with low magnetic permeability μ. In the two-dimensional case considered in this paper, μ turns out to be of order $(\mathrm{Re}_m)^{-1}$. However, the assertion regarding the diamagnetic properties of the turbulent motion is valid for the three-dimensional case as well. Here we find $\mu \sim (\mathrm{Re}_m)^{-1}$ (see review [2]). The next article in this collection is devoted to further development of the results obtained in this paper.

[1] *Cowling T. G., Phil D.*, Month. Notic. Roy. Astr. Soc. **94**, 39–48 (1933).
[2] *Vainshtein S. I., Zeldovich Ya. B.*—Uspekhi Fiz. Nauk **106**, 431-457 (1972).

8

The Magnetic Field in a
Conducting Fluid Moving in Two Dimensions*

With A. A. Ruzmaikin

§1. Introduction

The classical problem of the magnetic dynamo, whether amplification or maintenance of a magnetic field is possible in a moving conducting fluid, has received, as we know, an affirmative solution. Many concrete examples of magnetic dynamos have been constructed, and this has stimulated applications of the theory to the explanation of the origin and maintenance of magnetic fields of planets, stars, and galaxies. On the other hand, even in its simplest kinematic form (i.e., when the velocity field is supposed known), the theory of the magnetic dynamo remains incomplete since necessary and sufficient conditions for the action of a dynamo have never been fully elucidated. It is known that a dynamo is impossible when the magnetic field and the velocity field do not depend on one of the space coordinates for the cases of planar [1, 2] and axial [3, 4] symmetry. In general, all three velocity components may be non-zero here. However, if all components of the velocity are present, these results do not extend to the spherical case (see below) and to other geometries. For a detailed survey of anti-dynamo theorems see [5, 6].

We recall the essence of the proofs, using as an example the plane case with $\partial/\partial z = 0$. The two-component part of the magnetic field in the x,y-plane, $H_2 = (H_x, H_y)$, separates from the z-component of the field due to the indicated symmetry, and is expressed in terms of the single component A_z of the vector potential. It follows from the induction equation that the component A_z is subject to the same equation as the temperature in a moving heat-conducting medium. Consequently [7], A_z can only decay. The two-dimensional field H_2 may be amplified up to some limit due to the decrease of the scale of the field (entanglement of the force lines), however, in the long run it must also be damped out because of ohmic losses [1]. The growth of this field and its subsequent decay was recently traced numerically in the work of Pouque [8] for plane turbulent motion. For the remaining component H_z of the field, in the simplest case $v_z = 0$ [1], one also obtains

*Zhurnal eksperimentalnoĭ i teoretichiskoĭ fiziki **78** 3, 980–986 (1980).

an equation like that for heat conduction which leads to damping. If $v_z \neq 0$, but the motion still does not depend on z, then in the equation for H_z there appears a source depending on H_2 and v_z. However, the decay of H_2 guarantees the impossibility of unbounded growth of the z-component of the field: taking the electric resistivity into account in the absence of fields at infinity leads to the decay of H_z.

Completely abandoning translational symmetry ($\partial/\partial z = 0$) does not yet mean that a dynamo is possible. In reality, as will be shown in this paper, a dynamo may be absent even for fields which depend on all three coordinates if one of the components of the velocity of the fluid vanishes. The impossibility of a dynamo in the three-dimensional situation was first indicated in a paper by Bullard and Gellman [9] for the spherical case with $v_r = 0$ (see also [10, 11]); the plane case was discussed by Moffatt [6]. The situation is simplest in a plane geometry for a conducting fluid moving with $v_z = 0$.

It is obvious that in this motion the vertical distance $z_{12} = (\mathbf{r}_{12})_z$ between any pair of points 1, 2 remains constant. In the approximation of frozenness of the magnetic field, it follows that H_z is conserved in an incompressible fluid (or H_z/ρ is conserved in a compressible medium, where ρ is the density). When the finite conductivity is taken into account H_z decays. For H_x, H_y, v_x, v_y dependent not only on x, y, but also on z, the field H_z is an additional source of the components H_x and H_y, and their temporary growth, or even generation is quite possible. However, there is no feedback. A decaying source cannot ensure non-decaying, much less growing generation of the fields H_x, H_y. Thus, we are able to substantially generalize the anti-dynamo theorem for plane motion.

The result can be translated completely to the spherical case by replacing the condition $v_x = 0$ by $v_r = 0$ and H_z by rH_r. The plane and spherical cases are special because in these geometries the curvature of the surfaces along which the fluid flows is constant in all directions; therefore, in the equations for H_z or rH_r the diffusion term $(\Delta H)_{z,r}$ is expressed only in terms of the field H_z (rH_r) and its derivatives, and the other components do not enter. This is what allows us to obtain a separate equation for H_z (rH_r) which is similar to the heat conduction equation and leads to decay of these components.[1] In an ideally conducting medium the assertion that H_z is conserved for $v_x = 0$ can be generalized to more complicated two-dimensional flows. Indeed, if the motion takes place in two dimensions, along surfaces which do not intersect and and do not diverge unboundedly, it is clear that the projection onto the normal of the distance between two points moving along neighboring surfaces will be bounded. And consequently, by the condition that the magnetic field is frozen in, the magnetic field component normal to these surfaces must also be bounded. The two other field com-

---

[1] We note that the mean value of H_r over any closed spherical surface vanishes and, consequently, the volume average over any sphere will also vanish. Therefore, decay leads, not to $H_r = $ const, but to $H_r = 0$ as $t \to \infty$.

ponents, for which the normal component serves as a source, can increase; however, owing to the boundedness of the source, this growth will be slow (not exponential). It will be similar, for instance, to the growth of the azimuthal component of a field in differential rotation and in the presence of a radial field.

However, if the electric resistivity is taken into account, there appears in general a new, very fundamental circumstance: the normal component of the Laplacian of an arbitrary vector also depends on the tangential components of the vector. The magnetic viscosity creates feedback, due to which the tangential components are able to modify (in particular, to amplify) the normal component. Therefore, in principle, exponential, coordinated amplification of all fields also becomes possible, i.e., a dynamo! But this is a peculiar kind of dynamo that depends on the magnetic viscosity.

The plane and spherical cases turn out to be degenerate and special.

It would be interesting to investigate (in both the plane and spherical cases as well) the very difficult nonlinear problem of a fluid whose resistivity depends on the magnetic field. Here the resistivity should be considered as a tensor relating the electric field vector to the current vector.

In connection with what has been said, we can give a definite classification of dynamo-solutions as a function of the magnetic Reynolds number. For $R_m \gg 1$ and three-dimensional motion a fast dynamo is possible with a characteristic time $\tau \sim l/v$, where l, v are the scales of length and velocity. A constructive example of a fast dynamo is the so-called figure-eight (see [5, p. 433]): a torus with azimuthal field flux, which, after its length is increased by a factor of two with attendant thinning, is folded with a twist in such a way that the flux is doubled. The doubling time is of order l/v, so that the dynamo has an increment v/l (we neglect the difference between $0.7 = \ln 2$ and unity). As was noted in [5], if the field is strictly frozen-in, the topology of the force lines changes during each doubling cycle. The introduction of a finite ν_m allows us to get conservation of the whole picture of the field for a small change in the increment.

In two-dimensional motion, in the plane and spherical cases with scalar resistivity, a dynamo is impossible, and in the general two-dimensional case, if it is possible, it will prove to be slow (with a characteristic time tending to infinity for fixed l, v and for $R_m \to \infty$). For small R_m an effective dynamo is possible for two-dimensional motion as well (excluding the plane and spherical cases).

In order to attain maximum clarity, we will carry out below a complete proof of the anti-dynamo theorem for the case of plane motion with $v_x = 0$ in Cartesian coordinates (cf. [6]). Then we will discuss the general case more briefly.

§2. Plane Incompressible Flow

Let us consider the motion of a conducting fluid subject to the conditions

$$v_z = 0, \quad v_x = v_x(x, y, z), \quad v_y = v_y(x, y, z),$$

$$\operatorname{div} \mathbf{v} = \frac{\partial v_x}{\partial x} + \frac{\partial v_y}{\partial y} = 0, \tag{1}$$

and otherwise arbitrary. The initial field is arbitrary and may depend on all three coordinates. We note that the helicity of the velocity field (1), i.e., the scalar product

$$\mathbf{v} \operatorname{curl} \mathbf{v} = v_y \frac{\partial v_x}{\partial z} - v_x \frac{\partial v_y}{\partial z} = -v_x^2 \frac{\partial}{\partial z} \left(\frac{v_y}{v_x} \right),$$

is generally nonzero (for $\partial/\partial z \neq 0$).

The general evolution equation of the magnetic field in the moving conducting medium has the form

$$\frac{\partial \mathbf{H}}{\partial t} = \operatorname{curl} \left([\mathbf{v} \times \mathbf{H}] - \nu_m \operatorname{curl} \mathbf{H} \right), \tag{2}$$

where $\nu_m = c^2/4\pi\sigma$ is the magnetic viscosity of the medium (σ is the conductivity). For simplicity, we will restrict our attention to an unbounded medium and assume that

$$\mathbf{H} \to 0, \qquad |\mathbf{r}| \to \infty. \tag{3}$$

We write the z-component of equation (2) in a medium with isotropic conductivity moving according to equation (1):

$$\frac{dH_z}{dt} \equiv \frac{\partial H_z}{\partial t} + \left(v_x \frac{\partial}{\partial x} + v_y \frac{\partial}{\partial y} \right) H_z = \nu_m \Delta H_z, \tag{4}$$

where d/dt is the substantional derivative and Δ is the Laplacian operator. Thus, H_z is subject to the same equation as a scalar quantity—the temperature or the concentration of an impurity—in a given velocity field with diffusion taken into account. It follows from the equation that if $\nu_m = 0$, H_z is conserved in each fluid particle, so that in this case

$$\int H_z^2 \, dV = \text{const},$$

and for finite conductivity and condition (3), at infinity

$$\frac{d}{dt} \int H_z^2 \, dV = -2\nu_m \int (\nabla H_z)^2 \, dV, \tag{5}$$

i.e., the quantity H_z only decays.

The behavior of the two-dimensional field $\mathbf{H}_2 = (H_x, H_y)$ becomes considerably more complicated when all the quantities depend on z. We recall that for $\partial H_z/\partial z = 0$ the condition $\operatorname{div} \mathbf{H} = 0$ implied

$$\frac{\partial H_x}{\partial x} + \frac{\partial H_y}{\partial y} = 0$$

and consequently $\mathbf{H_2}$ could be represented as the curl of A_z. In the case here, when H_z depends on z, this is impossible.

We note that, even if at the initial instant the component H_z were to depend only on x and y, and not on z, a dependence on z would appear because the motion along x and y is accompanied by a translational deformation, $\partial v_x/\partial z \neq 0$, $\partial v_y/\partial z \neq 0$. Therefore, for $\partial H_z/\partial z = 0$ at the initial moment (in the absence of magnetic viscosity)

$$\frac{\partial}{\partial t}\frac{\partial H_z}{\partial z} = \frac{\partial v_x}{\partial z}\frac{\partial H_z}{\partial x} + \frac{\partial v_y}{\partial z}\frac{\partial H_z}{\partial y},$$

which is intuitively obvious if one thinks of the transport of H_z by the fluid particles.

Thus, for H_z we obtain an equation which does not depend on the other components and which describes the general decay of H_z and defines the quantity $-\partial H_z/\partial z \equiv q(x,y,z,t)$. Consequently, the two-dimensional vector $\mathbf{H_2}$ can be represented only as the sum of the vortical and potential components

$$H_x = \frac{\partial \Phi}{\partial y} + \frac{\partial \varphi}{\partial x}, \qquad H_y = -\frac{\partial \Phi}{\partial x} + \frac{\partial \varphi}{\partial y}. \tag{6}$$

We obtain from

$$\frac{\partial H_x}{\partial x} + \frac{\partial H_y}{\partial y} = -\frac{\partial H_z}{\partial z}$$

that

$$\Delta_2 \varphi = q, \qquad \Delta_2 = \frac{\partial^2}{\partial x^2} + \frac{\partial^2}{\partial y^2}. \tag{7}$$

Taking account of the boundary conditions, this equation can be integrated by elementary methods at each given instant and in a given layer. This determines the function φ. The decay of φ (caused by the decay of H_z) means that at a sufficiently late stage the plane field may be represented in the form $\mathbf{H_2} = \operatorname{curl}(\mathbf{n}\Phi)$, where $\mathbf{n} = (0,0,1)$. After this the function Φ (which now coincides with the vector potential component A_z) is also subject to an equation of the heat conduction type. Consequently, $\mathbf{H_2}$ decays asymptotically.

For the sake of clarity, and having in mind generalization to the case of motion along curved surfaces, we write out the equation for Φ before the decay of H_z and φ. From definition (6) it follows that

$$\Delta_2 \Phi = -\operatorname{curl}_z \mathbf{H}. \tag{8}$$

Therefore, in order to obtain the equation for Φ, it suffices to take the z-component of the curl of equation (2). We obtain

$$\partial \Phi/\partial t + \mathbf{v}\nabla\Phi = \nu_m \Delta\Phi + S, \tag{9}$$

where the source S has the form

$$S = [\mathbf{v} \times \nabla]_z \varphi + \Delta_2^{-1}\frac{\partial}{\partial z}\left(\frac{\partial v_x H_z}{\partial y} - \frac{\partial v_y H_z}{\partial x}\right). \tag{10}$$

Taking (7) into account, all terms in (10) depend linearly on H_z. The arbitrary function of z which may be present in the right-hand side of equation (9) is insignificant in the calculation of the integral of Φ^2 since Φ may be selected such that its average over the planes $z = \text{const}$ will vanish.

§3. The General Case

It is clear that in the limit of complete frozenness of the magnetic field ($\nu_m \to 0$), the results obtained can also be generalized to two-dimensional motions of a more general type. It is easy to check, for instance, that in motion along spherical or cylindrical surfaces the equation for the radial component is decoupled, and this field component is conserved in each fluid particle. It is simplest and most intuitive to use the property that, if the magnetic field is frozen into the fluid, H_r behaves like δr, which is conserved due to the condition $v_r = 0$.

In considering motion along a family of stationary, non-intersecting surfaces of a more general type[2] one has to bear in mind that these surfaces are not parallel and, therefore, the normal projection of an infinitesimal distance, $(\delta r)_n$, between two points on neighboring surfaces varies and may increase in time. However, if the neighboring surfaces do not diverge unboundedly, which will be the case, for instance, for a family of closed surfaces, it is clear that $(\delta r)_n$, and consequently H_r will be bounded.[3] In this case one can speak about the conservation of the product of H_r and a factor which measures the divergence of the surfaces (a Lamé coefficient).

When considering a two-dimensional field on a surface, one can make use of the decomposition of an arbitrary solenoidal field into a "poloidal" and a "toroidal" part (see, e.g., [6, 12]):

$$\mathbf{H} = \text{curl} \, (\, \text{curl} \, \mathbf{n} \, \Psi + \mathbf{n} \Phi) \equiv \text{curl} \, [\nabla \times \Psi \mathbf{n}] + [\nabla \times \Phi \mathbf{n}]. \qquad (11)$$

This implies

$$H_n = -[\mathbf{n} \times \nabla]^2 \Psi; \qquad \text{curl}_n \, \mathbf{H} = -[\mathbf{n} \times \nabla]^2 \Phi. \qquad (12)$$

For the function Φ we again obtain equation (9) with the source S determined by the time-bounded quantity $\nabla_n H_n$ and its derivatives and velocity. This implies that rapid (exponential) growth of the two-dimensional field component is impossible.

---

[2]Essential here is the condition of stationarity of the surfaces, allowing for a nonstationary potential velocity field on the surfaces. The possibility of taking an instantaneous family of surfaces over an instantaneous velocity field does not suffice for the validity of the assertions made in the present paper.

[3]The boundedness of $(\delta r)_n$ is to be understood in the sense that it remains an infinitesimal of the same order. We underscore the fact that, in the formulation of the frozenness theorem for $\nu_m = 0$, the field or H/ρ change proportionally to the *infinitesimal* vector distance between infinitely close fluid particles. Therefore, a rapid dynamo is possible even in the case when the motion occurs in a finite volume.

However, the conclusion of decay of H_n and its related source when finite conductivity is taken into account is, in general, not valid (see the Introduction). In the general case $(\Delta H)_n$ is expressed not only in terms of H_n and its derivatives, but also in terms of the other field components.[4] For example, in the cylindrical coordinates s, φ, z,

$$(\Delta H)_s = \frac{1}{s}\Delta(sH_s) + \frac{2}{s^2}\frac{\partial H_\varphi}{\partial \varphi}.$$

This allows one to draw two conclusions. First, the theorem of the impossibility of a dynamo with $v_n = 0$ for arbitrary magnetic Reynolds number R_m is valid only for plane and spherical geometries. Second, the increment of a dynamo which is possible in other geometries will be determined by the magnetic diffusion and, consequently, such dynamos are effective only for small magnetic Reynolds numbers.

Let us point out that the impossibility of a dynamo in the spherical and plane cases has already been noted in [6, 9, 10]. We have shown that these cases are the only ones (for a stationary dynamo this fact was noted in [11]), and have generalized the assertion of boundedness of one of the field components to the general case. As an illustration of the second conclusion we point out the dynamo effect of a helical flow with $v_s = 0$ in cylindrical coordinates [13, 14]. The leading term in the maximal increment of this dynamo is proportional to $\nu_m^{1/3}$; for $\nu_m = 0$ the dynamo effect is absent. Moreover, here the most effective generation is for angular harmonics with $\partial/\partial\varphi \neq 0$, which is an additional indication of the decisive role of the coupling term $2s^{-2}\partial H_\varphi/\partial\varphi$ in the equation for H_s.

§4. *The Role of Inhomogeneity and Anisotropy of the Conductivity and Compressibility*

We shall make some remarks regarding the cases when the magnetic viscosity and the density of the medium are functions of the coordinates. It is obvious that no anti-dynamo theorems are possible if ν_m depends on all three space coordinates. In such a medium solutions are possible which simulate the functioning of dynamo-machines constructed by means of insulated conductors. However, in the simplest case $\nu_m = \nu_m(z)$ or $\nu_m = \nu_m(r)$ we can prove the impossibility of a dynamo. The equation for H_z in this case, as may be easily verified, again has the form (4), but relation (5) will no longer hold. In its place we must multiply (4) by H_z/ν_m and then, making use of the relation $\mathbf{v}\nabla\nu_m = 0$, we obtain

$$\frac{d}{dt}\int \frac{H_z^2}{\nu}\,dV = -2\int (\nabla H_z)^2\,dV. \qquad (13)$$

---

[4]The peculiarity of the plane and spherical cases is related to the fact that the curvature of the surface is zero in the first case and is isotropic in the second.

Thus, a dynamo is impossible in the plane and spherical cases and will be slow in the general case.

The theorem is easily generalized also to the case when the fluid density is $\rho = \rho(z)$ or $\rho = \rho(r)$. Indeed, in this case, the continuity equation implies that for $v_z = 0$ we have div $\mathbf{v} = 0$, i.e., the problem reduces to that already considered. We note that in an ideally conducting medium, for arbitrary density, the quantity H_z/ρ is conserved (instead of H_z in the case of an incompressible fluid), since instead of (4) we have in this case the equation

$$\frac{\partial}{\partial t}\frac{H_z}{\rho} + (\mathbf{v}\nabla)\frac{H_z}{\rho} = \frac{\nu_m}{\rho}\Delta H_z. \tag{14}$$

Of particular interest is the case of anisotropic tensorial conductivity, which couples together the various field and current components. It is then, of course, impossible to obtain a separate equation for H_z and, in principle, a dynamo is possible with an increment determined by the off-diagonal (in x, y, z coordinates) components of the conductivity tensor. Usually the direction of anisotropy is related to the magnetic field, but then the whole problem becomes nonlinear.

Finally, the most important problem is whether one can draw any conclusions for the general three-dimensional problem. Can one, for example, in the absence of helicity transfer directly to the case of isotropic three-dimensional general turbulence the concepts of diamagnetism of a turbulent plasma and of temporary growth of fields with a simultaneous decrease of the scale. These concepts can be applied not only to an infinite medium, but also to situations with boundary conditions at finite distances. But this problem is only posed here—its solution is a matter for the future.

We are grateful to T. Cowling who brought his paper [11] to our attention.

M. V. Keldysh Institute of Applied Mathematics   *Received*
USSR Academy of Sciences, Moscow   *October 10, 1979*

References

1. *Zeldovich Ya. B.*—ZhETF **31**, 154 (1956); here art. 7.
2. *Lortz D.*—Phys. Fluids **11**, 913 (1968).
3. *Cowling T. G., Phil D.*—Month. Notic. Roy. Astron. Soc. **94**, 39 (1933).
4. *Braginskiĭ S. I.*—ZhETF **47**, 1084 (1964).
5. *Vainshtein S. I., Zeldovich Ya. B.*—Uspekhi fiz. nauk **106**, 431 (1972).
6. *Moffatt H. K.* Magnetic Field Generation in Electrically Conducting Fluids. Cambridge Univ. Press, p. 108–117 (1978).
7. *Zeldovich Ya. B.*—ZhETF **7**, 1466 (1937); here art. **4**.
8. *Pouque A.*—J. Fluid Mech. **88**, 1 (1978).
9. *Bullard E. C., Gellman H.*—Philos. Trans. Roy. Soc. London A **247**, 213 (1954).
10. *Backus G. E.*—Ann. Phys. **4**, 372 (1958).

11. *Cowling T. G.*—Quart. J. Mech. and Appl. Math. **10**, 129 (1957).
12. *Rädler K. H.*—Astron. Ztschr. Nachricht. **295**, 73 (1974).
13. *Ponomarenko Yu. B.*—Zhurn. prikl. mekhaniki i tekhn. fiziki **6**, 47 (1973).
14. *Gailitis A. K., Freinberg Ya. Zh.*—Magnit. gidrodinamika **2**, 3 (1976).

Commentary

In this paper substantial generalization of the results of the previous paper is accomplished. First of all, it is shown that magnetic fields decay even if the motion is such that the velocity vector is still located in the x,y-plane, but depends on all three coordinates and time. Furthermore, it appears that this statement can be proved even for the case in which the surfaces where the motion occurs are not planes but spheres. Then, for the velocity component normal to the surface, a diffusion-like equation which does not contain any other components is obtained.

A more complicated situation emerges in motion along nonintersecting surfaces with variable curvatures. If the distance between these surfaces remains finite everywhere, then the field lines do not expand infinitely in the directions normal to the surfaces. In the absence of dissipation this means that there is no unbounded growth of the normal field component. However, introduction of the finite conductivity yields an equation for the normal component which is not decoupled: it contains the contribution of the Laplacian of the remaining components. At the same time, it is possible for all other components to increase exponentially with an increment which depends on the conductivity and vanishes for infinite conductivity. The authors called this mechanism of field amplification a "slow dynamo," in contrast to the "fast dynamo" feasible in the three-dimensional case, i.e., the mechanism related only to infinite expansion of the field lines as, for example, in motion with magnetic field "loop doubling." In a fast dynamo the characteristic time of the field increase must be of the same order as the characteristic period of the motion's fundamental scale.

There are several known steady motions along nonintersecting surfaces in which the increase of the magnetic field is exponential, i.e., a "slow dynamo" occurs.

9

Gas Motion Under the Action
of Short-Duration Pressure (Impulse)[*]

§1. Statement of the Problem and the General Character of the Motion

Let us consider a half-space (the region $x > 0$) occupied by an ideal monatomic gas. At an initial moment $t = 0$ the gas is at rest, its density is the same everywhere ($\rho = \rho_0$ for $x > 0$; $\rho = 0$, $x < 0$), and the temperature and pressure are equal to zero. The outer surface of this gas is acted upon by a short pressure impulse, $\pi(t)$, where the function $\pi(t)$ is such that $\pi(t) = 0$ for $t < 0$, then it attains a maximum value P, after which $\pi(t)$ rapidly decreases. The time of the decrease is described by the quantity τ, so that, for $0 < t < \tau$, we can write $\pi = Pf(t/\tau)$, where f is a dimensionless function of the dimensionless variable t/τ.

We may imagine that a flat piston is located at the coordinate origin. At the initial time it begins to move into the gas so that $\pi(t)$ is the gas pressure at the piston's surface. Then, after a time of order τ, the piston is removed or begins to move left from the gas with a velocity exceeding the propagation velocity of the gas compressed by it. Another possible mechanism for generating this motion is the explosion of a thin, flat layer of explosive which lies on the surface of the cold gas.

A typical feature of the statement of the problem is the vacuum in the left half-space, $x < 0$, at the initial time. Gas particles located initially at the surface (at $x = 0$) fly out into the vacuum after the action of the external pressure is completed.

The problem consists in finding the motion of the gas and the distribution of density, pressure and other parameters after an elapsed time which is large compared to the effective duration of the pressure, τ.

Thus, we seek an asymptotic solution of the one-dimensional gasdynamic equations for a given pressure evolution curve, $f(t/\tau)$, which is characterized by a sufficiently rapid pressure decrease. In a slightly different way, we may formulate the problem thus: preserving the form of the dimensionless function $f(t/\tau)$, we let the pressure duration go to zero and the maximum pressure to infinity, and look for the asymptotic solution—the distribution of the velocity, pressure and other quantities—after a finite time t, at a finite distance x.

[*]Akusticheskiĭ zhurnal **2** 1, 28–38 (1956).

It is obvious that the solution of the problem should also give an answer to the question of *precisely how*, according to what law, should P increase with decreasing τ in order to ensure a given finite gas pressure after a finite time t. If, for example, the combination $P^k \tau^l$ enters into the solution, then P should increase as $\tau^{-1/k}$ as τ decreases.

The general character of the motion is shown in Fig. 1. A shock wave propagates in an unperturbed cold gas; the maximum compression, which is dependent on the adiabatic index of the gas, is reached at the front of this wave. The velocity of propagation of the shock wave and the mass velocity at the front are related in an elementary way to the pressure of the shock wave. Behind the front the pressure, density, and velocity decrease.[1]

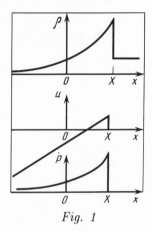

The amplitude of the shock wave decreases with time. The solution of the problem should provide an answer to the interesting question of the maximum possible rate of amplitude decay of a plane wave propagating in an ideal gas with constant initial density.

Fig. 1

This solution is related to the problem of an instantaneous pressure impulse ($\tau \to 0$); obviously, any extended application of pressure and any delay in the pressure decrease at the piston can only diminish the rate of the shock wave pressure decrease.

The character of the solution is independent of the concrete form of the pressure decrease law at the piston $\pi(t)$, described by the dimensionless function $f(t/\tau)$, only if f decreases fast enough[2] for large values of t/τ. In particular, it is unimportant whether f increases continuously or discontinuously at $t < \tau$. Even for a continuous increase of the gas pressure, a shock wave will form with a pressure amplitude of the order of the maximum pressure P at the piston. The shock wave velocity here is of order $D \approx \sqrt{P/\rho_0}$, so that some small mass of gas, of order $D\rho_0\tau \sim \tau\sqrt{P\rho_0}$, is subjected to a shock compression of amplitude P. After the pressure at the piston goes to zero this gas, extending into the vacuum, attains a velocity of order $u_0 \approx \sqrt{P/\rho_0}$, and at the time t it will be located at a point $x = u_0 t \approx t\sqrt{P/\rho_0}$. So, for example, if $f(t/\tau) = 1$ at $0 < t < t/\tau < 1$ and

---

[1]For the comparison of waves of different amplitudes at different gas densities, it is convenient to characterize the rate of pressure decay by a dimensionless quantity, the logarithmic derivative $d\ln p/d\ln t = (t/p)(dp/dt)$ or, equivalently, by the value of the exponent in the formula $p \approx t^{-g}$.

[2]We will be able to indicate specifically what we mean by "fast enough" only after the solution has been found.

$f = 0$ at $t/\tau > 1$, then for a diatomic gas we easily find u_0 as the difference between the dispersion speed, equal to $5c$, where c is the speed of sound in the compressed gas, and the mass speed in the wave, u. At the limit, for indefinitely increasing P, the absolute value of u_0 increases indefinitely, so that the distributions of ρ, p, and u in the required solution, shown in Fig. 1, should be imagined as extending indefinitely to the left to $x = -\infty$.

§2. *The Self-Similar Solution and the Laws of Conservation of Energy and Momentum*

The assumption of independence of the desired solution from the concrete form of $\pi(t)$ suggests that the solution, for example, the instantaneous pressure distribution $p(x, t)$ at $t \gg \tau$, should be sought in a self-similar form,

$$p(x, t) = AP^k \tau^l \rho_0^h t^{-g} \tilde{p}\left(\frac{x}{X}\right),$$

where $X = X(t)$ is the shock wave coordinate (see Fig. 1).

For the sake of definiteness, let $\tilde{p}(1) = 1$, so that the pressure at the front is $p(X, t) = AP^k \tau^l \rho_0^h t^{-g}$.

The form of the function $X(t)$ is easily found:

$$\frac{dX}{dt} = D \approx \sqrt{\frac{p(x, t)}{\rho_0}},$$

so that $x \approx t^{1-g/2}$; from the law of the wave amplitude decay as a function of time, $p \approx t^{-g}$, it is easy to pass to the decay law as a function of the path traveled by the wave:

$$p(x, t) \approx X^{-n}, \quad p(x, X) \approx BP^{k'} \tau^{l'} \rho_0^{h'} X^{-n}, \quad n = \frac{2g}{2 - g}.$$

The basic result of this paper is the determination of the exponents k, l, h, g, or k', l', h', g', as well as the dimensionless function $\tilde{p}$ and the analogous functions $\tilde{u}$, $\tilde{\rho}$ which describe the distribution of velocity, density and other quantities.

It is easy to verify that dimensional considerations establish only three relations between the four exponents (since all the quantities are expressed in a dimensional system with only three basic units: mass, length, and time).

The missing relation may be obtained by considering the ordinary differential equations which follow from the partial differential equations of gas dynamics written for the dimensionless functions $\tilde{p}$, $\tilde{\rho}$, $\tilde{u}$ of the single variable $\xi = x/X$.

In a number of problems, e.g., the well-known problem of localized explosion solved by Sedov [1], the exponents are found by elementary means from the conservation laws of mechanics (from the conservation of energy in Sedov's example).

What makes this problem different? Is it not possible to find the exponents from elementary considerations in the case of impulse?

After the pressure at the piston returns to zero, the gas is not acted upon by any external forces since the pressure in the cold gas is equal to zero both at $x \to -\infty$ (in the limit at the boundary of dispersion), and at $x \to \infty$ ($x > X$ is sufficient) in the cold gas. In the motion, therefore, the total momentum of the gas, $\int_{-\infty}^{\infty} \rho u \, dx$, is conserved. This momentum is equal to the pressure impulse of the piston,

$$T = \int_0^\infty \pi \, dt = P\tau \int_0^\infty f\left(\frac{t}{\tau}\right) \frac{dt}{\tau} = aP\tau$$

(all quantities are taken with respect to the unit area of intersection with the Y,Z-plane, i.e., to the unit area of the piston).

It is obvious that the bulk energy of the gas, equal to

$$\int_{-\infty}^{\infty} \left(\rho \frac{u^2}{2} + \frac{1}{\gamma - 1} p\right) dx,$$

is conserved as well, where γ is the adiabatic index. The energy must be equal to the work done by the piston during the action of the pressure.

The piston's velocity is completely determined when the pressure at the piston is given; however, to find $u(t)$ from a given function $\pi(t)$ requires solution of the gasdynamic equations. In any case, from dimensional considerations it is easy to establish that $E = bP^{3/2}\tau\rho_0^{-1/2}$, where b is a dimensionless number depending in some way on the dimensionless $f(t/\tau)$.

Formally, the integrals of momentum and energy of the gas are taken over the entire volume occupied by the gas. In practical terms, the gas in front of the shock front is motionless and cold, and contributes nothing to these integrals; without changing their values, then, we may set X as the upper limit of integration.

We substitute into the integrals the self-similar expressions

$$p(x, t) = \Pi \tilde{p}(x, X), \qquad (\Pi = BP^{k'} \tau^{l'} \rho_0^{h'} X^{-n});$$

$$u = \sqrt{\frac{\Pi}{\rho_0}} \tilde{u}\left(\frac{x}{X}\right); \qquad \rho = \tilde{\rho}_0 \tilde{\rho}\left(\frac{x}{X}\right),$$

taking into the account the relation between pressure and velocity at the shock wave front, as well as the constant value of the density at the front. We obtain

$$I = \int \rho u \, dx = \sqrt{\Pi \rho_0} X \int_{-\infty}^{1} \tilde{\rho}(\xi)\tilde{u}(\xi) \, d\xi = \text{const},$$

$$E = \int \left(\rho \frac{u^2}{2} + \frac{1}{\gamma - 1} p\right) dx = \Pi X \int_{-\infty}^{1} \left(\tilde{\rho}\frac{\tilde{u}^2}{2} + \frac{1}{\gamma - 1}\tilde{p}\right) d\xi = \text{const}.$$

It would be natural to consider the dimensionless integrals to be constants. Then each of the two conditions written above, taken separately, would allow us to establish the exponent n (from the momentum equation, $n = 2$, and from the energy equation, $n = 1$), and after this it would be elementary to find the rest of the exponents.

However, taken together, these two equations contradict one another! We are left with the impression that no self-similar solution can at once satisfy both conditions. Both necessarily follow from the equations of mechanics and, consequently, the problem does not have a self-similar solution at all.

Below, we shall find a specific self-similar solution of the power type. We will show that the actual exponent is always in the interval $1 < n < 2$; for a diatomic gas $n = 1.333 \approx 4/3$.

It is appropriate here, even before we find a concrete solution, to figure out just how this solution agrees with the conservation laws. It is difficult to imagine a reader, willing to follow the algebraic calculations and integration of the differential equations, while expecting to obtain a solution which satisfies neither the law of conservation of momentum, nor the law of conservation of energy!

The following complete expression for the pressure corresponds to the exponent $n \sim 4/3$ found by Adamskiĭ in [2]:

$$p = \left(\frac{\tau \sqrt{P/\rho_0}}{X} \right)^n BP\tilde{p} = BX^{-4/3} \rho_0^{-2/3} \tau^{4/3} P^{5/3} \tilde{p} \left(\frac{x}{X} \right),$$

so that to obtain a finite (not 0 and not ∞) pressure at a finite distance for $\tau \to 0$ and $P \to \infty$, it is necessary for P to increase as $\tau^{-4/5}$.

For P increasing according to this law, in the limit, the impulse of the external force causing the motion (the impact impulse) I goes to zero: $I \approx P\tau \approx \tau^{1/5}$. Consequently, the impulse equation for self-similar motion is written thus:

$$B'X^{1/3} \int_{-\infty}^{1} \tilde{\rho}(\xi)\tilde{u}(\xi) \, d\xi = 0.$$

This leads to a condition which should be satisfied by the dimensionless functions $\tilde{\rho}$, $\tilde{u}$: $\int_{-\infty}^{1} \tilde{\rho}\tilde{u} \, d\xi = 0$.

We turn now to the energy equation. As we have said, the work done by the piston (by impulse) on the gas, given the relation found between the pressure P and the time of impact τ is proportional to

$$E \approx P^{3/2}\tau \approx (\tau^{-4/5})^{3/2}\tau \approx \tau^{-1/5}.$$

In the limit as $\tau \to 0$, the work done by the impulse $E \to \infty$.

We turn to the energy of the gas, calculated according to the self-similar solution as the volume integral of the kinetic and thermal gas energies. The dimensional coefficient is proportional to $\Pi X = P^{3/2} \tau^{4/3} X^{-1/3}$. It decreases indefinitely with propagation of the wave, which at first glance appears to be impossible.

The paradox is resolved because the dimensionless integral in the expression for the energy

$$\int_{-\infty}^{1} \left(\tilde{\rho} \frac{\tilde{u}^2}{2} + \frac{\tilde{p}}{\gamma - 1} \right) d\xi$$

turns out to be divergent in the region of large negative ξ. The self-similar solution in this region has an asymptotic form: $\tilde{u} = k\xi$; $\rho = r(-\xi)^{-3/2}$.

In fact, the pattern of motion is such that a certain mass of the gas, located near the impacted surface experiences a pressure which differs from the self-similar value. For a pressure P, acting over a time τ, the shock wave travels a distance $\tau \sqrt{P/\rho_0}$ and compresses the gas mass by $m_0 \tau \sqrt{P\rho_0}$ (per unit surface).

Only for $t > \tau$ do we have the right to expect that the shock wave amplitude will approach the self-similar value which we have found. Due to the adiabatic character of the motion, the entropy of the mass m_0 always remains constant, and differs from the value which is obtained by extrapolation of the self-similar entropy distribution to this mass.

Convergence of the actual solution (for finite P and τ) to the self-similar solution as the wave propagates, i.e., as X increases, is non-uniform. As X increases, the pressure and all other quantities, over almost all of the gas mass involved in the motion, approach the values corresponding to the self-similar solution. At the same time, in a small mass m_0 of gas this convergence does not occur, however, with time this mass constitutes a smaller and smaller fraction of the total mass involved in the motion.[3] In the limit, this small mass can be disregarded in both the differential equations and in the convergent impulse integral; however, in the energy integral, because it diverges, it is essential that the integration be stopped in the region to which the self-similar solution is inapplicable.

We find the boundary of this region x_2, $\xi_2 = x_2/X$ ($x_2 < 0$, $\xi_2 < 0$), from the condition

$$m_0 \approx \tau \sqrt{P\rho_0} \approx \int_{-\infty}^{x_2} \rho \, dx = \rho_0 X \int_{-\infty}^{\xi_2} \tilde{\rho} \, d\xi \approx \rho_0 X (-\xi_2)^{-3/2}.$$

---

[3]Convergence of the actual solution to the self-similar one over time occurs in a way similar to convergence of a Fourier series to a discontinuous function as the number of terms in the sum increases: the well-known Gibbs phenomenon leads to the fact that, near the discontinuity, for any number of terms, the maximum difference between the series and the function does not approach zero; however, the width of the region in which the series differs noticeably from the function approaches zero as the number of terms increases.

The condition indicates that we take the self-similar solution and find within it the boundary of the region containing the mass m_0 (Fig. 2).

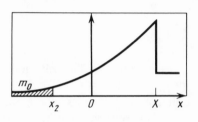

Fig. 2

This method should give the correct order of magnitude and the correct dependence of x and ξ on the parameters. We obtain

$$\xi_2 \approx -\left(\frac{X}{\tau\sqrt{P/\rho_0}}\right)^{2/3}$$

$$x_2 \approx X^{4/3}(\tau\sqrt{P/\rho_0})^{-2/3}.$$

As X increases, the region in which the self-similar solution is not applicable (shaded in Fig. 2) is pushed further and further to the left, away from the main region of motion. However, because the main portion of the kinetic energy is also concentrated at the left, in the zone of low density and rapid motion, the presence of the shaded region always (even for X indefinitely increasing and m_0/X indefinitely decreasing) remains significant for calculation of the energy integral.

We shall take this integral not from $-\infty$, but from x_2 or ξ_2:

$$E = \Pi X \int_{\xi_2}^1 \left(\tilde{\rho}\frac{\tilde{u}^2}{2} + \frac{1}{\gamma-1}\tilde{p}\right)\,d\xi.$$

Substituting $\rho \approx (-\xi)^{-3/2}$, $\tilde{u} \approx \xi$, we find that the value of the integral is determined by its lower limit, for $|\xi_2| \gg 1$, $\xi_2 < 0$,

$$\int_{\xi_2}^1 \left(\tilde{\rho}\frac{\tilde{u}^2}{2} + \frac{1}{\gamma-1}\tilde{P}\right)\,d\xi \to |\xi_2|^{1/2}.$$

Let us express Π and ξ_2 in terms of X: $\Pi \sim X^{4/3}$, $\xi_2 \sim X^{2/3}$. Substituting into the expression for the energy, we obtain

$$E \sim \Pi X|\xi_2|^{1/2} \sim X^{-4/3}X(X^{2/3})^{1/2} = X^0 = \text{const},$$

as in fact we should by the energy conservation law.

It is also easy to verify that, by using the above dependencies of Π and ξ_2 on P and τ, we obtain $E \sim P^{3/2}\tau$. The impulse integral converges, and we do not need to consider ξ_2.

Thus, the concept of zero total momentum, and of the existence of a small region which does not obey the self-similar solution, resolves the paradoxes which arose when the solution was compared with the conservation laws.

§3. General Properties of the Solution

The value of the exponent given above, $n = 1.333$, describing the dependence of the shock wave amplitude Π on the distance X traveled by the wave, $\Pi \sim X^{-4/3}$, yields the maximum rate of decay with distance of a plane shock wave in a diatomic gas.

As may be seen from the formula, the amplitude decreases by 2.5 times when the path length doubles.

It is also not difficult to find the dependence of the position of the front and its amplitude on time; we obtain $X \sim t^{3/5}$ and $\Pi \sim t^{-4/5}$. When the time doubles the wave amplitude decreases by 1.74 times.

Self-similarity of motion means that the distribution of all quantities at different times are similar. In particular, a quantity like the ratio of the mass between the origin and the shock wave to the total mass involved in the motion is a constant. For a diatomic gas, this quantity is .90 .

Thus, a tenth of the gas, as a result of compression and subsequent expansion, is thrown to the left of the initial position of the gas boundary (left of the coordinate origin). It is possible also to find the fraction of the gas which moves to the right, in the direction of the impulse. To this end we find ξ_1 such that $U(\xi_1) = 0$. The required fraction of the gas is expressed by the formula

$$\beta = \int_{\xi_1}^{1} \tilde{\rho} \, d\xi = 0.78.$$

It is not difficult to construct formulas for the momentum and energy of this fraction of gas:

$$I_1 = \rho_0 X \sqrt{\Pi} \int_{\xi_1}^{1} \tilde{p}\tilde{u} \, d\xi, \quad E_1 = \Pi X \int_{\xi_1}^{1} \left(\tilde{\rho} \frac{\tilde{u}^2}{2} + \frac{\tilde{p}}{\gamma - 1} \right) g\xi.$$

Both integrands are positive and (in the interval $\xi_1 < \xi < 1$, $\tilde{u} > 0$) are bounded such that both integrals are finite positive numbers.

Substituting $\Pi \sim X^{-n}$, we find that $I_1 \sim X^{1-n/2}$, and $E_1 \sim x^{1-n}$.

It is easy to see without calculation that, with time, I_1 should increase and E_2 should decrease. Thus, we are able in advance, before integration of the equations, and for any adiabatic index, to show that

$$1 - n/2 > 0, \quad 1 - n < 0, \quad 1 < n < 2.$$

To this end we note that the point $\xi = \xi_1$, i.e., the rest point at which the mass velocity of the gas is zero, moves from left to right, and $x_1 = \xi_1 X$ increases.

Since the gas at this point is at rest we may say that this point moves from left to right with respect to the gas as well.

We consider the volume of gas for which the momentum and energy equations are written—the volume contained between the rest point and the shock wave: $x_1 < x < X$.

Cold, unperturbed gas at zero pressure enters this volume from the right, through the shock wave surface. It is obvious that this gas does not bring into the volume either momentum or energy. Through the surface $x = x_1$ the gas leaves the volume with finite pressure and temperature and zero velocity.

Because the rest point was chosen as the second boundary of the volume, the momentum of the gas leaving the volume is zero.

The momentum of the gas inside the volume increases since the surface x_1 is acted upon by a non-zero pressure: $dI_1/dt = p_1 > 0$, hence $1 - n/2 > 0$.

The change in the energy contained in this volume is determined by the fact that the work done by the pressure forces $pu \, dt$ on this surface is equal to zero ($u = 0$). When the point moves to the right from the volume with respect to the gas, gas with nonzero thermal energy flows out of the volume (the kinetic energy of the gas is zero since $u = 0$). Finally,

$$\frac{dE_1}{dt} = -\frac{1}{\gamma - 1}\frac{p_1}{\rho_1}\rho_1\frac{dx_1}{dt} = -\frac{1}{\gamma - 1}p_1\xi_1\frac{dX}{dt} =$$
$$= \frac{d(aX^{1-n})}{dt} = (1 - n)\frac{E_1}{X}\frac{dX}{dt};$$

whence it follows that $n > 1$.

The overall picture of the motion is such that the energy contained in a particular portion of the mass (for example, in the region $x_1 < x < X$ or $0 < x < X$) continuously decreases, and there occurs an irreversible energy discharge into the zone of low-density propagating gas.

The shorter the impulse, the greater the decrease of the energy E_1 compared to the initial work done by the impacting piston. It is not surprising that, at the limit of the impulse duration, $\tau \to 0$, infinite energy of the piston is required.

The momentum of the gas as a whole is zero, or more precisely, in the actual, not the self-similar, solution, it is equal to the small initial momentum of the impulse. The momentum of a certain part of the gas (for example, the part which moves to the right), I_1, indefinitely increases with the motion. The momentum of this part of the gas is quite large compared to the initial impulse momentum, and it is balanced by the momentum ("recoil") of the gas flowing out to the left. We note that, by order of magnitude, we have the identity: $I_1 = \sqrt{M_1 E_1}$ (the "Stanyukovich formula"). However, due to the essentially nonuniform velocity distribution in the dispersing gas, this formula can not be applied to the dispersion zone where the energy is $E_{\text{disp}} = E_{\text{piston}} - E_1$ and the mass is $M_{\text{disp}} = M - M_1 = \rho_0 X - M_1$. In this zone,

$$\frac{I_{\text{disp}}}{\sqrt{M_{\text{disp}} E_{\text{disp}}}} \to 0.$$

In order to complete the solution of the problem of gas motion under the action of a short impulse, we must not only find the exponents and dimensionless functions, which is accomplished by integrating the ordinary differential equations. We must also determine the numerical coefficients A and B in the formulas.

The dimensionless numbers A and B depend on the adiabatic index and on the concrete form of the impulse pressure curve, $f(t/\tau)$. For them, however, we have not succeeded in obtaining closed-form expressions.

At present, we see only one route to complete solution of the problem: for $t \approx \tau$, for example, up to $t = 5\tau$ or 10τ, we solve the partial differential equations with a given pressure law at the piston; continuing the solution for $t \gg \tau$ in asymptotic (self-similar) form, we determine the constants A and B from the condition that the two solutions coincide at the extreme point to which the calculation is carried.

The law of pressure decay $f(t/\tau)$ is restricted by the obvious condition that the pressure at the piston must decrease faster than the pressure amplitude in the self-similar solution itself, i.e., faster (with a greater absolute value of the negative exponent) than $t^{-4/5}$.

§4. Explosion at the Boundary of Two Gases with Different Densities

Impact of a gas by a piston with subsequent gas dispersion into a vacuum is an extremely idealized statement of the problem. A more realistic problem is to consider a thin flat layer of explosive dividing two gas regions: one with initial density ρ_0 (in which we consider the motion) located to the right at $x > 0$, and one with initial density R, $R \ll \rho_0$ (Fig. 3–the explosive material is indicated by the *1*). The mass of the explosive per unit surface we denote by m, and its caloricity (specific energy in erg/g) by Q. Following M. A. Lavrentiev, we break the beginning of the process into stages.

At the moment of explosion the chemical energy is transformed into thermal and elastic energy, then the thermal and elastic energy are transformed into the kinetic energy of dispersion of the explosion products (EP). Only in the next stage do the EP slow down, giving up their energy to the gas.

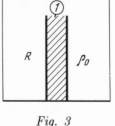

Fig. 3

The velocity of the EP is of order $w \sim \sqrt{Q}$. Their total energy is mQ, and momentum, $m\sqrt{Q}$. In a gas of density ρ_0 the EP form a shock wave with pressure of order $P = \rho_0 w^2 = \rho_0 Q$. From the condition that over the duration of the pressure P the EP must give up their energy and momentum to the gas we find the duration of the pressure action,

$$\tau = \frac{m\sqrt{Q}}{\rho_0 Q} = \frac{m}{\rho_0 \sqrt{Q}};$$

in this time the shock wave will travel in the gas a path of order m/ρ_0, and the mass of gas involved in the motion will be of order m. As we might expect, the proposed self-similar solution is applicable to the problem of the explosive's effect only after the mass of gas involved in the motion, $M = \rho_0 X$, exceeds the mass of the explosive.

Dropping numerical factors, we substitute our estimates of the magnitude and duration of the pressure into the self-similar expression for the shock wave amplitude. We obtain

$$\Pi = \rho_0^{-1/3} Q X^{-4/3} m^{2/3} = Q\rho_0 \left(\frac{m}{X\rho_0}\right)^{4/3} = Q\rho_0 \left(\frac{m}{M}\right)^{4/3}$$

for $M \gg m$.

When there is a gas, albeit one of lesser density ($R \ll \rho_0$) rather than a vacuum on the left, applicability of the solution is restricted in later stages of the process as well.

For the later stages, when the explosive mass is small compared to the mass of gases involved in the motion on both sides, there is every reason to apply to the solution of the problem the theory of strong explosion developed by Sedov. It is not difficult to generalize the theory of a one-dimensional plane explosion to that at the boundary of two gases of differing densities.

In each half-space we take Sedov's solution, which satisfies the condition of zero velocity at the coordinate origin, and impose on them the condition of pressure equality. Since the law of pressure decay with time in a strong explosion is independent of the gas density (in the plane case this law is $p \sim t^{-2/3}$), this joining of the two solutions is possible. At equal pressures, the ratios of the paths traveled by the two shock waves to the left, Y, and to the right, X, are inversely proportional to the ratio of the square roots of the gas densities: $Y/X \sim \sqrt{\rho_0/R}$.

From normalization by the total energy of the explosion we find the order of magnitude of the pressure amplitude Π_S in the shock wave (the index S denotes "pressure calculated according to Sedov's theory") at the moment when the wave has traveled a path X in the gas:

$$mQ = (Y + X)\Pi_S,$$

$$\Pi_S = \frac{mQ}{X(1 + \sqrt{\rho_0/R})} = \frac{mQ\sqrt{R/\rho_0}}{X} =$$

$$= Q\rho_0 \left(\frac{m}{\rho_0 X}\right)\sqrt{R/\rho_0} = Q\rho_0 \left(\frac{m}{M}\right)\sqrt{R/\rho_0}.$$

Thus, in the limit of large time and greater X and M, the presence of gas at the left (with small density $R \ll \rho_0$) results in a slower decrease in pressure (Π_S decreases as M^{-1} or X) than in the case considered earlier of dispersion into a vacuum, where Π decreases as $M^{-4/3}$ or $X^{-4/3}$.

We may expect that, beginning from the values M, X at which $\Pi = \Pi_S$, a greater value of Π will occur in subsequent motion. This criterion yields an upper bound of applicability of the self-similar solution from the condition

$$\rho_0 Q \left(\frac{m}{M}\right)^{4/3} = \rho_0 Q \left(\frac{m}{M}\right)\sqrt{R/\rho_0}, \qquad M = m \left(\frac{\rho_0}{R}\right)^{3/2}.$$

Thus the applicability of the solution is finally given by the inequality $m < M < m(\rho_0/R)^{3/2}$. For a large ratio ρ_0/R, this may be quite broad.

If we compare the pressures according to the proposed solution and the theory of strong explosion at identical times (and not at equal X, as was done above), then we obtain the same result.

The dimensionless numerical coefficients in the expressions for the pressure and the extent of applicability cannot be obtained without very time-consuming calculations requiring integration of partial differential equations, and without concrete assumptions regarding the equation of state of the EP and the explosion process.

§5. A Note on the Two Types of Self-Similar Solutions

It is well known that self-similar solutions are divided into two sharply differing types.

The first type, which includes, for example, the problem of strong explosion or propagation of heat in a medium with nonlinear thermal conductivity [3], is characterized by the fact that the exponents are found from physical considerations, from the conservation laws and their dimensionality. In addition, the exponents turn out to be rational numbers. The task of the calculation is to find the dimensionless functions by integration of ordinary differential equations. After this the problem is completely solved, since the numerical constants are determined by normalizing the solution to the conserved quantity (the total energy released in these examples).

The second type includes, for example, the problem investigated by Guderley (cited in [4]) of convergence of a cylindrical shock wave to a line, or of a spherical wave to a point. In this case, just finding the exponent requires integration of ordinary differential equations. The exponent is found from the condition that the integral curve passes through a singular point; without this it is impossible to satisfy the boundary conditions.

One would think that the type to which a problem belongs could easily be established in advance. Thus, in the examples given, the problem of a strong explosion is "closed"; the mass of the gas involved in the motion is finite, and is easily expressed in terms of the wave radius. In the problem of convergence of a shock wave, only a neighborhood of the origin, and only a small fraction of the gas involved in the motion are considered. In this case, we cannot pose the problem of motion of the whole gas. It is natural, therefore, that in the first case we may use the energy conservation equation, but cannot in the solution of the second problem; in the case of a converging wave the energy of the self-similar zone continuously decreases, and the exponent is found from other considerations.

The problem considered in this paper is of methodological interest because, from the way it was posed, it would seem to belong to the first type since, at each moment, the amount of matter involved in the motion is determined by the wave location, just as in the strong explosion problem. Therefore, the impossibility of determining the exponent by means of the energy conservation law is a non-trivial circumstance. As was explained in detail above, the energy integral in our case diverges, and the finite value of the energy is determined by a small region which bears the imprint of the initial conditions and in which the self-similar solution is inapplicable.

This example illustrates that it is necessary to exercise great care and thoroughness in finding self-similar solutions.

In conclusion, I use this occasion to express my gratitude to V. B. Adamskiĭ, who performed the calculations, and to N. A. Dmitriev, who took part in discussions of the work.

Conclusions

We consider the problem of motion of a gas adjacent to a vacuum under the influence of a short pressure impulse.

A self-similar solution is found in which the pressure Π at the shock wave front propagating in the gas decreases as a power function of the distance traveled, X: $\Pi \sim X^{-n}$, where $1 < n < 2$. In the general form, for any adiabatic index of the gas, it is proved that $1 < n < 2$; for $\gamma = 7/5$ the numerical value of n is 1.333. The law found yields the greatest possible rate of the plane shock wave decay under any circumstances.

The dependence of the shock wave pressure at a given distance on the duration τ and on the magnitude of the pressure P of the impulse is found. The shock wave pressure depends on the product $P\tau^{-k}$, where $1 < k < 3/2$, and k is expressed in terms of n.

In the limit of small τ the impulse pressure momentum goes to zero, and the work done by the impulse is infinite.

The gas momentum in self-similar motion is identically zero. The energy calculated from the self-similar solution is expressed by a divergent integral.

The energy conservation law cannot be used to determine the self-similarity exponent; conservation of energy is ensured by the departure of the actual solution from the self-similar one in a small (tending to zero as the process develops) part of the gas.

The question of the extent of validity of the solution is considered for the description of the propagation of a shock wave produced by an explosion on a plane surface separating two gases with different densities.

Institute of Chemical Physics   *Received*
USSR Academy of Science, Moscow   *June 11, 1955*

References

1. *Sedov L. I.* Metody teorii podobiĭa i razmernosti [Methods of the Theory of Similarity and Dimensionality]. Moscow: GTTI, 328 p. (1954).
2. *Adamskiĭ B. V.*—Akust. zhurn. **2**, 3 (1956).
3. *Zeldovich Ya. B., Kompaneets A. S.*—In: Sbornik, posvyashchennyĭ 70-letiĭu akademika A. F. Ioffe [Academician A. F. Ioffe's 70th Birthday Festschrift]. Moscow; Leningrad: Izd-vo AN SSSR, 61 p. (1950).
4. *Courant P., Fridrikhs K. O.* Sverkhzvukovoe techenie i udarnye volny [Supersonic Flow and Shock Waves]. Moscow, Leningrad: Isd-vo inostr. lit., 427 p. (1950).

Commentary

In writing this paper the author was unaware of the work by Weizsäcker[1] in which the same problem was studied less completely. Later, exhaustive references to publications by Weizsäcker and his followers were given in a monograph by Ya.B. and Yu. P. Raizer.[2] See also the commentary to article 28 in this volume.

The asymptotic analysis of the problem of an explosion at the boundary separating two media, given in this article by Ya.B., was subsequently confirmed in numerical calculations.[3] Later, the interesting question of possible generalization of the self-similar solution obtained by Ya.B. to the cases of two- and three-dimensional motion was examined.[4] It turned out that such solutions exist, but that they have singularities at the axis and the center of symmetry, respectively. Using the ideas developed by Ya.B., Yu. G. Maloma performed[5] a numericaι study of high speed impulse applied to meteorite impact.

The existence of two types of self-similar solutions, explicitly formulated by Ya.B. for the first time, stimulated extensive studies to clarify the general character of the difference between them and to apply the concept of self-similar solutions of the second kind to various problems in mathematical physics. The present state of the problem can be found in a monograph by G. I. Barenblatt.[6] We note also the existence of an exact analytic solution with a rational self-similarity exponent when the adiabatic index is equal to $7/5$.[7]

[1] *Weizsäcker C. F. von*—Ztschr. für Naturforsch. **9a**, 269–275 (1954)
[2] *Zeldovich Ya.B., Raizer Yu. P.* Fizika udarnykh voln i vysokotemperaturnykh gazodinamicheskikh yavlenii [Physics of Shock Waves and High-Temperature Gasdynamic Phenomena]. 2nd Ed., Moscow: Nauka, 686 p. (1966).
[3] *Vlasov I. O., Derzhavina A. I., Ryzhov O. S.*—Zhurn. vychisl. matematiki i mat. fiziki **14**, 1544–1552 (1974).
[4] *Parkhomenko V. P., Popov S. P., Ryzhov O. S.*—Zhurn. vychisl. matematiki i mat. fiziki **15**, 1325–1331 (1977).
[5] *Maloma Yu. G.* Izv. AN SSSR. MZhG **2**, 119–125 (1982).
[6] *Barenblatt G. I.* Podobie, avtomodelnost, promezhutochnaĭa asimptotika [Similarity, Self-Similarity and Intermediate Asymptotics]. 2nd Ed.. Moscow: Gidrometeoizdat, 426 p. (1982).
[7] *Adamskiĭ V. B.*—Akust. Zhurn. **2**, 3–15 (1956).

III

Phase Transitions

Molecular Physics

10
On the Theory of New Phase Formation.
Cavitation*

The critical nucleus of a new phase (Gibbs) is an activated complex (a transitory state) of a system. The motion of the system across the transitory state is the result of fluctuations and has the character of Brownian motion, in accordance with Kramers' theory, and in contrast to the inertial motion in Eyring's theory of chemical reactions. The relationship between the rate (probability) of the direct and reverse processes—the growth and the decrease of the nucleus—is determined from the condition of steadiness of the equilibrium distribution, which leads to an equation of the Fourier–Fick type (heat conduction or diffusion) in a rod of variable cross-section or in a stream of variable velocity. The magnitude of the diffusion coefficient is established by comparison with the macroscopic kinetics of the change of nuclei, which does not consider fluctuations (cf. Einstein's application of Stokes' law to diffusion). The steady rate of nucleus formation is calculated (the number of nuclei per cubic centimeter per second for a given supersaturation). For condensation of a vapor, the results do not differ from those of Volmer.

In the case of cavitation of a low-vapor-pressure fluid under high negative pressure, the rate of formation of nuclei is determined by the viscosity of the fluid (and not by its evaporation rate, as was supposed by Döring). In a

*Zhurnal eksperimentalnoĭ i teoretichiskoĭ fiziki **12** 11/12, 525–538 (1942).

similar way, we may consider the crystallization of a diluted solution, where the rate of formation of a nucleus is determined by diffusion. On the basis of an approximate solution of the non-steady equation, the dependence of the probability of cavitation on the duration of application of negative pressure (breaking stress) and on the volume of the region subjected to a negative pressure is investigated. Simultaneously, considerations regarding Frenkel's theory of heterophase fluctuations are given.

§1. The cavitation phenomenon offers at present considerable technical interest, first of all, in connection with the destruction of materials and energy losses that arise in hydraulic machines, propellers, etc., when the velocity is increased.

An investigation of cavitation naturally suggests several directions:

a. assuming the absence of cavitation, determination of the velocity field of the fluid and of the pressure field in the considered aggregate, especially determination of the negative pressures obtained (breaking stress), the duration of their effect, and the dimensions of the regions of negative pressure;

b. finding the probability of disruption of the fluid and of formation of a bubble in it depending on the factors enumerated above;

c. investigation of the effect of bubble formation on the motion of the fluid, in particular, determination of the power and energy losses, and also determination of the bubbles' fate, their motion and disappearance when the pressure is increased;

d. investigation of physico-chemical conditions in a contracting bubble and establishing the mechanism of destruction of materials in cavitation.

Items a and c are pure hydrodynamic problems; d is much closer to a physical chemist. However, in the present article, we shall consider only problem b, treating it as a particular case of the general theory of new phase formation.

§2. A fluid under a pressure lower than its vapor pressure, and especially under a negative pressure, is unstable thermodynamically with respect to formation of a bubble filled with vapors of the substance and even (in the case of negative pressure) of an empty one. A fluid subjected to negative pressure is completely analogous in this respect to a supersaturated vapor, unstable with respect to formation of a condensate, or to a supercooled liquid, unstable with respect to crystal formation.

However, as experiment shows, supersaturated systems under pure conditions, in the absence of the phase with respect to which the system is unstable (bubbles, condensate drops, small crystals in the examples given), turn out in practice to be quite stable within a rather broad range.

The general reason for this stability was first pointed out by Gibbs [1] and is related to the surface energy of the boundary separating the phase considered and the new phase being formed from it. The surface energy is particularly important for small-size particles, which demand expenditure of relatively large quantities of work for their formation. To Gibbs also belongs the concept of a "critical nucleus," i.e., of a particle of a new phase of a size for which the work of formation is maximum. The condition of maximum work coincides with the condition of equilibrium (chemical and mechanical) between the critical nucleus and the supersaturated medium surrounding it. This equilibrium is unstable, however; an increase in the size of the nucleus will lead to its further growth, a decrease—to its annihilation. The work of formation of the critical nucleus is a measure of the stability of a supersaturated state; a supersaturated state is stable with respect to nuclei of a new phase which are smaller than the critical one. The kinetics of formation of two phases from one under conditions when continuous transition, without formation of a critical nucleus of a new phase, is possible, was considered in a recent paper by Todes and the author [2]. In this case the time of formation depends on the saturation algebraically, instead of exponentially, and this does not allow determination of a practical boundary of stability. In the case of the cavitation, the critical nucleus means simply a bubble in which the surface tension exactly balances the negative pressure. We give for the case of water at room temperature (surface tension $\delta = 80\,\mathrm{dyn/cm^2}$) a table of values of the radius (x), volume (V) and work of formation (A) of a critical nucleus as functions of the magnitude of the negative pressure:

p, atm	x, cm	v, cm^3	A, erg
$-$ 0.1	1.6×10^{-3}	1.6×10^{-8}	8×10^{-4}
$-$ 10.0	1.6×10^{-5}	1.6×10^{-14}	8×10^{-8}
$-$ 1\,000.0	1.6×10^{-7}	1.6×10^{-20}	8×10^{-12}
$-$ 10\,000.0	1.6×10^{-8}	1.6×10^{-23}	8×10^{-14}

The radius is calculated according to the formula $p + 2\delta x^{-1} = 0$. The work of formation or, more precisely, the part not compensated by the work of pressure forces, is equal to one third of the surface energy of the bubble (Gibbs). For the sake of comparison we point out that the value of kT, which characterizes the energy of thermal motion, is of order 4×10^{-14} erg.

The size of the critical bubble approaches the size of a molecule, and the work of its formation approaches the molecular energy when the negative pressure, at any rate by order of magnitude, approaches the molecular strength of the fluid with respect to disruption.

§**3.** Volmer [3] was the first to pose the question of the consequences of Gibbs' ideas for the kinetics of new phase formation. The most important conclusion is that the probability of formation of a nucleus depends on the supersaturation as $\exp(-A/kT)$. For condensation of a vapor, Volmer chose as a factor the number of collisions of vapor molecules (per unit time in a unit volume). In more recent papers by Farkas [4], Becker and Döring

[5], the formation of the critical drop is treated more precisely, and the whole chain of successive reactions leading from a single molecule through aggregates of 2, 3, 4, 5, ... molecules to the critical nucleus is considered. Together with annexation of molecules, the reverse process of disintegration of the aggregates is considered, which is especially important before they have reached the critical size. A special method is developed for finding the steady solution of a system of kinetic equations (the number of equations is equal to the number of different aggregates, i.e., is of the order of the number of molecules in the critical nucleus). The results differ little from Volmer's simplest assumption. Since the expression in the exponent is known with far from sufficient precision, further precision is practically unrealistic.

Further, Döring[1] treats cavitation in a similar way, supposing that the size of the bubble (before reaching critical conditions) at each moment corresponds to the condition of mechanical equilibrium between the external pressure, the force of surface tension, and the vapor pressure in the bubble. It is thus supposed that the change of the volume of the bubble is determined by the kinetics of evaporation of the liquid into the bubble. Meanwhile, it is physically obvious that under negative pressures such as those at which cavitation arises, the vapor tension in the bubble ($p = 0.02$ atm. for saturated water vapors at room temperature) does not play any role. The frequency of fluctuation and the rate of variation of the bubble volume in this case is determined by the fluid viscosity. Below we shall develop a general method for considering the transition across the state of the critical nucleus. The most important feature of the method is that it is combined with determination of a process which limits the velocity—be it viscosity, heat conduction or diffusion—by macroscopic analysis (without regard to fluctuations and the molecular mechanism).

§4. In Eyring's theory of chemical reactions (see, e.g., [6]), it is supposed that the motion of the system across the transitory state takes place according to the laws of classical mechanics, without any "friction"; in particular, the inertial motion leads to the independence of the flow from the extent of the intermediate state in the direction of the reaction path.

Recently, Kramers [7] has generalized the theory of the intermediate state (activated complex) by considering systems which over the entire course of the transition are subjected to random exterior forces, so that the motion acquires the character of Brownian motion of a particle in a field of forces. In view of the high generality of Kramers' derivations and their complexity, and for the sake of completeness of the present article, we shall give a simplified derivation of the equations,[2] bearing in mind the processes of new phase formation in which we are interested.

---

[1] Unpublished work whose content is presented in Volmer's book [3].

[2] Our calculations are close to the diffusional treatment of chemical reaction by Christiansen [8].

Let us consider a system in which the nuclei of a new phase, e.g., drops or bubbles, may be of different sizes. If we plot on some axis the dimension of the nucleus (its radius, volume, or the number of particles) then motion of a point along that axis will correspond to variation of the dimensions. Plotting the free energy as a function of the coordinate, we shall determine the critical size as a coordinate corresponding to the maximum free energy, and the critical region as a region in which the values of the free energy differ from the maximum by no more than kT (Fig. 1—the critical region is denoted by *1*). We shall be able to deduce the desired equation very simply by supposing that motion along the x-axis takes place in discrete jumps of $\pm\lambda$ (the length of the free path). The frequency of jumps is determined by the probabilities q_+ and q_- of a jump to the right and left [over a distance $+\lambda$ or $-\lambda$ (Fig. 2)]. λ has the dimensions of the value plotted on the x-axis, q—of inverse time. In certain cases, λ has a completely obvious physical meaning, and its introduction and calculation do not cause any difficulties. Thus, in the process of drop formation, the number of molecules in a drop may vary by only ±1, the volume of the drop by $\pm v_1$, where v_1 is the volume of a single molecule.[3] In the general case (in particular, when a bubble is formed in a liquid), the introduction of λ means that the process is schematized, since in a number of cases in reality the variable plotted on the axis may change continuously and the motion is continuous. The meaning of the path length (jump) λ in this case is the distance over which, on average, the effect of random forces changes the velocity by a quantity of the order of the mean velocity. If λ is small compared to lengths of interest on the axis, transition to the critical region takes place through many jumps—we have what Kramers called the case of large viscosity.

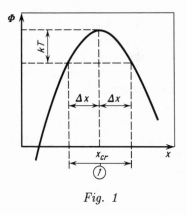

Fig. 1

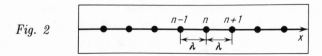

Fig. 2

---

[3]It may be shown that the collision of two aggregates, in which the number of molecules changes by several units at once, is much less probable and may be disregarded.

Let us construct an equation, considering schematically that all the nuclei are discretely distributed according to their sizes over the nodes of the axis at a distance λ from each other, and let us renumber these nodes. The change in the number of nuclei $Z(n)$ at the nth node may be easily expressed using the probabilities of transition

$$\frac{\partial Z(n)}{\partial t} = -Z(n)[q_+(n)+q_-(n)]+Z(n-1)q_+(n-1)+Z(n+1)q_-(n+1). \quad (1)$$

Designating by $b(n)$ the equilibrium number of nuclei we may, from the principle of detailed balance, establish the relationship between the probabilities of opposite transitions, for instance, $n \rightarrow n+1$, or $q_-(n)$, and $n+1 \rightarrow n$, or $q_-(n+1)$:

$$b(n)q_+(n) = b(n+1)q_-(n+1),$$
$$b(n-1)q_+(n-1) = b(n)q_-(n). \quad (2)$$

With the aid of the relations (2) we shall eliminate from equation (1) the quantities q_-; the remaining quantity q_+ we shall henceforth designate simply by q, omitting the index:

$$\begin{aligned}
\partial Z(n)/\partial t &= q(n)[-Z(n) + Z(n+1)b(n)/b(n+1)] \\
&\quad + q(n-1)[Z(n-1) - Z(n)b(n-1)/b(n)] \\
&= q(n)b(n)[Z(n+1)/b(n+1) - Z(n)/b(n)] \\
&\quad - q(n-1)b(n-1)[Z(n)/b(n) - Z(n-1)/b(n-1)]. \quad (3)
\end{aligned}$$

We turn from consideration of the nodes to a continuous distribution with density $Z(x)$. Here $Z(n) = \lambda Z(x)$, $Z(n+1) = \lambda Z(x+\lambda)$, etc. (it does not seem reasonable to introduce for a smooth function of the density new notations, since the node functions $Z(n)$, $b(n)$, will not be encountered any more below). Assuming λ to be constant, we expand in a power series of λ, confining ourselves everywhere to the first non-vanishing term—we suppose the functions Z, b, q to vary little over the length λ, so that $Z > \lambda Z' > \lambda^2 Z'' > \dots$. Thus, we obtain the fundamental equation

$$\frac{\partial Z}{\partial t} = \frac{\partial}{\partial x}\left(q\lambda^2\right)b\frac{\partial}{\partial x}\left(\frac{D}{b}\right) = \frac{\partial}{\partial x}Db\frac{\partial}{\partial x}\left(\frac{Z}{b}\right). \quad (4)$$

The quantity $q\lambda^2 = D$ plays the role of the diffusion coefficient; in the particular case when $b = $ const, i.e., when all the points of the axis are equivalent, as is the case for the motion of particles in space without exterior forces, we get Fick's law,

$$\frac{\partial Z}{\partial t} = D\frac{\partial^2 Z}{\partial x^2}. \quad (5)$$

The main problem is the evaluation of the quantity D. Kramers expresses it in terms of the average values of the force acting on the system, the square of the force, the cube of the force and similar expressions whose practical evaluation is very difficult.

We have taken a different route. Similar to the way we found the relationship between q_+ and q_- by comparing equation (1) with the thermodynamics, we shall find the absolute value of D by comparing (5) with the macroscopic rate of change of the system, in the spirit of the general principle of correspondence of theoretical physics.

Let the system contain at the initial time a certain number N of nuclei of exactly the same size. The distribution function Z is equal to zero everywhere except at one specific point (curve *1*, Fig. 3). In the macroscopic theory, each nucleus changes with time in a quite definite way, depending on its size and external conditions; N nuclei which were identical at the initial moment will remain identical even after a certain time interval t, and curve *1* will be shifted as a whole to another place (curve *2*) that corresponds to the change in the size of the nucleus, in accordance with the kinetics equation of the form

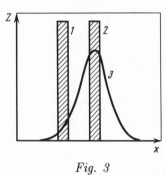

Fig. 3

$$\frac{\partial x}{\partial t} = w(x, T, \ldots). \tag{6}$$

Meanwhile, in a more exact theory which takes fluctuations into account, the theory that led us to equation (4), the distribution depicted by curve *1*, after a time t, will be transformed into the distribution corresponding to curve *3*; as a consequence of fluctuations the nuclei, which initially did not differ from each other, will become somewhat different; the distribution will become diffuse. Together with the general increase in x in this example, we see from curve *3* that there will also be a very small number of nuclei which have accidentally shifted to the left and decreased the value of x. There are very few such nuclei, but in some problems—especially in the theory of a new phase formation—it is precisely these nuclei that reach a thermodynamically unfavorable critical size that are essential. Here, however, we shall restrict ourselves to the statement that the center of gravity of the distribution of the curve moves in the same way as the center of curve *2*, Fig. 3. In other words, in taking account of fluctuations, the average rate of variation of x must correspond to the rate of variation of x evaluated macroscopically. This statement follows from the condition that our theory, which accounts for fluctuations, links with the usual calculation methods, i.e., from the principle of correspondence. Let us construct the quantity $\bar{x}$ for distribution changing with time according to (4). When the time interval is short, the variation of x is not great, and the overall number of nuclei is conserved, $M = \int Zx\,dx = \text{const}$. By definition (all the integrals are taken over the entire range of definition of x, for instance, from $-\infty$ up to $+\infty$),

$$\bar{x} = \frac{\int Zx\,dx}{\int Z\,dx} = \frac{1}{M}\int Zx\,dx, \tag{7}$$

hence

$$\frac{d\bar{x}}{dt} = \frac{d}{dt}\frac{1}{M}\int Zx\,dx = \frac{1}{M}\int \frac{\partial Z}{\partial t}x\,dx. \tag{8}$$

We substitute $\partial Z/\partial t$ from equation (4) and integrate by parts twice:

$$M\frac{d\bar{x}}{dt} = \int\left[\frac{\partial}{\partial x}Db\frac{\partial}{\partial x}\left(\frac{Z}{b}\right)\right]x\,dx = Dbx\frac{\partial}{\partial x}\left(\frac{Z}{b}\right)\Big|_{-\infty}^{\infty} - \int Db\frac{\partial}{\partial x}\left(\frac{Z}{b}\right)dx$$

$$= \left[Dbx\frac{\partial}{\partial x}\left(\frac{Z}{b}\right) - DZ\right]\Big|_{-\infty}^{\infty} + \int\left(\frac{Z}{b}\right)\frac{d}{dx}(Db)\,dx. \tag{9}$$

For the functions Z, corresponding to the distributions under consideration (Fig. 3), the quantities in square brackets vanish at both limits. In the remaining integral, Z is non-zero only in a narrow interval. Further, in the expression $d(Db)/dx$, the dependence of D on x may be neglected in comparison with the dependence of b on x, since b contains the factor $\exp(-\Phi(x)/kT)$. Finally, we find

$$\frac{d\bar{x}}{dt} = Db^{-1}\frac{db}{dx} = D\frac{d\ln b}{dx} = -\frac{D}{kT}\frac{d\Phi}{dx}. \tag{10}$$

Comparing with (6), we find the expression

$$D = \frac{kTw}{-d\Phi/dx}. \tag{11}$$

Here w always has the sign opposite that of $d\Phi/dx$ so that D is always positive; formula (11) represents a natural generalization of the expression first found by Albert Einstein [9], for the diffusion of a dissolved substance in a liquid. In our problem of formation of a new phase, $\Phi(x)$ has a maximum which corresponds to an extremely sharp minimum[4] of b.

Expanding Φ in a series,[5] we find

$$b = b(x_{\text{crit}})\exp\left[\frac{-d^2\Phi}{dx^2}\frac{x - x_{\text{crit}}^2}{2kT}\right]_{\text{crit}}. \tag{12}$$

It is not difficult to find the steady solution of equation (4),

$$0 = \frac{\partial Z}{\partial t} = \frac{\partial}{\partial x}Db\frac{\partial}{\partial x}\frac{Z}{b}, \qquad Db\frac{\partial}{\partial x}\frac{Z}{b} = \text{const} = N, \tag{13}$$

$$\frac{Z}{b} = N\int\frac{dx}{Db} + \text{const}. \tag{14}$$

We find the values of the constants in equation (14) from the boundary conditions: if the transition occurs from left to right, then in the initial state

---

[4]At the minimum of b both w and $d\Phi/dx$ vanish, while D suffers no change.

[5]Note that $d^2\Phi/dx^2 < 0$.

at $x = 0$ or $x = -\infty$ (depending on the definition of x), complete equilibrium takes place, $Z/b = 1$; in contrast, the number of nuclei greater than the critical size which grows further is small, especially small as compared with the very large equilibrium number of large nuclei, so that for $x \to +\infty$, $Z/b \to 0$. We satisfy the boundary conditions by taking

$$\frac{Z}{b} \int_x^\infty b^{-1}\, dx \bigg/ \int_{-\infty}^\infty (Db)^{-1}\, dx. \tag{15}$$

The integrated function has a sharp maximum at $x = x_{\text{crit}}$. Making use of the expansion (12), we find the expression for diffusive flow, i.e., for the number N of nuclei formed per unit time; here only those formations are called "nuclei" which exceed the critical size, and which, with a probability that is practically identical to unity, will grow to macroscopic size

$$N = b(x_{\text{crit}})D\sqrt{\frac{-d^2\Phi/dx^2}{kT}}. \tag{16}$$

Before turning to consideration of non-steady solutions of the fundamental equation (4), we will apply the information obtained to the problem of cavitation.

§**5.** Let us construct Φ for a bubble

$$\Phi = \delta S + pv = 4\pi x^2 \delta + \frac{4\pi x^3 p}{3}. \tag{17}$$

Supposing $d\Phi/dx = 0$, we find the expression for the critical size cited above:

$$8\pi x\delta + 4\pi x^2 p = 0; \quad x_{\text{crit}} = \frac{-2\delta}{p};$$

$$\Phi(x_{\text{crit}}) = A = \frac{4\pi\delta^2 x_{\text{crit}}}{3} = \frac{16\pi\delta^3}{2p^2};$$

$$\frac{d^2\Phi}{dx^2} = -8\pi\delta. \tag{18}$$

In order to construct the expression for the equilibrium number of nuclei in a unit volume (the dimension of $b(x)\, dx$ is cm^{-3}, the dimension of $b(x)$, when x is defined as the radius, is cm^{-2}), we must multiply the exponent $\exp(-\Phi/kT)$, where Φ is determined by (17), by a quantity of dimension cm^{-2}. Exact evaluation of a pre-exponential factor is presently an unsolved problem of statistical mechanics. From dimensional considerations we may propose d^{-2} or x^{-2}, where d is the linear size of a molecule of liquid and x is the radius of a bubble. In the present problem of evaluating the critical (i.e., minimum) value of the equilibrium concentration, we are dealing with a region where the factor in the exponent is large and exact evaluation of the pre-exponential factor is not actually necessary.

The value of the pre-exponential factor is very important in Frenkel's theory of pre-transition states and heterophase fluctuations [10]. However, simple qualitative arguments appear to show that the effects considered by him are extremely small. The quantity which characterizes the role of het-erophase fluctuations is represented by the integral

$$J = \int \alpha(x) \, b(x) \, dx, \tag{19}$$

if $\alpha(x)$ is the dependence of the property of interest (for instance, the total volume) on the coordinate x for an individual nucleus of the new phase ($\alpha = 4\pi x^3/3$ for a bubble in this example). Up to the point of transition, as long as the phase is thermodynamically stable, the integral J converges and is finite. Beyond the point of transition, the integral diverges; the quantity under the integral has a sharp minimum near $x = x_{crit}$. The divergence of the integral is related to the fact that it incorporates the formation of the new phase in macroscopic quantities. To study the properties of a supersaturated phase, it is necessary to break off the integration upon reaching the critical size: only smaller formations may be called fluctuations; formations exceeding the critical size are nuclei of a new phase capable of growth and leading to an irreversible phase transition. Having thus determined J, we see that this quantity increases rapidly upon approaching the conditions of equilibrium of the two phases and continues to grow after the equilibrium conditions have been crossed in the supersaturated phase.

It has been established experimentally that all thermodynamic properties of supersaturated phases (a supersaturated vapor, a liquid heated beyond the boiling point, or a liquid cooled down below the melting point) are in no way remarkable and fail to show any substantial deviations that would point to strong heterophase fluctuations. All the more, *à fortióri*, it may hence be concluded that before the transition point and at the point itself the quantity J is negligibly small.

The melting of a crystalline substance deserves special attention. At equal distances from the melting point (but from different sides), the prop-erties of the crystal and liquid with respect to heterophase fluctuations and pre-transition phenomena are completely symmetrical; in both cases we are dealing with the same liquid–crystal surface tension, while the difference of thermodynamic potentials of the crystal and liquid passes linearly through a value which is equal to zero at the melting temperature. The experimen-tally established absence of noticeable heterophase phenomena in a liquid near the melting temperature demonstrates the absence of pre-melting phe-nomena (in Frenkel's sense) in a crystal. At first glance, the symmetry of transition properties of a crystal and liquid contradicts commonly known facts: innumerable supercooled liquids are known, but there are no super-heated crystals. However, this non-symmetry is related to the properties of surfaces of separation from the gas phase: melting wets the surface of the crystal, which indicates an unsymmetric relation

$$\sigma_{sol-gas} > \sigma_{liq-gas} + \sigma_{sol-liq}. \tag{20}$$

As a consequence of this last relation, formation of a film of the liquid on the free surface is thermodynamically favorable, and occurs without an energy barrier and without any delay at the melting temperature; in contrast, on a liquid–gas surface the formation of a solid film would require expenditure of energy equal to

$$\sigma_{\text{sol}-\text{gas}} + \sigma_{\text{sol}-\text{liq}} - \sigma_{\text{liq}-\text{gas}} > 2\sigma_{\text{sol}-\text{liq}} \tag{21}$$

[the inequality follows from the preceding formula (20)]. Indeed, Volmer and Schmidt [11] have succeeded in observing insignificant superheating of mercury crystals heated from the surface; the authors relate the superheating to incomplete wetting of the crystal by liquid mercury; in this case they observed a finite angle of contact of melted drops on the surface of solid mercury. Khaikin and Bene [12], in heating a monocrystal of lead with an electric current and cooling it from the surface, were able to observe considerable superheating without melting in the body of the monocrystal.

Thus, melting of a crystalline substance without superheating is a superficial effect. Pre-melting phenomena are apparently also related to the formation of liquid films on the surfaces of crystals, if not to other incidental causes (for example, impurities), and are not pertinent to Frenkel's theory. Heterophase fluctuations are quite large where the difference between two phases and the surface tension between them tend to zero—near the critical point and near the Curie point. The first case is commonly known, the second was earlier investigated quantitatively in Landau's fine work [13, 14].

It also seems to me that attempts to build a theory of melting from the condition that the coefficient of elasticity of a crystal vanishes [15], or from other conditions which relate solely to properties of the solid phase, are fundamentally mistaken. The melting point for a given pressure (or melting-line in the pressure–temperature plane) is determined at the intersection of the surfaces of the solid- and liquid-phase thermodynamic potentials, but does not correspond to any features of either surface, either phase, taken separately. We are convinced of this by the possibility of a shift of the melting point by a change in the properties of the liquid phase alone, for example, by dissolving an impurity in the liquid, and the possibility of superheating a crystal under conditions which exclude the formation of the nucleus of a new phase.

Let us find the magnitude D of formulas (4)–(16) for the case of cavitation. We shall consider the motion of an incompressible fluid which arises with a change in the radius of a bubble, and construct an expression for the energy, which is transformed by viscosity forces into heat per unit time. The energy dissipation in a unit volume per unit time is equal to

$$\eta \left[2 \left(\frac{\partial u_1}{\partial y_1} \right)^2 + \ldots + \left(\frac{\partial u_3}{\partial y_1} + \frac{\partial u_1}{\partial y_3} \right)^2 + \ldots \right], \tag{22}$$

where η is the viscosity of the fluid; y_1, y_2, y_3 are the three space coordinates;

u_1, u_2, u_3 are the components of the velocity of the fluid. Near the bubble all the derivatives in (22) are of order w/x, where $w = dx/dt$ and x is the radius of the bubble. As the distance from the bubble is increased, the dissipation quickly subsides; the characteristic distance of the decrease is of the order of the bubble radius, x. Finally, the volume integral of expression (22), accurate to within a numerical factor of order 1, is equal to

$$H = \frac{\eta w^2 x^3}{x^2} = \eta w^2 x. \tag{23}$$

Equating the dissipated energy to the work performed per unit time by the forces of surface tension and by external forces causing stretching of the liquid, we find the rate of variation of the bubble radius:

$$H = -p\frac{dv}{dt} - \sigma\frac{ds}{dt} = -\frac{d\Phi}{dx}\frac{dx}{dt} = -\frac{d\Phi}{dx}w, \tag{24}$$

$$w = \frac{dx}{dt} = \frac{d\Phi/dx}{\eta x} = \frac{-8\pi x\sigma + 4\pi x^2 p}{\eta x}. \tag{25}$$

Comparing with equation (11), we find

$$D = \frac{kTw}{-d\Phi/dx} = \frac{kT}{\eta x}. \tag{26}$$

Finally, in accordance with (16), we obtain the following expression for the number of bubbles formed in 1 cm^3 in 1 sec (we have chosen d^{-2} as a pre-exponential factor in the expression for σ, and for x we have substituted in equation (23) the expression for x_{crit}):[6]

$$N = (-p)kTe^{-16\pi\sigma^3/3p^2 kT}\frac{\sqrt{\sigma/kT}}{\sigma\eta d^2}. \tag{27}$$

Substituting numerical values for water at room temperature, we find (expressing p in atmospheres)

$$\log N = 6.1 \times 10^7/p^2 + \log(-p) + 15. \tag{28}$$

Setting $N = 1\,\text{cm}^{-3}\cdot\text{sec}^{-1}$, we find

$$p = -1800\,\text{atm}; \qquad x_{\text{crit}} = 8 \times 10^{-8}; \qquad D = 5 \times 10^{-5}\,\text{cm}^2/\text{sec} \tag{29}$$

and the equilibrium number of nuclei in the transition region (cf. Fig. 1)

$$b(x)\Delta x = b(x_{\text{crit}})\frac{1}{\sqrt{\frac{-d^2\Phi/dx^2}{kT}}} = 10^{-12}\,\text{cm}^{-3}, \tag{30}$$

accordingly, the time spent by a nucleus in the transition state is of order 10^{-12} sec. The evaluation has been performed under the assumption of high viscosity.

Let us check whether this assumption is fulfilled in the case of disruption of water. We express the diffusion coefficient in terms of the mean square of

---

[6]We recall that $p < 0$.

the (fluctuating) velocity,

$$D = \lambda \sqrt{\bar{w}^2}, \tag{31}$$

and determine the latter in the critical region from the equipartition of energy. The kinetic energy of motion of the liquid when the radius of a bubble is changed corresponds to a reduced mass of order ρx^3, where ρ is the density of the fluid,

$$\frac{m_{\text{eff}} \bar{w}^2}{2} = \frac{\rho x^2 \bar{w}^2}{2} = \frac{kT}{2}. \tag{32}$$

For water [see the values in (29)] we find $(\bar{w}^2)^{1/2} \simeq 10^4 \text{cm/sec}$, $\lambda = 5 \times 10^{-9}$, $\Delta x = 10^{-8}$; the calculations are still applicable. In the opposite case we should have had to apply the ordinary theory of an activated complex that assumes inertial motion across the barrier with an average velocity $(\bar{w}^2)^{1/2}$, which leads to the following expressions [instead of the general formula (16) and the expression for cavitation (27)]:

$$N = b(x_{\text{crit}})\sqrt{\bar{w}^2} = b(x_{\text{crit}})\sqrt{\frac{kT}{m_{\text{eff}}}} = e^{-16\pi\sigma^3/3p^2kT}\sqrt{\frac{kT}{\rho x^3}}\frac{1}{d^2}. \tag{33}$$

For more viscous fluids and for glasses, the assumption $\lambda < \Delta x$, which serves as the foundation for (27), is fulfilled even better than for water.

Let us return to discussion of the results obtained for water.

The dependence of p on the choice of a particular value of N, i.e., on the duration of the experiment in which we determine the "tensile strength" of water, and on the volume subject to the tension, is very slight; choosing $N = 10^3$, 1 and 10^{-3}, we find in (28) $p = 1640$, 1800, 2000. The dependence of the breaking stress on the viscosity may become essential only in the region of glassification of the fluid, where the viscosity varies with temperature very strongly (exponentially with E/kT). Defining the limit of the strength as the tension corresponding to a specific value of N, we find

$$C_1 - \frac{C_2}{p^2 T} - \frac{E}{kT} = 0; \quad p = -\frac{1}{\sqrt{C_3 T + \frac{E}{kC_2}}}, \tag{34}$$

where C_1, C_2, C_3 are constants.

The value of the breaking stress found by us is quite high; experiments carried out under the purest conditions with all precautions taken give values which are considerably smaller, although they agree qualitatively with (28) and (29). Thus, for ethyl ether, Meyer [16] reached $p = -170$ atm, while calculation according to (27), (28) gives about 250 atm. It is obvious, however, that for the critical bubble size found, evaluation with a constant value of the surface tension (entering into the exponent) can give no more than the order of magnitude. Even the experimentally achieved values of the tension stress themselves, found under pure conditions, show that cavitation in propellers, in a water jet in a Venturi tube, in a super-sonic wave, and under other similar conditions, where the stress does not exceed a few dozen atmospheres, is related to the effect of factors not pre-viously taken into account. It does not seem possible to account for the influence of suspended solid particles in a general way, since the size, shape, conditions of wetting and similar concrete properties enter into considera-tion.

The question of the influence of surface-active substances on the condi-tions of cavitation is relatively easy and lies wholly within the scope of the developed theory. The breaking stress depends basically on the expression in the exponent, so that we may expect that in the presence of surface-active substances

$$-p \sim \sigma^{3/2}. \tag{35}$$

Because of the small overall surface of nuclei, and due to their small equi-librium number, we may expect a very strong effect from insignificant traces of surface-active substances whose discovery by other methods would be very difficult. For a significant effect from these traces, low solubil-ity of the active substance is particularly essential, since at a given vol-ume concentration a less soluble substance would more willingly concen-trate on the surface, thus lowering the surface tension. When there are traces of a surface-active substance present in the solution, and when the tension is insufficient for disruption of the pure fluid, diffusion of the ac-tive substance to the surface of a bubble may become the process that determines the rate. From equations (26) and (29) it is evident that the rate at which mechanical equilibrium is established is relatively high. For a given total quantity of the active substance m in the surface layer of a bubble, its concentration per unit surface is $c = m/4\pi x^2$; the sur-face tension σ depends on the concentration c (this dependence—the so-called equation of state of an adsorbed substance—has often been stud-ied):

$$\sigma = \sigma(c) = \sigma(m/4\pi x^2). \tag{36}$$

For a given quantity of an active substance m, the size of the bubble cor-responding to mechanical equilibrium (the critical size) is determined by solution of the equation

$$x = \frac{2\sigma}{-p} = \frac{2\sigma(m/4\pi x_2)}{-p}. \tag{37}$$

Equation (37) has either one or three solutions for x, at given m and p. De-pending on the number of solutions, the form of the surface $\Phi(m, x)$ varies, as does the character of the solution of the diffusion equation of type (4), generalized to the case of two coordinates, m, x. However, in view of the present state of experimental data, a full analysis of all these questions seems premature.

§6. Non-steady solutions of the equation for the probability of formation of a nucleus. Above we sought a steady solution of the fundamental equation (4), which gave us the number of "virile" (exceeding the critical size) nuclei which form in a unit volume per unit time under constant conditions. As is obvious from the form of solution (15), in a steady state the number of formations of subcritical size does not differ from the equilibrium number:

$$z \simeq b, \quad x < x_{\text{crit}}; \qquad z = \frac{b}{2}, \quad x = x_{\text{crit}}; \qquad z \ll b, \quad x > x_{\text{crit}}. \quad (38)$$

Establishing such a steady state is a necessary condition for correctness of the number N obtained [formula (6)]. Let us clarify the question of the *time* required to establish the steady state. This question corresponds to the following experimental regime: by prolonged maintenance of a state in which there is no supersaturation (or it is small, so that N is very small), we establish the equilibrium state[7] $Z = b_1(x)$. At a certain moment of time $t = 0$, we instantaneously establish (or increase) supersaturation, and we further observe the number of nuclei formed in the system as a function of time. The character of the curve is given in Fig. 4. At first, as long as the distribution for $x < x_{\text{crit}}$ is closer to b_1 than to b [cf. (38)], it is obvious that the number of centers formed per unit time will be considerably less than N_{steady}. If the initial distribution b_1 differs from the steady one, beginning from a certain value $x = x_1$,

$$b(x) - b_1(x) < b(x), \quad x < x_1; \qquad b(x) \gg b_1(x), \quad x > x_1, \quad (39)$$

and in the non-steady solution the primary term will be the expression

$$e^{-(x-x_1)^2/4Dt}.$$

Consideration of the exact solution found by Ornstein and Uhlenbeck [17] in one case of diffusion in a field of forces shows that on the segment from x_1 to x_{crit}, the non-steady solution sought has a form close to

$$Z(x,t) = Z_{\text{steady}}(x)e^{-(x-x_1)^2/4Dt}. \quad (40)$$

The quantity $Z_{\text{steady}}(x)$ was given above [equation (15)]. The time τ (see Fig. 4) by order of magnitude is here equal to

$$\tau \simeq \frac{(x_{\text{crit}} - x_1)^2}{4D}, \qquad N = N_{\text{stat}}e^{-\tau/t}. \quad (41)$$

The total number ν of nuclei formed during the time τ is not greater than $N_{\text{stat}}\tau$ by order of magnitude (see Fig. 4)

$$\nu = N_{\text{stat}}\tau = b(x_{\text{crit}})\frac{D(x_{\text{crit}} - x_1)^2}{\Delta x 4D} = b(x_{\text{crit}})\Delta x \left[\frac{(x_{\text{crit}} - x_1)^2}{2\Delta x}\right]^2, \quad (42)$$

where Δx is the width of the critical region [cf. equation (30) and Fig. 1]. From equation (42) it is evident that the non-steady calculation can not be of

---

[7]For all x when supersaturation is absent, but only for $x < x_{\text{crit}}$ when supersaturation is low.

any significance for Christiansen's [8] diffusion theory of chemical reactions: in the theory of chemical reactions we are interested in the transformation of a number of molecules which is comparable to the overall number of reacting molecules, while the time required to establish the steady rate corresponds to insignificant transformation, to transformation of a quantity of molecules not much greater than the number of particles in the transition state, i.e., it corresponds to a time which is very small compared to the overall time of reaction.

In the theory of formation of a new phase, in the case of cavitation in pure water to which the values (28)–(30) pertain, evaluation of (41) yields a very small quantity, of order 10^{-10}sec. The diffusion of surface-active substances, by lowering the breaking stress, may increase τ considerably. The role that the non-steady part of the curve in Fig. 4 may, in principle, play in the theory of new phase formation is defined by the fact that, in this case, the total number of nuclei that should form is not given. In contrast to a chemical reaction, where the number of molecules that have reacted must be

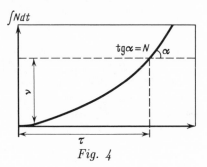

Fig. 4

large, in the case of formation of a new phase a single successfully grown nucleus reaches macroscopic size and removes supersaturation from the entire system. The term "successfully" is used intentionally, since expression (40) in fact represents the probability of a fortunate coincidence, in which over a time t the number of fluctuations necessary to attain x_{crit} occurs. The natural unit for the ordinate scale in Fig. 4 in the theory of formation of a new phase is $1/V$, where V is the volume of the system; the value $\int fN\,dt = 1/V$ corresponds to the formation of a single nucleus. The role of the non-steady part of the curve is determined by the relation between ν and $1/V$.

In reality we are interested not in the formation of a nucleus, even of a virile (supercritical) one, but in the formation of a macroscopic quantity of a new phase which will remove the supersaturation. The growth of the nucleus after crossing the barrier at $x > x_{\text{crit}}$ is governed by the kinetics equation (6). The role of fluctuations in this region is negligibly small. Taking a definite value x_2, which we will call "large" or "macroscopic," we find the time of growth

$$t_2 = \int_{x' \simeq x_{\text{crit}}}^{x_2} dx/w(x). \tag{43}$$

For $x' = x_{\text{crit}}$ and $w(x') = 0$ we shall get a definite value of t_2, taking at the lower limit $x' = x_{\text{crit}} + \Delta x$ (the right edge of the critical region). Comparison of τ and t_2 gives the second criterion for the importance of the non-steady solution (the first was the relation between ν and $1/V$). Using the connection between w and D [equation (11)], we find

$$t_2 = \int_{x'}^{x_2} -kT\, dx/(D d\Phi/dx). \tag{44}$$

Since $d\Phi/dx$ increases proportionally to the distance from the critical size,

$$\frac{d\Phi}{dx} = \frac{x - x_{\text{crit}}}{d^2\Phi/dx^2} = -kT\frac{x - x_{\text{crit}}}{(\Delta x)^2} \tag{45}$$

(in accordance with the definition of the width of the critical region, cf. Fig. 1), with logarithmic accuracy we have

$$t_2 = \ln\frac{x_2}{\Delta x}\frac{(\Delta x)^2}{D}, \tag{46}$$

from which it is clear that t_2 may be smaller than τ, which was determined by (41). If at the same time $\nu > 1/V$, consideration of the non-steady process leads to non-trivial formulas for the decrease in the probability of formation of a new phase for brief supersaturation. A special case is the formula for the increase in the resistance of a material under a brief loading of the form, for example,

$$\frac{a}{p^2} + \frac{\tau}{t} = C \tag{47}$$

(a and C are constants).

In conclusion, I should like to express my gratitude to I. I. Gurevich, O. M. Todes and D. A. Frank-Kamenetskiĭ for their valuable suggestions in discussions of this work.

Institute of Chemical Physics   *Received*
USSR Academy of Sciences, Moscow   *April 2, 1941*

References

1. *Gibbs W.*—Sci. Pap. **1**, 15 (1899).
2. *Zeldovich Ya. B., Todes O. M.*—ZhETF **10**, 1441 (1940).
3. *Volmer M., Weber N.*—Ztschr. phys. Chem. **119**, 277 (1926); *Volmer M.* Kinetik der Phasenbildung, Ed. by T. Steinkopf. Dresden (1939).
4. *Farkas L.*—Ztschr. phys. Chem. **125**, 236 (1927).
5. *Becker R., Döring W.*—Ann. Phys. **24**, 719 (1935).
6. *Eyring H.*—J. Chem. Phys. **3**, 107 (1935).
7. *Kramers H. A.*—Physica **7**, 107 (1940).
8. *Christiansen J. A., Kramers H. A.*—Ztschr. phys. Chem. **104**, 451 (1923); **B33**, 145 (1936).
9. *Einstein A.*—Ztschr. Elektrochem. **14**, 235 (1908).
10. *Frenkel Ya. I.*—ZhETF **9**, 199, 952 (1939).
11. *Volmer M., Schmidt O.*—Ztschr. phys. Chem. **35**, 467 (1937).

12. *Khaikin S. E., Bene N. P.*—Dokl. AN SSSR **23**, 31 (1939).
13. *Landau L. D.*—ZhETF **7**, 1232 (1937).
14. *Landau L. D.*—ZhETF **7**, 19 (1937).
15. *Born M.*—J. Chem. Phys. **7**, 591 (1939).
16. *Meyer J.*—Abh. Bunsen Ges. **6**, 130 (1911).
17. *Ornstein G. E., Uhlenbeck L. S.*—Phys. Rev. **36**, 823 (1930).

Commentary

This paper by Ya.B. helped lay the foundation for the study of the kinetics of phase transitions of the first kind. It considers the fluctuational formation and subsequent growth of vapor bubbles in a fluid at negative pressures. It is assumed that the fluid state is far from the boundary of metastability and that the volume of the bubbles formed is still small in comparison with the overall volume of the fluid. The first assumption ensures slowness of the process: the time of transition to another phase is large compared to the relaxation times of the fluid *per se.* This allows the application of the Fokker-Planck equation in the space of embryo dimensions to describe the growth of the embryos.

The second assumption allows us to seek steady solutions of this equation. It was established that the process rate is determined with the exponential accuracy by its behavior near the critical region, a certain "bottleneck" in the space. It is very significant that the embryo size which corresponds to this region turns out to be a macroscopic quantity so that, in principle, the diffusion coefficient can be found from the macroscopic equations describing the embryo's growth. The exact form of the boundary condition for large sizes of the embryos, where we have a significantly non-equilibrium situation, is not important due to the narrowness of the critical region. It was later found that these features are typical for a large number of problems in the kinetics of slow decay of non-equilibrium states. The mathematical technique of this paper is transferable to these problems without modification.

The course of the process at a later stage, where the second assumption is not satisfied, was studied by I. M. Lifshitz and V. V. Slezov.[1] The kinetics of phase transitions of the first kind near absolute zero, where fluctuations have a quantum character, were described by I. M. Lifshitz and Yu. M. Kagan[2] and by S. V. Iordanskiĭ and A. M. Finkelshtein.[3] In these works the ideas of Ya.B.'s paper also play an important role.

[1] *Lifshitz I. M., Slezov V. V.*—ZhETF **35**, 478–492 (1958).
[2] *Lifshitz I. M., Kagan Yu. M.*—ZhETF **62**, 385–403 (1972).
[3] *Iordanskiĭ S. V., Finkelshtein A. M.*—ZhETF **62**, 403–411 (1972); J. Low Temp. Phys. **10**, 423–432 (1973).

11

Theory of Interaction Between an Atom and a Metal*

The problem of the interaction of an atomic system with the surface of a metal at large distances is of significant interest for the theory of gas and vapor adsorption on solids surfaces. Just as in the interaction of two atomic systems, the universal attraction at large distances for neutral and non-polar systems may be obtained only in the second approximation of perturbation theory [1]. Hitherto, only one attempt has been made in this direction, but the untenability of the assumptions, methods, and results of this attempt were obvious [2].

The perturbation energy may be written in the form

$$V = \mu e \sum_i \frac{1}{r_i^2}, \tag{1}$$

where μ is the dipole moment of the atom, e is the electron charge, r is the distance from the center of the atom to an electron in a metal. The summation is carried out over all the electrons in the metal. The presence of ions in the metal equal in number to the electrons is accounted for by considering only the charge density corresponding to transitions of electrons from one stationary state to another. We restrict ourselves to the case in which the stationary states of all the electrons together can be constructed from the eigenfunctions of individual electrons. Then, from Pauli's principle it follows directly that the matrix perturbation energies, which correspond to the simultaneous processes of optically active transition of an atom and the jump of a single electron to an unoccupied level, are nonzero. We note that, due to the large energy of optically active atomic transitions, the problem is non-degenerate. Examination of the interaction of a free charge with the surface of a metal by means of the method used below would be impossible.

We feel it reasonable to seek only the order of magnitude and distance-dependence of the forces with which we are concerned. Therefore, in what follows we will systematically omit dimensionless numerical factors of order unity.

We first consider the one-dimensional problem or, more precisely, the interaction of a single atom with a linear chain of metal atoms. For the

*Zhurnal eksperimentalnoĭ i teoreticheskoĭ fiziki **5** 1, 22–27 (1935).

eigenfunctions of the electrons in the metal we take

$$\psi = (h')^{-1/2} \sin \frac{px}{h'} \qquad (2)$$

(normalized to a continuous spectrum), which corresponds to Sommerfeld's metal. At large distances, $d \gg a$, where d is the distance of our atom from the "surface" of the metal and a is the metal lattice constant, the functions we have chosen are fully equivalent to Bloch's functions.

The matrix element of the interaction energy is

$$(op'|V|kp'') = \frac{\mu_{0ke}}{h'} \int_0^\infty \frac{\sin(p'x/h')\sin(p''x/h')}{(x+d)^2}\, dx. \qquad (3)$$

We denote

$$I(p',p'') = \int_0^\infty \frac{\sin(p'x/h')\sin(p''x/h')}{(x+d)^2}\, dx. \qquad (4)$$

Choosing for the electrons a Fermi distribution with values of p' from 0 to p_{max}, we obtain for the total energy of interaction the following cumbersome expression:

$$U = -\frac{e^2}{(h')^2} \int_0^{p_{max}} dp' \int_{p_{max}}^\infty dp'' \sum_k \frac{\mu_{0k}^2}{E_k - E_0 + (p'')^2/2m - (p')^2/2m}[I(p',p'')]^2. \qquad (5)$$

It will be shown below that, for $|(p'-p'')/h'| > 1/d$, $I(p',p'')$ behaves as

$$\frac{(h')^2}{(p'-p'')^2 d^3} - \frac{(h')^2}{(p'+p'')^2 d^3}.$$

Due to this rapid decrease of the integral, the only part of significance is that where p' and p'' differ little from one another (and, hence, from p_{max} as well). This allows us to disregard $((p'')^2/2m) - ((p')^2/2m)$ in the denominator compared to $E_k - E_0$. Then, in the expression for U, the following sum is factored out:

$$\sum \frac{\mu_{0k}^2}{E_k - E_0} = \alpha, \qquad (6)$$

where α is the polarizability of the atom.

It remains for us to calculate

$$\overset{p_{max}}{\inf}\, dp' \int_{p_{max}}^\infty dp''[I(p',p'')]^2,$$

$$I(p',p'') = \int_0^\infty \frac{\sin(p'x/h')\sin(p''x/h)}{(x+d)^2}\, dx =$$

$$= \frac{1}{2}\int_0^\infty \frac{\cos[(p'-p'')x/h']}{(x+d)^2}\, dx - \frac{1}{2}\int_0^\infty \frac{\cos[(p'+p'')x/h']}{(x+d)^2}\, dx.$$

The term containing the cosine of the addition is always much less than the first term, and after integration over p' and p'' it yields components which decrease with the distance faster than $1/d^4$.

Neglecting these terms we assume roughly that

$$I(p', p'') = 1/d \quad \text{at} \quad -p'/h' < 1/d,$$
$$I(p', p'') = (h')^2/[(p' - p'')^2 d^3] \quad \text{at} \quad (p'' - p')/h' > 1/d. \qquad (7)$$

The last expression is obtained by reducing the integral to integral trigonometric functions and to the asymptotic expansions in a series of these. We restrict ourselves here to the first non-zero term.

In Fig. 1 the domain of integration is the upper rectangular strip extending to infinity. We divide it into three parts, a, b, and c, by the lines along which $I(p', p'')$, which depends only on $p' - p''$, is constant.

Here a is separated from b by the line $(p' - p'')/h' = 1/d$. The integral over a gives a quantity proportional to $(h')^2/d^4$. Introducing the variable $p'' - p' = \xi$, we take the integral

Fig. 1

$$\int_{h'/d}^{p_{max}} \left[\frac{(h')^2}{\xi^2 d^3} \right]^2 \xi\, d\xi = \frac{(h')^2}{d^4} - \frac{(h')^4}{p_{max}^2 d^6}.$$

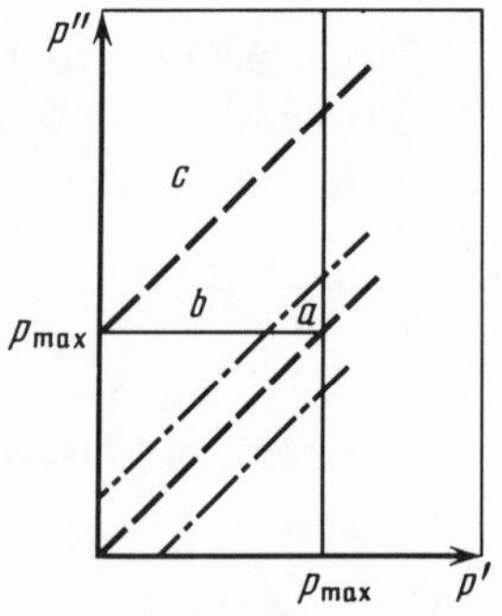

The second term, being of higher order in $1/d$, we throw out.

Integration over c yields only the term with $1/d^5$. Finally we obtain for the energy of interaction between a single atom and a linear metal at large distances

$$U = \frac{-\alpha e^2}{d^4}, \qquad (8)$$

accurate to within dimensionless factors of order unity.

In the interaction of an atom and a linear chain of individual atoms we would have a decrease of the interaction energy proportional to $1/d^5$.

The slower decrease of the forces in interaction with free electrons could be expected on the basis of physical considerations as well.

We note that we can easily find $\int_0^\infty [I(p', p'')]^2\, dp''$ exactly using the completeness theorem or the formula for matrix multiplication [3].

Indeed,

$$\frac{1}{h'} I(p', p'') = \frac{1}{h'} \int_0^\infty \frac{\sin(p'x/h') \sin(p''x/h')}{(x+d)^2}\, dx = p' \left| \frac{1}{(x+d)^2} \right| p'',$$

$$\frac{1}{h'} \int_0^\infty [I(p', p'')]^2\, dp'' = \int_0^\infty \left(p' \left| \frac{1}{(x+d)^2} \right| p'' \right) \left(p'' \left| \frac{1}{(x+d)^2} \right| p' \right) dp''$$

$$= p' \left| \frac{1}{(x+d)^2} \right| p'' = \frac{1}{h'} \int_0^\infty \frac{\sin^2(p'x/h')}{(x+d)^4}\, dx.$$

With accuracy to within higher order terms, we obtain $1/(6h'd^3)$. Thus, ignoring the Pauli principle in the calculation of the interaction $\int_0^\infty dp''$, we would have obtained for the energy

$$U = \frac{-\alpha e^2}{p_{\max} d^{-3}(h')^{-1}}.$$

Comparing this expression with the actual one, we can say that, because of the Pauli principle, only a thin layer of electrons in the Fermi distribution $(p_{\max} - p') > h'/d$, rather than all of the electrons from 0 to $p_{\max}$, takes part in the interaction or "works." The feasibility of this formulation is related to the form given above for $I(p', p'')$ as a function of $(p' - p'')$.

A direct, complete calculation of the interaction of an individual atom with a three-dimensional piece of metal presents great mathematical difficulties. However, this calculation is hardly of much interest since both the dependence of the interaction energy on the distance and its order of magnitude can be found (with accuracy to within dimensionless factors of order unity) from simpler considerations. We may now define the function $I(\mathbf{p'}, \mathbf{p''})$ as

$$\int \frac{\sin \frac{p'_x x}{h} \sin \frac{p'_y y}{h'} \sin \frac{p'_z z}{h'} \sin \frac{p''_x x}{h} \sin \frac{p''_y y}{h'} \sin \frac{p''_z z}{h'}}{(x+d)^2 + y^2 + z^2} \, dx \, dy \, dz,$$

or, equivalently,

$$I(\mathbf{p'}, \mathbf{p''}) = \int \frac{\exp[i(\mathbf{p'} - \mathbf{p''})\mathbf{r}/h']}{(x+d)^2 + y^2 + z^2} \, dx \, dy \, dz.$$

Just as in the one-dimensional case, the choice between the product of sines and the exponent does not influence the result.

As before, $I(\mathbf{p'}, \mathbf{p''})$ decreases very rapidly when $|(\mathbf{p'} - \mathbf{p''})/h'|$ increases to more than $1/d$. In addition, for $|(\mathbf{p'} - \mathbf{p''})/h'| > 1/d$, $I(\mathbf{p'}, \mathbf{p''})$ contains only terms which decrease rapidly with distance. We may again assume that only the nearest level transitions of electrons in the metal are important for the interaction:

$$\left| \frac{\mathbf{p'} - \mathbf{p''}}{h'} \right| < \frac{1}{d}.$$

The quantity

$$\int \int_0^\infty \int [I(\mathbf{p'}, \mathbf{p''})]^2 \, dp''_x \, dp''_y \, dp''_z$$

is determined from the completeness theorem. It turns out to be equal to $(h')^3/d$.

The Pauli principle and the Fermi distribution for electrons in the metal lead to the fact that in the interaction corresponding to this value of

$$\int [I(\mathbf{p'}, \mathbf{p''})]^2 \, d\tau''_p,$$

only electrons with $(p_{\max} |\mathbf{p'}|) < h'/d$ participate.

The volume occupied by the "interacting" electrons in the momentum space is equal (to within a factor) to $p_{\max}^2 h'/d$.

Finally, taking into account the new normalization of the eigenfunctions for the three-dimensional case (the normalization factor is $(h')^{-3/2}$), we obtain

$$U = -\alpha e^2 p_{\max}^2 d^{-2}(h')^{-2},$$

or, noting that $h'/p'_{\max}$ is of the order of magnitude of the metal lattice constant a,

$$U = -\alpha e^2 d^{-2} a^{-2}.$$

In the interaction of an atom with a nonmetallic solid body, as we know, the interaction energy decreases as $1/d^3$.

To experimentally determine the law of the adsorption potential decrease with distance, we may propose measurement of the thickness of a fluid layer wetting a solid object as a function of the elevation above the fluid level in a vessel (Fig. 2).

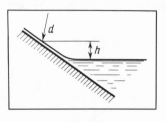

Fig. 2

The above calculations do not allow determination of the absolute value of U at atomic distances, i.e., of the actually measurable heat of adsorption of a gas on a metal, since for $d \cong a$, $h'/d \cong p_{\max}$ the use of a continuous spectrum and an infinitely large piece of metal no longer makes sense. The required numerical value of U can be easily found with sufficient accuracy if we take a piece of metal with a finite (small) number of atoms and, hence, with a small number of eigenfunctions, and directly sum the general expression for

$$U = -\sum \frac{|V_{0m}|^2}{E_m - E_0'},$$

where, as before, V is equal to $-\mu e \sum 1/r_i^2$, with the summation carried out over all excited states of the system m.

This calculation, however, makes sense only for such metals as sodium and potassium, where we know precisely the properties of the positive ions which form the metal lattice. The absence of appropriate experimental data makes the calculation useless at present.

The primary task for the future theoretical study of the problem is to use phenomenological properties of the metal rather than various approximate formulations from the modern theory of metals.

Laboratory of Catalysis. Theoretical Section   *Received*
Institute of Chemical Physics   *October 9, 1934*
USSR Academy of Sciences, Leningrad

References

1. *London H., Eisenschitz D.*—Ztschr. Phys. **60**, 491 (1910); *London H.*—Ztschr. Phys. **63**, 245 (1930).
2. *Jones L.*—Trans. Faraday Soc. **28**, 333 (1932).
3. *Dirac P., Harding F.*—Proc. Cambridge Philos. Soc. **28**, 209 (1932).

Commentary

This paper by Ya.B. is the first published work to study the interaction of an isolated atom with the surface of a metal. The author arrived at the conclusion that the interaction energy decreases with distance as $1/r^2$. Later, H. Casimir and D. Polder[1] considered this problem again using a different method, and arrived at a different result—the interaction energy is proportional to $1/r^3$ at distances $r \ll c/\omega_0$ (where ω_0 is some characteristic frequency of the absorption spectrum of the atom and metal), and $U \sim 1/r^4$ at distances much greater than c/ω_0. (E. M. Lifshitz[2] arrived at the same result.) The difference between these results and those in Ya.B.'s paper is related, however, to differences in the approximations made and in the areas of applicability of the formulas obtained.

Ya.B. applied formal perturbation theory to the interaction of an atom with the electrons of a metal, where the latter are assumed to be free. Meanwhile, Casimir and Polder and Lifshitz neglected the spatial dispersion of the dielectric permittivity of the metal. Therefore, in the region of small distances, frequencies of order ω_0 are important at "small" distances in the sense indicated above, as are arbitrarily small frequencies at large distances. In both limits the dielectric permittivity of the metal is not at all close to one. Meanwhile, the perturbation theory used by Ya.B. corresponds formally to an expansion in powers of $\epsilon - 1$ and is therefore not applicable in this case. Neglecting the spatial dispersion is valid, however, only at distances $r \gg a$ (a is the Debye radius in the metal) of the atom from the surface. At the opposite extreme, $r \ll a$, the wave vectors $\mathbf{k}_1 \sim 1/r \gg a \sim v_F/\omega_0$ are of importance (v_F is the electron speed at the Fermi boundary). In this region of strong spatial dispersion perturbation theory can be applied, and the r-dependence satisfies Zeldovich's law.

[1] *Casimir H. B. C., Polder D.*—Phys. Rev. **73**, 300–306 (1948).
[2] *Lifshitz E. M.*—ZhETF **29**, 94–110 (1965).

12

Proof of the Uniqueness of the Solution
of the Equations of the Law of Mass Action*

Intuitively, the uniqueness of the chemical equilibrium state of a mixture of reacting gases is more or less obvious. However, it may be of some interest to rigorously prove that the system of equations of the law of mass action (LMA), together with the imposed conditions of conservation of matter for given T and v or T and p, has one and only one real-valued and positive solution.

To carry out the proof, we note that the LMA equations are equivalent to imposing an extremum on some function of the concentrations—the free energy F for $v = $ const or the thermodynamic potential Φ for $p = $ const. It is in precisely this way, as is well known, that the LMA may be derived from general thermodynamic principles. We will solve the problem if we prove that the surface F or Φ, under the conditions imposed on the concentration and for constant v or p, has one and only one minimum, and does not have either maxima or any other critical points (so-called minimax, or saddle points).

First we consider the case $v = $ const.

In an ideal system—and it is only for such a system that LMA holds—our proof can be carried out

$$f = \frac{F}{kT} = \sum_{i=1,\ldots,m} N_i \lg\left(\frac{N_i}{v}\right) + a_i' N_i = \sum_{i=1,\ldots,m} N_i \lg N_i + a_i N_i, \quad (1)$$

where the N_i are the total amounts of various components and $a_i = a_i' - \lg v$.

The conditions of conservation of matter are written in the form

$$\sum_{i=1,\ldots,m} B_{ij} N_i = C_j \qquad j = 1,\ldots,n \tag{2}$$

for the number of elements in the system, n; here all $B_{ij} > 0$. Thus, for dissociation of water vapor at high temperatures, we have in the system the compounds H_2O, H_2, O_2, OH, H, O.

The conservation of the matter yields two conditions:

$$2H_2O + 2H_2 + OH + H = H°,$$
$$H_2O + 2O_2 + OH + O = O°, \tag{2a}$$

*Zhurnal fizicheskoĭ khimii **11** 5, 685–687 (1938).

where H° and O° denote the overall number of hydrogen and oxygen atoms in the system. For the n linear equations (2) relating the m variables $N_i, \ldots, N_m$, we may choose $m - n$ independent variables $\xi_1, \ldots, \xi_{m-n}$ in terms of which all concentrations N_i will be expressed linearly:

$$N_i = \sum_{k=1,\ldots,m-n} D_{ik}\xi_k + d_i, \qquad i = 1, \ldots, m \tag{3}$$

(the D_{ik} can be both positive and negative). Successively we find

$$\partial f/\partial \xi_l = \sum_i D_{il}(\lg N_i + a_i + 1). \tag{4}$$

Equating $\partial f/\partial \xi_l = 0$ we obtain precisely $m - n$ equations for the LMA:

$$\sum_i D_{il}(\lg N_i + a_i + 1) = 0, \tag{5}$$

$$\lg \prod N_i^{D_{il}} = -\sum D_{il}(a_i + 1) = \lg K_l, \tag{6}$$

where K_l is the reaction equilibrium constant corresponding to variation of one variable ξ_l while all the other $m - n - 1$ variables ξ_k are held constant.

The second derivatives are

$$\frac{\partial^2 f}{\partial \xi_k \partial \xi_l} = \sum_i \frac{D_{ik}D_{il}}{N_i}. \tag{7}$$

The quadratic form $\partial^2 f/\partial \xi_k \partial \xi_l$ is positive definite since, for any set of values $z_1, \ldots, z_{m-n}$, the quantity

$$\sum_k \sum_l \frac{\partial^2 f}{\partial \xi_k \partial \xi_l} z_k z_l = \sum_k \sum_l \sum_i \frac{D_{il}D_{ik}}{N_i} z_k z_l = \sum_i \frac{(\sum_k D_{ik}z_k)^2}{N_i} > 0, \tag{8}$$

because, obviously, all the amounts $N_i > 0$. The surface f in the variables ξ is concave down throughout the domain of ξ in which all $N_i > 0$. From this we immediately see that the minimum is unique, if it exists.[1]

The existence of the minimum is clear from the fact that at all boundaries of its definition domain f, as it approaches the boundary, behaves as $x \lg x$ for x approaching zero. Therefore, f cannot reach its minimum at the very edge of the region since it certainly increases as it approaches the edge. Consequently, the minimum is reached inside the region. The uniqueness of this minimum was proved above.

The proof for $p = \text{const}$ is a little bit more complicated. Here

$$\Phi' = \frac{\Phi}{kT} = \sum_i N_i \lg \frac{N_i}{v} + a_i' N_i. \tag{9}$$

---

[1] It is easy to show that in the variables ξ the definition domain of f (all $N_i > 0$) is convex.

But v is more variable, and a'_i cannot contain $\lg v$. Now,

$$v = \frac{kT}{p \sum N_i}. \tag{10}$$

Substituting, we find

$$\Phi' = -\left(\sum N_j\right) \lg N_j + \sum N_i \lg N_i + a_i N_i. \tag{11}$$

Note that a_i and a'_i are different from those in (1). The conditions (2) for N have not changed since, as we have already said, the N_i are not concentrations but the total amounts of components in the system, which is of importance for a variable volume proportional to $\sum N_j$.

Introducing the new variables $\xi_1, \ldots, \xi_{m-n}$ just as before and constructing $\partial \Phi / \partial \xi_k$ and $\partial^2 \Phi / \partial \xi_k \partial \xi_l$, we obtain the LMA in the form

$$-\sum D_{ik} \lg \left(\sum_j N_j\right) + \sum D_{ik} \lg N_i = \lg \prod_i \left(\frac{N_i}{\sum_j N_j}\right)^{D_{ik}}$$

$$= \lg \Pi \left(\frac{N_i}{v}\right)^{D_{ik}} + A = B, \tag{12}$$

where A and B are constants. Because the volume varies, it is essential that the concentrations N_i/v, and not the actual amounts N_i, should enter into the LMA. From the second derivatives we obtain

$$\sum_k \sum_l \frac{\partial^2 f}{\partial \xi_k \partial \xi_l} z_k z_l = \sum_i \left(\sum_k D_{ik} z_k\right)^2 \left(\frac{1}{N_i} - \frac{1}{\sum N_j}\right) > 0,$$

since all

$$N_i > 0, \qquad N_i < \sum N_j, \qquad 1/N_i > 1/\sum N_j. \tag{13}$$

With this we also establish the uniqueness of the solution of the LMA equations for an ideal system at constant pressure.

It is clear that the study of negative, complex-valued solutions of the LMA equations, their bifurcation points, etc. [1], is devoid of any physical sense.

Physico-Chemical Laboratory   *Received*
USSR Academy of Sciences, Leningrad   *January 13, 1938*

References

1. *Stepanov B. M.*—Uspekhi khimii **5**, 972 (1936).

Commentary

This paper shows that the conditions of thermodynamic equilibrium in a mixture of chemically reacting ideal gases always have a solution for the concentrations of the mixture components and that this solution is unique. The paper has acquired special significance in the last few years in connection with the intensive study of systems in which this uniqueness does not occur. Such anomalies may be related either to nonideal components, or to treatment of stationary states, rather than truly equilibrium ones, in which the system exchanges matter or energy with the surrounding medium.

The behavior of a mixture is determined by a system of ordinary differential equations, while the required state, either equilibrium or stationary, is determined by a time-independent system of algebraic equations. Therefore, at first glance one would not expect any qualitative difference between the equilibrium and stationary states. Ya.B. shows that in the equilibrium case, even for an ideal system, a variational principle exists which guarantees uniqueness. Such a principle cannot be formulated for the case of an open system with influx of matter and/or energy.

13

On the Relation Between
Liquid and Gaseous States of Metals*

With L. D. Landau

General considerations regarding the character of the transition of a substance from a metal to a dielectric state lead to the conclusion that such a transition occurs as a normal phase transition even up to high temperatures. For mercury and other low-boiling metals the critical point of transition from a liquid to a gaseous state probably corresponds to a lower temperature. One should expect the existence in some region of two separate (at different pressures and temperatures) transitions, from a metallic to a nonmetallic state, and from a liquid to a gaseous state, i.e., the existence of a liquid nonmetallic phase which transforms into a metal with increased pressure, and into a gas with decreased pressure.

A metal differs sharply from a dielectric by its electron energy spectrum at absolute zero. The basic state of a metal is contiguous to a continuous spectrum of states. For this reason, an arbitrarily weak electric field causes an electric current in the metal which depends on transition of the system to states which are arbitrarily close in energy to the basic state. On the other hand, the electron energy spectrum of a dielectric is characterized by the existence of a finite "gap," a certain difference in energies between the basic state with minimum energy (in which there is no current) and adjacent excited states in which one of the electrons of the dielectric becomes free and electrical conductivity appears.

We note that it is not possible to define a metal as a substance with a continuum of states contiguous to the basic state without additional conditions; in fact, any paramagnetic, e.g., liquid oxygen or gadolinium sulfate, has a continuous spectrum of states corresponding to different values of the magnetic moment; indeed, it is due to this that variation of the moment, typical for a paramagnetic substance, occurs in a weak magnetic field. However, liquid oxygen and gadolinium sulfate are not metals and do not conduct current. The existence of a continuous spectrum is a necessary but not sufficient condition for a substance to be a metal. For electrical conductivity the excited states contiguous to the basic state must have the property of electrical charge transfer.

*Zhurnal eksperimentalnoĭ i teoreticheskoĭ fiziki **14** 1/2, 32–34 (1944).

The assumption has been repeatedly made, and seems plausible (although it has never been proved in a general form), that at sufficiently high compression any substance will be transformed into a metal. An illustration of this is the transformation of phosphorus into its conducting variant at very high pressure—the Bridgman black phosphorus.

At absolute zero a metal and a dielectric are qualitatively different, and it is always possible to determine just which one we are dealing with; there is a definite point of transition.[1]

A dielectric differs from a metal by the existence of an energy gap in the electron spectrum. But can the gap width go to zero when the point of transition from a dielectric to a metal is approached? In this case we would have transition without latent heat, and without a jump in volume and other properties. Peierls indicated that a continuous transition in this sense is impossible. Let us consider an excited state of a dielectric in which it conducts current. An electron leaves its position and moves in the crystalline lattice, uncovering at another lattice position a positive charge. When the electron is far from the positive charge, the electron of course experiences a Coulomb attraction which tries to move it back. In the Coulomb field of attraction there always exist discrete states with negative energy corresponding to bonding of the electron. Consequently, the excited and conducting states of a dielectric are always separated by a gap of finite width from the basic state in which the electrons are bound.

If, even at 0°K, the transformation of a metal into a dielectric is a phase transition of the first kind (i.e., with latent heat of transition and a discontinuous change in properties), then obviously this transformation will occur as a phase transition of the first kind at low, non-zero temperatures as well. Continuous transition is possible only at high temperatures when excitation and conductivity of the dielectric become significant. The excitation energy is of the order of the ionization energy, i.e., a few volts, or at worst 1 volt. Thus the line of thermodynamic equilibrium between the metal and dielectric phases can end up in a critical point only at very high temperature (on the order of a volt, i.e., of order 10,000 degrees) and accordingly at enormous pressure. Both of these phases at high temperatures are noncrystalline (the melting of a metal in no way deprives it of its metallic properties). The question then arises of the relation between the line of transition from a metallic to a dielectric state and the line of transition between the liquid and gas states in metals. It is perfectly obvious that, at low pressure, a substance with low density (an ideal gas in the extreme case) is a nonconductor, a nonmetal. For mercury the energy spectrum of the gas is discrete, and for

---

[1] At non-zero temperatures there occurs, in principle, in any dielectric a certain excitation; some fraction, even if only an infinitesimal one, of the electrons are in an excited state, so that the electrical conductivity is non-zero and the system as a whole is in a state of continuous spectrum. Therefore we may completely rigorously distinguish a metal from a dielectric only at absolute zero.

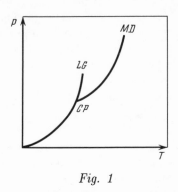

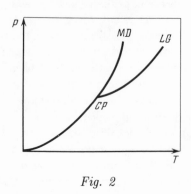

Fig. 1   Fig. 2

paramagnetic vapors of sodium we have a continuous spectrum which, however (as in the case of oxygen), is not directly related to metallic properties
and conductivity.

In principle, three cases are possible.[2] Transition from a metallic to a
dielectric state is always accompanied by transition from a fluid state to a
gaseous one; there is a single common curve, and one critical point which
is reached at very high temperatures. This case perhaps occurs for non-
volatile metals. For metals with a low heat of evaporation (for example,
mercury), one may expect the liquid–gas critical point (LG) to be at a
temperature substantially lower than the metal–dielectric critical point of
transition (MD). Here the second and the third cases appear (Figs. 1 and 2).

In the second case, heating of a liquid metal at high pressure results in a
density jump along the CP–LG line. However, the lower-density phase is also
a metal ("metallic gas"). Transition to a normal gas occurs along the line
CP–MD. This case is quite improbable. In the third case, we expect that in a
certain pressure interval at high temperature the liquid metal is transformed
into a liquid non-conducting phase (on the line CP–MD), and that only after
this is the non-conducting phase transformed into a gas along the line CP–
LG. Loss of metallic properties also occurs via metal–gas phase transition
at temperatures and pressures much higher than those corresponding to the
liquid–gas critical point.

In both of the latter cases, the appearance of a triple point, CP, of coexis-
tence is characteristic—in the second case, of two metallic and one dielectric
phase, and in the third of a metal and two dielectric (liquid and gaseous)
phases.

In the case of mercury the comparatively low evaporation heat indicates
that the point LG is located not far away, at 1000–1500°K by different
estimates, whereas the point MD is probably not even accessible now to

[2]We do not consider crystalline phases which exist at low temperatures. The corresponding transitions are not relevant to our topic.

experimental study. From our considerations it follows that, in this case, it is apparently the third case that occurs. Our physical predictions may be summarized thus: (1) a nonconducting liquid phase exists and (2) at temperatures and pressures greater than critical, phase transition occurs with discontinuous changes in electrical conductivity, volume, and other properties.

Institute of Chemical Physics   *Received*
USSR Academy of Sciences   *June 15, 1943*
Institute of Physical Problems
USSR Academy of Sciences

Commentary

This article discusses the question of the character of the phase diagram in the transition of a liquid metal to a dielectric state. It formulates for the first time the important hypothesis that it is impossible, strictly speaking, to distinguish a metal from a dielectric at finite temperatures, and that therefore the transition between the dielectric and metallic phases at finite temperatures may occur discontinuously. This means that the transition curve, which at sufficiently low temperatures is a transition of the first kind, must end at a critical point. Here, in principle, two outcomes are feasible. Either the critical point of metal–dielectric transition coincides with the critical point of liquid–vapor transition (in this case the transition curves themselves also coincide), or these transformations have different critical points. The paper generated enormous literature and prompted numerous experimental studies. Nonetheless the experimental situation is unclear even today, and this is related primarily to the difficulty of accurate measurements at high pressures and temperatures. The most detailed studies have been carried out for cesium and mercury.[1] In cesium, significant changes in the electrical properties occur in the critical region of liquid–vapor transition which makes probable the assumption of the existence of a common critical point. In mercury the changes in electrical properties occur at a density approximately two times greater than the critical one. It is not excluded that this may indicate the existence of two critical points, i.e., that the second of the possibilities indicated in the paper is realized.

In the exposition of the problem, this paper presents a good deal of discussion on the problem of the conductivity decrease at $T = 0$ with the increase of the distance r between fixed sodium atoms, where the zone theory yields a continuous conductivity decay to zero according to $\exp(-r/r_0)$ while in fact Mott's transition takes place.

[1] *Alekseev V. A., Andreev A. A., Sadovskiĭ M. V.*—Uspekhi Fiz. Nauk **132**, 47–90 (1980).

IV

Theory of Shock Waves

14

On the Possibility of

Rarefaction Shock Waves[*]

A shock wave is the surface of a sudden very sharp variation in the motion and state (pressure, density, etc.) of a material, which moves with respect to this material.

The relations between the state of the material before the wave passes (I), the properties of the wave, and the state after the wave passes (II) are easily obtained from the laws of conservation of mass, momentum, and energy. These relations are symmetrical with respect to the quantities describing I and II, and are equally suited to description of the transition from I to II and the reverse transition from II to I.

Zemplen [1], considering an ideal gas with constant specific heat, showed that entropy is not conserved in a shock wave: it rises with increased pressure and falls with decreased pressure. From this follows the so-called Zemplen theorem of the impossibility of rarefaction shock waves.

In fact, just one year earlier Jouguet [2] gave an expression for the entropy change in a small-amplitude shock wave,

$$S_{II} - S_I = \frac{1}{12} \frac{1}{T} \left(\frac{\partial^2 v}{\partial p^2} \right)_S (p_{II} - p_I),$$

whence follows, for an ideal gas and for all substances with a positive second derivative $(\partial^2 v / \partial p^2)_S$, the assertion of the impossibility of rarefaction shock waves.

[*]Zhurnal eksperimentalnoĭ i teoreticheskoĭ fiziki **16** 4, 363–364 (1946).

E. Jouguet also proved in a general form that this thermodynamic criterion of the possibility of compressive shock waves

$$S_{II} > S_I \quad \text{for} \quad p_{II} > p_I,$$

and the impossibility of rarefaction shock waves

$$S_{II} < S_I \quad \text{for} \quad p_{II} < p_I,$$

is identically related to the relation between the speed of sound before and after passage of a wave and the speed of the wave relative to the corresponding states ($c_I > D_I$, $c_{II} > D_{II}$ for $p_{II} > p_I$ and $c_I > D_I$, $c_{II} < D_{II}$ for $p_{II} < p_I$). This interrelation may be viewed as a mechanical condition of the possibility of compression shock waves and the impossibility of rarefaction shock waves.

Our point is that, without a doubt, there exist substances in nature for which, under certain conditions of pressure and temperature, $(\partial^2 p/\partial v^2)_S < 0$ holds.

In this case, it follows from the general theory that in such a state (sharp) rarefaction shock waves will exist, while compression shock waves will be diffuse and ill-defined, with a width proportional to the path traveled by the wave.

For proof we note that the isotherm has an inflection point at the critical point, i.e., $(\partial^2 p/\partial v^2)_T = 0$ for

$$p = p_{cr}, \qquad v = v_{cr}, \qquad T = T_{cr}.$$

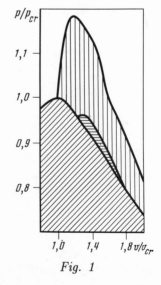

It is easy to verify that on the isotherm corresponding to the critical temperature $T = T_{cr}$ there is a finite region $(\partial^2 p/\partial v^2)_T < 0$ bounded on one side by the critical point and on the other by some point at a lower pressure and higher specific volume. This region, like the entire isotherm at $T = T_{cr}$, lies completely outside the condensation region. Analogous regions are present on other isotherms which are close to the critical one for $T < T_{cr}$ and

Fig. 1

$T > T_{cr}$; taken together, they form in the p,v-plane a region in which $(\partial^2 p/\partial v^2)_T < 0$. Part of this region obviously lies outside the condensation region since the entire isotherm $T = T_{cr}$ lies outside the condensation. In this region, in which $(\partial^2 p/\partial v^2)_T < 0$, the quantity $(\partial^2 v/\partial p^2)_T$ is also negative since

$$\left(\frac{\partial^2 v}{\partial p^2}\right)_T = -\left(\frac{\partial v}{\partial p}\right)_T^3 \left(\frac{\partial^2 p}{\partial v^2}\right)_T$$

In the stable phases, $(\partial v/\partial p)_T < 0$, and the signs of $(\partial^2 p/\partial v^2)_T$ and $(\partial^2 v/\partial p^2)_T$ coincide.

For the problem in question, it is the isentropic derivative, $(\partial^2 v/\partial p^2)_S < 0$, that is important, not the isothermic one.

However, the greater the internal molecular specific heat, the closer the adiabate is to the isotherm. We may therefore assert that for sufficiently large specific heat outside the condensation region, the region of interest, $(\partial^2 v/\partial p^2)_S < 0$, probably exists. The arguments given above are completely general, but at the same time they do not allow us to determine the required value of the specific heat without experiment.

Using the very rough assumptions of a constant specific heat c_v and strict accuracy of the Van-der-Waals equations, numerical calculations were performed which implied that the minimum value of the specific heat for which the desired region appears is equal to $c_v =20$ cal/mole $\cdot$ deg. The figure above shows in the p,v-plane the condensation region (diagonal lines), the region $(\partial^2 v/\partial p^2)_T < 0$ (vertical lines), and the region $(\partial^2 v/\partial p^2)_S < 0$ at $c_v = 40$ cal/mole $\cdot$ deg (horizontal lines).

F. E. Yudin participated in the calculations.

This problem is considered in detail in the monograph "Theory of Shock Waves and Introduction to Gasdynamics" [3].

Institute of Chemical Physics   *Received*
USSR Academy of Sciences, Moscow   *September 5, 1945*

References

1. *Zemplen G.*—C. r. Bull. chim. Soc. **141**, 712 (1905); **142**, 142 (1906).
2. *Jouguet E.*—C. r. Bull. chim. Soc. **138**, 1685 (1904); **138**, 786 (1904).
3. *Zeldovich Ya. B.* Teoriĭa udarnykh voln i vvedenie v gazodinamiku [Theory of Shock Waves and Introduction to Gasdynamics]. Moscow, Leningrad: Izd-vo AN SSSR, 186 p. (1946).

Commentary

It is shown in this paper that, contrary to previous common belief, under certain special, but still fully feasible experimental conditions rarefaction shock waves can exist. In particular, this situation should certainly occur in gases with a sufficiently large specific heat c_v near the critical point of fluid–vapor transition. In recent years the prediction made by Ya.B. has been conclusively confirmed by experiment.[1] Later Ya.B. considered the peculiarities of the state near the critical point which may occur in a rapid, "shock" expansion.[2]

[1] *Kutateladze S. S., Borisov Al. A., Borisov A. A., Nakoryakov V. E.*—Doklady AN SSSR **252**, 595–598 (1978)
[2] *Zeldovich Ya. B.*—ZhETF **80**, 2111–2112 (1981).

15

On the Propagation of Shock Waves in a Gas with Reversible Chemical Reactions*

The acoustics of a gas in which occur reversible chemical reactions whose equilibrium shifts with pressure and temperature variations in the acoustic wave was studied by Albert Einstein [1]. His results were later applied to the study of the very fast processes encountered at the boundary between physics and chemistry—the transition of molecules from one vibrational state to another [2].

The propagation of shock waves accompanied by an irreversible chemical reaction with substantial release of heat is the subject of the theory of detonation [3–6].

Below we consider the question of shock wave propagation in a gas with a reversible chemical reaction or with delayed transition of molecules from one vibrational state to another.

In contrast to an explosive capable of detonating, this gas is in a state of complete thermodynamic equilibrium. Propagation of a wave occurs only as a result of external action, for example, the motion of a piston in a tube filled with gas.

In the first moment after motion begins, the chemical reaction has not yet occurred, and the wave propagates as it would in a normal gas. However, the state attained in this compression is certainly not an equilibrium one. The change in pressure and temperature which has occurred in the wave causes a change in the state of the chemical equilibrium and, after the compression, a reaction takes place which pushes the system toward a new state of equilibrium. This reaction, in turn, causes a change in the pressure and density. In the one-dimensional case of a plane wave or a wave in a tube, in a normal gas with a piston moving at constant velocity instantaneously acquired at the initial moment of motion, the wave amplitude and the state of the gas after the wave has passed remain unchanged through the entire duration of the piston's motion; from what we have said, it is clear that this is not so for a gas with a reversible reaction.

Without considering transient regimes, we find the asymptotic laws which are satisfied by the wave and the state of gas compressed by it during the motion of the piston (which is long compared to the time of the chemical

*Zhurnal eksperimentalnoĭ i teoreticheskoĭ fiziki **16** 4, 365–368 (1946).

reaction). It is obvious that the gas compressed by the wave will arrive at a
new state of chemical equilibrium.

To find the final state of the gas after compression by the wave we
must construct three equations—momentum, and conservation of mass and
energy—and add to them the equation of chemical equilibrium.

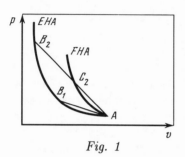

Fig. 1

Taken together, these equations determine in the p,v-plane the curve of possible final states after compression—the "equilibrium Hugoniot adiabate" (Fig. 1). Near the point A, which describes the initial state, this curve coincides with the isentropic curve, i.e., with the curve of the change of state in very slow adiabatic compression (not shown in the figure). The equilibrium Hugoniot adiabate (EHA) gives relations between all the wave parameters: a specific pressure p_1 on the EHA corresponds to a given specific volume v_1; the wave velocity is equal to

$$D = V_0 \sqrt{\frac{p_1 - p_0}{v_0 - v_1}},\qquad(1)$$

the velocity of the compressed matter and, consequently, the velocity of the
piston necessary to generate the wave are equal to

$$U = \sqrt{\frac{p_1 - p_0}{v_0 - v_1}}.\qquad(2)$$

Einstein showed that when a reversible reaction is present sound dispersion
occurs: at low frequency the equilibrium is shifted within the time of oscillation, the effective specific heat is at a maximum, and the speed of sound c_0 is
at a minimum. At high frequency the oscillations occur so rapidly that the
equilibrium has no time to shift (it is "frozen"). The corresponding Hugoniot adiabate (FHA) is shown in the figure. Here the effective heat capacity
is minimal, the speed of sound c_∞ is maximal: $c_\infty > c_0$. From consideration
of the final state and the theory of shock waves it follows that $D \gg c_0$.

Equality occurs in the limit of very small shock wave amplitude; as the
amplitude increases, D increases indefinitely. The structure of the wave front
substantially depends on the relation between the wave velocity D and the
speed of sound in the frozen equilibrium, c_∞.

To determine the front structure we observe that from the equations of
momentum and conservation of energy it follows that all interim states between the initial state A and the final state B_1 (at small amplitude) or B_2
(at large amplitude) are located in the p,v-plane on the corresponding lines
AB_1 or AB_2 (cf. the theory of detonation).

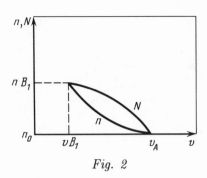

Fig. 2

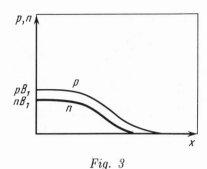

Fig. 3

The equation of the straight line is

$$p = p_0 + D^2 \frac{v_0 - v}{v_0^2}. \tag{3}$$

Because of this the problem of calculating the wave front structure becomes an elementary one. The chemical reaction is the slow process. Along the line AB_1 or AB_2 (depending on the wave amplitude) we calculate at each point the actual reaction state, which we describe by a quantity n—the concentration of some product of the chemical reaction. This quantity we find from the energy equation for the point under consideration

$$E(p, v, n) = E_0 + \frac{(v_0 - v)(p_0 + p)}{2}; \tag{4}$$

v is arbitrary, p is found from the straight line AB_1 or AB_2, and the index "zero" denotes quantities relating to the initial state. The velocities D and U which enter into the energy equation above are expressed in terms of p_1, v_1, p_0 and v_0.

Further, for every point on the line we also find the quantity N—the equilibrium concentration of the chemical reaction product corresponding to the temperature and pressure which is determined by the product pv at that point:

$$N = N(p, v). \tag{5}$$

The equation of the reaction kinetics is characterized by the approach of the concentration toward its equilibrium value and, in a first approximation, may be written as:

$$\frac{dn}{dt} = \frac{N - n}{\tau}, \tag{6}$$

where τ is the time required to establish equilibrium.

In a steadily propagating wave

$$\frac{dn}{dt} = (D - U) \frac{dn}{dx}$$

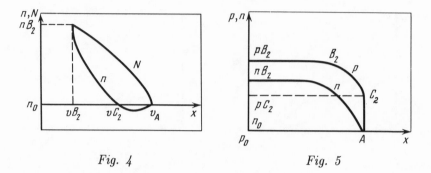

Fig. 4   Fig. 5

With the help of (3)–(5) we express n, N, and U in terms of v, after which the distributions of v, of the concentration n, and of other quantities in space are obtained as the quadratures of one ordinary differential equation.

For small amplitude (final state B_1 with $D < c_\infty$) the curves N and n are located as shown in Fig. 2, and solution of the equation yields a diffuse wave front (Fig. 3) whose width is proportional to the product $D\tau$ and inversely proportional to the amplitude of the wave.[1]

In Fig. 3 the pressure and concentration (n) variations in the wave are shown. The diffuse structure in this case may be explained qualitatively by the fact that the high-frequency waves necessary to generate a steep front propagate with a speed greater than D and, moving forward away from the wave, are damped (absorbed). As Einstein showed, waves with an oscillation period less than τ decrease in amplitude by e times at a distance of order $c\tau$.

For large-amplitude waves (final state B_2, with $D > c_\infty$, see Fig. 1) the curves N and n are as shown in Fig. 4. Here it is obvious that a continuous solution of the type in Fig. 3 is not possible since, as before, $N > n$, and the chemical reaction can not induce a decrease in n near v_0. In this case the structure of the wave has the form shown in Fig. 5: the unperturbed gas A is sharply compressed by the shock wave AB_2, after which the state of the gas changes smoothly until it reaches the final state B_2.

We find the speed of propagation D by knowing the overall amplitude from the EHA curve and determining the point B_2. The parameters of the sharp wave AB_2 running along the wave front are found with the help of the curve FHA for a given speed D from the condition that in the wave AB_2 the concentration n does not change. Therefore, the wave AB_2, as in the ordinary theory of shock waves, has a very sharp front: its width for large-amplitude waves is of the order of the mean free path. The subsequent

---

[1]The overall change of n from n_0 to nB_1 is proportional to the amplitude, and the maximum difference $N - n$ is proportional to the square of the amplitude. From this it is clear that the time of reaction is inversely proportional to the amplitude.

compression from C_2 to B_2 occurs in the reaction time τ and occupies a width of order $D\tau$. It is obvious that all we have said may be directly transferred to the case of slow transition of molecules from one state to another, i.e., to the case of delayed excitation of part of the specific heat of the gas.

The widening of the shock wave front in this case can also be foreseen from the viewpoint of Leontovich and Mandelstam [7] who indicate that delayed excitation corresponds to an anomalously large second gas viscosity coefficient. However, this approach is approximate: being qualitatively valid for small-amplitude waves (of the type in Fig. 3), the concept of the viscosity coefficient is inappropriate to describe the more complicated structure of large-amplitude waves (of the type in Fig. 5).

Direct observation of the shock wave structure, for example, by the semi-shadow method of Toepler, is, in our opinion, a very convenient method for the detection and study of phenomena of delayed energy transfer in gases. According to Kneser [2], in a calculation with an excitation time of $\tau = 10\,s$, wave amplitude $p = 0.07p_0$, $D = 1.03$, and $c_0 = .99c_\infty$, the width of the wave in carbonic acid (which shows delayed excitation of the specific heat) should be 180 times greater than in air, reaching 1.45 mm at atmospheric pressure and 10 mm at $p_0 = 100$ mm of mercury. The problem was first considered in the author's monograph, "Fundamentals of Gasdynamics and Theory of Shock Waves," which is in press at the USSR Academy of Science Publishing House.*

Institute of Chemical Physics
USSR Academy of Sciences, Moscow

Received
September 5, 1945

References

1. *Einstein A.*—S.-Ber. Berlin Akad. Wiss. **380** (1920).
2. *Kneser H.*—Ann. Phys. **11**, 761 (1931).
3. *Jouguet E.* Mécanique des explosifs. Paris: Dion (1917).
4. *Zeldovich Ya. B.*—ZhETF **10**, 542 (1940).
5. *Zeldovich Ya. B.* Teoriĭa goreniĭa i detonatsii gazov [Theory of Combustion and Detonation of Gases]. Moscow: Izd-vo AN SSSR, 71 p. (1944).
6. *Rankine W. J. M.*—Phfl. Transact. **160**, 277 (1870).
7. *Leontovich M. A., Mandelstam L. I.*—ZhETF **7**, 438 (1937).

Zeldovich Ya. B. Teoriĭa udarnykh voln i vvedenie v gazodinamiku [Theory of Shock Waves and Introduction to Gasdynamics]. Moscow, Leningrad: Izd-vo AN SSSR, 186 p. (1946).—*Editor's note.*

АКАДЕМИЯ НАУК СОЮЗА ССР

ИНСТИТУТ ХИМИЧЕСКОЙ ФИЗИКИ

Я. Б. ЗЕЛЬДОВИЧ

ТЕОРИЯ ГОРЕНИЯ
и
ДЕТОНАЦИИ ГАЗОВ

ИЗДАТЕЛЬСТВО АКАДЕМИИ НАУК СССР
Москва   1944   Ленинград

PART TWO

THEORY OF COMBUSTION AND DETONATION

16

Theory of Combustion
and Detonation of Gases*

Chapter I. Theory of Gas Combustion

§I.1 Introduction and Survey of Results

At every stage in the development of science the study of combustion has been very closely related to general and physical chemistry.

The first period in the development of combustion science was a period of determination of the basic chemical facts; to this period belong the refutation of the phlogiston theory and the discovery of oxygen, the discovery and study of the properties of carbon monoxide and carbon dioxide, and the so-called "pneumatic chemistry"—the investigation of various gases and determination of the stoichiometric laws (1650–1820).

Later, the energy aspects of combustion were studied, beginning with determination of the heat of combustion of various compounds and the pressures and temperatures which develop in explosions, and ending with the application of chemical thermodynamics to questions of equilibrium, dissociation and completeness of combustion in flames (1820–1900).

During this same period the basic theoretical concepts were formed. Properties common to all explosive mixtures—the ability to ignite when heated, the ability to propagate flame after local ignition—were explained on the basis of common characteristics of explosive mixtures—the presence in them of a large supply of chemical energy. For a qualitative explanation of combustion phenomena it is sufficient to know that at low temperatures explosive compounds are inert, and no heat of reaction is released; rapid chemical reaction begins only at elevated temperatures.

During this period detailed investigation of the phenomena of combustion took place, primarily in mixtures of combustible gases (hydrogen, carbon monoxide, methane) with air and oxygen.

At the beginning of the twentieth century attempts were made to measure the rate of chemical reaction in these systems. But here they were unable to arrive at clear results.

*Zeldovich Ya. B. Teoriĭa goreniĭa i detonatsii gazov [Theory of Combustion and Detonation of Gases]. Ed. by N. N. Semenov. Moscow, Leningrad: Izd-vo AN SSSR, 71 p. (1944).

The kinetics of combustion reactions turn out to be quite complicated; they do not satisfy the classical law of mass action and its kinetic formulation. Neither did Duhem's formal conceptions of the existence of regions of false equilibria and of a special "chemical friction," which ignores the molecular mechanism of chemical reactions, correspond to reality.

Moreover, the phenomena of combustion themselves prove to be more complicated. For a long period the study of combustion broke away from chemical kinetics and set itself its own specific tasks. These included especially studies of the influence of instrumental parameters on ignition, flame propagation and limits, i.e., the influence of the diameter and length of tubes, the form of vessels, the direction of propagation, etc.

Over the last 20 years chemical kinetics, and especially the theory of chain reactions, have achieved major successes. A theory of ignition of heated explosive mixtures has been created. However, attempts to directly explain the propagation of flame as the diffusion of active centers, or to explain the limits of propagation by the conditions of chain breaking fail to yield positive results.

What problems face the theory of combustion? The theory of combustion must be transformed into a chapter of physical chemistry. Basic questions must be answered: will a compound of a given composition be combustible, what will be the rate of combustion of an explosive mixture, what peculiarities and shapes of flames should we expect? We shall not be satisfied with an answer based on analogy with other known cases of combustion. The phenomena must be reduced to their original causes. Such original causes for combustion are chemical reaction, heat transfer, transport of matter by diffusion, and gas motion. A direct calculation of flame velocity using data on elementary chemical reaction events and thermal constants was first carried out for the reaction of hydrogen with bromine in 1942. The problem of the possibility of combustion (the concentration limit) was reduced for the first time to thermal calculations for mixtures of carbon monoxide with air. Peculiar forms of propagation near boundaries which arise when normal combustion is precluded or unstable were explained in terms of the physical characteristics of mixtures.

Laws were found for the combustion of gases in laminar or turbulent flow. Similarity laws were formulated for flame propagation under conditions in which the difference in density of a substance before and after combustion causes convective motion of the gas and flame.[1]

For the overwhelming majority of combustion reactions we do not have reliable quantitative kinetic schemes. Therefore, instead of precalculation of the flame velocity, the task of using combustion as a kinetic experiment becomes the first priority. The theory of combustion allows us to use data

---

[1]This last group of studies (combustion in a flow and the influence of convection) is closely related to thermal technology and will not be treated in the present physicochemical treatise.

on the combustion rate and limits of propagation to determine the kinetics of chemical reactions which last no more than a thousandth of a second at 2000–3000°K. Thus our goal is not only to transform combustion science into a branch of physical chemistry, but also to enrich chemical kinetics with new methods and new data.

In industrial applications, in the vast majority of cases, fuel and air are supplied separately to the furnace. Under these conditions the chemical reaction rate and heat release rate are determined primarily by the process of fuel and air mixing and depend little on the kinetics of the chemical reaction.

In the present work we consider only the combustion and detonation of gaseous explosive mixtures. The practical significance of these problems is related to accident prevention in working with explosive mixtures and to the combustion of gasoline-air mixtures in engines. In the latter case, incidentally, there are a number of peculiarities which we do not consider here.

Due to the fact that mixing of the fuel and air is carried out in advance, the rate of reaction in a gaseous mixture is basically determined by thermal factors and by the kinetics of the chemical reaction. It is precisely the clarification of the role and laws of chemical reaction in a flame and in an explosion that constitutes our task.

§I.2 Conditions of Chemical Reaction in a Flame

A flame propagating in an explosive mixture represents a thin zone which divides the cold, unreacted mixture in its original state from the reaction products in which all of the chemical energy has gone to thermal energy. We are given the state of the original mixture directly; the state of the products upon completion of the chemical reactions, restricted by mobile equilibria, may be calculated without any particular difficulty although, in practical terms, in the case of significant dissociation the calculation may be laborious.

How does the transition from one state to another occur in a flame? What is the width of the transition zone, and how long does the reacting substance remain in it? These quantities depend on the physical constants of the substance with which we are working, and on the velocity of propagation of the flame with respect to the gas.

Let us begin by determining the order of magnitude of the heating zone width.

Heinrich Hertz long ago solved the problem of a thermal wave in front of a heated surface which is moving with constant velocity. This solution, first applied to a flame by Michelsohn, has the form

$$T = T_0 + Ne^{-ux/\kappa}, \tag{I.2.1}$$

where T_0 is the temperature of the substance unperturbed by the heat wave, u is the velocity of the surface (the flame velocity in this case), x is the coordinate perpendicular to the flame front, κ is the coefficient of thermal diffusivity, i.e., the ratio of the thermal conductivity η, to the volumetric specific heat ρc_p,

$$\kappa = \frac{\eta}{\rho c_p}, \qquad (\text{I.2.2})$$

and N is a constant determined from the boundary conditions.

A solution of the form (I.2.1) is itself applicable only to that region of the "flame" in which the chemical reaction does not occur and there is no release of heat (Fig. 1—the regions indicated by the numbers are **1**: original mixture, **2**: heating zone, **3**: reaction zone, **4**: combustion products).

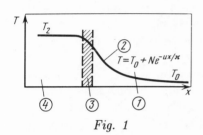

Fig. 1

As a scale for the zone width l we may take the distance at which the heating increases by $e \approx 2.7$ times:

$$l = \frac{\kappa}{u} = \frac{\eta}{\rho c_p u}. \qquad (\text{I.2.3})$$

Let us calculate this quantity for the most slowly burning mixtures with $u = 5$ cm/sec (6% methane, 94% air), and the fastest burning mixtures with $u = 1000$ cm/sec (the detonating mixture $2H_2 + O_2$); substituting constants for a temperature of about $500°\,C$, where the velocity is higher as well, we find

$$l = 0.06 \text{ cm } (CH_4); \qquad l = 0.0003 \text{ cm } (2H_2 + O_2).$$

In both cases the zone width is many times greater than the mean free path. This is as it should be, since the velocity is small compared to the speed of sound; as proof it is sufficient to substitute an expression from kinetic theory,

$$\kappa = \frac{\lambda c}{3}, \qquad (\text{I.2.4})$$

where c is the velocity of molecules, of the order of the speed of sound, and λ is the mean free path. This also proves the applicability of the differential equations for molecular transfer to a flame, which will prove to be of importance in what follows.

We obtain the order of magnitude of the time spent by the substance in the flame, τ_{Fl}, by division of the zone width, found above, by the flame velocity:

$$\tau_{Fl} = 4 \cdot 10^{-3} \text{ sec } \quad (CH_4); \qquad \tau_{Fl} = 10^{-7} \text{ sec } \quad (2H_2 + O_2).$$

This time also exceeds by many times the mean free time of molecules in the gas in accordance with the fact that any reaction, whether because of a significant heat of activation or because of a complicated mechanism, requires a large average number of collisions for one effective collision leading to reaction.

Comparison of the equations for thermal conductivity and diffusion shows that the temperature and composition change simultaneously as a result of the chemical reaction. But they also change simultaneously in the zone where there is no chemical reaction: the temperature changes due to thermal conduction, and the composition—to diffusion. If the physical properties (molecular weight) of the reagents and reaction products are close to one another, the kinetic theory of gases predicts that the coefficient of diffusion will be close in magnitude to the thermal diffusivity, and the relation between the temperature and the composition will prove to be especially simple: in a flame this relation is the same as if the reaction ran adiabatically, without heat or matter exchange with neighboring layers.

In particular, it is obvious that this relation implies conservation of the bulk energy of a unit of mass.

In the flame zone the increase in energy by heating of the mixture is compensated by the loss of the chemical energy carried away by diffusion with the original substances, the carriers of the chemical energy.

This statement was first made in 1934 by Lewis and Elbe in the form of a hypothesis that the sum of the thermal and chemical energy is constant.

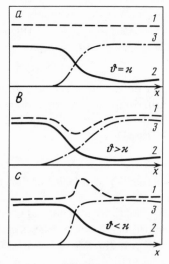

Fig. 2

They, however, associated it with the diffusion of active centers for a chain reaction whereas, in fact, variation of the sum of the thermal and chemical energy depends not on the reaction mechanism, but on the interrelation between the thermal diffusivity and the diffusion of the basic components participating in the reaction.

Therefore, only now, when we know the meaning of the law of constant total energy, do we know the conditions under which it is valid, as well as its exceptions. Thus, in a lean mixture of hydrogen with air, the diffusion coefficient of hydrogen (which may naturally be considered the carrier of the chemical energy) exceeds the thermal diffusivity of the mixture: diffusion carries away hydrogen from those layers which have not yet been heated. The energies of the initial and final states are equal by the energy conservation law (in these states the thermal and diffusion fluxes are absent), and in the intermediate states of a hydrogen–air mixture the bulk energy is at a minimum.

In contrast, in a mixture of a high-molecular-weight hydrocarbon, for ex-

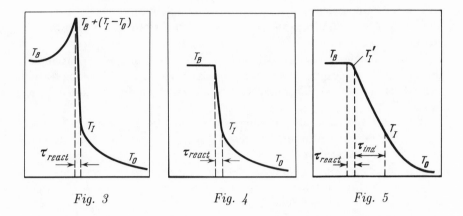

Fig. 3   Fig. 4   Fig. 5

ample benzene, with air, the thermal diffusivity is larger than the diffusion coefficient of the benzene vapors, and heating of the mixture occurs more intensively than the supply of benzene by diffusion to the flame zone; the energy is thus maximal in the combustion zone. The interrelation between concentration, temperature and energy in three typical cases is graphically presented in Fig. 2a–c (where 1 is the energy, 2 is the temperature and 3 is the concentration of the fuel).

We analyzed above the interrelation between changes in the temperature and composition. By what law does each of these quantities change individually? In the old literature the question of the temperature distribution in a flame was widely discussed. Michelsohn considered a chemical reaction which, below a certain temperature T_I (the ignition temperature), does not run at all, but beginning with this temperature it runs extraordinarily rapidly. He supposed that in such a reaction the temperature distribution shown in Fig. 3 would apply: the mixture is heated from T_0 to T_I; it then reacts rapidly, developing a temperature which corresponds to combustion of a mixture heated to T_I, i.e., a temperature which is higher than the final combustion temperature T_B. Haber disputed Michelsohn's arguments: heating from T_0 to T_I occurs due to the reaction heat and, therefore, in the reaction of the heated mixture part of the heat is in turn spent on heating the next layer of the mixture to T_I. The loss of thermal energy to the heating of the layers ahead balances the gain from combustion of a preheated mixture. In Haber's opinion, the combustion temperature is not exceeded and the distribution has the form of Fig. 4.

Haber is right regarding the maximum temperature, and Fig. 4 is closer to the truth than Fig. 3. The correction which Fig. 4 requires is related to the fact that rapid heat release leads not to a rapid change in the thermal

flux, i.e., not to a large temperature gradient, but to a large second derivative of the temperature. An approximate distribution is shown in Fig. 5.
The place where rapid reaction occurs is characterized by a sharp bend in
the temperature curve.[2]

Thus, a chemical reaction in a flame occurs in a mixture which has already
been heated by thermal conduction and which has changed its composition:
it has been diluted by the combustion products which diffuse from adjacent
layers.

A distinctive feature of the kinetics of a reaction in a flame is the freedom of the reaction from any delaying stages: whereas in self-ignition the
stages of accumulation of heat in the
system and of active centers delay development of the process, in a flame
there is no slow accumulation of heat.
In the temperature interval in which
heating of the mixture by the reaction heat would occur slowly, the thermal flux from neighboring layers in fact
causes a rapid rise in the temperature.
Just as for the heat, active centers diffuse from layer to layer as well. In a
flame the entire interval of temperatures is realized, from the initial temperature to the temperature of combustion, and the reagent concentration
varies simultaneously with the temperature. It is extremely important for us
to trace here the variation of the chemical reaction rate.

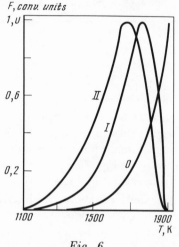

Fig. 6

In Fig. 6 the dependence of the reaction rate on the temperature is
presented for a typical case: $T_0 = 300°K$, $T_B = 1900°K$, activation heat
$A = 50,000$ cal/mole. The three curves correspond to three different assumptions about the dependence of the reaction rate on the concentration:
independence *(0)*, direct proportionality—a monomolecular reaction *(I)*,
and proportionality to the square of the concentration—a bimolecular reaction *(II)*. As is clear from the drawing, in all the cases the reaction rate is
large for temperatures close to the combustion temperature. This circumstance will form part of the foundation of a theory of flame velocity.

---

[2]Neither Haber nor Michelsohn took into account the fact that the concept of "instantaneous" reaction following immediately after the ignition temperature is attained is
incompatible at $T_I < T_B$ with a finite flame propagation velocity. In Fig. 5 the temperature distribution is given for a rapid chemical reaction which occurs over a very small time
τ_R. In order to reconcile the small reaction time with a finite flame velocity, we have to
assume that rapid chemical reaction begins only after a certain period of induction τ_{Ind}
following attainment of the ignition temperature T_I, or else that the ignition temperature
T_I' is very close to the combustion temperature (see Fig. 5).

There is a certain difficulty in understanding a basic fact of combustion theory, that the reaction always occurs primarily at a temperature close to the combustion temperature.

Considering a mixture which attains 2000°K during combustion, we assert that for a certain A noticeable reaction occurs in the interval 1700–2000°K. Then, for example, by raising the initial temperature we achieve an increase in the combustion temperature to 3000°K, and we shall assert that the reaction now occurs in the temperature interval 2300–3000°K, and ignore any reaction at temperatures under 2300°K. On what basis, in a reaction which develops a temperature of 3000°K, do we now ignore chemical reaction at 1700–2000°K, which we earlier (for a mixture which develops 2000°K) considered to be rapid?

A higher combustion temperature, while it does not reduce the rate of reaction at 1700–2000°K, creates the opportunity for even faster reaction at higher temperatures. As a result the propagation velocity of the flame increases significantly and the time during which the temperature varies in the interval 1700–2000°K is sharply reduced. In a slowly burning mixture (at low combustion temperature) this time is sufficient for transformation of 100% of the mixture; in a rapidly burning mixture, the time during which the mixture is at 1700–2000°K is reduced so much that only a small part of the mixture reacts: the flame velocity corresponds to a more rapid reaction at the higher temperature, 2300–3000°K.

§I.3 A Theory of the Flame Velocity

The velocity of a flame with respect to a gas is a basic characteristic of the combustion process, and calculation of the velocity or analysis of the relation of the flame velocity to the chemical reaction rate and the thermal properties of the mixture are very important tasks of the theory.

The general equation of thermal conduction in a system of coordinates associated with the flame may be written as follows:

$$-\rho u c_p \frac{dT}{dx} = \frac{d(\eta dT/dx)}{dx} + F, \qquad (I.3.1)$$

where T is the temperature, η is the coefficient of thermal conductivity, and F is the volumetric rate of the heat release. Relating the reagent concentration to the temperature, we may represent the reaction rate and the rate of heat release as functions of the temperature alone. We then obtain a second-order equation which is nonlinear due to the nonlinear dependence $F(T)$: the combustion rate u enters the equation as a parameter, and only for one value of this parameter is it possible to satisfy both boundary conditions at $x = \pm\infty$.

Attempts at solution reported in the literature are based on linearization of the equation. An expression for the reaction rate is chosen such that the equation becomes linear and allows a simple solution. This requires that the Arrhenius dependence of the reaction rate on the temperature be abandoned.

We have taken a different approach, using precisely the property of the Arrhenius law which is typical for a chemical reaction.

As may be seen from Fig. 6, rapid chemical reaction occurs in a narrow temperature interval adjacent to the combustion temperature; we will denote the width of this interval by θ. Thus, the interval of interest extends from T_B to $T_B - \theta$. The heat released in the chemical reaction is partially spent on heating the reacting mixture itself, and partially on heating the unreacted mixture from the initial temperature to the temperature at which the reaction begins, i.e., from T_0 to $T_B - \theta$.

The two terms of the differential equation correspond precisely to these two parts of the energy release: the term $-\rho u c_p dT/dx$ represents the energy spent on heating the reacting mixture, the term $d(\eta dT/dx)/dx$—the energy carried away by heat conduction. Because the interval in which the reaction occurs is narrow,

$$\theta < (T_B - T_0), \tag{I.3.2}$$

we may neglect the energy spent on heating the reacting mixture in the reaction zone and omit the corresponding term in equation (I.3.1):

$$\frac{d(\eta dT/dx)}{dx} + F = 0; \qquad T_B > T > T_B - \theta. \tag{I.3.3}$$

Solution of this equation (the thermal conductivity η may be considered practically constant in the reaction zone) yields an expression for the heat flux,

$$\eta\left(\frac{dT}{dx}\right) = \left(2\eta \int F \, dT\right)^{1/2}. \tag{I.3.4}$$

If we neglect the thermal energy spent on heating the reacting mixture itself, then all the heat of reaction is carried away by thermal conduction, and we must equate the heat flux to the overall amount of heat released in the flame per unit time, i.e., to the product of the volumetric heat capacity of the mixture, ρQ, and the flame velocity:

$$u\rho Q = \left(2\eta \int F \, dT\right)^{1/2}; \qquad u = \frac{\left(2\eta \int F \, dT\right)^{1/2}}{\rho q}. \tag{I.3.5}$$

The integral of the heat release rate with respect to temperature is taken over the entire region in which the reaction rate is non-zero. In fact, due to the rapid decrease in the reaction rate as the temperature falls, the basic contribution to the integral is made by a comparatively narrow temperature interval adjacent to the combustion temperature. Substituting the appropriate law for the reaction rate (mono- or bimolecular, etc.), we obtain the corresponding, rather cumbersome expressions for the flame velocity which we do not give here. The basic formula (I.3.5) allows us to analyze the influence of various parameters of the mixture on the velocity.

At first glance it is surprising that the heat capacity of the mixture, Q, enters the denominator in the formula for the velocity since we know that as the heat capacity increases, so does the flame velocity. The fact is that for a constant initial temperature, growth of the heat capacity of the mixture induces an increase in the combustion temperature, but the increase of the integral of the heat release rate outweighs the increase in Q as the temperature is raised.

In unpublished experiments by the late P. Ya. Sadovnikov (Institute of Chemical Physics), the combustion velocities of explosive mixtures of carbon monoxide with air, diluted by the combustion products, were compared. The diluted mixtures were preheated so that their combustion temperatures[3] did not differ from the combustion temperature of the undiluted mixture. These experiments confirmed with sufficient accuracy the relation required by the theory

$$u\rho Q = \text{const} \tag{I.3.6}$$

for $T_B = \text{const}$.

The influence of the initial temperature of the mixture on the combustion rate depends on the change in the combustion temperature and the heat of activation of the reaction occurring in the flame: if the rate of heat release depends on the temperature according to the Arrhenius law,

$$F = fe^{-A/RT}, \tag{I.3.7}$$

then for the flame velocity we obtain the expression,

$$u = \left(\frac{G}{\rho Q}\right)e^{-A/2RT_B}. \tag{I.3.8}$$

This rational formula replaces formulas proposed at various times which originate from the concept of an ignition temperature T_I and which disagree with experiment:

$$u \sim \frac{T_B - T_I}{T_I - T_0} \qquad \text{(Mallard and Le Chatelier),} \tag{I.3.9a}$$

$$u \sim \left(\frac{T_B - T_I}{T_I - T_0}\right)^{1/2} \qquad \text{(Jouguet),} \tag{I.3.9b}$$

and the empirical formulas,

$$u \sim T^2 \qquad \text{(Passauer),} \tag{I.3.9c}$$

$$u = hg^{T^{3/4}} \qquad \text{(Tamman and Thiele).} \tag{I.3.9d}$$

Belyaev's experiments on the combustion of methylnitrate have confirmed formula (I.3.8) over a broad temperature interval, and the activation heat calculated in the dependence of the combustion velocity on the temperature

---

[3]Since we are located far from the boundary we may neglect the influence of heat transfer in the process. We equate the combustion temperature T_B with the so-called theoretical combustion temperature T_T, calculated from thermodynamic data under the assumption of adiabatic combustion. For more detail, see § I.4 and I.5.

coincides with the activation heat found by Apin in his study of the kinetics of decomposition of methylnitrate at lower temperatures.

Table 1   Influence of Combustion Products on Flame Velocity

Products, %	Mixture, %	Velocity, cm/sec
Mixture I (50% CO, 50% air, 2% H_2O)		
0	100	62
10	90	50
20	80	40
30	70	25
40	60	17
Mixture II (30% CO, 70% air, 2% H_2O)		
0	100	40.5
10	90	32
30	70	13

For gaseous mixtures a methodologically convenient technique for varying the combustion temperature in order to determine the activation heat is the dilution of the mixture by its combustion products. Here a change in the combustion temperature is achieved without changing the composition of the combustion products. Table 1 shows the data obtained by Sadovnikov. This data is presented graphically in Fig. 7.

The activation heats found are equal to 25,000 cal/mole (Mixture I) and 38,000 cal/mole (Mixture II). The dependence of the activation on the composition indicates that we are dealing with a complex chemical reaction. It is characteristic that admixture of the reaction products produces no specific catalytic effect (positive or negative) on the flame velocity, although in experiments on self-ignition such effects have been observed. The reason is that in a flame the reaction occurs in a zone where in all cases (including combustion of an undiluted mixture) a high concentration of the combustion products is attained.

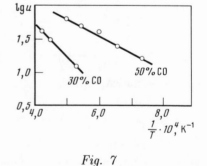

Fig. 7

Knowledge of the heat of activation allows us to determine the temperature interval in which the chemical reaction occurs and the order of magnitude of the volumetric rate of heat release in the zone of highest reaction rate.

This last quantity is of particular interest for thermal engineering since we determine the upper intensification limit of fuel-burning devices depending on the chemical reaction rate. Barskiĭ and Zeldovich determined this quantity for a number of carbon monoxide mixtures: the heat release rate exceeds 10^9 kcal/hr · m^3. For purposes of comparison we note that in present-day industrial furnaces and reactors, the heat release rate taken with respect to the entire volume of the furnace wavers between $2 \cdot 10^5$ and $5 \cdot 10^6$ kcal/hr·m^3.

Finally, most interesting is the dependence of the flame velocity on the composition of the mixture. In the literature one encounters simple formulas whose authors proceed from naive conceptions of the chemical kinetics of combustion reactions and do not take into account that the combustion temperature also depends on the composition. The formula of Stevens may serve as an example: $u \approx [CO]^2[O_2]$.

In reality, in studying the dependence of the flame velocity on composition, it is necessary to distinguish the influence of components whose concentration exceeds or falls short of the stoichiometric amount.

Thus, comparing the flame velocity in two mixtures—I: 50% CO, 10% O_2, 40% N_2 and II: 50% CO, 20% O_2, 30% N_2—we will not be able to determine the influence of the oxygen concentration on the reaction rate since the increase in oxygen concentration will cause a very large increase in the combustion temperature.

We may determine the influence of the oxygen concentration by comparing mixtures in which oxygen is not the deficient component, for example, I: 20% CO, 10% O_2, 70% N_2;   II: 20% CO, 30% O_2, 50% N_2; III: 20% CO, 80% O_2. In this series of compositions the combustion temperature is constant, as is the carbon monoxide content in the reaction zone.

Barskiĭ's experiments showed that the flame velocity, and consequently the reaction rate, are practically independent of the oxygen concentration: in these experiments the oxygen concentration in the reaction zone varied from 2% (I) to 72% (III), while the flame velocity varied by less than a factor of 1.5.

Analogous experiments with mixtures of identical combustion temperature containing various excess amounts of carbon monoxide showed that the flame velocity is proportional to $[CO^*]^{1/2}$, where $[CO^*]$ is the carbon monoxide concentration in the reaction zone. From this it follows that the chemical reaction in a flame is first order in carbon monoxide. The role of water in the combustion of carbon monoxide is well known. Analysis of available data shows that the flame velocity is proportional to $[H_2O]^{1/2}$, i.e., the reaction is first order in water vapor content. The influence on combustion of such flegmatizers as CCl_4 may be ascribed to the binding of hydrogen by halogen with the formation of a molecule of HCl, which is dissociable only with difficulty. However, the latest experiments by Kokochashvili in our laboratory show that the influence of the halogens is more complex: Kokochashvili showed that addition of HCl also noticeably lowers the flame velocity. In this case we cannot speak of hydrogen binding; the role of HCl can only lie in the transformation of active forms of hydrogen to less active ones, for example, by the reaction

$$H + HCl = H_2 + Cl - 1.2\,kcal,$$
$$OH + Cl = O + HCL + 2.7\,kcal.$$

On the other hand, in mixtures which contain neither water vapor nor hydrogen, HCl plays the role of a hydrogen supplier: without HCl the mixture does not burn, and the reaction rate at low content grows with an increase in the concentration of HCl. The activation heat of the combustion reaction of such mixtures, determined from the influence of the reaction products on the flame velocity, proves to be magnified in comparison to mixtures containing H_2O or H_2, in accordance with the more difficult dissociation of HCl.

Thus, measurement of the flame velocity allows us to establish, step by step, the kinetic patterns of a chemical combustion reaction, while the theory of combustion serves as a link between the measurement and its kinetic explanation.

Compared to ordinary kinetic experiments the study of flame velocity has disadvantages: we cannot simultaneously and independently vary the temperature and concentration of all the reacting components; peculiar conditions interrelate, for example, the temperature and concentration of a deficient component. However these disadvantages are expiated by the tremendous expansion of the field of kinetic studies: in a flame we study reactions which occur at 1500–3000°K and which are so rapid that the reaction time lies within the limits 10^{-3}–10^{-6} sec. This expansion of the kinetic experimental region, undoubtedly, will yield a number of new laws of particular interest in connection with the fact that at high temperatures the equilibrium concentration of active centers of the reaction becomes significant.

Let us mention certain transformations of the expressions for the propagation rate (3.5). Since, by the Arrhenius law, F depends exponentially on the temperature, the integral in (I.3.5) may be replaced in order of magnitude by the product of the maximum value of F, which we denote by F_{max}, and the effective width of the temperature interval θ_{eff}. This latter, in the case of Arrhenius dependence on the temperature, is expressed by the formula

$$\theta_{eff} = \frac{RT_T^2}{A}. \tag{I.3.10}$$

Thus we obtain the simplified equation

$$u\rho Q = T_T \left(2\eta F_{max}\frac{R}{A}\right)^{1/2}. \tag{I.3.11}$$

Using this formula, in a joint paper with Barskiĭ on the basis of measurements of the flame velocity and activation heat determined from dilution by the reaction products, we found the value F_{max} for several carbon monoxide mixtures (Table 2).

Table 2

Mixture Composition	u, m/sec	T_T, °K	θ, deg	η, kcal/ /hr $\cdot$ m $\cdot$ deg	A, kcal/mole	F_{max}, 10^9 kcal/h $\cdot$ m^3
I	0.74	3000	240	0.089	75	140
II	0.26	2370	250	0.079	46	15
III	0.27	2030	420	0.072	20	5

Mixture I: $2CO + O_2$, 2% H_2O
Mixture II: (29% CO, 71% air, 2% H_2O)
Mixture III: (50% CO, 50% air, 2% H_2O)

In the calculation of T_T and F_{max} for stoichiometric mixtures the dissociation is taken into account. The values u, ρ and Q are taken with respect to the original mixture; the product $u\rho$ and the quantity Q do not depend on whether they are taken with respect to the original mixture at room temperature or to the hot and expanding reaction products. The quantities η and F_{max} are given in technical units; they are related to the state of the mixture in the reaction zone, i.e., to temperatures close to T_T. In order to understand the quantity F_{max} which is found, we note that in modern boilers which burn powdered coal the average value of the heat release rate in the furnace is $F = 300,000\,kcal/h \cdot m^3$, and in the most forced furnaces using liquid fuel it may reach $F = (10\text{--}20) \cdot 10^6\,kcal/h \cdot m^3$. Comparison of these values shows the enormous intensity of chemical reaction in a flame. Existing fuel-burning devices lag far behind this intensity—only a miniscule part of the furnace volume is occupied by actually reacting gas. The chemical kinetics of combustion reactions give heat engineers practically unlimited possibilities for further intensification which in reality is basically limited by other factors.

From the technology of combustion we move to the molecular mechanism of flame propagation. We shall give a molecular-kinetic expression for the heat release rate by calculating the frequency ν of collisions of fuel molecules with other molecules (ν is proportional to the molecular velocity and inversely proportional to the mean free path), further taking into account that only a small $(1/\nu)$ part of all collisions are effective. The quantity $1/\nu$—the probability of reaction taken with respect to a single collision—depends on the activation heat of an elementary reaction event, as well as on the fraction of all molecules comprised of those radicals or atoms by means of which the reaction occurs. The molecular-kinetic expression for the coefficient of thermal conductivity follows from formulas (I.2.4) and (I.2.3).

After simple transformations we obtain the result in the following form:

$$u = \frac{c\varphi}{\sqrt{\nu_{\min}}}, \tag{I.3.12}$$

$$l = \frac{\lambda\sqrt{\nu_{\min}}}{3\varphi}. \tag{I.3.13}$$

We recall that c is the velocity of the molecules. The index on ν means that we calculate the number of collisions necessary for reaction in the part of the zone where the reaction rate is highest and conditions are most conducive, so that $\nu_{\min}$ is the minimum value of ν. Finally, φ is a dimensionless quantity of order (but less than) unity, algebraically (but not exponentially) dependent on the reaction mechanism, the activation heat, the temperatures T_0 and T_B, and the reagent concentrations. From the formula it is obvious first of all that u is always many times smaller than c, and less than the speed of sound. This fact will be important for the theory of detonation (Part II).

Turning from the factor φ, we recognize in (I.3.12) the Einsteinian expression for the average velocity of a molecule undergoing $\nu_{\min}$ collisions, and in (I.3.13)—the Einsteinian expression for the mean path traveled by the molecule over $\nu_{\min}$ free paths. This result is in complete agreement with the molecular picture of flame propagation. A molecule of fuel sustains $\nu_{\min}$ collisions in the reaction zone; over this time it travels a path l with an average velocity u. At the end of this path the fuel molecule reacts and thereby raises the temperature, making it possible for the next fuel molecules to react, and so on.

The average velocity of a certain number of given "marked" molecules decreases over time, approaching zero as $t^{-1/2}$, according to Einstein, since the number of collisions sustained by the molecules grows proportionally to t.

A constant, non-decaying velocity of flame propagation is ensured by the relay organization of the process: each molecule of fuel undergoes, on the average, $\nu_{\min}$ collisions in the reaction zone, after which it reacts and "passes the baton" to other molecules, which begin the path with fresh strength.[4]

A crucial quantitative verification of the theory of flame velocity was performed in 1942 in processing data from the dissertation of Kokochashvili who was studying the combustion of mixtures of hydrogen and bromide in our laboratory.

The mechanism of the reaction of hydrogen with bromide is known today in complete detail: all the individual stages of the reaction, their probabilities, and activation heats are known. Because of this, it was possible to calculate the flame propagation velocity from kinetic data after the combustion temperature had been found by thermochemical computations, and with

---

[4]In an unpublished paper by Skalov (Institute of Chemical Physics) similar considerations were applied earlier to the theory of propagation of an auto-catalytic reaction by diffusion of the active product, without accounting for the reaction heat.

the aid of the kinetic theory of gases the heat conductivity of the reacting mixture was found.

The calculation gave a combustion rate which differed by 15% from the experimental one. Thus, for the first time, the combustion rate was calculated from independent data, and for the first time in this example the practical possibility was demonstrated of reducing the laws of flame propagation to the laws of phenomena which lie at the basis of the process, i.e., to the laws of chemical kinetics and of heat conductivity and diffusion.

§I.4 Limits of Propagation and Critical Diameter. History of the Modern Theory of Combustion

Together with the theory of the flame propagation velocity, we are also faced with the very important question of limits, i.e., whether a given mixture under particular conditions will burn and what will determine the combustibility of the mixture?

We shall begin analysis of this question with the most fundamentally simple phenomenon, that of the critical diameter. The same mixture may burn in a wide tube, but not support flame propagation in a narrow tube whose diameter is less than a certain *critical diameter*. This phenomenon was discovered by Humphrey Davy in 1816 and became the basis for construction of a safe miner's lamp in which a copper mesh with small openings prevents the possibility of flame propagation from the inside of the lamp to the atmosphere of the mine. In this case the mixture in the narrow tube is combustible (an experiment in a wide tube confirms this), and if combustion is nevertheless impossible, then this is so because of the increase in heat loss with the decrease in the diameter of the tube. By heat loss we mean heat carried to the walls of the tube from the reaction zone and the heating zone of the mixture. The heat transferred from the reaction zone into the unburned mixture cannot be considered a loss (as was done by Holm) since this heat returns to the reaction zone in the form of thermal energy of the heated mixture.

The transfer of heat from layer to layer is a natural mechanism of flame propagation. The cooling of the combustion products which follows combustion does not by itself inhibit combustion.

By heat losses capable of stopping propagation we mean those processes which lower the combustion temperature. These include: (1) heat transfer from the heated explosive mixture to the tube walls, (2) evacuation of heat from the reaction zone itself to the walls of the container and (3) evacuation of heat from the reaction zone to the combustion products which depends on the cooling of the reaction products—cooling of the products creates a drop in temperature as one moves away from the reaction zone and a corresponding longitudinal (in the direction opposite the direction of propagation) heat flux.

Study of these sources of heat loss leads to the conclusion that the decrease in the combustion temperature T_B compared to the theoretical combustion temperature T_T depends on the flame velocity: the lower the flame velocity, the more the temperature falls for a given heat transfer rate. As it happens,

$$T_B = T_T - \frac{a}{u^2}, \tag{I.4.1}$$

where u is the flame velocity and a is a constant which depends on the conditions of heat transfer (tube diameter, heat conductivity of the gas). Combining (I.4.1) with the known dependence of the flame velocity on the combustion temperature,

$$u = be^{-A/2RT_B} \tag{I.4.2}$$

(which follows from the idea that the chemical reaction runs basically at a temperature close to T_B), allows us to construct a simple theory of the limit of flame propagation. It turns out that the two equations (I.4.1) and (I.4.2), by means of which we must determine the two unknown quantities—the combustion rate and the combustion temperature T_B—have a real solution only if the intensity of heat transfer (characterized by the coefficient a) does not exceed a certain value,

$$a < \left(\frac{RT_T^2}{Ae}\right) b^2 e^{-A/RT_T} = a_{\text{crit}} \tag{I.4.3}$$

If the inequality (I.4.3) does not hold, combustion is impossible. If (I.4.3) does hold, solution of equations (I.4.1) and (I.4.2) gives a completely definite decrease in the temperature and combustion rate. It is convenient to introduce a theoretical combustion rate—that which would occur in combustion without losses at $a = 0$, for which $T_B = T_T$,

$$u_T = be^{-A/2RT_T}. \tag{I.4.4}$$

It turns out that throughout the region of variation of the coefficient a from 0 to its critical value a_{crit} the combustion rate falls from

$$u = u_T \tag{I.4.5a}$$

(for $a = 0$) to the minimum value

$$u = \frac{u_T}{\sqrt{e}} = 0.61\, u_T \quad \text{for} \quad a = a_{\text{crit}}. \tag{I.4.5b}$$

Further, using the dependence of the coefficient a on the tube diameter, we obtain the dependence of the critical diameter on the properties of the explosive mixture,

$$d_{\text{crit}} = \frac{\text{const}}{u_T p} \tag{I.4.6}$$

These conclusions agree qualitatively with experiment: the critical diameter (as was shown by experiments carried out in our laboratory by Shaulov) is smaller for rapidly burning mixtures.

As the heat transfer increases, the flame velocity decreases, but before the decrease becomes significant propagation stops. Quantitative verification is hampered by complicating factors: convective motion of the gas, which bends the flame front, causes a difference between the theoretical rate of flame motion with respect to the gas and the measured rate of flame motion with respect to the walls.

In the present physico-chemical study we do not discuss our results relating to the theory of flame motion in the presence of convection.

Results pertinent to the theory of critical diameter are contained for the most part in earlier works by English authors. Despite his erroneous assumptions, Holm obtained the correct relation between the critical diameter and the flame velocity (I.4.6). The remarkable work by Daniell on the theory of flame propagation contains an analysis of the influence of heat losses. The losses enter directly into the equation describing the temperature distribution in the flame zone. A solution exists only for heat losses which do not exceed a certain limit, and under critical conditions (at the limit of propagation), the flame velocity drops to a certain fraction (40–50%) of the theoretical flame velocity. Daniell was also the first to indicate definitely that the flame velocity cannot be constructed from thermal quantities alone and by dimensional considerations must be proportional to the square root of the reaction rate.

A general flaw in all of Daniell's work is his neglect of diffusion and an absolutely implausible assumption about the chemical reaction rate: it is assumed that, above the ignition temperature, the reaction rate does not depend on the temperature, but depends only on the reaction time.

In fact the interval of variation of the reaction rate in a flame is so large that the formulas into which the reaction rate enters do not even determine the order of magnitude of the flame velocity and critical diameter until the conditions—temperature and concentration—to which the reaction rate relates are indicated. By entering the dependence of the reaction rate on the temperature into Daniell's equations, we render them unsolvable.

Turning to older works, we note the little-known, but wonderful studies by Taffanel, carried out in the years preceding the First World War at an experimental mining research station in Levene (France), and published in the form of several short notes in "Comptes Rendus Aca. Sci."

In these notes we find the conditions of self-ignition for an explosive mixture introduced into a heated vessel. Taffanel compared the rate of heat release in a chemical reaction of a mixture with the rate of heat transfer by a gas mixture to the vessel walls. Both quantities depend on the temperature (the heat transfer is proportional to the difference between the temperatures of the gas and vessel). In order for self-ignition to occur, it is necessary that the curve of the heat release remain always above the line of heat transfer. The same interpretation, leaving far behind the vague notions of Vańt Hoff, is given in a monograph by Jouguet in which he cites the lectures of Le Chatelier, lithographed in 1912.

Semenov arrived at an analogous interpretation of self-ignition independently, starting from a thermal theory of electric breakdown which he developed. The success and broad recognition of Semenov's work, which became the point of departure for Soviet work on combustion, arose from the fact that a correct fundamental scheme was brought together with general kinetic dependences (Arrhenius's law of the chemical reaction rate) and with the subsequent intensive work of Semenov's school on self-ignition phenomena of the most important gas systems.

Taffanel used measurements of the chemical reaction rate at temperatures lower than the temperature of self-ignition, and measurements of the time of self-ignition at a higher temperature, in order to determine the dependence of the heat release rate on the temperature and concentration. Further, Taffanel introduced measurements of the flame propagation velocity. He compared experimental data with the theoretical calculation, carried out under the assumptions of a constant chemical reaction rate in the interval from T_B to $T_B - \theta$ and the absence of chemical reaction at all lower temperatures, also ignoring the Arrhenius dependence of the reaction rate on the temperature and the variation of the concentration.

Taffanel's results are summarized in Fig. 8, borrowed from his paper. The temperature is plotted on the abscissa; along the ordinate, on a logarithmic scale, a quantity equal in our notation to $RT/273$ is plotted, where R is the volumetric rate of heat release in the reaction of methane and oxygen and T is the absolute temperature of the reaction.

In his last note Taffanel approximates the law found for the reaction rate by the exponential (non-Arrhenian) expression $\exp(T/a)$ and finds the propagation velocity corresponding to these reaction kinetics. His formulas are close to our formulas in § I.3.[5] Taffanel notes that, due to the rapid decrease of the exponential function in the integration of the reaction rate, only the high-temperature region is significant. Taffanel left unsolved the problem of the concentration of reagents in the reaction zone, and did not even raise the question of diffusion in a flame.

Taffanel held the view that the limit of propagation is caused by the existence of thermal losses and that, as a result, a certain minimum chemical reaction rate is required in order for combustion to occur. He further showed that under certain instrumental conditions, as the initial temperature is raised, increasingly diluted mixtures are capable of burning, and that in all the cases at the limit the combustion temperatures are practically the same. We have no information on the external reasons why Taffanel's work stopped. The internal, fundamental reasons which prevented him from developing an orderly and complete theory of combustion processes lay in the undeveloped

---

[5]The author found Taffanel's work after publication of [1].

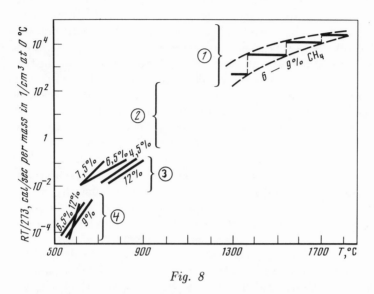

Fig. 8

Methane and air — **1**: Studies based on propagation velocity, **2**: Not studied, **3**: Studies based on ignition delays, **4**: Slow reaction without self-ignition (direct measurements).

state of chemical kinetics at the time.

Taffanel's indisputable achievement is his statement of the problem of flame velocity in a mixture characterized by a specific smooth dependence of the reaction rate on the temperature, whereas many authors, both before and after Taffanel, based their calculations on the concept of an ignition temperature at which there was supposedly a jump in the reaction rate. This achievement is a consequence of the theory of self-ignition which Taffanel developed in which the temperature of ignition depends on the interrelation between the continuously varying heat release rate and the conditions of heat transfer.

In 1934 Lewis and Elbe again proclaimed rejection of the use of the ignition temperature in the theory of flame propagation. Above, in §2, we mentioned these authors' hypothesis regarding concentration propagation in a flame, which was correct in the case of a mixture of gases of the same weight. Independently of the concrete analytical calculations, this paper played a significant role in the creation of the combustion theory.

It would seem that the concept of ignition temperature as a basic kinetic characteristic of a mixture, and a crucial one in the propagation of the flame as well, is revived in the theory developed by Semenov of branching chain reactions. We recall that in this theory at temperatures below the ignition temperature chain breaking prevails over chain branching, and reaction is completely absent. Above the ignition temperature branching prevails, and the number of active centers and the reaction rate grow uncontrollably and lead to ignition. At the so-called upper limit, where both branching and breaking of chains depend on homogeneous chemical reactions, the ignition temperature is indeed a physico-chemical (kinetic) constant of the explosive mixture which is independent of instrumental conditions. Semenov was able to explain quantitatively and in complete detail the remarkable phenomena which he and his collaborators observed in the ignition of sulphur, phosphorus, hydrogen, carbon monoxide and many other substances.

Detailed analysis shows that above the ignition temperature the increase in the concentration of active centers occurs the more rapidly the farther away we are from the limit, i.e., from the ignition temperature at which branching and breaking are exactly balanced. In a flame the time that the mixture remains in each particular temperature interval is limited. Therefore it is not the critical conditions, at which slow growth of chains becomes possible, but conditions at much higher temperatures, close to the combustion temperature at which chain branching occurs at an enormous rate, that determine the reaction rate in the flame and the flame propagation velocity.

Extensive research on the kinetics of gas reactions in various countries, illuminated by Semenov's chain theory, showed that the great majority of reactions, particularly those in systems where combustion is possible, occur according to a chain mechanism. Very recently N. N. Semenov has confirmed, using simple, convincing arguments, that the chain mechanism in the majority of reactions is not accidental, but lies in the nature of things, and depends on the deepest and most general interrelations between the energy of chemical bonding and the heat and activation energy of reactions.

We again note that the thesis of the decisive role of the chemical reaction in the zone of highest temperatures, which lies at the basis of the combustion theory developed here, remains fully applicable to chain reactions with both branching and non-branching chains. It is precisely this fact which has allowed us to successfully apply the general laws of the theory of combustion to interesting, non-artificial, and practically important systems.

§I.5   Concentration Limits

We considered above the limit of propagation which depends on the heat transfer by molecular heat conduction to the walls of the tube in which the combustion is being studied, and so depends on the diameter of the tube.

Experiment shows that as the diameter of the tube is increased, propagation is enhanced, but only to a certain limit which cannot be extended by further increase of the diameter.

Thus we establish the concept of the concentration limit.[6] Investigation of these limits was carried out on a broad scale in connection with the requirements of accident prevention, so that the region of concentrations in which mixtures of gases and vapors with oxygen and air are explosive has been established for a large number of substances.

However, no theory of the concentration limit has existed until recently. The opinion was expressed that the limits correspond to the composition at which the flame velocity goes to zero (Bunte, Jahn) since the combustion temperature is equal to the ignition temperature (Jouguet).

The theory of flame propagation developed above leads us to an expression for the flame velocity which does not go to zero even for very diluted mixtures; calculations for definitely incombustible mixtures lead to a very small, but nonetheless non-zero propagation velocity.

Experiments by Payman showed that the most diluted mixtures which are still combustible yield a flame propagation velocity in a pipe of order 12–20 cm/sec, and Wheeler's empirical rule gives a velocity of 15 cm/sec. The corresponding flame propagation velocity for a gas (after accounting for the influence of convection) is of order 3–8 cm/sec. Thus, naive notions of the concentration limits are refuted by both theory and experiment.

Just what is the reason for the impossibility of propagation of a slower flame in more diluted mixtures? In our opinion, heat loss by radiation is such a reason common to different chemical substances.

Chemoluminiscent radiation as shown by spectroscopic analysis of a flame is so small that in calculations of thermal losses it may be neglected;[7] we will be interested in thermal radiation, whose intensity is determined by the composition and temperature of the gas. Transparent, colorless gases radiate in the infrared part of the spectrum. Significant and well-studied radiation is given off by water vapors and carbon dioxide.

In a first approximation, for the dimensions of the radiating layers with which we must work, we may consider that each element of volume radiates independently of the others, and we thus have the conditions to which the calculations of the previous section refer. For mixtures of carbon monoxide, in which the flame velocity has been studied in detail, we carried out detailed calculations of the influence of radiation on the combustion temperature.

These calculations, necessarily rough, showed that the heat transfer by

[6]We distinguish here a lower limit, which depends on the fuel deficiency, and an upper limit, which depends on oxygen deficiency in mixtures with an overabundance of fuel. For fluids we also give the temperature at which the saturating vapor tension becomes sufficient to form an explosive mixture at atmospheric pressure.

[7]We are interested in the combustion of ordinary explosive mixtures at pressures on the order of one atmosphere. We do not consider here rarified flames, for example, the combustion of sodium in chlorine, etc., where the quantum output of radiation may be significant.

radiation makes flame propagation with a velocity of less than 2 cm/sec impossible. In evaluating the agreement with Wheeler's rule and Payman's data (3–8 cm/sec), one should keep in mind that our result relates to a specific mixture: the minimum velocity depends on thermal and kinetic parameters of the reacting mixture. In mixtures of carbon monoxide with air we were able to carry the calculations up to determination of the concentration limits and obtained (at a humidity corresponding to room temperature) a lower limit of 10–13.5% CO, and an upper limit of 81–87.5% CO. The values given in Landolt's reference book vary within the limits 12.5–13.6% CO for the lower and 70–85% CO for the upper limit.

Thus, considering the low accuracy of the calculations and the inexactness of experimental determination of concentration limits, the agreement must be deemed satisfactory. In evaluating the character of the agreement, it should be kept in mind that we calculate the concentration limits from completely independent data on the combustion velocity, gas radiation and thermo-chemical constants. Meanwhile, to date, calculations have not gone beyond determination of limits for mixtures of several fuels on the basis of the limits of the individual fuels (Le Chatelier's rule).

How are the basic dependencies of the concentration limits to be interpreted from our point of view?

The influence of a number of chemical substances on these limits is known and has been studied in extraordinary detail: the most active influence is exerted by the halides, whose admixture in small amounts makes mixtures of carbon monoxide, hydrogen and hydrocarbons with air incombustible. Yet the thermal properties of the mixtures change comparatively little. In our laboratory it was shown that the influence of admixtures on the limits is a result of their influence on the flame velocity: the addition of tetrachloride stannic to hydrogen–air mixtures (Sadovnikov), or the addition of chlorine or tetrachloride carbon to mixtures of carbon monoxide with air and oxygen (Barskiĭ, Drozdov) in amounts insufficient to fully flegmatize the mixture significantly decrease the flame velocity.

Thus, in a stoichiometric mixture of $2CO + O_2$ with 2% water content, the flame velocity is 90 cm/sec; addition of 1.8% CCl_4 decreases the flame velocity to 40 cm/sec. At the limit, with a 4.5% CCl_4 content, the minimum velocity is 4 cm/sec, in good agreement with absolute calculations and with the minimum velocities in other carbon monoxide mixtures. Drozdov showed experimentally that any factor which increases the flame velocity (an increase in the carbon monoxide and water vapor content, an increase in the combustion temperature) simultaneously increases the amount of flegmatizor (tetrachloride carbon) which is necessary to render the mixture incombustible.

The theory of the upper limit of hydrocarbons presents great interest. In mixtures of the higher hydrocarbons, for example, pentane, heptane and so on, with air the phenomena of the so-called "cold" flame occur, which is related to the formation of intermediate oxidation products; here we do not consider this phenomenon, which has been studied in detail by Neumann and his associates at the Institute of Chemical Physics. The simplest hydrocarbon, methane, does not yield these phenomena, and either burns at a relatively high temperature, or does not burn at all.

Comparison of the data available in the literature[8] shows that the addition of an excess amount of methane has a much stronger flegmatizing effect on a stoichiometric mixture of methane and air than does addition of inert nitrogen.

The limiting compositions:

I. 14.8% CH_4 + 85.2% air (20.9% O_2, 79.1% N_2;
II. 6.4% CH_4 + 93.6% of a mixture with (13.7% O_2, 86.3% N_2)

may be represented in the following form:

I. 94.1% stoichiometric mixture (9.5% CH_4, 90.5% air) + 5.9% CH_4;
II. 67.4% stoichiometric mixture (9.5% CH_4, 90.5% air) + 32.6% N_2.

The flegmatizing influence of excess methane is five times stronger than the action of nitrogen, although the specific heat of methane does not exceed that of nitrogen by more than two times in the temperature interval of interest, 300–1700°K.

The strong action of methane may only be explained by the fact that methane at high temperatures enters into endothermic chemical reactions with the combustion products of the stoichiometric mixture, for example, according to the equation

$$CH_4 + CO_2 = 2CO + 2H_2 - 56 \, kcal,$$

and, as a result, lowers the flame temperature much more strongly than an inert diluent of equal specific heat.

Calculations which relate the concentration limits to losses by thermal radiation lead to the conclusion that, as the pressure is raised, the minimum possible propagation rate decreases proportionally to $p^{-1/2}$. Unfortunately, we do not have the data necessary to compare this assertion with experiment. Reliable measurements of the flame velocity (especially for slow flames) at non-atmospheric pressures are rarely encountered.

The shift of the concentration limit with a change in pressure is related to the dependence of the flame velocity on pressure. In particular, if at constant

---

[8]The properties of methane mixtures have been carefully investigated, from the papers of H. Davy up to the present day, since in coal mines the danger of explosion is directly related to the release of methane.

composition the flame velocity is proportional to $p^{-1/2}$ (as it should be for a chemical reaction which is first order in pressure), the concentration at the limit does not depend on the pressure.

Bone's data shows that the concentration limits of mixtures of hydrogen with air do not, within the limits of experimental error, depend on the pressure in the interval 1–125 atm. The limits of mixtures of carbon monoxide even narrow somewhat, which may be related to the decrease in the percent composition of water vapors as pressure increases: the combustion rate of carbon monoxide mixtures depends substantially on the concentration of water or hydrogen. The lower limit of mixtures of methane with air

Table 3

| Pressure, atm | Methane mixed with air, % | |
	Lower limit	Upper limit
1	5.6	14.3 (14.8)*
10	5.9	17.2
50	5.4	29.4
125	5.7	45.7

*The second number (14.8) is the average of the data of Coward, Payman and Iorissen.

is practically independent of the pressure. In this respect the extraordinarily sharp enhancement of combustion at high pressure of mixtures containing excess methane is particularly interesting. Table 3 shows the experimental data of Bone, Newitt and Smith.

The mixture composition at the upper limit at 125 atm may be represented in the following form:

III. 60% stoichiometric mixture (9.5% CH_4, 90.5% air) + 40% CH_4.

The mixture composition at the lower limit (5.7% CH_4) may be represented in the same form:

IV. 60% stoichiometric mixture (9.5% CH_4, 90.5% air) + 40% air.

At high pressure methane becomes thermodynamically stable and does not enter into endothermic reactions which are accompanied by an increase in the number of molecules. Under these conditions the flegmatizing action of excess methane proves weak. Evidently, the higher heat capacity of methane is compensated for by the increase in the chemical reaction rate for increased methane concentration.

Thus we have been able, not only to establish the physical nature of the concentration limit, but also to relate it in a number of cases to the chemistry of combustion reactions.

§I.6   *Diffusion Phenomena at the Limits of Propagation*

Experiment shows that near the concentration limits of flame propagation extremely interesting features of propagation phenomena are observed in a number of cases. A dependence arises of the concentration limit on the direction of flame propagation.

The difference in the limits for upward and downward propagation was first noticed by White in hydrogen–air mixtures.

Later, the difference in the upward and downward limits under otherwise equivalent conditions (atmospheric pressure, a tube of 2 cm diameter) was determined by Clusius and Cutschmid for a number of mixtures.

Table 4 shows data from this work for mixtures of hydrogen and deuterium with various gases.

In the first column the mixture composition is given; in the second—the limiting fuel concentration for upward propagation; and in the third—the same for downward propagation; the last column shows the ratio of one value of the limiting fuel concentration to the other. As we see, this ratio reaches 2–2.5, with the ratio for deuterium smaller than for hydrogen.

In contrast, in mixtures with excess hydrogen and deficient oxygen, the concentration ratio of the deficient component (oxygen) at both limits (upward and downward) is close to unity.

According to Drozdov and Zeldovich, the upper limit for hydrogen–air mixtures for upward propagation is 73.5% H_2, 26.5% air, and for downward propagation it is 72.6% H_2, 27.4% air; the ratio is $z = 27.4/26.5 = 1.03$.

In our laboratory, Kokochashvili showed that the behavior of mixtures of hydrogen with bromine bears the same qualitative character: in mixtures with deficient hydrogen the ratio of the limiting concentrations is quite large. In mixtures with excess hydrogen and deficient bromine the two limits practically coincide.

At a pressure of 200 mm at the upper limit (with excess hydrogen) the mixture composition for upward propagation is 50% H_2, 50% Br_2, and for downward propagation—60% H_2, 40% Br_2; $z = 1.2$.

At the lower limit, i.e., for deficient hydrogen, the mixture composition for upward propagation is 8% H_2, 92% Br_2, and for downward propagation—40% H_2, 60% Br_2; $z = 5.0$.

The ratio $z = 5$, which characterizes the influence of the direction of propagation, exceeds all values of z known to date.[9]

In order to understand how small the hydrogen concentration is at the limit of upward propagation (92% Br_2, 8% H_2) it is enough to calculate the maximum increase in temperature which will occur in complete combustion of such a mixture in the absence of heat losses:

$$T_T - T_0 = \frac{8.2 \cdot 12\,000}{100 \cdot 8.7} = 220°.$$

The theoretical combustion temperature is thus only 240°C. The minimum temperature of self-ignition, meanwhile, recorded in experiments by

---

[9]We note that all data on limits given in the previous section were related to downward propagation. We shall see below just why this limit should be considered the undistorted one.

Table 4 Dependence of the Limiting Concentration of Fuel on the Direction of Flame Propagation in Mixtures with a Deficiency of Fuel

Mixture composition	Flame propagation		Ratio of limiting concentrations z
	Upward	Downward	
Concentration limit of mixtures of *hydrogen* with various gases			
100% O_2	3.8	9.5	2.5
20% O_2, 80% N_2	3.9	9.6	2.46
20% O_2, 80% He	5.7	8.0	1.40
20% O_2, 80% Ne	3.5	7.0	2.00
20% O_2, 80% Ar	2.7	7.1	2.63
Concentration limit of mixtures of *deuterium* with various gases			
100% O_2	5.6	11.1	1.98
20% O_2, 80% N_2	5.6	11.0	1.96
20% O_2, 80% He	7.4	8.3	1.12
20% O_2, 80% Ne	4.2	7.7	1.83
20% O_2, 80% Ar	3.7	7.7	2.08

Kokochashvili, is 425°C.

At a temperature of 250°C in experiments by Bodenstein and Lind as well only a very slow chemical reaction occurred, requiring long hours to complete.

Comparison of these temperatures shows that the observed upward flame propagation in lean hydrogen mixtures requires special explanation. Taken together, the data on the influence on this effect by admixtures of various inert gases, from argon to helium, and by replacement of hydrogen by deuterium, shows that the ratio of the molecular weight of the fuel (hydrogen, deuterium) to that of the mixture is of importance for this effect. The large molecular weight of bromine explains the particular sharpness of the effect in the latter case.

Gartek and Goldman relate the expansion of the concentration limit in upward propagation in lean hydrogen mixtures to the fact that "due to the large coefficient of diffusion in the mixture, high temperatures may be achieved on the surfaces of particles of the catalyst." Comparison of hydrogen and deuterium confirms this point of view. In addition, a number of facts show that it is not the absolute value of the diffusion coefficient, but its ratio to some other value that is important. Replacing oxygen with helium increases the diffusion coefficient of hydrogen, while replacing with bromine decreases it; the ratio of concentrations at the limit in these mixtures, meanwhile, changes in the opposite order: it is greatest in mixtures of hydrogen with bromine and least in mixtures containing helium. The coefficients of mutual diffusion of the gases are equal to one another: the coefficient of diffusion of hydrogen into oxygen in lean hydrogen mixtures is no larger than the coefficient of diffusion of oxygen into hydrogen at the upper concentration limit. Meanwhile, at the upper limit, there is no difference between the upward and downward propagation limits, z is close to 1. Thus the ideas of Gartek and Goldman in their original form are insufficient.

Let us consider the thermal balance of a flame pellet at rest with respect to an explosive mixture on whose surface a chemical reaction is occurring. The reagents are transported to the surface of this pellet by diffusion, and the reaction products are also carried away from the surface by diffusion. The heat of the reaction is carried away from the surface of the flame by heat conduction.

If we are dealing with mutual diffusion of gases which are close in molecular weight (e.g., carbon monoxide and air), it may be shown that the temperature of the flame pellet will prove to be equal to the theoretical combustion temperature of the mixture. This equality depends on the existence in the kinetic theory of gases of a simple relation between the diffusion coefficient (on which the supply of reagents and heat release rate depend) and the thermal conductivity (on which the heat evacuation depends).

However, in the case of mixtures with low hydrogen content this relation is sharply violated: the diffusion coefficient of light hydrogen in such a mixture is large, while the thermal conductivity of the mixture, which is comprised primarily of heavy molecules, is small. For this reason the combustion temperature rises sharply. Roughly speaking, the flame pellet in such a mixture collects fuel (hydrogen) from a large volume, and gives off heat to a small volume. It is for this reason that a temperature increase many times greater than the "theoretical" one is achieved: for ordinary calculations of the theoretical temperature we assume that the reaction heat of the fuel contained in a certain volume goes to heat the entire mass of the combustion products which form from the portion of mixture which is located in this volume. Thus it is not the diffusion coefficient, but the ratio of the diffusion coefficient to the thermal diffusivity[10] that determines the difference between the up- and downward limits of propagation.

From this point of view, all the dependencies are explained naturally: in a series of mixtures containing helium, oxygen or nitrogen, and bromine, the diffusion coefficient of the hydrogen falls from the first to the last, but the thermal diffusivity falls even more sharply, ϑ/κ grows and, therefore, the

[10]See §I.2 for a definition of thermal diffusivity.

ratio z of the concentrations at the up- and downward limits of propagation increase.

At the upper limit in mixtures containing excess hydrogen, the diffusion coefficient of oxygen is large, but the thermal diffusivity of a mixture rich in oxygen is still larger; the ratio $\vartheta/\kappa < 1$ and there is no noticeable difference between the up- and downward propagation limits; z is close to unity.

When $\vartheta/\kappa < 1$, conditions in the flame pellet, which receives fuel from the surrounding medium by diffusion, are even less favorable than for normal flame propagation with respect to the gas. Therefore the flame moves normally with respect to the gas in both directions, upward and downward, and identical mechanisms yield practically identical limits; z is close to unity.

In lean hydrogen mixtures the flame temperature in diffusive combustion is higher than in normal propagation and so, in a wide region of concentrations (from 4 to 9% H_2 in a mixture of hydrogen with air), normal propagation is impossible, only "diffusive" combustion is possible. Our point of view is in accord with the observed properties of flame propagation in this concentration interval.

1. If diffusive combustion occurs and the flame does not move with respect to the gas, then convective flows will carry it upward; such a flame can not propagate downward. Therefore we relate the dependence of the limits on the direction of propagation to the possibility of a diffusive mechanism, and we compare the theory of limits of normal propagation (§ I.4 and I.5) with data relating to downward propagation.

2. An increased temperature is achieved in diffusive combustion due to the fact that only part of the fuel is burned, but the heat propagates in an even smaller part of the mixture. According to the laws of diffusion and heat conduction, complete extraction of the fuel by diffusion would lead to distribution of the heat throughout the entire mixture, and the diffusive mechanism would not yield a temperature increase compared to normal propagation.

Experiment indeed shows that at a concentration such that the flame propagates only upward, the completeness of the combustion when ignited from below is not high. As confirmation, we cite the data of Coward and Brinsley:

H_2 in mixture, %	4.35	4.7	5.1	7.2	8.0	9.1	10.0
Fraction of hydrogen that burns	0.11	0.28	0.49	0.87	0.92	0.97	1.0

3. In a mixture at rest, a diffusive flame would take the form of a sphere; convective motion of the gas leads to the flame taking the form of a bent cap. When a lean mixture is ignited with an electric spark, one or several such caps form and rise slowly upward to the end of the tube. An elementary calculation of diffusive combustion in a gas at rest yields a combustion temperature which depends only on the ratio ϑ/κ, and not on the radius

of the sphere; however, the intensity of combustion with respect to a unit surface of the diffusive flame is inversely proportional to the radius. Consideration of heat lost on radiation and of the influence of convection lead to the conclusion that individual pellets or caps of flame cannot be large and yields a reasonable size for them—from several millimeters to 1–2 cm.

On the basis of the theory developed above we foresaw and realized experimentally (in Drozdov's work in our laboratory) a new kind of mixture which exhibits a significant dependence of the limit on the direction of propagation; such mixtures do not contain hydrogen, and the diffusion mechanism in them facilitates combustion due to kinetic, rather than thermal factors.

In air and oxygen mixtures of carbon monoxide, the flegmatizing concentration of CCl_4 turned out to be significantly different for ignition from above and ignition from below.

We give a table for stoichiometric mixtures (Table 5).

Table 5   Flegmatizing Concentration of CCl_4

CO, %	O_2, %	N_2, %	CCl_4, %		Ratio of concentrations z
			Ignition from above	Ignition from below	
20	10	70	0.36	0.58	0.62
23.5	11.7	64.8	0.58	1.03	0.57
29	14.8	56.2	1.16	2.05	0.57
40	20.4	39.6	2.14	3.75	0.58
48	24	28	2.72	4.73	0.56
66.7	33.3	0	4.55	–*	–

*The flegmatizing concentration exceeds the vapor tension of CCl_4 at room temperature. We note that all mixtures also contain 1.8% water vapor above 100%.

The ratio z in this case is less than unity since we are considering the concentration of a substance which inhibits combustion, whereas before we calculated z on the basis of the concentration of the combustible material.

Calculations show that addition of CCl_4 has practically no influence on the combustion temperature and that its flegmatizing action depends on its influence on the chemical reaction rate.

Whatever the specific mechanism of the influence of CCl_4, it is clear that the intensity of its flegmatizing action depends on the interrelation between the amounts of flegmatizer and combustible mixture entering the flame.

For normal propagation, the flegmatizer and reagents enter the flame in the same ratio as they are taken in the preparation of the mixture.

In a diffusive combustion mechanism the ratio of the amounts of flegmatizer and reagents entering the flame also depends on the rate of diffusion to the flame surface. If the molecular weight of the flegmatizer is significantly larger than the molecular weight of the reacting substances (CCl_4–154, CO–28; O_2–32), combustion conditions in a diffusive mechanism turn out to be more favorable than in a normal mechanism. This explains the presence of a concentration interval for the flegmatizer for which normal propagation is no longer possible, while the diffusive mechanism allows upward flame propagation.

Flame propagation in this concentration interval is similar in every respect to combustion of lean hydrogen mixtures; the flame propagates in the form of individual caps or pellets of diameter 5–8 mm, behind which stretches, in mixtures which develop a high combustion temperature, a narrowing luminescent trail about 30–40 mm long. The flame surface does not cover the full cross-section of the tube so that complete combustion cannot be expected.

Finally, in mixtures of hydrogen with bromine in which the difference in molecular weights is particularly large, Kokochashvili observed a new, never before described phenomenon of downward flame propagation in the form of individual caps. Using military terminology (flame front), we may say that in this case we have an attack by individual isolated columns or "wedges," instead of the attack as a solid front in the normal mechanism. This phenomenon was observed and reproduced numerous times in mixtures containing 35–40% hydrogen and 65–60% bromine at pressures above 200 mm.[11] This form of flame propagation cannot be reduced to a diffusive mechanism since the caps of flame move downward; downward motion cannot be explained by convection either since the flame truly moves with respect to the gas.

In connection with Kokochashvili's observations there arises an important fundamental question about the stability of normal propagation of a continuous plane flame front. We must analyze the influence of convexity and concavity of the flame front on the propagation velocity. In mixtures in which the diffusion coefficient is equal to or less than the thermal diffusivity, a convexity (in the direction of propagation) decreases and a concavity increases the flame velocity. The increase in the velocity is explained by the fact that the mixture, enveloped by the concave flame from all sides, heats up more rapidly.[12]

It is obvious that in this case a flat front is stable. Let us imagine a

[11] We mentioned in §I.3 the agreement achieved between the calculated and measured flame velocity; this agreement is pertinent to mixtures containing 50–55% H_2 in which the flame propagates as a solid front.

[12] The velocity increase in a concave flame was noted long ago in connection with the theory of the Bunsen burner: it is on this that the rounding of the apex of the flame cone depends.

wave-like flame front. If the velocity of the concave areas is greater than that of the convex ones, the front will straighten out and become flatter.

In lean hydrogen mixtures, in which the diffusion coefficient is greater than the thermal diffusivity, the influence of the curvature on the flame velocity changes sign: in a convex area the increase in fuel supply by diffusion is more significant than the increase in thermal conduction of the unburned mixture, and the combustion velocity increases. In contrast, a mixture enveloped by a concave front loses fuel by diffusion more rapidly than it is heated, and the flame velocity falls. A flat flame front becomes unstable: for a small initial curvature of the front the velocity of the protruding convexities grows and the velocity of the concave regions decreases. They lag further behind the convex parts and the curvature

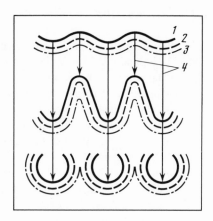

Fig. 9

increases. Finally, the convex tongues of flame which have gone forward make the mixture remaining in the openings between them so lean that it becomes incombustible, and the flame front breaks up into separate, unconnected convex areas, as shown in Fig. 9 (where *1* is the reaction zone, *2* is the heating zone, *3* is the depletion zone and *4* is the combustion rate). This is precisely the picture of flame propagation in mixtures of 35–40% H_2, 60–65% Br_2.

We have thus succeeded in providing a theoretical explanation of the peculiar phenomena which arise in combustion of mixtures of gases of different molecular weights. We began with an analysis of the reasons for the influence of the direction of flame propagation on the limits of lean hydrogen mixtures. Similar phenomena were observed in our laboratory for explosive mixtures which were flegmatized by substances of large molecular weight. Finally, in mixtures of hydrogen with bromine we found a new form of flame propagation which testifies to the instability in this case of a continuous plane flame front.

Chapter II.   Theory of Gas Detonation

§II.1   Introduction

The phenomenon of gas detonation was discovered comparatively late, in 1881, independently by Mallard and Le Chatelier and by Berthelot and Vieille in the course of work on flame propagation in tubes commissioned by the Mine-Industrial Committee. These studies followed the horrible catastrophes in the mines of France and Belgium. Using photography, flame propagation at enormous velocities (from 1.5–2 to 4–5 km/sec) was discovered. Originally, detonation of gas mixtures was discovered when they were ignited with a charge of explosive material (detonating mercury), whereas thermal ignition, with a spark or flame, caused slow combustion of the mixture (so-called deflagration, with normal propagation according to the theory given above). Soon, however, it was shown that for flame propagation in sufficiently long tubes the flame accelerates, oscillations in the flame velocity frequently appear, and finally, detonational combustion of the mixture occurs.

The very first experiments revealed the distinctive features of detonation:

1. The stability of the velocity. On a moving piece of photographic paper detonation leaves a perfectly straight, distinct track, in contrast to the curved, often oscillating track of slow propagation. The detonation velocity is not only constant in a given experiment, but is constant also for a given mixture in tubes of different diameters; it does not depend on the means of ignition, and depends very weakly on the initial pressure of the explosive mixture and on any admixtures, including chemically (catalytically) active ones.

2. the large mechanical effects of a detonation wave which indicate that significant pressures are reached in this regime. Due to the high velocity of detonation, the pressure which is developed in an open tube decreases compared to a closed one.

The subject of our study will first of all be the theory of the detonation velocity, the mechanical effects of a detonation wave and, finally and most importantly, the conditions of chemical reaction in the wave. A short summary of the results obtained by the author are given at the end of the book.

§II.2   The Classical Theory of Detonation

The authors who discovered detonation noted that in order of magnitude the detonation velocity is close to the molecular velocity of the reaction products, to the velocity of sound in the products, and to the propagation velocity of strong explosive waves. These comprise, in embryo form, the ideas which were later developed by many different scientists.

Chapmann developed a hydrodynamic theory of detonation along the lines used earlier by Riemann, Rankine and Hugoniot to construct a theory of shock (explosive) waves in a chemically inert gas.

Taking as a basis the experimental fact that the process is stable and strictly steady, Chapmann compared the states of the initial mixture and the reaction products. He did not consider the reaction zone itself, but, however the reaction runs inside this zone, the amount of mixture entering the zone is equal to the amount of the products output, and the energy of the mixture entering the reaction zone (containing the still unreleased chemical energy) is equal to the energy of the reaction products leaving the reaction zone. In the products the chemical energy has been transformed primarily into thermal energy[1] (Fig. 10). Finally, by Newton's second law, the change in the velocity of the substance depends on the pressure difference across the reaction zone,

$$\frac{D}{v_0} = (D - w)v, \tag{II.2.1}$$

$$E_0 + p_0 v_0 + \frac{D^2}{2} = E + pv + \frac{(D-w)^2}{2}, \tag{II.2.2}$$

$$\frac{Dw}{v_0} = (D - w)\frac{w}{v} = p - p_0. \tag{II.2.3}$$

We must consider the internal energy E of the combustion products, which enters into the second equation, as a known function of the pressure p and the volume v.

For a given initial state of the mixture these three equations are insufficient to determine the four variables: the detonation velocity, i.e., the velocity D of the reaction zone itself with respect to the mixture; the specific volume v; the pressure p of the products leaving the reaction zone; and the velocity w acquired by them.[2]

However, it is enough to assign a specific value to one of the four variables p, v, w, D in order to determine the other three variables from the three equations.

---

[1] The energy balance also includes the kinetic energy and the work of the pressure forces.

[2] In a coordinate system in which the initial mixture is at rest, the velocity of the products is denoted by w; in a system in which the reaction zone (the "wave") is at rest, the velocity of the original mixture is D and the velocity of the products is $D - w$ (see the arrows in the upper and lower parts of Fig. 10).

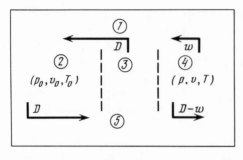

Fig. 10   Fig. 11

1: System at rest in which the mixture is not   1: Hugoniot adiabate.
moving, **2**: explosive mixture, **3**: reaction zone,
4: reaction products, **5**: system in motion in
which the zone is not moving.

To each value of v thus corresponds a definite p. Taken together they form a curve in the p, v-plane—the so-called Hugoniot adiabate (Fig. 11). We find the value D corresponding to each v by the formula

$$D^2 = v_0^2 \frac{p - p_0}{v_0 - v},   \tag{II.2.4}$$

which has a simple geometric interpretation—D depends on the slope of the line[3] which joins the points p, v and p_0, v_0:

$$D \sim (\tan \alpha)^{1/2}   \tag{II.2.5}$$

The quantity D reaches a minimum in the state B, which corresponds in the drawing to the point of tangency of the line AB, drawn from A, (p_0, v_0). Chapmann asserts that it is just this minimum value of the detonation velocity that is realized in experiment. Comparison with measurements fully confirms this hypothesis. In this same way definite values are established for the volume and pressure of the products. The substance is compressed by almost two times, and a temperature is achieved which slightly exceeds the explosion temperature. The reaction products acquire a velocity close to $D/2$.

The pressure reaches double the pressure of explosion. Sharp braking when the detonation wave encounters an obstacle (a wall or the closed end of the pipe) can lead to tripling of the pressure, which reaches $6p_{\text{expl}}$. This

[3]We note that the use of the line drawn from the point A to one of the points on the Hugoniot adiabate CBG, e.g., the line AB, to geometrically determine the detonation velocity D is not related to any assumptions about the intermediate states of the substance in the wave or about the trajectory of the point which describes the state of the substance in the p, v-plane. On this see § II.4.

explains the increased destructiveness of detonation, even compared to explosion in a closed volume.

The kinetic energy of the products reaches 15% of the reaction heat, and the total energy reaches 150% of the reaction heat. The question arises, where does the additional energy, over 100%, come from and might we use it? In reality the excess energy of the reaction products at the moment of their formation is due to expansion, cooling and braking of previously reacting layers.

For detonation to propagate it is necessary that part of the energy continue to be transmitted forward. In this way the first law of thermodynamics is not violated and the energy use of detonation, if we ignore constructive problems, does not promise substantial gain.

The state which corresponds to the minimum value of the detonation velocity possesses a number of remarkable properties: in this state the extremum of the entropy is reached—a minimum on the Hugoniot adiabate and a maximum on the line joining the corresponding point with the initial one, p_0, v_0, in the p, v plane. The detonation velocity in this state is equal to the sum of the velocity of the products and the velocity of sound in them.

§II.3 The Motion of the Reaction Products

The motion of the reaction products as they exit the reaction zone obeys the usual equations of gasdynamics. It is comparatively easy to construct distribution curves of the pressure, velocity, density, and other quantities which characterize the state of the reaction products in the one-dimensional problem (propagation of detonation in a tube), when losses are ignored. In Fig. 12a we show a solution for detonation propagation from the closed end of a tube for both the state B, occurring in reality with a minimum velocity (bold line), and two other states, C and G, corresponding to a higher detonation velocity (see Fig. 11). The density is plotted on the ordinate axis; the density of the original mixture ρ_0 is shown by the dotted line. The coordinate x, measured from the closed end of the tube, is plotted on the abscissa.

In Fig. 12b the velocity of the substance is plotted on the ordinate. As is clear from Fig. 12, in the regime which occurs in experiment, the almost doubled density of the products at the wave front is compensated for by a density at the end of the tube which is lowered to almost 75% of ρ_0. Rarefaction and braking begin immediately after completion of the reaction. About 60% of the total mass of reacted matter is in motion, and it occupies 50% of the length traveled by the wave.

With time the distribution curves, without changing in form, are stretched out along the length.

In a rarefaction wave, BP, the propagation velocity of each state is equal

to the sum of the velocity of the gas and the velocity of sound in it; this
relation is valid for the reaction products at the moment of their formation,

$$D = w + c,$$

and remains valid subsequently in the course of the expansion:

$$x = (w + c)t$$

(where w and c denote their current values).

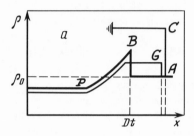
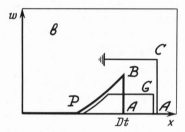

Fig. 12

For a greater detonation velocity two regimes are possible, corresponding to the two points, G and C, of intersection with the line AGC and the
Hugoniot adiabate (see Fig. 11).

In the first regime, which corresponds to a smaller pressure $D > w + c$,
the gasdynamic equations allow a solution which satisfies the boundary condition. In the second regime, the only solution, $v = $ const, $w = $ const, does
not satisfy the boundary condition. The corresponding state of motion satisfies only the boundary condition which occurs for a piston moving with a
velocity[4] w.

The author also studied for the first time spherical propagation of a detonation wave (1942). In Fig. 13 (where *1* denotes the original mixture and
2—the reaction products) the distribution of the pressure and velocity in the
case of minimum detonation velocity is shown. This distribution exhibits interesting concrete features: immediately after completion of the reaction, an
extraordinarily sharp, then slowing, expansion takes place; in a rarefaction
wave, $x > (w + c)t$.

92% of the reacting matter is involved in the motion.

Consideration of the two other regimes leads to conclusions analogous to
the one-dimensional case, in particular, the state C again yields a solution
which does not satisfy the condition $w = 0$ at the center of the sphere.

[4]The patterns of motion (just as the incompatibility of the state C with a rarefaction
wave) have long been known (see Jouguet, Crussard). Recently they were studied again
by Grib in an unpublished dissertation (Leningrad Mining Institute, 1940), as the author
of the dissertation kindly informed me.

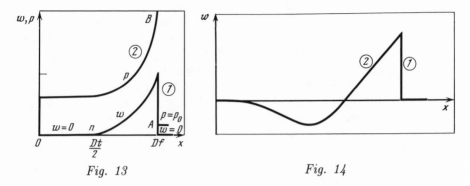

Fig. 13   Fig. 14

The author also studied the propagation of detonation in a long tube in which, due to heat transfer and friction against the walls of the tube, a steady state is achieved which moves together with the wave front. In Fig. 14 (again *1* denotes the original mixture and *2*—the reaction products) the velocity distribution found in the case of minimum detonation velocity is shown. Of interest here is the change of sign of the velocity in the course of expansion.

The change in the direction of motion may be clearly seen in Dixon's photographs and previously had no explanation.

The final state of the products when friction and heat transfer are taken into account is naturally achieved when the products are fully braked and cooled to the temperature of the walls.

The distance at which this state is achieved depends on the intensity of the heat transfer and friction; in accordance with the laws of turbulent flow it reaches a value several dozen times the tube diameter.

Again, study of the equations shows that the regime C is incompatible with conditions at infinity, while the solution for the regime G differs little from the solution for regime B, illustrated in Fig. 14.

§II.4   *The Chemical Reaction Mechanism and the Selection of the Detonation Velocity*

Up to now we have studied the state B, corresponding to the minimum velocity, considering its occurrence to be an experimental fact. It is obvious that a rigorous justification of the necessity of this state is a most important task of the theory.

The exclusion of states which lie above the point B follows from the impossibility, demonstrated above, of constructing for them a regime which satisfies the boundary conditions; the physical reason is that at $D < w+c$ the rarefaction wave in the products overtakes the detonation wave and weakens it.

However, all attempts encountered in the literature to exclude states of type G using either hydrodynamic or thermodynamic arguments are unconvincing and incorrect. Their fundamental error is obvious from the regimes which we have constructed which satisfy all the equations and conditions. For consecutive ignition of one layer of the mixture after another by an external ignition source with a velocity exceeding the minimum velocity, it is precisely a regime of type G that is realized. An explanation of the fact that, in the distribution of detonation (without any external ignition source), these regimes do not occur must in fact follow from a consideration of the reaction mechanism. This justification was given for the first time by the author.

Views on the mechanism of reaction in the wave which are encountered in the literature are marked by great diversity.

There exists a common belief that the high linear velocity of detonation propagation, which is thousands of times higher than the normal velocity of flame propagation, indicates a rapid chemical reaction rate.

A number of authors consider that the reaction is caused by compression of the original substance by a shock wave (Vieille, Nernst and Wendtlandt, Jouguet, Sokolik). Becker, in analogy with shock wave theory, considers that heat conduction is involved in the transfer of the reaction from layer to layer, and he argues against the proposition of ignition by a shock wave.

In a shock wave, as is known, a change of state occurs on the mean free path of molecules of the gas. The outward analogy between the theory of detonation and the shock wave theory prompted many authors to consider that a detonation front is just as sharp as the front of a shock wave. Jouguet spoke in favor of "instantaneous" reaction.

But occurrence of a chemical reaction along one or even several mean free paths is quite surprising, and to explain such a large reaction rate and large velocity of propagation, these authors resorted to electrons and radiation, quantum-mechanical resonance of collectively moving particles and direct impact of rapid active centers of a chain reaction.

Jost recognized the necessity of at least thousands of molecule collisions for reaction and hastily concluded from this that steady propagation of detonation is impossible (1939).

Without going into a detailed critique of these points of view, we shall lay out our conceptions of how the reaction occurs.

The chemical reaction is caused by sudden and energetic heating of the mixture when it is compressed by the shock wave which is part of the complex called the "detonation wave." Compression by the shock wave occurs so rapidly that during the compression the chemical composition is unable to change at all. In the compressed gas the chemical reaction is accompanied by expansion and a pressure change.

All the processes occurring in the detonation wave we subordinate to the condition of its (the wave's) steadiness: the structure should be such that the wave as a whole moves in space without deformation.

The expression for the propagation velocity of a detonation wave with respect to the motionless mixture was given earlier (II.2.4). This expression does not contain the reaction heat; it follows from the equations of momentum and conservation of matter.

Expression (II.2.4) is therefore also applicable to any intermediate state in which only a part of the energy has been released or only compression has occurred. In §II.2 we applied (II.2.4) to the final reaction products. The condition of steady propagation means that the velocities at which any of the intermediate states propagate are equal. Hence follows the elementary relation:

$$p = p_0 + \left(\frac{D}{v_0}\right)^2 (v_0 - v). \tag{II.4.1}$$

For D constant (the same for all states) this equality is the equation of a straight line in the coordinates p, v which passes through the initial point, p_0, v_0. As we have said, this equation is satisfied not only by the final state of the reaction products, but also by any intermediate state.[5]

In particular, in the diagram of p, v the state of the mixture which has just been compressed by the shock wave and has not yet started to react is represented by the points Y or S (Fig. 15; here 1 denotes the original mixture and 2—the reaction products) (for two different values of the detonation velocity) of the Hugoniot adiabate, AYS, constructed without taking account of heat release.

[5]Earnshaw was the first to note that an acoustic wave of arbitrary form can propagate without deformation (so that all states propagate with the same velocity, i.e., steadily) only in the case of a linear relation between the pressure and specific volume. Actually, the adiabatic relation between the pressure and volume is not linear, and this is the reason for the deformation of acoustic waves, discovered by Poisson and Stokes and studied by Riemann, which leads to the formation of shock waves.

Rankine assumes that there is a linear relation of pressure and volume in the change of state in a shock wave. The deviation from the adiabatic law, $pv^\kappa = \text{const}$, is due to the effect of heat conduction at the shock wave front. In fact, in a shock wave both heat conduction and viscosity act simultaneously so that the law of state change differs both from Poisson's adiabate and from a linear relation. In addition, the very question of a relation between pressure and volume for compression in a shock wave makes sense only for small-amplitude waves; for a large amplitude the change of state takes place over a time of the order of the molecular mean-free time. For further details see the author's monograph [9].

In the theory of detonation Michelsohn was the first to propose the assumption of a linear law (II.4.1) for the change of state during a chemical reaction. Todes and Izmailov, unaware of his work, expressed the same point of view in an unpublished paper. In both cases it was assumed that in the p, v-plane, along the line passing through the point A, (p_0, v_0) and satisfying (II.4.1), the motion begins from the point A; this would mean a chemical reaction in the initial state at room temperature. Our understanding of the path of motion along the line is presented below. In the author's first paper [7] the line corresponding to (II.4.1) was called the "Todes line." In the following paper [10] it was indicated that it would be more correct to call it the "Michelsohn line."

In the initial state A the mixture (usually at room temperature) is chemically inert. Compression to the state Y or S heats the mixture to 1000–2000°C (typical values for detonating mixtures). At this temperature the chemical reaction begins and heat is released, similar to the way a chemical reaction begins when a mixture is introduced into a heated vessel. As heat is released the mixture moves to states which are intermediate in composition between the initial and final states. In Fig. 15 these states are represented by the points between Y and B or S and C (for a different detonation velocity), respectively. We recall that the curve AYS describes the states of the compressed but unreacted mixture, and the curve CBG—the states of the final reaction products.

We see directly that the reaction caused by compression in the shock wave at a large detonation velocity (corresponding to the line $AGCS$) leads

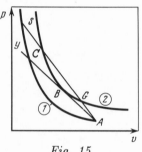

Fig. 15

to the state C on the upper part of the curve CBG. We saw that the state C is incompatible with the boundary conditions, since in this state the detonation velocity is less than the propagation velocity of a perturbation in the reaction products. At the same velocity which corresponds to C, another state, G, is possible as well, one with lower compression and a smaller velocity of the reaction products: in the state G the velocity of a perturbation is less than the detonation velocity, and therefore the state G is completely compatible with the subsequent expansion of the reaction products and with the boundary conditions.

However this state G is unattainable for a substance which is chemically inert at the initial state A (e.g., at room temperature and atmospheric pressure). This substance enters into a chemical reaction only when it is compressed by a shock wave; no other sufficiently powerful perturbation propagates along the continuous substance with the necessary velocity. Compression by a shock wave of the velocity which corresponds to the state G implies a transition from the initial state A to the point S which lies on the continuation of the line AG: the fact that the points A, G, S lie on a single line in fact indicates that the velocity of the wave AS is equal to the propagation velocity of the state G with respect to A. The point S describes a state of compressed and heated, but unreacted mixture. The chemical reaction and heat release begin, the mixture expands, and the pressure falls; the change of state is described by motion along the line SA of Fig. 15 from the point S to the right and down. In this motion, by the time all the reaction heat has been released, we end up precisely at the point C since the point C lies on the Hugoniot adiabate of the reaction products (satisfies the conservation equations, taking into account the release of all the heat of reaction). The point G satisfies these equations and does so at the same value of the velocity. But, in order to arrive at G from S, it would be necessary, after completion of the reaction (C), to add some additional amount of heat to reach the middle of the segment CG, and then to remove this heat. The path from Y to B or from S to C passes completely through the states which are intermediate in terms of the amount of energy released between the original mixture and its reaction products. In the course of the reaction all of these intermediate states are realized. The path from S to G lies partially (in the section CG) in the region of states whose realization requires release of an amount of energy which exceeds the reaction heat of the mixture; the corresponding states are unattainable in the course of the reaction, and therefore the state G cannot occur in the reaction of a mixture compressed by a shock wave.

Only the tangent point B satisfies both conditions at once: it may be attained in the process of chemical reaction of a gas which is compressed by a shock wave (the jump HY and the drop YB), and at the same time the state B is compatible with the conditions of expansion of the detonation products upon completion of the reaction (the detonation velocity in state B is exactly equal to the velocity of propagation of a perturbation in the reaction products).

Thus, selection of a single specific value for the detonation velocity and a single possible state of the reaction products depends on the adopted mechanism of the chemical reaction induced by compression in a shock wave.

We give a rigorous analytical proof of the assertions

1. that states of type C are incompatible with the boundary conditions,
2. that states of type G cannot occur in the reaction of a substance compressed by a shock wave,
3. that the state of the reaction products in which the detonation velocity is at a minimum is compatible with the boundary conditions of motion of the products of the reaction after its completion, and that this state may be achieved in a chemical reaction of a substance compressed by a shock wave

for three typical cases: propagation of detonation in a tube without losses; detonation propagation in an infinitely long tube, taking into account heat transfer and friction of the reaction products against the walls of the tube; and, finally, for spherical propagation of detonation with central ignition and no losses. Here we pause on the mathematical details of the proof, the basic ideas and results of which are presented above.[6]

[6]The fact that in reality it is precisely the state B that occurs in the detonation of gaseous mixtures was discovered simultaneously with the formulation of the conservation equations for a substance in which a chemical reaction is propagating (Chapmann, Jouguet).

Chapmann considered that the state B corresponded to maximum entropy, and

We are now able to clearly picture the interrelation between shock wave theory and detonation theory.

In both theories we begin from the conservation equations, which we apply to a comparison of the initial state of the substance (before compression or reaction) and the final state of the (compressed or reacted) substance.

Writing the conservation equations does not in itself presuppose instantaneous transition from one state to another. The only condition is that the transition not be stretched out over an amount of time such that external losses noticeably decrease the heat release and velocity. In the propagation of a perturbation in a tube, in order for the conservation equations to be satisfied we need only for the reaction not to take longer than the time required for the moving gas to travel a distance of several tube diameters.

In a shock wave the compression is tied to a change in entropy, the only source of which are the dissipative forces—viscosity and heat conductivity. In the calculation we obtain a negligible front depth and compression time in the shock wave. We emphasize that this is a result of the calculation, not an assumption necessary to write the conservation equations.

In a detonation wave the change of state—after equally rapid compression—depends on the process of chemical reaction and is extended in accordance with the kinetics of the reaction. The only restriction is that the wave (reaction zone) not be extended to a length which is many times larger than the tube diameter. Comparison with a shock wave shows only that the role of heat conduction and diffusion of active centers in a detonation wave is negligible.[7] But they are not needed: the mixture, which has been heated to a high temperature, enters the reaction and reacts under the influence of active centers created by the thermal motion and multiplying in the course of the reaction. Each layer reacts without exchanging heat or centers with other layers.

It is significant that, throughout the reaction zone, the velocity of transport of an acoustic perturbation $w + c$ exceeds the detonation velocity: expansion of one layer, caused by release of reaction heat, is conveyed to another and beyond to the original mixture. The mechanism of detonation

---

considered this fact sufficient to justify the selection of the state B. Jouguet and Crussard showed that on the curve CBG at the point B the entropy is at a minimum. So the question of the selection of B arose anew. The incompatibility of a state of type C with the boundary conditions—with a rarefaction wave—is obvious and generally known (Jouguet, Becker, Jost, Grib).

Essentially new is the author's proof above of the impossibility of states of type G, which completes the theory of the detonation velocity.

[7]The thermal flux is proportional to the temperature gradient, i.e., inversely proportional to the distance at which the temperature changes. For a very sharp shock wave of large amplitude the thermal flux is of the order of the energy of the mixture itself. The width of a detonation wave is greater than that of a shock wave by approximately ν times, where ν is the average number of collisions necessary for reaction. The thermal flux is correspondingly ν times smaller. Since usually $\nu \gg 10^3$, the flux, transported by thermal conduction, comprises less than one thousandth part of the energy of the mixture so that it may be completely neglected.

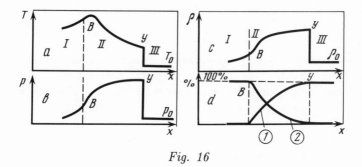

Fig. 16

propagation lies precisely in the transport of pressure from layer to layer, just as the mechanism of slow combustion lies in the transport of heat from layers heated by the reaction to subsequent ones.

Direct experimental proof of the possibility of rapid gas ignition under compression by a shock wave was given by Leipunskiĭ and the author. Compression by the shock wave was accomplished by shooting a fast bullet from a special small-caliber rifle into the explosive mixture. In front of the body, which was flying with supersonic velocity, a steady shock wave formed whose velocity with respect to the gas was equal to the velocity of the body. This condition determined the pressure and temperature amplitudes in the wave. The duration of the compressed state of the gas did not exceed 10^{-5} sec. For a bullet velocity of 1700–2000 m/sec, ignition of the mixture $2H_2 + O_2 + 5Ar$ was observed.

§II.5   Conditions of the Chemical Reaction

If we compare detonation with other combustion phenomena we find more similarity to self-ignition than to flame propagation: in a detonation wave the original mixture enters the reaction undiluted by products.

The reaction is accompanied by a change of state according to an unusual law: the reaction runs not with constant volume or constant pressure, but with a constant propagation velocity, which leads to a linear relation between pressure and volume.

The release of heat is accompanied by expansion, and this expansion is so significant that the pressure falls. Over the course of a large part of the reaction the release of heat is accompanied by growth of the temperature.

Somewhat before the end of the reaction the temperature reaches a maximum, and the release of the last portions of the heat is accompanied by such a strong expansion that the temperature falls somewhat (by only 100–200°): the specific heat of the system, taken at a constant velocity c_D (similar to the way we speak of c_p or c_v), changes sign, and is negative in the narrow region of states close to the final one.

It is significant, however, that the entropy during heat release invariably grows, reaching a maximum just at the moment of completion of the reaction.[8]

As a result of the temperature growth and accumulation of active centers we can expect growth of the reaction rate in the initial stage. The reaction rate decreases and approaches zero only when the reagent concentration approaches zero or thermodynamic equilibrium (in the case of significant dissociation).

In Fig. 16 (where I denotes the reaction products, II—the reaction zone, and III—the original mixture and the lines indicated by 1 and 2 are the percentages of original material and the reaction products, respectively) we show the approximate distribution curves of the temperature (a), pressure (b), density (c), and percentage content of the original and final reaction products (d), constructed with our earlier comments taken into account.

The change of state in the shock wave Y may be considered absolutely abrupt (we note that the composition here does not change). The change of state from Y to B which accompanies the chemical reaction occupies a depth in space which is equal to the product of the velocity of the compressed gas with respect to the wave and the duration of the chemical reaction. The duration of the chemical reaction is primarily determined by its slowest stage of initial acceleration and, consequently, depends on the ability of the mixture to self-ignite and on the temperature achieved in the shock wave.

Taking losses into account (§II.7) leads to the fact that the duration of the chemical reaction in the wave cannot be arbitrary; in order for detonation of the mixture to be possible, the duration of the reaction must not exceed a certain quantity. The maximum allowable reaction time increases for decreasing external losses. Thus, under given conditions, the induction period of self-ignition determines the mixture's ability to propagate detonation. This last assertion, together with incorrect ideas of the magnitude of T_Y (the temperature at which self-ignition occurs) and of the allowable magnitude of the induction period, may be found in papers by Sokolik, Rivin and others (Institute of Chemical Physics). Their calculations of the absolute quantities have no value, but comparative data on the influence of small concentrations of admixtures on the ability to detonate and self-ignite

---

[8]Returning to Fig. 15, we emphasize that the entropy is at a maximum at the point B, in the state which corresponds to completion of the reaction, compared to other states which propagate at the same velocity (points on the line ABY), i.e., compared to all other states in the given detonation wave in which part of the heat has not yet been released or has already been dissipated into the surrounding medium. Jouguet and Becker note that, relative to the states of the reaction products which satisfy the conservation equations (points on the Hugoniot adiabatic curve CBG), the entropy at B is at a minimum. But the other points in this case correspond to a different velocity and a different regime (not to other states in the same regime), and comparison with them does not allow us to draw any physical conclusions.

of inert mixtures of carbon monoxide and methane are of very great interest and are a valuable verification of the views presented above. It is quite instructive that the admixtures which do not influence the detonation velocity in any way noticeably expand the limits (of concentration and pressure) in which propagation of detonation is possible. These same admixtures expand the limits and accelerate thermal self-ignition of the mixtures, as is correctly pointed out by the authors cited.

§II.6 Application to High-Temperature Chemistry

The propagation velocity of detonation in the absence of losses and after selection of a specific state of the reaction products turns out to be dependent on the thermodynamic properties of the original mixture: the reaction heat, the change in the number of molecules during the reaction, the specific heat and dissociation at the temperatures which develop.

The first studies of the behavior of gases at high temperatures were done by Bunsen and Le Chatelier using explosions. Without quantitative information on the dissociation, Le Chatelier arrived at expressions for the specific heat which differ significantly at high temperatures from the true values known to us today (e.g., for CO_2 at $3000°K$, $c_v = 26.7$ kcal/mole instead of the true value of 13.1).

Jouguet's first calculations, carried out using these values for the specific heat, yield satisfactory agreement. In fact, the reason for this is that the specific heat and dissociation, each incorrect on its own, were selected by Le Chatelier in accord with the maximum pressure of explosion in the same mixtures in which the detonation velocity was measured. The agreement of the velocity with that calculated by Jouguet does not prove the validity of all the individual assumptions in his calculation; approximately the same agreement may be obtained by applying a formula which directly relates the velocity of detonation with the explosion pressure of a given mixture in a closed volume:

$$D^2 = 4.44 p_{\text{expl}} v_0$$

(all quantities are in absolute units: D—in cm/sec, p—in dyne/cm^2, v—in cm^3/g).

Today our knowledge of the specific heat and dissociation of gases at high temperatures has made extraordinary progress, primarily through the work of American scientists. However, even now, in order to solve a number of problems in high-temperature chemistry, analysis is needed of experimental data on phenomena of combustion, explosion and detonation.

Determination of the detonation velocity offers a number of advantages, both practical and fundamental, over measurement of the pressure of explosion: the equipment is simpler, no mechanical system is required and the experimental accuracy is higher.

True, calculations related to a detonation wave are more complex than the elementary calculation of an explosion, but in fact the phenomenon of explosion is complicated by non-uniformity of the temperature of the explosion products.

We have used data available in the literature on detonation velocity to solve several problems in high-temperature chemistry (the calculations were carried out in collaboration with a post-graduate student, S. B. Ratner).

In detonation of a mixture of dicyanogen with oxygen, extremely high temperatures develop as a result of the exothermic character of dicyanogen: according to Michelsohn's approximate evaluation (1920) the temperature exceeds 6000°K.

Calculation of the detonation velocity of the mixture $C_2N_2 + O_2$ is related to the calculation of the dissociation of carbon monoxide with the formation of a free carbon atom. The chemical constants of the substances which participate in the reaction (CO, C, O, O_2) are sufficiently well known. However, there still exists great uncertainty about the magnitude of the dissociation energy of carbon monoxide, and discussion continues even now.* The magnitude of the dissociation heat of carbon monoxide is also of much interest because it is related by simple thermochemical relations to the evaporation heat of carbon and to the energy necessary to break up any organic molecules into their component atoms.

In the literature the following values for the heat of dissociation of carbon monoxide are cited (kcal/mole): 166—Schmid (1938), 210—Herzberg (1937), 256—Kohn (1920), Kinch and Penney (1941). In his last note, Herzberg concluded that the dissociation heat is not more than 221 kcal/mole.

We shall show how different assumptions about the dissociation heat influence the composition of the reaction products at identical temperatures (Table 6).

Results of the calculation of the detonation velocity under the three assumptions are given in Table 7 (the first column gives the adopted value of the heat of dissociation Y, the second—the temperature of the reaction products, and the third—the detonation velocity D).

Dixon experimentally determined the detonation velocity of an equimolecular mixture as 2728 m/sec (1903). In a later paper, Campbell (1922) found a velocity of 2667 m/sec.

The observed value of the detonation velocity best agrees with the assumption of maximum dissociation heat and a minimum amount of atomic carbon in the reaction products. It should be noted that the lower the amount of atomic carbon, the less sensitive is the detonation velocity to the choice of the dissociation heat: a change in Y of 44 kcal (from 166 to 210) changes the velocity by 170 m/sec, while a change in Y of 46 kcal (from 210

*See: *Termodinamicheskie svoistva individualnyx veshchestv, Spravochnik* [*Handbook of Thermodynamic Properties of Individual Substances*]. Moscow: Nauka, 1979, V. 2, Book 1, p.25.—Editor's note.

Table 6

Influence of the Dissociation Heat on Composition of the Reaction Products (in %)

Q, kcal/mole	CO	N_2	C	O	N	NO	CN	CO_2
				$T = 4000°K$				
166	59.0	32.3	4.1	0.8	0.7	0.6	0.0	2.5
				$T = 5000°K$				
166	45.6	26.3	12.8	$8.1 + 1\% O_2$	3.9	2.2	0.1	1.1
210	52.6	29.6	1.2	1.2	4.3	0.3	0.6	0.2
256	64.0	29.7	0.0	0.8	4.3	0.2	1.0	0.0
				$T = 6000°K$				
210	51.4	21.0	5.9	5.5	13.7	1.3	1.1	0.1
256	57.8	20.5	0.4	2.7	14.9	0.6	3.1	0.0

Table 7
Calculation of the Detonation Velocity

Y, kcal/sec	T, °K	D, m/sec
166	4640	2440
210	5420	2610
256	6610	2660

Table 8

Composition	Assumption	T, °K	Calculated D, m/sec	Measured D, m/sec Dixon	Measured D, m/sec Campbel
I	First	3655	2040	2110	—
	Second	4095	2135		
II	—	4395	2265	2166	2230

I: $C_2N_2 + 3O_2$
II: $C_2N_2 + O_2 + 2N_2$

to 256) changes the velocity by only 50 m/sec.

Realistically evaluating the accuracy of the calculation and experiment, we may conclude that the data on the detonation velocity in any case exclude the possibility of a dissociation heat smaller than 210 kcal. Thus, these data prove to be a valuable supplement to Herzberg's considerations and, taken together, determine:

$$Y = 220 \text{ kcal/mole}, \qquad L = 134 \text{ kcal/atom}$$

(L is the sublimation heat of a carbon atom).

We paused to consider this example in detail because of a natural interest in the highest temperatures which develop in a chemical reaction. No less interesting are the results of an investigation of diluted mixtures of dicyanogen with oxygen. Dixon's data, according to which the velocity is greater in mixtures diluted with nitrogen than in mixtures diluted with oxygen, are cited everywhere as proof that the oxidation of CO to carbonic acid does not have time to occur in the wave.

Our calculations disprove this point of view. The fact is that at high temperatures oxygen intensively consumes energy for dissociation; without the reaction of carbon monoxide with oxygen, the velocity in a mixture containing excess O_2 would be even smaller than the experimental value. On the other hand, the dissociation of carbonic acid limits the completeness of the carbon monoxide reaction.

In Table 8 we show a comparison with experiment of the calculated detonation velocity under each of two assumptions: (1) the usual assumption that the velocity of the reaction $2CO + O_2 = 2CO_2$ is so small that CO_2 does not form in the wave at all and should not be considered in the calculation of the velocity; or (2) the opposite assumption that the reversible reaction $2CO + O_2 = 2CO_2$ occurs so rapidly that a thermodynamic equilibrium is established in the wave which corresponds to the pressure and temperature of the products.

Also shown are data for a mixture diluted with nitrogen in which the formation of CO_2 cannot occur because of insufficient oxygen, and both assumptions yield a single result.

The second assumption yields significantly better agreement with experiment than the first.

Thus, experimental data, when analyzed in quantity, lead to conclusions which are diametrically opposed to those reached by the experimenter, and other authors after him.

How the reaction mechanism is reflected in the detonation velocity may be seen from the following comparison of two reactions: hydrogen with oxygen and hydrogen with chlorine. The detonation velocity, calculated under the assumption that complete chemical equilibrium is achieved, depends equally in both cases on the pressure. In the case of hydrogen with oxygen, the dissociation of H_2O in either direction of the reaction

$$2H_2O = 2H_2 + O_2, \quad 2H_2O = H_2 + 2OH,$$
$$H_2O = H + OH, \quad H_2O = O + H_2,$$

takes place with an increase in the number of molecules and, consequently, is suppressed when the pressure is increased.

In the case of hydrogen with chlorine, it would appear that dissociation by the equation

$$2HCl = H_2 + Cl_2 \qquad\qquad (I)$$

leads to independence of the composition, temperature and velocity from the pressure. In fact, the dissociation basically runs according to the equation

$$2HCl = H_2 + 2Cl \qquad\qquad (II)$$

and, consequently, depends on the pressure. Thus, for initial atmospheric pressure, and taking into account all possible reactions, the calculation yields the following equilibrium state: temperature—3130°K; pressure—21 atm; composition—HCl 80.1%, H_2 5.8%, Cl_2O 2%, H 1.4%, Cl 12.5%.

The concentration of molecular chlorine which is obtained in reaction (I) is small compared to the concentration of atomic chlorine which forms from

Table 9

Mixture composition	p_0, mm	Calculated		Measured			
				Dixon		ICP	
		D, m/sec	Δ, %*	D, m/sec	Δ, %*	D, m/sec	Δ, %*
$H_2 + O_2$	200	2760	2.5	2627		2685	7.1
	760	2820		2821	6.9	2835	
$H_2 + Cl_2$	200	1680	2.3	—	—	1729	0.7
	760	1720		—	—	1741	

$$* \ \Delta, \% = \frac{(D_{760} - D_{200})}{D_{760}} 100\%.$$

HCl in reaction (II) and partly in the reaction $HCl = H + Cl$ with an even greater expansion of the volume.

In Table 9 we compare the results of thermodynamic calculations and measured values of the detonation velocity for initial pressures of 200 and 760 mm [data are from Dixon (1903) and the Institute of Chemical Physics (1934)].

While the pressure dependence of the calculated velocity in both mixtures is the same, the dependence found in experiment is significantly different. How can this difference be explained?

The reaction of hydrogen with oxygen runs according to a branching chain mechanism:

$$H + O_2 = OH + O,$$
$$O + H_2 = H + OH,$$
$$OH + H_2 = H_2O + H.$$

If we sum these three reactions, we obtain

$$H + O + OH + O_2 + 2H_2 = H_2O + O + 2OH + 2H.$$

Eliminating terms common to both sides, we find

$$O_2 + 2H_2 = H_2O + H + OH.$$

This is the result of one complete link of the chain: by the formation of one molecule of water two new active centers are obtained. The reaction is practically thermoneutral. In the first stage of the reaction large numbers of centers are accumulated; it is through recombination of these that the reaction heat is released. Recombination ceases when an equilibrium concentration of centers is reached. The recombination rate is proportional to the cube of the pressure.

In the reaction of hydrogen with oxygen equilibrium is approached from the side of excess atoms and radicals, and the amount of reaction heat released asymptotically and gradually approaches the thermodynamic limit.

Delayed heat release at low pressure increases losses and causes an additional (beyond that depending on the equilibrium dissociation) decrease in the velocity (see § II.7 below).

The situation is different for a mixture of hydrogen with chlorine. The reaction mechanism is known—the Nernst chain; subsidiary reactions with admixtures and recombinations become of secondary importance at high temperatures in a detonation wave:

$$H + Cl_2 = HCl + Cl,$$

$$Cl_2 + H_2 = HCl + H.$$

The result is the formation of two molecules of HCl from H_2 and Cl_2 without a change in the number of active centers H and Cl.

Thus, the reaction

$$H_2 + Cl_2 \rightleftharpoons 2HCl$$

may run to equilibrium at any concentration of active centers and, conversely, the formation of hydrogen chloride does not itself cause an increase in the number of centers.

The activation heat of the reactions $Cl + H_2$ and $H + Cl_2$, comprising the essence of the chain, is quite small: the activation heat of the dissociation reaction of chlorine, which supplies active centers, is significantly greater. Therefore formation of a large amount of hydrogen chloride is quite possible before an equilibrium amount of chlorine atoms is formed.

Heat release in the first stage of the reaction is replaced by the endothermic reaction of dissociation of Cl_2 with the formation of atomic chlorine. Our theory leads to the conclusion that in the general case the detonation velocity is determined not by the final state of complete equilibrium, but by the state in which the maximum amount of heat is released. For hydrogen with oxygen the heat release, at first small, continues until equilibrium is reached. For hydrogen with chlorine the possibility of the release of an excess amount of heat which is subsequently absorbed (approach to equilibrium from the other side) was shown above.

Thus the fundamental possibility is discovered of explaining the independence observed for hydrogen with chlorine of the velocity from the pressure and the greater-than-calculated value of the velocity.[9]

The material presented contains neither a full analysis of the kinetics of individual chemical reactions in a wave, nor certainly proof of any particular mechanism based on data on the detonation of the corresponding mixtures.

[9]If we were considering the reaction of hydrogen with chlorine in a vessel with constant volume, the maximum pressure would be achieved not in the equilibrium mixture, but in the mixture in which formation of hydrogen chloride had occurred, but its dissociation to chlorine atoms had not; the value of the maximum pressure would exceed that calculated for the equilibrium state. We noted the relation between the pressure and the detonation velocity at the beginning of this section. Compare also the conditions in the system $H_2 + Cl_2$ with those necessary for occurrence of the state G (see Fig. 15, §II.4) with an increased detonation velocity.

Our goal was to analyze, using the examples given, the kinds of complications and peculiarities of detonation which may be expected in complex reactions, i.e., in the vast majority of combustion reactions.

§II.7 Losses and the Propagation Limit

In the first studies of detonation the efforts of scientists were directed toward proving general agreement between the calculated and measured values of the detonation velocity. Later, interest was concentrated precisely on the small differences between the calculated and measured values of the velocity. This interest was based on the idea that the lag of the measured value from the calculated one was due to the fact that part of the reaction energy did not have time to be released in the wave and was released somewhere later, no longer influencing the detonation velocity.

However, attempts to formulate a condition on which the use of some part of the heat in a detonation wave depended were unsuccessful: its formulation was in fact impossible without establishing the reaction mechanism in the wave and without giving a fundamental analysis of all the questions of the theory of the detonation velocity.

According to our theory, for steady propagation of detonation in a tube, a reacting substance compressed by a shock wave moves such that the propagation velocity of a perturbation in it exceeds the detonation velocity (the segment YB of the line ABY of Fig. 15),

$$D < w + c, \tag{II.7.1}$$

where w is the velocity of the substance, c is the velocity of sound in it, and D is the detonation velocity.

After completion of the reaction, braking and cooling of the reaction products occurs. However, these processes now run in the state in which

$$D > w + c. \tag{II.7.2}$$

The course of dissipative processes after the reaction therefore no longer has any influence on the process in the wave, in particular, on the detonation velocity.

Thus, the boundary of the "wave" should be considered in fact to be that state B in which

$$D = w + c. \tag{II.7.3}$$

In a quite general form we have been able to show that this state is achieved at the moment when the substance contains the maximum amount of heat, i.e., when the release of reaction heat is just balanced by the heat transfer and braking, so that beyond this point the heat transfer[10] outweighs

[10]More precisely, heat transfer and braking by friction against the tube walls; the influence of braking with a particular transfer coefficient is equivalent to the influence of heat transfer, as a detailed mathematical analysis shows. Below, when we speak of heat

the heat release.

Thus we answer the question, at what reaction rate is the reaction heat used in the wave? It is necessary that the rate of heat release exceed the rate of heat transfer.

Thus the reaction rate sought is not constant and, for example, in very wide tubes (underground vaults) the combustion energy of coal dust may be used in the wave, although in laboratory equipment combustion of dust proves so slow that detonation is out of the question.

We return to the detonation velocity. The condition at the boundary of the wave in state B requires that a certain reaction rate be maintained; this is accompanied by a certain incompleteness of combustion, an unreacted remainder which is proportional to the intensity of the heat transfer and inversely proportional to the specific rate of the reaction. This is the first source of energy loss. More significant in reality are losses in the first stage, during the reaction on the line YB. The losses here are less than the heat release, but must nevertheless be considered in the energy balance. To find the detonation velocity we construct the equations determining the state B.[11] The losses here must be accounted for in constructing the momentum equation (in it the forces of friction against the walls must be taken into account), and in the energy equation where, besides the chemical energy remainder in the reaction products in the state B, we must account for heat losses to the walls of the tube during compression of an element of gas until it attains the state B.

The absolute quantity of the losses and the loss-dependent decrease in the detonation velocity are inversely proportional to the reaction rate. In a narrow tube, where the intensity of heat transfer is greater, the decrease in the detonation velocity is greater, etc.

An extremely valuable confirmation of these conceptions of the character of the losses is provided by the data of Shchelkin (Institute of Chemical Physics) who studied detonation of gaseous mixtures in tubes whose internal surface was made very rough.

These experiments were carried out in connection with Shchelkin's conception of the role of gas turbulization in the appearance of detonation. However, they simultaneously provided valuable material on the influence of external conditions on the steady propagation of detonation. The detonation velocity fell noticeably in rough tubes compared with smooth ones,

---

transfer, we mean both factors. On the other hand, together with the release of reaction heat, the change in the number of molecules in the reaction also enters into the expression (in approximately the same combination as that of the reaction heat and the number of molecules in the expression for the pressure of explosion).

[11] As a result of losses the state which occurs in fact differs from the point B on the Hugoniot adiabate calculated without losses: we should have called it B'. When drag is taken into account the line of state change in the reaction YB differs from a straight line. However, in the state B' the basic property which distinguished the state B in the theory without losses is preserved, namely the condition of equality of the detonation velocity and perturbation velocity, $D = w + c$.

which proved the role of the hydraulic drag of the tube and the role of losses to braking of the gas in the reaction process in a detonation wave.[12]

Accounting for losses makes it possible to understand the factors which in some explosive mixtures allow propagation of a detonation wave, while in others they do not, and a wave, once initiated, decays. In other words, we may now construct a theory of the limit of propagation of detonation.

The greater the duration of a chemical reaction, the more the detonation velocity decreases.

The detonation velocity in turn influences the pressure and temperature achieved during compression of the gas by a shock wave of a given velocity (the position of the point Y). A decrease in the detonation velocity brings with it a decrease in the temperature at which the reaction should start. For a reaction which runs most slowly at the beginning, the temperature achieved in the shock wave in fact basically determines the duration of the reaction. Thus, in an elementary formulation, the theory of the limit is reduced to three equations:

$$D = D_{\text{T}} - \beta\tau, \tag{II.7.4}$$

$$\tau = \gamma e^{A/RT_Y}, \tag{II.7.5}$$

$$T_Y = MD^2, \tag{II.7.6}$$

where D is the detonation velocity, D_{T} is the "theoretical" detonation velocity calculated in the absence of losses, τ is the duration of the reaction, β is a coefficient which characterizes the losses; it is calculated from the theory of turbulent heat transfer and drag and is proportional to the tube diameter. γ is a constant which is related to the composition of the mixture and characterizes the reaction rate, A is the activation heat of the reaction;

---

[12]We speak of losses (convective heat transfer and friction against the walls of the tube) for a very rapid process which ends before the substance has time to move a distance of several tube diameters in the tube: we measure this distance from the point at which the shock wave sharply sets the gas in motion and heats it to the point where a given element of the gas attains the state B. It is known that, over a small distance, the cooling and braking effect of the wall propagates directly to a comparatively narrow layer adjacent to the wall and does not cover the full cross-section of the tube. Does this not refute our conceptions? No, since the change in velocity and density (which depends on the temperature) of the gas in a layer adjacent to the tube wall will invariably influence the velocity in the central core of the flow as well, because the magnitude of the total matter flow over the entire cross-section of the tube is given and cannot change. In order to satisfy the condition of constant flow and to change the velocity in the core accordingly, the pressure in the given cross-section must change. In a flow in which the propagation velocity of a perturbation exceeds the detonation velocity (as occurs for states in a wave), a change in pressure in one cross-section will cause some restructuring of the entire flow; in particular, it will cause a change in the amplitude and velocity of the wave itself. Thus, a local direct effect of the wall on the material in the wave does not preclude the influence of corresponding heat and momentum losses on the entire regime and on the propagation velocity of detonation. Compared to the simplified one-dimensional calculation, in which it is assumed that heat transfer and friction are uniformly distributed over the entire cross-section, a fairly complicated calculation of the true picture, accounting for a non-uniform distribution over the cross-section, yields results for the detonation velocity which differ only by numeric factors.

T_Y is the temperature of compression in a shock wave of velocity D, M is a constant related to the molecular weight and specific heat of the mixture and is calculated from the theory of shock waves.

The first equation describes the decrease in the detonation velocity due to losses during the reaction. The second equation represents the dependence of the reaction time on the initial temperature. The third equation is the limiting dependence of the temperature of the gas compressed by a shock wave on the velocity of the shock wave; the formula takes a simple form for the case of strong shock wave, $D \gg c_0$ (c_0 is the speed of sound), $p_Y \gg p_0$: for detonation these relations are satisfied.

Study of this system shows that it certainly does not always have a real solution; a condition which is both necessary and sufficient for the existence of a solution may be written thus:

$$\frac{2e\beta\gamma MRD_{\mathrm{T}}}{A}e^{A/RMD_{\mathrm{T}}^2} \leq 1. \qquad (\mathrm{II.7.7})$$

It turns out here that for all possible regimes a small change in the velocity is characteristic. The stronger the dependence of the reaction rate on the temperature and the higher the activation heat, the smaller is this change:

$$\frac{D_{\mathrm{T}} - D}{D_{\mathrm{T}}} \leq \frac{RT_Y}{2A} = \frac{RMD_{\mathrm{T}}^2}{2A}. \qquad (\mathrm{II.7.8})$$

This conclusion is indeed in satisfactory agreement with the whole body of experimental material. In all cases we are dealing either with comparatively slow or non-steady flame propagation, or with detonation, whose velocity differs little from the calculated value. As conditions change in the direction inhibiting detonation (decrease in tube diameter, decrease in pressure, dilution of the explosive mixture with inert gases), the velocity falls and the difference between the calculated and measured detonation velocities grows. However, before this difference becomes significant and exceeds 10–15% of D, the limit is reached and steady propagation of detonation ceases altogether (Wendtlandt). Such losses in a tube of diameter 10–20 mm correspond to a reaction time which does not exceed $(2–3) \cdot 10^{-4}$ sec.

This quantity, small compared to induction periods measured by ordinary methods at lower temperatures, is quite large in comparison to the compression time in the shock wave: 10^{-10}–10^{-11} sec.

§II.8   Propagation of Detonation in Rough Tubes

We referred above to experiments by Shchelkin on detonation in rough tubes: spirals made of wire were inserted into glass tubes. The spirals were attached to the tube walls so that the tube surface become rough, with the magnitude of the roughness dictated by the diameter of the wire from which the spiral was made. In these experiments a significant decrease in the detonation velocity was observed compared to the calculated value and to the velocity observed in a smooth tube, other conditions being equal. The decrease in the velocity is natural in the sense that, for an equal reaction time, losses increase as a result of the increased drag coefficient. However, from the standpoint of the theory of limits we must now consider surprising, and seek special explanation for the fact that the increased losses do not lead to termination of detonation.[13]

In the published experiments of Shchelkin the velocity fell to 60–50% of D_T, which is incompatible with the theory of limits developed above. The behavior of detonation in rough tubes in this respect differs from the behavior of detonation in smooth tubes. We could increase the losses in the smooth tube too by decreasing its diameter, but then, instead of steady propagation with a lower velocity, we observe termination of the propagation of detonation.

How do we explain this special behavior of rough tubes? On the one hand, in a rough tube in which the drag dominates, after compression of the original mixture in the shock wave the pressure and temperature continue to increase. This is the difference between a rough tube and a smooth one of smaller diameter in which both the drag and the heat transfer are increased. The reaction time in a rough tube should be taken with respect, not to the temperature in the shock wave, $T_Y = MD^2$, but to another, higher temperature. As a result the reaction time is more weakly dependent on the velocity, which in fact gives a qualitative explanation for the larger possible decrease in the velocity.

On the other hand, in a rough tube one may imagine a fundamentally different mechanism of detonation propagation (we shall abbreviate it SM, for second mechanism). Let the detonation velocity be small, and let the temperature achieved in a shock wave of a given velocity be insufficient to bring about the chemical reaction, the ignition of the compressed mixture. In a smooth tube this is sufficient to terminate detonation. In a rough tube, the reflection of the shock wave off each irregularity is accompanied by a local increase in the pressure and temperature, and creates local foci of ignition of the mixture. Ignition near the surface sets off propagation of the flame from the periphery toward the center of the tube. While this combustion is taking place the shock wave moves forward, so that the combustion ends with a certain lag behind the shock wave. The surface of the flame (Fig. 17 [a–original mixture, b–compressed mixture, c–reaction products]) forms a cone with its base adjacent to the flat surface of the shock wave and its apex pointing in the direction opposite that of propagation (1 denotes the ignition

---

[13]If detonation does not propagate with the calculated velocity D_T, the reaction at the corresponding temperature MD_T^2 runs insufficiently rapidly. But then it is surprising that the reaction runs at a much lower temperature $T_Y = MD^2$, corresponding to the decreased velocity D.

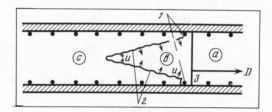

Fig. 17

foci, 2 is the flame front, 3 is the shock wave front).

Let us compare the macroscopic theory of such a regime with the theory of the normal mechanism (NM) of detonation, presented in §II.4 and II.5. In both cases a substance which is compressed by a shock wave and whose parameters are determined by its velocity enters into the reaction.

In the normal mechanism the reaction runs simultaneously over the entire cross-section of the tube: the curves presented in §II.5 illustrate the change in pressure, temperature and composition. We are fully justified in using an approach in which we consider all quantities characterizing the state to be dependent only on the distance of the point from the shock wave front. In the case of the SM, in the mechanism which we have proposed here for rough tubes, in each intermediate cross-section part of the substance has not reacted at all (the core of the flow) and part of the substance has completely reacted (the peripheral layers); the states of the two parts— composition, temperature, specific volume—are sharply different. The only common element is the pressure, which is practically identical in a given cross-section in the two parts of the flow (in the compressed, but unreacted mixture and in the combustion products), but which changes as combustion progresses from one cross-section to another.

The mathematical formulation of the theory becomes drastically more complicated; however, the physical conclusions in the part of the curve relating to the pressure change during the reaction, the selection principle, and the calculation of the detonation velocity and the effect of external losses on the detonation velocity remain practically unchanged. As was to be expected, a theory of pressure and velocity of a detonation wave based on the general conservation laws proves not very sensitive to the mechanism of chemical reaction.

Let us analyze another aspect of the problem and compare the reaction conditions themselves. In our presentation of the NM we emphasized that at enormous velocities neither thermal conduction nor diffusivity play a role; each element of the substance reacts adiabatically and is related to other elements only by the pressure. In the case of SM, meanwhile, we consider that the mixture enters the reaction due to flame propagation, i.e., to a process which depends on thermal conduction and diffusivity. Why, in this case, does the role of thermal conduction and diffusivity become noticeable and significant? The result has to do with rejection of one-dimensionality of the theory.

In the one-dimensional theory of NM we can imagine only a flat flame front; the temperature varies only as a function of the coordinate along which the flame propagates, and the direction of the temperature gradient coincides with the direction of propagation. The gradient is small, as is the surface through which heat is transferred (it is equal to the tube cross-section).

In the theory of the SM we reject one-dimensionality. The greatest temperature differences occur in each cross-section between the burnt and unburned substance.

The local temperature gradient in the flame zone is now much larger than that corresponding to a longitudinal change in the average temperature.

Moreover, the surface through which the thermal flux passes increases sharply as well, in the ratio of the cone surface to the surface of its base.[14] These reasons fully explain the role of thermal conduction in the SM; similar arguments may be applied to diffusion.

Why can't the SM occur in a smooth tube? In this case, after identical compression in the shock wave over the entire cross-section, cooling of the substance in the peripheral layer occurs. Therefore, either the temperature of compression is sufficient and the reaction takes place over the entire cross-section (NM), or the reaction does not occur at all. In the theory of the SM, flame propagation ensures combustion of the mixture over the entire cross-section of the tube after ignition at the walls; however, a mechanism of ignition which would follow the propagation of the shock wave is also necessary. For this the propagation speed of the flame is insufficient; in rough tubes this mechanism is the ignition when the shock wave is reflected off irregularities.

The overall reaction time, which determines the losses, for the SM is the time the flame propagates from the periphery to the center. This time depends only weakly on the temperature achieved in the shock wave, whence follows the possibility (from the standpoint of the theory of detonation limits) of significant losses and a significant decrease in the detonation velocity for the SM.

§II.9 Theory of the Origination of Detonation

Until very recently the origination of detonation in the thermal ignition of a mixture, i.e., the transition from slow combustion to detonation, remained completely unexplained.

---

[14]The effect of unevenness of the cone surface, due to turbulization of the compressed gas flow, is added here as well.

Detonation is the propagation of a reaction preceded by compression of the gas by a shock wave of enormous amplitude; the pressure in this shock wave reaches dozens of atmospheres. How does such a wave form in the propagation of a flame when the flame velocity with respect to the gas in the most rapidly burning mixtures does not exceed 10 m/sec? The transition from combustion to detonation takes place through acceleration of the flame, whose velocity passes through intermediate values of 300, 500 m/sec. What is the mechanism of the motion of a flame at such velocities, too large for propagation of thermal conduction, and too small for us to assume ignition by a shock wave?

In flame propagation in tubes, as a rule, one observes a definite stage of uniform flame distribution (the "uniform movement" of English authors) at a low velocity. What are the reasons for the acceleration of the flame?

Alongside the difficult physical problems, there were formal difficulties as well. The three equations of steady propagation of a reaction through a mixture in a given initial state—the equation of matter conservation, the momentum equation and the energy balance—have two groups of solutions. One group corresponds to a propagation velocity ranging from 0 to 100–200 m/sec, with combustion accompanied by expansion and an insignificant drop in pressure; the other group describes propagation of detonation at a velocity of 2000–4000 m/sec with a decrease in the specific volume and multiple pressure increases.

First of all we must imagine a mechanical picture of flame propagation with an intermediate velocity that does not decrease in the theory of the Hugionot adiabate created by Chapmann and Jouguet.

We consider some concrete experimental set-up, for example, flame propagation in a long tube with a constant cross-sectional area and ignition at the closed end. Combustion of the mixture and the gas expansion which results cause motion of the explosive mixture before the flame front. If the average velocity of the flame with respect to the original mixture is equal to $\bar{u}$, then in unit time a volume $\bar{u}S$ of the mixture burns, where S is the tube cross-section. The volume of the combustion products formed is $n\bar{u}S$ where n, a coefficient characterizing the change in volume during combustion, depends on the ratio of the temperature and number of molecules before and after the reaction; usually n lies between 5 and 10.

When the volume $\bar{u}S$ of the original mixture is replaced by volume $n\bar{u}S$ of the hot reaction products, the unburned mixture must move so as to free a volume equal to $(n-1)\bar{u}S$. Thus, the motion of the flame from the closed end at a velocity $\bar{u}$ results in motion of the original mixture with a velocity of $(n-1)\bar{u}$, i.e., 4–11 times greater. The flame acts as a moving piston causing the gas before it to move.

This motion does not immediately encompass the entire gas: the perturbation propagates at low amplitude with the velocity of sound. In Fig. 18

(where *1* denotes the original mixture, *2*—the compressed, moving mixture, and *3*—the reaction products) we show the pressure distribution *t* sec after ignition (ignoring losses by heat transfer and braking). The segment *0–I* is occupied by combustion products at rest and the segment *I–II*, by the original mixture moving at a velocity of $(n-1)\bar{u}$. To the right of *II* stretches the unperturbed original mixture at rest; *I* is the flame front, moving at a velocity $n\bar{u}$ with respect to a motionless coordinate system and at a velocity $\bar{u}$, in accord with the condition, with respect to the moving mixture; *II* is the shock wave front.

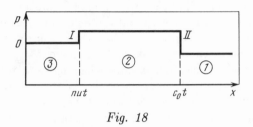

Fig. 18

In this formulation of the problem we may construct a regime with any flame speed with respect to the unperturbed—far from the flame front—mixture. In particular, all intermediate values of the velocity between slow combustion and detonation are possible. Why have they now become possible, while in the equations of the Hugoniot adiabate no real solutions corresponded to intermediate values of the velocity?

If we consider in the regime described in Fig. 18 the propagation of a flame with respect to the original mixture at rest, we see that it has become non-steady: over time the zone occupied by compressed but unreacted gas, *I–II* in Fig. 18, grows. On the other hand, we see that the flame now directly propagates not through the motionless mixture whose state was given, but through the compressed and moving mixture whose state itself depends on the flame velocity.[15] As $\bar{u}$ increases, the difference between the shock wave velocity and the propagation velocity of the reaction decreases and, finally, at the detonation velocity they coincide; the reaction steadily, without falling behind, follows the compression in the shock wave, forming a steady, non-deteriorating complex. This takes care of the mechanical possibility of intermediate velocities and the character of the motion in regimes between slow combustion and detonation. We are now prepared to analyze the physical causes of rapid combustion and flame acceleration.

The flame, like a piston, moves and compresses the gas before it. To flame propagation at normal velocity in a detonation mixture of $2H_2 + O_2$

[15]The author investigated, with similar results, propagation regimes for all intermediate values of the velocity in other cases as well, in particular, for the case of steady propagation in a cylindrical tube, accounting for losses to heat transfer and friction.

(about 10 m/sec) corresponds a gas velocity up to 70 m/sec; this velocity is small compared to the velocity of sound in this mixture (500 m/sec). In a shock wave caused by flame propagation, the pressure increases from 1 to 1.5 atm, and the temperature grows by 12°. This change of state in the burning mixture only insignificantly changes the flame velocity and cannot explain the acceleration of the flame.

A very important conclusion is that the acceleration of the flame depends on the gas motion itself; as long as all the gas moves as a whole at constant velocity over the entire cross-section, propagation of the flame through the moving gas does not differ from flame propagation through a gas at rest. But, as the gas contained in a tube with immobile walls moves, the gas layers adjacent to the wall are slowed and the motion of the gas in the center of the tube accelerates accordingly. The distribution of the velocity over the cross-section becomes non-uniform. There appear streams of gas whose velocity is less than average, and streams which move more rapidly.

Under these conditions the velocity of the flame with respect to the gas (the difference between the flame velocity and the average velocity of the gas) increases and the amount of gas which burns in unit time increases; the motion of the flame is determined by the maximum velocity of the gas stream,

$$\bar{u} = u_{\mathrm{N}} + w_{\max} - \bar{w},$$

where $\bar{u}$ is the flame velocity with respect to the average motion of the gas, u_{N} is the normal flame velocity (with respect to the motionless gas), $w_{\max}$ is the maximum and $\bar{w}$ the average velocity of the gas. The corresponding increase in the overall amount of gas which burns in unit time is explained by the fact that in a stream with variable velocity across the cross-section the flame front bends; as a result its surface increases and the amount of the substance burned increases proportionally.

K. I. Shchelkin ascribed the acceleration of the flame to turbulization of the moving gas. However, evaluations based on the theory of turbulent combustion show that the influence of turbulization is smaller than the influence of the velocity profile, which should be considered the basic one.

Thus the picture of the origination of detonation reduces to the following: combustion causes motion of the gas before the flame front; braking of a layer of gas adjacent to the walls leads to restructuring of the velocity profile; the change in the profile causes an increase in the combustion rate, which in turn leads to an increase in the velocity of the gas; corresponding to the increased velocity is an increased flame velocity, and so on. As the flame accelerates the amplitude of the shock wave grows as well. Finally, the compression temperature attains the self-ignition temperature of the mixture. The most difficult phase of the initial acceleration of the flame to explain is now complete. It may be easily proved that the subsequent development of the phenomenon after self-ignition leads to propagation of detonation. Indeed, self-ignition is characterized by a specific value of the induction period; each particle of the gas will enter into the reaction within a certain time after compression by the shock wave. The motion regime will change until the velocity of the shock wave attains the steady velocity of detonation. Here the state of the compressed gas during the induction period is described by the point Y and the state of the reaction products by the point B (see Fig. 15). Only in such a regime can the propagation of the shock wave and the chemical reaction which follows it with a certain constant lag-time become steady.

To what extent does this conception agree with the basic facts?

Our conception ascribes an essential role to braking of the gas by the tube walls. Experiment shows that as the tube diameter increases the distance at which detonation arises also increases. In spherical bombs with central ignition occurrence of detonation has never yet been observed.

The distance at which the transition from combustion to detonation takes place reaches several dozen tube diameters. This distance coincides in order of magnitude with the length over which, according to hydrodynamic data, the velocity profile is established.[16]

The extraordinarily striking fact that transition from combustion to detonation is eased in rough tubes was discovered by Shchelkin, who was led in this by certain ideas about the role of gas turbulization.

In order to interpret this recently discovered, but absolutely fundamental fact, we shall consider more carefully the conditions of the gas motion. The flame functions as a piston, and the dependence written above of the gas velocity on the flame velocity, $w = (n-1)\bar{u}$, is valid insofar as the combustion products do not cool. Therefore, for detonation to occur the ratio of the drag and heat transfer is of particular importance. It is precisely in rough tubes that conditions are most favorable: the increased drag accelerates the establishment of the velocity profile, while the heat transfer remains practically unchanged by the introduction of roughness.

We presented the theory as applied to ignition of a mixture at the closed end of a tube. Ignition at some distance from the closed end increases the amount of the substance which burns in unit time (since combustion will propagate in both directions from the point of ignition), and will accelerate the gas motion which depends on its expansion during combustion. And indeed, experiment shows some decrease in the distance at which detonation appears.

In contrast, for ignition at the open end of a tube the expanding combustion products flow out into the atmosphere and create much lower compression and gas motion before the flame front. However, as the flame moves

[16]The pressure and temperature change in a compression wave corresponding to a change in the combustion rate occurs extraordinarily rapidly. If this change were the cause of the flame acceleration, the transition from combustion to detonation would occur at a distance which exceeds the width of the reaction zone in the flame by only a few times, i.e., at a distance of not more than a few millimeters.

further into the tube, the outflow of combustion products is made more difficult as a result of the hydraulic drag of the tube segment already traveled, and the compression and velocity of combustion-induced motion increase. We approach the conditions of steady flame propagation in a long tube (see footnote 16). Thus, ignition at the open end hinders the occurrence of detonation, but does not make it impossible. This conclusion agrees with experiment. It is also completely natural that the introduction of roughness eases the transition from combustion to detonation in the case of ignition at the open end as well.

In order for detonation to arise, it is necessary that upon achievement of a certain amplitude of the shock wave, self-ignition of the mixture occur: this self-ignition, at some distance before the flame front, i.e., where direct thermal or diffusive effects by the flame are excluded, was noted by Bone and Becker in some cases.

However, we know that the kinetic characteristics on which the flame velocity depends differ from the factors which determine the conditions of self-ignition: in particular, self-ignition is relatively hindered, while flame propagation is eased in the case of an autocatalytic reaction or a reaction with branching chains.

The initial acceleration of the flame, which determines the impossibility of its propagation at a constant low velocity, depends, as we have seen, on the flame velocity and on the hydrodynamic conditions of propagation.

It is quite possible here that in the acceleration of the flame, despite the amplitude increase of the shock wave preceding the flame, the conditions of

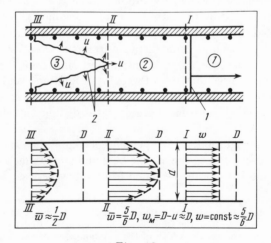

Fig. 19

self-ignition will not be achieved and self-ignition will not occur. We obtain a peculiar steady regime of rapid flame propagation. Its velocity with respect to the motionless gas (which is far ahead of it) can exceed the velocity of sound. In fact, the flame propagates in this case through a mixture which is compressed by the shock wave, so that the flame velocity with respect to the compressed gas is subsonic, though of the same order as the velocity of sound. The true (normal) propagation velocity of the flame with respect to the gas always comprises only a small fraction of the velocity of sound (see §I.3). In accord with the theory developed above for flame acceleration, a high flame velocity is possible only due to braking of the gas streams adjacent to the walls and to the related acceleration of the central streams which carry the flame forward. In Fig. 19 we show the location of the shock wave (*1*) and the flame (*2*). The circled numbers *1–3* represent the original mixture, the compressed mixture and the reaction products, respectively. In the lower part of the drawing we show the form of the velocity profile in three cross-sections. We shall call such a propagation regime TM (third mechanism, or braking mechanism). An exact calculation of the properties of the TM requires a difficult, detailed analysis of the concrete hydrodynamic conditions.

General assertions which may be made concerning the TM reduce to the following.

1. In order for propagation to occur at a high velocity which far exceeds the normal flame velocity, it is necessary that the braking (friction of the gas against the walls) be significantly more intensive than the heat exchange between the gas and the wall. This relation in fact occurs in rough tubes.

2. The propagation velocity of the TM does not exceed 85–87% of the theoretical detonation velocity D_T. Calculation of D_T is carried out under the assumption of a chemical reaction which runs after compression by the shock wave without any thermal or hydrodynamic losses. In the case of the TM, meanwhile, the very possibility of propagation of a fast flame with the velocity of the shock wave depends on a velocity redistribution as a result of braking of the layers adjacent to the wall. In constructing the equations for the motion as a whole, braking plays the role of a loss which reduces the velocity. In fact, the velocity will be even smaller than the value cited; besides the losses in the hydrodynamic preparation zone (the zone of velocity redistribution between the shock wave front and the forward point of the flame front, zone *I–II* in Fig. 19) we must add the losses in the combustion zone (from the forward point of the flame front to the cross-section in which combustion has ended, zone *II–III* in Fig. 19).

3. In its external manifestations—the pressure developed, destructive action, product temperature and similar characteristics—the TM differs little from detonation in the strict sense, i.e., from the NM.

The concept of the TM is necessary to describe experimental regimes of flame propagation (observed by Ditzen and Shchelkin[17]) in mixtures of carbon monoxide with air and mixtures of ether with air in long tubes with a high degree of roughness. In these experiments stable, steady flame propagation at a constant velocity of order 300–500 m/sec was observed (with the velocity depending on the mixture composition and on the diameter and roughness of the tube). At this velocity neither the compression in the shock wave, nor the additional compression by reflection of the shock wave off irregularities is sufficient to cause ignition of the mixture.

A puzzling problem is that of detonation spin—the propagation of detonation along a spiral path—discovered by Campbell in 1922.[18] Theoretical attempts to describe the spin as periodic propagation are at variance with the observed picture of the phenomenon. Apparently, we are dealing with a phenomenon of the same character as the SM or TM, in which simultaneous self-ignition over the entire cross-section does not occur. More favorable conditions for ignition or propagation occur at a specific point which moves in a spiral; the reasons for this are unknown today. We may assert only that these reasons are internal and depend on a peculiar distribution of the pressure and motion of the gas, not on interaction with the wall, which was a key factor in the cases of the SM and TM in rough tubes. This is confirmed by the results of a calculation by Ditzen and Ratner,[19] who showed that the velocity of propagation of detonation in a mixture of $2CO + O_2$ in the presence of a sharply defined spin does not differ from that calculated theoretically under elementary assumptions without accounting for losses.

We now complete our presentation of the results of work in the theory of propagation and the origination of detonation. Briefly recounted, the results include the completion of the classical theory of Chapmann and Jouguet, analysis of the conditions of chemical reaction in a wave and the relation of the reaction mechanism to the theory of the detonation velocity. The role of losses and the conditions which determine the possibility of detonation combustion of a given mixture were analyzed, a theory was given for the origination of detonation, and the detonation theory was applied to a number of problems in high-temperature chemistry. This development of the theory took place in a period of intense multiplication of new experimental facts which did not fit the classical interpretations, a period when expansion of the narrow boundaries of the classical theory, which ignores all questions of chemical kinetics, became an urgent necessity.

[17]Unpublished paper of the Institute of Chemical Physics, 1940.

[18]In 1903 Dixon noted a "periodic distribution of luminescence, too regular to be accidental." In a series of elegant experiments, for example, by synchronous photography in two perpendicular planes and photography from the end of the tube, Campbell showed that, in fact, it is a helical, spiral motion that occurs which only appears periodic in one projection.

[19]Unpublished paper of the Institute of Chemical Physics, 1940.

In turn, the concepts developed in the present work relate the phenomenon of detonation to broad areas of gasdynamics and chemical kinetics and advance new experimental problems whose solution belongs to the future.

The basic ideas of this monograph were conceived, theoretically developed and subjected to experimental testing in the process of creative work by the collective at the combustion laboratory of the Institute of Chemical Physics, and I consider it my pleasant duty to note particularly the participation of G. A. Barskiĭ, N. P. Drozdov, V. I. Kokochashvili, O. I. Leipunskiĭ, P. Ya. Sadovnikov, D. A. Frank-Kamenetskiĭ and K. I. Shchelkin and I thank them here for permission to use the results they obtained.

In concluding this work I wish to express my sincere gratitude to the Director of the AS USSR Institute of Chemical Physics, Academician N. N. Semenov, for his lively interest and valuable discussions, which in large measure determined the direction of our work in a period when the fundamental thermal, diffusive and hydrodynamic constructions were being overgrown with studies of specific chemical systems. I would also like to thank Academician A. F. Ioffe for his attention to the author's work.

References

I. Theory of Gas Combustion

Belyaev A. F.—Zhurn. Fiz. Khimii **12**, 92 (1938); **14**, 1000 (1940).
Kokochashvili V. I. Candidate Dissertation in Phys./Math. Sci., Moscow: AS USSR Institute of Chemical Physics, Tbilisi Univ. (1942).
Michelsohn V. A. Sobr. Soch. Moscow: Novyĭ Agronom **1**, 56 (1930).
Sadovnikov P. Ya.—Nauch.-Tekh. Byul. Ins-ta Khim. Fiziki AN SSSR, 32 (1941).
Semenov N. N. Tsepnye reaktsii [Chain Reactions]. Leningrad: Goskhimtekhizdat, 555 p. (1934).
Clusius K., Cutschmid G.—Ztschr. Elektrochem. **42**, 498 (1936).
Coward H. F., Brinsley F.—J. Chem. Soc. **105**, 1859 (1914).
Daniell P.—Proc. Roy. Soc. London A **126**, 393 (1930).
Goldmann R.—Ztschr. Phys. Chem. B **5**, 307 (1929).
Haber F. Thermodynamik technischer Gasreaktionen. Leipzig, 263 p. (1905).
Holm J.—Philos. Mag. **14**, 18 (1932).
Jahn G. Der Zündvorgang in Gasgemischen. Berlin (1934).
Jouguet E. Méchanique des explosifs. Paris (1917).
Lewis B., Elbe G.—J. Chem. Phys. **2**, 537 (1934).
Mallard E., Le Chatelier A.—Ann. Mines **4**, 274 (1883).
Passauer H.—Gas. und Wasserfach. **73**, 313 (1930).
Payman W.—Industr. and Eng. Chem. **20**, 1026 (1928).
Stevens B.—Industr. and Eng. Chem. **20**, 1028 (1928).
Taffanel J.—C. R. Bull. Chim. Soc. **157**, 714 (1913); **158**, 42 (1913).

Taffanel J., Le Floch B.—C. R. Bull. Chim. Soc. **156**, 1544 (1912).
Tamman G., Thiele H.—Ztschr. Anal. Chem. **192**, 68 (1930).
White A.—J. Chem. Soc. **121**, 1268 (1922).

II. Theory of Gas Detonation

Appin A. Ya., Khariton Yu. B.—In: Sbornik Referatov Rabot Khim. Otd. AN SSSR (1940).
Grib A. A. Candidate Dissertation in Phys.-Math. Sci. Leningrad: Leningrad Mining Institute (1940).
Michelsohn V. A. Sobr. Soch. Moscow: Novyĭ Agronom **1**, 115 (1930).
Rivin M. A., Sokolik A. S.—Zhurn. Fiz. Khimii **7**, 571 (1936).
Sokolik A. S.—Uspekhi Khimii **7**, 976 (1938).
Sokolik A. S., Shchelkin K. I.—Zhurn. Fiz. Khimii **5**, 1459 (1934).
Todes O. M., Izmailov S. V.—Dokl. Nauchn.-Tekh. Soveta In-ta Khim. Fiziki AN SSSR (1935).
Shchelkin K. I.—Dokl. AN SSSR **23**, 636 (1939); ZhETF **10**, 823 (1940).
Berthelot E., Vieille P.—C. R. Bull. Chim. Soc. **993**, 18 (1881).
Becker R.—Ztschr. Phys. **8**, 321 (1922).
Campbell C.—J. Chem. Soc. **121**, 2483 (1922).
Chapmann D.—Philos. Mag. **47**, 90 (1899).
Dixon H. D.—Philos. Trans. Roy. Soc. London A **184**, 27 (1893); **200**, 315 (1903).
Herzberg G.—Chem. Rev. **20**, 145 (1937); J. Chem. Phys. **10**, 306 (1942).
Jost W.—Ztschr. Phys. Chem. B **42**, 136 (1939).
Jouguet E. Méchanique des explosifs. Paris (1917).
Kinch B., Penney K.—Proc. Roy. Soc. London A **179**, 214 (1941).
Kohn H.—Ztschr. Phys. **3**, 145 (1920).
Lewis B.—J. Amer. Chem. Soc. **52**, 3120 (1930).
Mallard E., Le Chatelier A.—C. R. Bull. Chim. Soc. **93**, 145 (1881).
Schmid P.—Proc. Phys. Soc. **50**, 283 (1938).
Wendtlandt R.—Ztschr. Phys. Chem. **110**, 637 (1934); **116**, 227 (1925).

Cited Works by the Author

I. Theory of Gas Combustion

1. *Zeldovich Ya. B., Frank-Kamenetskiĭ D. A.* Teoriĭa teplovogo rasprostraneniĭa plameni [Theory of Thermal Propagation of a Flame].—Zhurn. fiz. khimii **12**, 100 (1938); here art. 19.
2. *Zeldovich Ya. B., Semenov N. N.* Kinetika khimicheskikh reaktsiĭ v plamenakh [Kinetics of Chemical Reactions in Flames].—ZhETF **10**, 1116 (1940).
3. *Zeldovich Ya. B.* Teoriĭa predela rasprostraneniĭa tikhogo plameni [Theory of the Propagation Limit for a Slow Flame].—ZhETF **11**, 159 (1941); here art. 20.

4. *Barskiĭ G. A., Zeldovich Ya. B.* Teplonapryazhennost khimicheskoĭ reaktsii v plameni [Thermal Stress of a Chemical Reaction in a Flame].—Nauch.-tekhn. otchet In-ta khim. fiziki AN USSR (1941).

5. *Barskiĭ G. A., Zeldovich Ya. B., Orlova L. I., Yudin F. E.* Teplota aktivatsii goreniĭa okisi ugleroda [The Activation Heat of Combustion of Carbon Monoxide].—Nauch.-tekhn. otchet In-ta khim. fiziki AN USSR (1941).

6. *Drozdov N. P. Zeldovich Ya. B.* Diffuzionnye yavleniĭa u predelov rasprostraneniĭa plameni [Diffusion Phenomena at the Propagation Limit of a Flame].—Zhurn. fiz. khimii **17**, 134 (1943); here art. 21.

II. Theory of Gas Detonation

7. *Zeldovich Ya. B.* K teorii rasprostraneniĭa detonatsii v gazoobraznykh sistemakh [On the Theory of Propagation of Detonation in Gas Systems].—ZhETF **10**, 542 (1940); here art. 27.

8. *Zeldovich Ya. B.* Ob energeticheskom ispolzovanii detonatsionnogo goreniĭa [On the Power Application of Detonation Combustion].—Zhurn. tekhn. fiziki **10**, 1454 (1940).

9. *Zeldovich Ya. B., Ratner S. B.* Raschet skorosti detonatsii v gazakh [Calculation of the Detonation Velocity in Gases].—ZhETF **11**, 170 (1941).

10. *Zeldovich Ya. B.* O raspredelenii davleniĭa i skorosti v produktakh detonatsionnogo vzryva [On the Pressure and Velocity Distribution in the Products of a Detonation Explosion].—ZhETF **12**, 389 (1942).

11. *Zeldovich Ya. B., Leipunskiĭ O. I.*—Zhurn. fiz. khimii (in press) [Acta physicochim. URSS **18**, 167 (1943)].

12. *Zeldovich Ya. B.* Teoriĭa udarnykh voln i vvedenie v gazodinamiku [Theory of Shock Waves and Introduction to Gas Dynamics] (in press). Moscow, Leningrad: Izd-vo AN-SSSR, 186 p. (1946).

Commentary

This monograph was one of the first in Soviet scientific literature to consider fundamental problems of combustion and detonation. Together with the book by N. N. Semenov,[1] it became a point of departure for modern voluminous publications on combustion. Ya.B. himself later wrote several books. Together with V. V. Voevodskiĭ, he published in monograph form the lecture notes, "Thermal Explosion and Flame Propagation in Gases,"[2] and with D. A. Frank-Kamenetskiĭ, "Turbulent and Heterogeneous Combustion."[3] Working with A. S. Kompaneets, he prepared and published "The Theory of Detonation."[4]

[1] *Semenov N. N.* Tsepnye reaktsii [Chain Reactions]. Leningrad: Goskhimtekhizdat, 555 p. (1934)

[2] *Zeldovich Ya. B., Voevodskiĭ V. V.* Teplovoi vzryv i rasprostranenie plameni v gazakh [Thermal Explosion and Flame Propagation in Gases]. Moscow: Mosk. Mekh. In-t, 294 p. (1947).

[3] *Zeldovich Ya. B., Frank-Kamenetskiĭ D. A.* Turbulentnoe i geterogennoe gorenie [Turbulent and Heterogeneous Combustion]. Moscow: Mosk. Mekh. In-t, 251 p. (1947).

[4] *Zeldovich Ya. B., Kompaneets A. S.* Teoriĭa detonatsii [The Theory of Detonation]. Moscow: Gostekhizdat, 268 p. (1955).

In 1980 he published "Mathematical Theory of Combustion and Explosion."[5] These books directly develop the ideas of the 1944 monograph. Besides these, Ya.B. wrote monographs on the combustion of powders and condensed explosives and on the oxidation of nitrogen (see the commentaries to the appropriate articles).

Modern Soviet scientific literature is distinguished by the excellent books of K. I. Shchelkin and Ya. K. Troshin,[6] E. S. Shchetinkov,[7] D. A. Frank-Kamenetskiĭ,[8] K. K. Andreev,[9] L. N. Khitrin,[10] R. I. Soloukhin,[11] L. A. Vulis,[12] A. S. Sokolik,[13] and others. All of them were directly influenced by Ya.B.'s little 1944 monograph. Many sections of the book are reviewed below in connection with the appropriate articles.

Let us begin with flame as an object of physico-chemical experiment which allows us to obtain information on the kinetics of exothermic chemical reactions at high temperatures. This direction, reflected in detail in this book, was initiated by the works of Ya.B. and D. A. Frank-Kamenetskiĭ,[14] and especially of Ya.B. and N. N. Semenov,[15] which investigate the kinetics of chemical reactions in flames, in particular, for a combustible mixture of carbon monoxide with air. Later, together with G. A. Barskiĭ, Ya.B. continued his study of this mixture.[16]   Another composition which attracted close attention was the mixture of hydrogen with oxygen. A. B. Nalbandyan and V. V. Voevodskiĭ published a monograph, "The Mechanism of Oxidation and Combustion of Hydrogen,"[17] and for the experimental study of intermediate reaction products, methods of electron-paramagnetic resonance were successfully

[5] *Zeldovich Ya. B., Barenblatt G. I., Librovich V. B., Makhviladze G. M.* Matematicheskaĭa teoriĭa goreniĭa i vzryva [Mathematical Theory of Combustion and Explosion]. Moscow: Nauka, 484 p. (1980).

[6] *Shchelkin K. I., Troshin Ya. K.* Gazodinamika goreniĭa [Gasdynamics of Combustion]. Moscow: Izd-vo AN SSSR, 256 p. (1963).

[7] *Shchetinkov E. S.* Fizika goreniĭa gazov [Physics of Gas Combustion].   Moscow: Nauka, 739 p. (1965).

[8] *Frank-Kamenetskiĭ D. A.* Diffuziĭa i teploperedacha v khimicheskoĭ kinetike [Diffusion and Heat Transfer in Chemical Kinetics]. 2nd Ed. Moscow: Nauka, 491 p. (1967).

[9] *Andreev K. K.* Termicheskoe razlozhenie i gorenie vzryvchatykh veshchestv [Thermal Decomposition and Combustion of Explosives]. Moscow: Nauka, 346 p. (1966).

[10] *Khitrin L. N.* Fizika goreniĭa i vzryva [Physics of Combustion and Explosion]. Moscow: Izd-vo MGU, 442 p. (1957).

[11] *Soloukhin R. I.* Udarnye volny i detonatsiĭa v gazakh [Shock Waves and Detonation in Gases]. Moscow: Fizmatgiz, 175 p. (1963).

[12] *Vulis L. A.* Teplovoi rezhim goreniĭa [The Thermal Regime of Combustion]. Moscow, Leningrad: Gosenergoizdat, 288 p. (1954).

[13] *Sokolik A. S.* Samovosplamenenie, plamĭa i detonatsiĭa v gasakh [Self-Ignition, Flame and Detonation in Gases]. Moscow: Izd-vo AN SSSR, 427 p. (1960).

[14] *Zeldovich Ya. B., Frank-Kamenetskiĭ D. A.*—Dokl. AN SSSR **19**, 693–698 (1938); Zhurn. fiz. khimii **12**, 100–105 (1938); here, art. 19; Acta physicochim. URSS **9**, 341–350 (1938).

[15] *Zeldovich Ya. B., Semenov N. N.*—ZhETF **10**, 1116–1136, 1427–1440 (1940).

[16] *Barskiĭ G. A., Zeldovich Ya. B.*—Zhurn. fiz. khimii **25**, 523–537 (1951).

[17] *Nalbandyan A. B., Voevodskiĭ V. V.* Mekhanizm okisleniĭa i goreniĭa vodoroda [The Mechanism of Oxidation and Combustion of Hydrogen]. Moscow-Leningrad: Izd-vo AN SSSR, 180 p. (1949).

applied.[18]

Papers on the relation between the detailed chemical mechanism and the properties of flames have become an important part of chemical monographic literature.[19] In a report by T. von Karman,[20] it was shown that the formula of Y. B. Zeldovich and D. A. Frank-Kamenetskiĭ is the limiting value of the flame velocity for a single chemical reaction with a large activation energy. In complex cases the method proposed by Ya.B. and G. I. Barenblatt for the solution of the problem of establishment of the regime is the most economic one. In 1961 Ya.B. extended the methods of the analytical study of flames with a single, simple, irreversible chemical reaction to multi-stage branching and non-branching chain transformations.[21] A. G. Istratov and V. B. Librovich applied these methods to calculate the flame velocity in mixtures of hydrogen with chlorine,[22] D. B. Spalding and P. L. Stephenson carried out the numerical integration of a system of equations of the combustion theory for a mixture of hydrogen with bromine,[23] and V. I. Golovichev[24] carried out a detailed study of the same mixtures.

The development of numerical methods allowed calculation of the structure and propagation velocity of a plane laminar flame for the most complex transformation schemes, encompassing up to seventy elementary chemical events. We note the works of G. Dixon-Lewis with S. M. Islam[25] and with F. Gramarossa,[26] which studied mixtures of hydrogen with air, methane with air, and flame in ozone decomposition.

The theoretical calculations agree well with experimental observations. Among Soviet studies we note the work of V. Ya. Basevich, S. M. Kogarko and V. S. Posvyanskiĭ on a methane-oxygen flame.[27] We should also mention papers on the propagation of a cold flame in mixtures of carbon bisulfate with air which were initiated by N. N. Semenov's ideas on chain reactions and the experimental investigations of V. G. Voronkov and N. N. Semenov.[28] B. V. Novozhilov and V. S. Posvyanskiĭ[29] carried out a detailed numerical calculation of the flame velocity in these mixtures in which interest had recently grown in connection with the

[18] *Balakhnin V. P., Gershenzon Yu. M., Kondratiev V. N., Nalbandyan A. B.*—Dokl. AN SSSR **154**, 1142–1144 (1964).

[19] *Kondratiev V. N., Nikitin E. E.* Kinetika i mekhanizm gazofaznykh reaktsiĭ [The Kinetics and Mechanism of Gas-Phase Reactions]. Moscow: Nauka, 558 p. (1974).

[20] *Karman T. von*—In: 6th Intern. Symp. on Combustion. New York: Reinhold; London: Chapman and Hall, 116 p. (1957).

[21] *Zeldovich Ya. B.*—Kinetika i kataliz **2**, 305–318 (1961).

[22] *Istratov A. G., Librovich V. B.*—Dokl. AN SSSR **143**, 1380–1388 (1962); Zhurn. prikl. mekhaniki i tekhn. fiziki **1**, 68–75 (1962).

[23] *Spalding D. B., Stephenson P. L.*—Proc. Roy. Soc. London A **324**, 315–337 (1971).

[24] *Golovichev V. I.*-In: Aerofizicheskie issledovaniia [Aerophysics Studies]. Novosibirsk. **5**, 80–81 (1975).

[25] *Dixon-Lewis G., Islam S. M.*—In: 19th Intern. Symp. on Combustion. Haifa, 314–330 (1982).

[26] *Dixon-Lewis G., Gramarossa F.*—Combust. and Flame **16**, 243–251 (1971).

[27] *Basevich V. Ya., Kogarko S. M., Posvyanskiĭ V. S.*—Fizika goreniia i vzryva **11**, 242–247 (1975).

[28] *Voronkov V. G., Semenov N. N.*—Zhurn. fiz. khimii **13**, 1695–1727 (1939).

[29] *Novozhilov B. V., Posvyanskiĭ V. S.*—Fizika goreniia i vzryva **9**, 225–230 (1973); **10**, 94–98 (1974).

application of carbon bisulfate in chemical combustion lasers.[30]

As was said in the scientific-biographical essay at the beginning of the book, as a whole the applicability of combustion processes has expanded tremendously. New directions in the technological use of combustion—for obtaining chemical compounds, soot, and for intensification of chemical-energetic processes—are well covered in the collection of articles, "Combustion Processes in Chemical Technology and Metallurgy."[31] Let us mention also the study of combustion waves in hydrogen which are maintained by laser radiation in the combustion chamber of a rocket engine. Such a chamber is very promising for future space equipment.[32] The study of complex chemical transformations in flames has led to the refinement of approximate asymptotic research methods.[33,34,35]

[30] *Dudkin V. A., Librovich V. B., Rukhin V. B.*—Fizika goreniĭa i vzryva **14**, 141–142 (1978).

[31] Protsessy goreniĭa v khimicheskoĭ tekhnologii i metallurgii: Sb. stateĭ [Combustion Processes in Chemical Technology and Metallurgy: A Collection of Articles]. Ed. by A. G. Merzhanov. Chernogolovka, 215 p. (1975).

[32] *Kemp N. H., Ruth R. G.*—J. Energy **3**, 133–141 (1979).

[33] *Fendell F. E.*—J. Fluid Mech. **56** 1, 81–89 (1972).

[34] *Berman V. S., Ryazantsev Yu. S.*—PMM **36**, 659–666 (1972); **37**, 1049–1058 (1973); **39**, 306–315 (1975).

[35] *Margolis S. B., Matkowsky B. J.*—SIAM J. Appl. Math. **42**, 314–321 (1982).

I

Ignition and Thermal Explosion

17

On the Theory of Thermal Intensity. Exothermic Reaction in a Jet I*

§1. Introduction

In 1908 an elementary theory of chemical reactions in a jet was given in a paper by Bodenstein and Wohlgast [1].

Only in the case when the reaction vessel is a long tube do the laws of chemical reaction in a jet approach the laws of chemical kinetics in a closed vessel uncomplicated by diffusion exchange, i.e., the laws which are obtained by integration of chemical kinetics equations of the form

$$W = \frac{dc}{dt} = f(c). \tag{1}$$

For the simplest cases, for example, a monomolecular reaction,

$$f(c) = kc, \tag{2}$$

$$c = c_0 e^{-kt}, \tag{3}$$

they were found in integral form [e.g., (3)] as long ago as 1850 by Wilhelmie.

In contrast, Bodenstein and Wohlgast consider the other extreme of a reaction vessel in which mixing and diffusion exchange occur quite intensively and the formation of noticeable differences in concentration within the reaction vessel, such as those which arise along the jet in a tubular vessel, is impossible.

*Zhurnal tekhnicheskoĭ fiziki **11** 6, 493–500 (1941).

In the case considered by Bodenstein and Wohlgast, the (volume) concentration of the reacting substance in the vessel, which coincides with the concentration in the incoming gas, is determined by the algebraic (not differential, as before) equation,

$$uc_0 = uc_1 + vf(c_1). \tag{4}$$

In equation (4) v is the vessel volume (cm^3) and u is the volume rate of the jet (cm^3/sec). The left-hand side of (4) is the rate of supply of the (reacting) material into the reaction vessel; on the right-hand side this rate is equated to the sum of the consumption rate of the substance during the chemical reaction in a vessel of a given volume at a concentration c_1 and the rate of output of the substance with the exiting gases (we have neglected the change in volume in the reaction).

The ratio of the volume of the vessel to the space velocity of the jet represents the time that the substance remains in the reaction vessel,

$$t = \frac{v}{u}. \tag{5}$$

For a given kinetics of the chemical reaction, the relation between the concentrations at entry and exit (i.e., before and after the chemical reaction) depends only on the time t. However, the form of the dependence, obtained by solution of equation (4), is itself substantially different from the integral of equation (1) for identical chemical kinetic laws [identical functions $f(c)$]. So, for a monomolecular reaction, instead of (3) we find

$$c = \frac{c_0}{1 + kt}. \tag{6}$$

Accordingly, for a bimolecular reaction, Bodenstein and Wohlgast find

$$c_0 = c + ktc^2; \qquad c = \frac{\sqrt{1 + 4ktc_0} - 1}{2kt} \quad \text{at } ktc_0 \gg 1; \qquad c \cong \sqrt{c_0/kt}, \tag{7}$$

instead of the usual, classical expression,

$$\frac{dc}{dt} = -kc^2; \qquad c = \frac{c_0}{1 + ktc_0} \quad \text{at } ktc_0 \gg 1, \qquad c \cong \frac{1}{kt}. \tag{8}$$

Over the last 10–15 years, interest has grown significantly in the kinetics of combustion and explosion reactions, which are characterized by the presence of some mechanism of acceleration of the reaction. This acceleration, which leads to ignition, may be related either to the accumulation of active products which catalyse the reaction, the chain carriers (autocatalysis, chain explosion), or to an increase in the temperature of the mixture due to heat release in an exothermic chemical reaction (thermal explosion).

Fundamental theoretical and experimental research on systems capable of ignition has been done by Semenov and his school, especially on the critical conditions of thermal and chain ignition. Later papers studied the process in time of chain (Semenov [2]) and thermal (Todes [3]) ignition by integration of equations and systems of the form (1).

In this paper we study theoretically ignition and combustion reactions under conditions of a jet in a reaction vessel with complete mixing, i.e., under the conditions described by equations of the form (4).

Partly anticipating the results of the calculations below, we note that this research is of interest from several standpoints:

1. Running a reaction in a jet is the most convenient and natural means of experimentally studying very rapid chemical reactions.

2. The high rate of explosive reactions is the result of long accumulation in the system of active products or heat; therefore, realization of the reaction under conditions of complete mixing, which can also replace a long period of accumulation, is for these reactions the best means of carrying out the reaction with maximum intensity over a minimum time in a minimum volume, which in a number of cases is of practical interest.

3. The study of a reaction in a steady jet with mixing is logically fundamentally simpler than the study of a single run of a chemical reaction in time in a closed volume. In our case we were able to absolutely rigorously introduce a number of very important concepts (such as the explosion limit) which in the case of a closed volume are of an approximate nature, although the approximation is in fact usually quite good.

4. Under the conditions considered here it is possible to foresee the peculiar phenomenon of quenching, or disruption of intensive combustion as the reaction time is decreased, similar to the extinction of a heterogeneous reaction (Frank-Kamenetskiĭ [4]); in contrast, in the kinetics of reaction in a closed volume this phenomenon finds no close analogy.

§2. Adiabatic Combustion and Autocatalysis by the End Products (In the Case of a Single Variable)

The rate of a chemical reaction is naturally related to the concentration of the reacting substance, and in most cases the reaction rate falls as the concentration is decreased. In most cases acceleration of an explosive reaction occurs not because the reagent concentration decreases: if we take a smaller initial concentration we would not obtain a large rate. Self-acceleration of a reaction occurs, as a rule, as a result of changes in other conditions, for example, an increase in the temperature or accumulation of active centers, which are only indirectly related to consumption of the original reacting substance.

To describe an explosive reaction in which these factors are significant a system of at least two equations is needed (for the concentration of the original substance and for the accelerating factor). However, in at least two very important cases the accelerating factor turns out to be identically related to the concentration of the original substance, and we may consider only one unknown and one equation.

These cases are:

1. Auto-catalysis by the stable end products of the chemical reaction under consideration. The concentration of these catalyzing products is related by the stoichiometric equation to the decrease in the concentration of the original substance. Thus, for a reaction equation,

$$mC + \ldots = nA + \ldots, \tag{9}$$

for given values of C and A in the jet,

$$C = c_0, \qquad A = 0, \tag{10}$$

we obtain for any reaction time, volume, rate, and so on

$$A = \frac{n}{m}(c_0 - c). \tag{11}$$

2. The most important case of adiabatic flow of an exothermic reaction—the condition of constant enthalpy,

$$J = E + pv, \tag{12}$$

under the simplest assumptions yields a relation between temperature and concentration,

$$c_p(T - T_0) = Q(c_0 - c), \tag{12a}$$

where c_0, T_0 are the concentration and temperature of the gas flowing in, and c, T are the same in the reaction vessel and exiting gas; Q is the reaction heat; c_p is the specific heat.

Here we always assume the presence of only one chemical reaction, and the absence of parallel reactions with different stoichiometric equations, heats, etc.

When only one reaction is present into which several initial substances enter (rather than C only, as we wrote above), all of their concentrations are identically related and the number of variables does not increase.

The identity relation between the concentration of the catalyzing product or the temperature and the concentration of the initial substance or product allows us to express the chemical reaction rate, which depends on several variables (c, A or T), as a function of only the concentration of the initial substance for a given initial concentration.

In Figs. 1–5, several such curves are represented by solid lines:

$$W = f[c, T(c_0, c)]; \quad \alpha(c_0, c) = \psi(c, c_0). \tag{13}$$

Fig. 1 relates to a classical isothermic monomolecular reaction:

$$W = kc. \tag{14}$$

Fig. 2 represents auto-catalysis of the first order:

$$W = kca; \quad W = kc(c_0 - c). \tag{15}$$

Fig. 3 represents auto-catalysis of the second order:

$$W = kca^2 = kc(c_0 - c)^2. \tag{16}$$

In both cases of auto-catalysis the initial concentration of the catalyzing product is taken equal to zero.

Fig. 4 shows a monomolecular reaction with the heat effect of a 7000 cal/mole mixture, which raises (for complete combustion) the temperature from 1000 to 2000°K. The activation heat of the reaction is 30,000 cal/mole:

$$W = k_0 c e^{-30\,000/RT}; \qquad 7(T - 1000) = 7000 \frac{c_0 - c}{c_0}. \tag{17}$$

Fig. 5 shows a monomolecular exothermic reaction with the same activation heat, at the same temperature of the incoming jet, 1000°K, but with a smaller heat effect—1400 cal/mole:[1]

$$W = k_0 c e^{-30\,000/RT}, \qquad 7(T - 1000) = 1400 \frac{c_0 - c}{c_0}. \tag{18}$$

Now it is not difficult to study graphically the different cases of solution of an arbitrary kinetic equation (4), which is represented in the form

$$W = \frac{c_0 - c}{t} = f(c) \tag{19}$$

The graphic solution reduces to searching in the c,W-plane for the intersection of the curves ψ (13) with lines drawn from the point $(c_0, 0)$, whose slope depends on the mean residence time of the substance in the vessel, i.e., on the ratio of the volume to the velocity of the jet.

In the case of a classical isothermic reaction (see Fig. 1), for any residence time, we find one and only one solution corresponding (as will be seen from the criteria given below) to a stable reaction regime; the concentration of the original substance in the jet leaving the vessel monotonically decreases as the residence time is increased. We now consider a typical case of non-isothermic reaction [equation (17), Fig. 4].

In a broad range of variation of the slope of the intersecting line which depends on the residence time t, each line has three points of intersection with the reaction rate curve, and equation (19) has three solutions.

The question of the stability of the regime corresponding to a given solution is easily analyzed. It turns out that the two extreme intersection points, A and C, correspond to stable regimes, while the middle point, B, corresponds to an unstable regime which is transformed to the respective stable regime A or C after small deviations in one or the other direction. The region A–B, which is closest in composition and temperature to the gas entering the vessel, recalls Semenov's theory of thermal explosion. Just as in the theory of thermal explosion we encounter here a steady, stable regime

---

[1] In expressions (17) and (18) the factor of the expression in parentheses $(T - 1000)$ is the specific heat of one mole (cal/mole · deg).

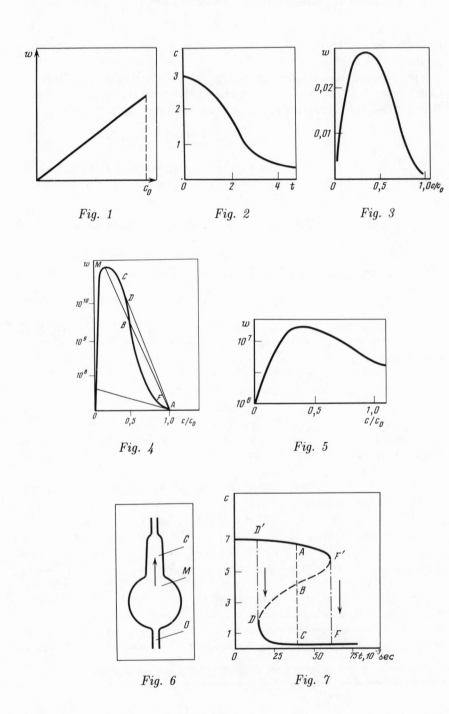

Fig. 1

Fig. 2

Fig. 3

Fig. 4

Fig. 5

Fig. 6

Fig. 7

with low heating which does not exceed the characteristic temperature difference,

$$\theta_0 = \frac{RT_0^2}{E},\qquad(20)$$

and an unstable regime B, below which the system returns to the steady state A with low heating, and above which (for a given residence time) rapid growth of the temperature occurs all the way up to the state C (see Fig. 4).

As the residence time is increased (which is equivalent in Semenov's theory to a decrease in heat transfer), the points A and B approach one another and, finally, after they contact at the point F, the steady regime with low heating disappears completely; in other words, self-ignition occurs. The point of contact F describes a state at the boundary of self-ignition. Alongside this similarity, there are also significant differences between self-ignition in a jet and in a closed vessel. In Semenov's theory, the presence or absence of self-ignition is due to the heat transfer of the reacting substance to the outside. In the case of a jet, however, the reacting substance's own specific heat (evacuation of heat with the reaction products exiting the vessel) causes the absence of self-ignition for a large jet velocity (small residence time) even in the case of adiabatic reaction under consideration. As a result, the conditions found for the limits do not contain any instrumental constants. In Semenov's theory the consumption of the substance in the course of the reaction, heat release and temperature increase were not taken into account.

The effect of consumption was investigated by Todes and Melentiev [5] in a paper which studied the temporal evolution of self-ignition. In principle, when consumption is taken into account, the transition from a reaction with low heating (below the limit) to ignition, i.e., to a reaction during which high temperatures develop, turns out to be gradual; in practical terms, for significant activation and reaction heats, the transition occurs with a very small, but still finite change in the parameters.

In our investigation consumption was taken into account in advance in the construction of the velocity curve in the form of relation (12), (12a) between the concentration and temperature. However little we might exceed the limit, self-ignition will occur in the jet. The boundary between self-ignition and the steady regime is absolutely abrupt; physically, this is related to the fact that in a jet with continuous supply of the substance we may allow the system an indefinitely long time, unlimited by consumption, for a steady or non-steady regime to be manifested.

Let us turn to consideration of the third intersection point, C, located in the high temperature region,[2] i.e., the steady regime which is obtained as a result of ignition. As the residence time in the reaction vessel increases, the

[2]This point has nothing in common with the third point of intersection in Semenov's theory. Cf. ref. [6] and Semenov's reply [7, p. 119].

completeness of combustion continuously increases and the temperature approaches the theoretical combustion temperature. On the branch to which the point C belongs there is also another point M which corresponds to the maximum (for given initial conditions and an adiabatic relation between temperature and concentration) chemical reaction rate.

In a monomolecular reaction with large activation heat E and reaction heat Q at the point M,

$$T_M - T_{\text{TC}} = \frac{-RT_{\text{TC}}^2}{E}, \tag{21}$$

$$c_M = \frac{RT_{\text{TC}}^2 c_0}{E(T_M - T_{\text{TC}})} = c_p \frac{RT_{\text{TC}}}{EQ}, \tag{22}$$

where T_{TC} is the theoretical combustion temperature.

The rate of the chemical reaction at the point M is quite large, and the work in this regime yields the best use of the volume. Thus, in the example of Fig. 4, if we take the volume necessary for realization of the point M at a given jet velocity as unity, then the volume necessary to attain the state in the jet without prolonged mixing turns out to be equal to 4 (see equations (1) and (13), and also Todes' article, "Adiabatic Thermal Explosion" [8]).

Physically, this means that mixing with the reaction products is a very effective means of overcoming the delay of an autocatalytic or exothermic reaction related to accumulation in the system of the active product or to heating. As may also be easily shown, in the case of exothermic reaction the maximum rate corresponds to very significant dilution by the reaction products—up to 70–80%.

In contrast, for classical reactions with a decreasing reaction rate, the volume is better used when mixing is absent since in this case the mean concentration always exceeds the output concentration. The most advantageous means of achieving a given completeness of combustion is shown in Fig. 6: the fresh mixture enters the reaction vessel (with mixing) which is filled with matter in the state M; subsequent reaction from M to C is carried out in the elongated vessel.

A classical technique of chemical technology is the use of the reaction heat in a reverse heat exchanger. For reactions in which the rate increase is related to heating, but not to the accumulation of chemically active catalytic products, heating of the mixture without dilution will yield even better results. However, for such very important chemical reactions as fuel combustion, selection of the material for heat exchange at high temperatures is quite difficult. In all cases the heat exchanger makes the construction more complex and heavier. Bearing in mind also the probable chain nature of these reactions, we must consider careful use of mixing of the incoming mixture with the hot combustion products the simplest and most effective means of intensive combustion. We cannot take the time here to analyze existing furnace constructions from this standpoint.

We note, finally, the close relation between chemical reaction conditions on the MCD branch (see Fig. 4) and conditions in a propagating flame.

As was shown in a paper by Frank-Kamenetskiĭ and the author [9], simultaneous diffusion and heat exchange between the fresh mixture and the hot reaction products under the simplest assumptions leads to the same relation between the concentration and temperature as their direct mixing. Since in a propagating flame the entire range of temperatures from the initial one to the combustion temperature of the given mixture is simultaneously realized, it is obvious that the bulk of the substance reacts precisely in the temperature region in which the chemical reaction rate is close to a maximum.

Thus, the study of chemical kinetics in a jet in the MCD sector of the curve in Fig. 4 will be extremely significant for the theory of flame propagation.

Disruption of the combustion regime when the blast rate is increased and the residence time of the reacting substance in the vessel is decreased is quite remarkable. It is obvious that this disruption occurs when the intersection point C moves to the tangent point D (see Fig. 4) in the direct vicinity of point M.

The study of kinetics at high temperatures should be begun precisely from the determination of the minimum reaction time corresponding to disruption of the combustion regime ("quenching," cf. [4]). Plotting the concentration of the initial substance in the reaction products as a function of the reaction time, we obtain for a typical exothermic reaction a dependence of the form in Fig. 7.

In a wide interval, $t_1 - t_2$, three steady states exist: for example, A–B–C, of which only A and C are realized. Which of the two regimes, A or C, will be realized under given conditions depends on the history of the system. Over time the point describing the system traces a hysteretic loop, $D'DF'F$; the direction of the jumps along the lines FF' and DD' is indicated by arrows. When the heat effect of the reaction is reduced, the area of the loop decreases. Finally, in a sufficiently diluted mixture, we do not have a multivalued curve $c(t)$ at all, nor a finite (FF' or DD') change in the regime for a small parameter change (e.g., at the time t; see Fig. 5).

This also establishes the absolute concentration limit of ignitibility of the mixture. Together with Todes, we define ignition (and its opposite, extinction) precisely as a finite change for an infinitely small change in a parameter. Under the simplest assumptions, for a monomolecular reaction, we find at the concentration limit

$$\Delta T = \frac{Q_{\lim}}{c} = \frac{R(T_c + T_0)^2}{E}.$$

We note that the fundamental difficulties associated with consumption did not allow Todes to determine the concentration limit in the theory of explosion in a closed vessel.

A similar analysis of reactions with autocatalysis by the final products [equations (15) and (16), see Figs. 2 and 3] is significantly simpler and we leave it for the reader. A simpler, but also weaker algebraic (rather than exponential, in accordance with Arrhenius' law) dependence of the reaction rate on the concentration in an isothermic autocatalytic reaction causes much less complete combustion at the limit of extinction.

Combustion Laboratory. Institute of Chemical Physics   *Received*
USSR Academy of Sciences, Leningrad   *October 26, 1940*

References

1. *Bodenstein M., Wohlgast K.*—Ztschr. phys. Chem. **61**, 722 (1907).
2. *Semenov N. N.*—Ztschr. Phys. B **42**, 571 (1929); **11**, 464 (1930).
3. *Todes O. M.*—Acta physicochim. URSS **5**, 785 (1930); **11**, 153 (1939).
4. *Frank-Kamenetskiĭ D. A.*—Zhurn. tekhn. fiziki **9**, 1457 (1939).
5. *Todes O. M., Melentiev P. K.*—Acta physicochim. URSS **11**, 153 (1939).
6. *Gururaja D.*—J. Ind. Chem. Soc. **10**, 57 (1933).
7. *Semenov N. N.* Tsepnye reaktsii [Chain Reactions]. Leningrad: Goskhimtekhizdat, 555 p. (1934).
8. *Todes O. M.*—Zhurn. fiz. khimii **4**, 71, 78 (1933).
9. *Zeldovich Ya. B., Frank-Kamenetskiĭ D. A.*—Zhurn. fiz. khimii **12**, 100 (1938); here art. 19.

17a

On the Theory of Thermal Intensity.
Exothermic Reaction in a Jet
II. Consideration of Heat Transfer
in the Reaction*

With Yu. A. Zysin

The theory of adiabatic reaction developed in the previous article is here generalized to the case when heat transfer is present. Consideration of the heat transfer leads to the appearance of new features in the "consumption–time" kinetic curves, specifically, the possibility of extinction as the residence time is increased and of self-ignition when the reaction time is decreased (in the previous article, in the adiabatic case, extinction occurred only for a decrease in the reaction time, and self-ignition only for an increase).

In a number of regions of the values of the initial concentration and of the constant which characterizes the heat transfer, the kinetic curves take a quite peculiar form. We find the conditions leading to particular forms of the kinetic curves.

The properties and limits of the combustion regime are studied in detail and expressions are constructed for the maximum and minimum thermal intensity.

§1. Types of Kinetic Curves

In the previous article [1], for an adiabatic reaction the thermal balance equation led to an identity relation between consumption and reaction temperature. We now write the thermal balance equation with heat transfer:

$$uc_pT_0 + uc_0Q = uc_pT + ucQ + \alpha S(T - T_0), \tag{1}$$

where α is the heat transfer coefficient, S is the surface of the vessel, and the other notations are as in [1].

Assigning a specific value to the concentration, i.e., a percentage consumption for a particular residence time and for a particular jet velocity, we

*Zhurnal tekhnicheskoĭ fiziki 11 6, 501–508 (1941).

easily find the temperature which is attained:

$$T = T_0 + \frac{u(c - c_0)Q}{uc_p + \alpha S} = T_0 + \frac{(c_0 - c)Q}{c_p + (\alpha S/v)(v/u)}$$

$$= T_0 + \frac{(c_0 - c)Q}{c_p + (\alpha S/v)t}. \tag{2}$$

As is clear from the formula, the temperature is again linearly related to the percentage consumption. For a given residence time taking the heat transfer into account is completely equivalent to increasing the effective specific heat of the reacting substance if the temperature of the incoming mixture does not differ from the temperature of the surrounding space, as we assumed in writing (1). This follows from the fact that both the heat spent on heating the reacting substance, which depends on the actual specific heat of the reacting substance, and the heat spent on thermal losses at a given jet velocity are proportional to the temperature difference, $T - T_0$, which in fact allows us to combine the two terms by introducing the concept of effective specific heat.

The effective specific heat which enters into (2) turns out to depend on the jet velocity. The greater the velocity of the jet, the less the role of thermal losses and the closer the effective specific heat is to the actual specific heat. However, for a given jet velocity and a given reaction time, the effective specific heat is constant, and we may make full use of the results of the previous article. In particular, just as in the previous article, we may state that for a given residence time either one solution of the equations is possible, or three solutions, of which one in this case turns out to be unstable; we shall call the other two "slow reaction" and "combustion."

A new circumstance arises in the transition from one residence time to another. The accompanying change in the effective specific heat will lead to a change in the maximum temperature achievable for complete consumption, i.e., will be equivalent to a continuous change in the heat effect.

For an adiabatic reaction we found two kinds of features in the curves of the dependence of the percentage consumption on the chemical reaction time. In stable regimes an increase in the time facilitates an increase in the percentage of reacted substance and an increase in the temperature. With an increase in the residence time, for sufficient initial concentration, self-ignition of the mixture is possible; in contrast, extinction occurs only when the reaction time is reduced.

When heat transfer is present we can expect the presence of new features. The decrease in temperature when the reaction time is increased, due to the growing relative role of heat transfer, can lead to extinction. For the same reasons, self-ignition is also possible for decreased time as a result of improvement of the thermal balance.

The corresponding features of the curves are portrayed graphically in Fig. 1*a,b,c,d*, where *a* and *b* are adiabatic ignition and extinction, respectively, and *c* and *d* are the new kinds of features just described (the abscissa plots time, the ordinate—the percentage of the reacting substance, which falls when consumption is increased).

Combining all four possible kinds of features of the curves *a*, *b*, *c*, *d* in various ways and taking into account the condition stated above that we must always have either one solution or three, we obtain the possible types of kinetic curves illustrated in Figs. 2–6. The dotted line on all of the curves indicates the region of unstable solutions.

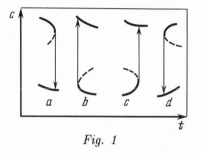

Fig. 1

We note that the indicated dependence of the temperature on the reaction time, related to the heat transfer, can lead also to a decrease in the consumption with increasing time on stable lines of the stable regime. A concrete calculation of the curves in Figs. 2–6 is given in §2.

Of particular interest are the curves in Fig. 5. This form corresponds to the most common case when the temperature T_0 is so low that, even when very little heat transfer is present, self-ignition is completely out of the question. For all ordinary explosive mixtures of hydrogen, carbon monoxide, methane and other industrial combustible gases with air, at initial room temperature, self-ignition is completely impossible. However, for sufficient fuel concentration and sufficient calorific value of the mixture, the temperature which develops in combustion may be large enough to ensure a very rapid chemical reaction or to ensure the possibility of steady combustion (branch AB of the curve).

The steady combustion depicted on the branch AB can be extinguished both by an increase in the velocity of the jet and a decrease in the residence time (point A), and by a decrease in the jet velocity and increase in the residence time, leading to growth of heat losses (B).

In Fig. 7 the initial concentration is plotted on the abscissa, and a quantity proportional to the heat transfer—along the ordinate. The numbers show under what conditions, initial concentration, and heat transfer each form of the kinetic curve in Figs. 2–6 occurs.

For a small initial concentration a reaction which is close to isothermic is always realized, i.e., a curve of type *2*. Increasing the initial concentration, with small heat transfer, we move into the region of curves of type *3*, analyzed earlier in the theory of adiabatic reaction.

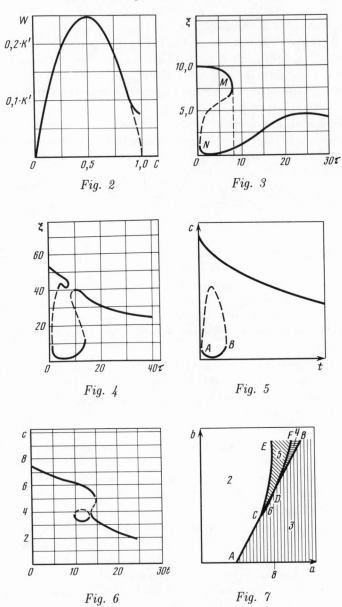

Fig. 2

Fig. 3

Fig. 4

Fig. 5

Fig. 6

Fig. 7

For increased heat transfer there appears, besides adiabatic self-ignition and extinction, a new type of self-ignition and extinction. Along the line DB we move into a region where the curve depicted in Fig. 4 is realized. When the initial concentration is decreased, crossing the line CD, we eliminate the possibility of self-ignition and move into region *5* of curves of type *5* with two extinctions, but no self-ignition. Finally, as the initial concentration is further reduced at constant heat transfer, combustion becomes completely impossible; the closed line of Fig. 5 contracts into a point and we again return to the isothermic curve of Fig. 6.

In the next section we shall present the mathematical part of the work, all the calculations which led to construction of the diagram in Fig. 7.

Finally, in §3 we discuss the combustion regime and its gradations in detail and, returning from dimensionless variables to the usual, well-known quantities of concentration, temperature and time, we formulate the basic quantitative results of the calculations.

§2. Analytical Theory

We take the heat transfer coefficient α to be independent of the jet velocity and of the residence time in the vessel. Physically, this assumption together with the assumption of complete mixing of the substance in the reaction vessel and of a constant mean temperature throughout the vessel corresponds to the idea that for heat transfer the governing factor is the thermal resistance from the internal wall of the vessel to the outside space in which the temperature is kept at T_0. In other words, our assumption corresponds to the concept of a vessel which is thermally insulated from outside.

For the sake of definiteness we consider a monomolecular reaction:

$$W = ck(T); \qquad k = Ae^{-E/RT}. \tag{3}$$

We expand the Arrhenius dependence of the reaction rate on the temperature in an exponential series using Frank-Kamenetskiĭ's method [2],

$$k = Ae^{-E/R[T_0+(T-T_0)]} = Ae^{-E/RT_0}e^{E(T-T_0)/RT_0^2}$$
$$= k(T_0)e^{E(T-T_0)/RT_0^2}. \tag{4}$$

We introduce dimensionless variables:

$$a = \frac{c_0 EQ}{c_p RT_0^2}; \qquad \xi = \frac{cEQ}{c_p RT_0^2};$$

$$b = \frac{\alpha S}{v c_p k(T_0)}; \qquad \tau = tk(T_0) = \left(\frac{v}{u}\right)k(T_0). \tag{5}$$

As the unit concentration we choose the concentration which, reacting adiabatically, leads to an increase in temperature that increases the chemical reaction rate by e times. The dimensionless initial concentration is denoted by a, and the current value by ξ. As the unit time we take the characteristic reaction time at the initial temperature; the dimensionless time is denoted by τ. Finally, as the unit heat transfer in constructing the dimensionless parameters we take the value of the heat transfer which halves the heating for a residence time equal to the characteristic reaction time. The dimensionless heat transfer is denoted by b.

In dimensionless variables the basic equation of the monomolecular reaction is written as follows:

$$a - \xi = \xi \tau e^{(a-\xi)/(1+b\tau)}, \tag{6}$$

$$\frac{\xi \tau e^{(a-\xi)/(1+b\tau)}}{a - \xi} - 1 = \varphi(\xi, \tau, a, b) = 0. \tag{6a}$$

This transcendental equation cannot be solved in elementary functions.[1] However, we may easily construct the desired curves of the dependence of ξ on τ for given values of a and b by introducing the parameter θ, which has the simple physical meaning of the temperature:

$$\theta = \frac{a - \xi}{1 + b\tau}; \qquad T = T_0 + \left(\frac{RT_0^2}{E}\right)\theta. \tag{7}$$

The quantities of interest, ξ and τ, may be found as functions of θ by solution of the quadratic equation.

We provide the final formulas:

$$\tau = \frac{a - b - \beta \pm \sqrt{\gamma}}{2b\theta}; \qquad \xi = \frac{a - \theta + \beta \pm \sqrt{\gamma}}{2}, \tag{8}$$

where we used the notations

$$\beta = b\theta e^{-\theta}; \qquad \gamma = a^2 + \theta^2 + \beta^2 - 2a\theta - 2a\beta - 2\theta\beta. \tag{8a}$$

With the help of these formulas, by forcing θ to range over all values from 0 to a $(T_0 < T < T_0 + c_0 Q/c_p)$, we exhaust all the points on the curve $\xi - \tau$. In the case of a curve like that in Fig. 5, for certain intermediate values of θ there are no real solutions (at $\gamma < 0$).

In a calculation using formulas (8) and (8a) it is necessary to consider both solutions of the quadratic equation corresponding to the two signs $\pm$ in (8) and (8a).

In order to construct the diagram in Fig. 7, which shows the regions of occurrence of the various curves, we must find the equation of the lines dividing the regions, i.e., of the lines on which transition from one type to another occurs. For this we shall now consider the form of the transition curves themselves in the $\xi - \tau$ planes. The first type of transition curve corresponds to simultaneous appearance or destruction of one extinction

---

[1]In the adiabatic case, equation (19) of the previous article always has an elementary solution with respect to the time $t = c_0 - c/f(c)$, and construction of the curves $c - t$, corresponding to $\xi - \tau$, does not pose any difficulties.

and one ignition; at the moment of their appearance the points of extinction and ignition are located at an infinitesimal distance from one another. In the limit, the coincident maximum and minimum of the curve $\tau(\xi)$ form an inflection point characterized by the conditions [cf. (6)]:

$$\tau = \tau(\xi); \quad \frac{d\tau}{d\xi} = 0; \quad \frac{d^2\tau}{d\xi^2} = 0. \tag{9}$$

Whether we are dealing with the occurrence of "adiabatic" extinction and ignition (see Fig. 1a,b) or with self-ignition and extinction which are essentially related to the heat transfer, as in Fig. 1c,d, obviously depends on the sign of the third derivative, $d^3\tau/d\xi^3$, at the inflection point.

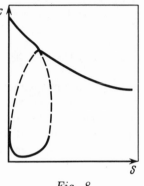

In the a,b-plane it is not difficult to find that the conditions (9) yield the equation of a straight line,

$$b = (a - 4)e^2 = 7.39(a - 4). \tag{10}$$

In Fig. 7 the line (10) is depicted by the line ADB. Here, on the segment AD adjacent to the abscissa the third derivative is negative (adiabatic case); on the segment

Fig. 8

from D up the third derivative is positive, which corresponds to the appearance of another type of extinction and ignition.

The second type of transition curves is related to the appearance or disappearance of a closed line. A closed line may disappear either by contracting into a point or by joining with the rest of the curve. In the first case the intermediate type we seek will be the kinetic curve of the form in Fig. 8. In both the case of a single isolated point and that of a point of self-intersection of the curve, we have the following conditions[2] [cf. the form of equation (6a)]:

$$\varphi = 0; \quad \frac{\partial\varphi}{\partial\tau} = \varphi'_\xi = 0, \quad \frac{\partial\varphi}{\partial\tau} = \varphi'_\tau = 0. \tag{11}$$

In order to distinguish just what we are dealing with, we must find the sign of the quantity $\varphi''_{\xi\xi}\varphi''_{\tau\tau} - \varphi''^2_{\xi\tau}$. At an isolated singular point this quantity is positive, at a self-intersection point it is negative. The three equations (11)

$$\xi\tau e^\theta + \xi - a = 0,$$

$$\xi e^\theta - \frac{\xi\tau(a - \xi)}{(1 + b\tau)^2}e^\theta = 0, \tag{11a}$$

$$\tau e^\theta - \frac{\xi\tau}{1 + b\tau}e^\theta + 1 = 0,$$

[2]See, e.g., *Courant R.* Differential and Integral Calculus, V. 2., transl. by E. J. McShane. New York: Interscience (1947).

may be solved for a in parametric form by the introduction of a parameter $y = b\tau$:

$$a = \frac{(1+y)^3}{y}; \qquad b = y^2 e^{1+1/y}, \tag{11b}$$

and we obtain a curve in the a,b-plane (see Fig. 7) with a cusp point at C: $a = 6.75$, $b = 5$ for $y = 0.5$.

By considering the sign of $\varphi''_{\xi\xi}\varphi''_{\tau\tau} - \varphi''^2_{\xi\tau}$ we establish that, on the branch EC at $y < 0.5$, contraction of the closed line of the kinetic curve into an isolated point occurs, and on the branch CF, the closed line of the kinetic curve joins with the basic curve at a point of self-intersection for $y > 0.5$.

§3. The Combustion Regime

Let us derive approximate relations for the combustion regime, taking $a \gg 1$. As the following calculations confirm, in combustion the temperature and completeness of the reaction are high so that $\xi \ll a$.

In the expression for the reaction rate constant, expansion in an exponential series is most appropriately done near the theoretical temperature [3]:

$$T_T = T_0 + Q\frac{c_0}{c_p}.$$

Just as in (5) we introduce

$$a = \frac{c_0 EQ}{c_p RT_T^2}; \qquad \xi_T = \frac{cEQ}{c_p RT_T^2};$$

$$b = \frac{\alpha S}{v c_p k(T_T)}; \qquad \tau_T = tk(T_T) = \left(\frac{v}{u}\right)k(T_T). \tag{12}$$

Neglecting ξ wherever this quantity does not enter into the exponent and is not a factor, we simplify the basic equation

$$tck(T) = c_0 - c \cong c_0, \tag{13}$$

which is written with dimensionless variables thus:

$$\tau_T \xi_T e^{-(\xi_T + a_T b_T \tau_T)/(1 + b_T \tau_T)} = a_T. \tag{14}$$

The exponent (14) describes the drop in the chemical reaction rate due to the drop in the temperature below the theoretical temperature of combustion.

If the total temperature drop, $(\xi_T + a_T b_T \tau_T)/(1 + b_T \tau_T)$, is of order unity (in dimensionless variables), then for $a \gg 1$ we may assert that $b_T \tau_T \ll 1$; the kinetics equation is further simplified:

$$\tau_T \xi_T e^{-\xi_T - a_T b_T \tau_T} = a_T. \tag{14a}$$

ξ_T is the temperature drop that results from incompleteness of the reaction, and $a_T b_T \tau_T$ is that from the heat transfer. At the limit of extinction (see

Fig. 1, *b* and *c*)

$$\frac{d\tau_T}{d\xi_T} = 0. \tag{15}$$

Let us take the derivative of (14) with respect to ξ_T, immediately taking (15) into account; we find

$$\xi_T = 1. \tag{16}$$

Physically this means that the temperature decrease due to incomplete combustion cannot exceed the characteristic quantity RT_T^2/E without full extinction or disruption of combustion also taking place. Substituting (16) into (14a) we obtain an equation for the critical reaction time at which extinction occurs:

$$\tau_T e^{-a_T b_T \tau_T - 1} = a_T. \tag{17}$$

If the heat transfer is large, $b_T > e^{-2} a_T^{-2}$, equation (17) has no solutions and the combustion regime is not possible at all (the kinetic curve of Fig. 2). If the heat transfer is small, $b_T < e^{-2} a_T^{-2}$, (17) has two solutions:

$$\tau_T^{(1)} = a_T e, \qquad a_T b_T \tau_T^{(1)} < 1, \qquad \xi^{(1)} = 1; \tag{18}$$

$$\tau_T^{(2)} \cong \frac{\ln\left(1/a_T^2 b_T\right)}{a_T b_T}, \qquad a_T b_T \tau_T^{(2)} \simeq \ln\left(1/a_T^2 b_T\right) \geq 1, \qquad \xi_T \geq 1. \tag{19}$$

Taking into account $b_T < a_T^{-2}$, $a_T > 1$, we note that $\tau_T^{(1)} < \tau_T^{(2)}$. Solution (18) represents the adiabatic extinction for a small reaction time and small heat transfer considered earlier.

In contrast, solution (19) represents the condition of extinction for a large reaction time. The temperature drop due to heat transfer only leads to disruption of the regime when it (the decrease in T) reduces the reaction rate so much that incompleteness of the chemical reaction becomes noticeable, as a result of which the temperature falls further and so on.

Returning from dimensionless variables to the conventional dimensional unburned fuel concentration c, temperature T, chemical reaction rate k, and time t, we rewrite the conditions of extinction at the limit of "extinction due to intensification of combustion" (disruption of combustion when the jet velocity is increased and τ is decreased—adiabatic extinction):

$$c = \frac{c_p RT_T^2}{EQ}; \qquad T = T_T - \frac{RT_T^2}{E};$$

$$k(T) = \frac{k(T_T)}{e}; \qquad t = \frac{ec_0 EQ}{k(T_T)c_p RT_T^2}. \tag{18a}$$

At the limit of "extinction due to low capacity" (disruption of combustion due to a decrease in the jet velocity and an increase in the time of disruption, related to the heat transfer) we obtain

$$c = \frac{c_p RT_T^2}{EQ}, \qquad T = T_T - \frac{(\ln + 1)RT_T^2}{E},$$

$$k(T) = \frac{\alpha S}{v} \frac{E^2 Q^2 c_0^2}{c_p^3 R^2 T_T^4 \ln}, \qquad (19a)$$

$$t = \frac{v}{\alpha S} \frac{c_p^2 R T_T^2 \ln}{c_0 E Q},$$

where

$$\ln = \ln\left[\frac{k(T_T)}{k(T)}\right] = \ln\left[\frac{v k(T_T) c_p^3 R^2 T_T^4}{\alpha S E^2 Q^2 c_0^2}\right].$$

Equation (18a) does not contain quantities characterizing the heat transfer; in contrast, (19a) depends only very weakly (logarithmically) on the chemical reaction rate.

The terms "intensification" and "low capacity" introduced above are related to the expression for the thermal loading of a unit of volume

$$W = \frac{c_0 Q}{t}. \qquad (20)$$

In accordance with our results, this quantity may vary in an interval whose upper bound is determined by the kinetics of the chemical reaction and lower bound by the heat transfer:

$$W_{max} = \frac{k(T_T) c_p R T_T^2}{eE} > W > \frac{\alpha S}{v} \frac{c_0^2 E Q^2}{c_p^2 R T_T^2 \ln} = W_{min}. \qquad (21)$$

Incompleteness of combustion is maximal at the boundaries of the interval (21) and reaches a minimum inside this interval at the thermal intensity at which the heat transfer lowers the combustion temperature by the quantity RT_T^2/E.

Finally, in order for combustion to be possible at all, it is necessary that the heat transfer coefficient and degree of cooling of the furnace volume not exceed a certain value which we find by narrowing the interval (21):

$$W max = W_{min}; \qquad \frac{\alpha S}{v} = \frac{k(T_T) c_p^3 R^2 T_T^4}{e^2 c_0^2 E^2 Q^2}. \qquad (22)$$

Condition (22) is identical to the equation of the line (11) in Fig. 7 for large t. Analysis of the working conditions of current furnace chamber constructions falls outside the scope of the present paper. There are a number of indications that only in a small part of their total volume does intensive chemical reaction occur since it is only in this small part that necessary conditions of good mixing of air, fuel and hot reaction products are created. We may expect that, under these conditions, with the thermal intensity referred to the entire volume, the observed maximum and minimum values of the thermal intensity will prove to have been underestimated.

Combustion Laboratory   *Received*
Institute of Chemical Physics   *October 26, 1940*
USSR Academy of Sciences, Leningrad

References

1. *Zeldovich Ya. B.*—Zhurn. tekhn. fiziki **11**, 493 (1941); here art. 17.
2. *Frank-Kamenetskiĭ D. A.*—Zhurn. fiz. khimii **13**, 738 (1939).
3. *Zeldovich Ya. B., Frank-Kamenetskiĭ D. A.*—Zhurn. fiz. khimii **12**, 100 (1938); here art. 19.

Commentary

The theoretical questions which are posed and solved in these papers by Ya.B. and by Ya.B. with Yu. A. Zysin (articles 17 and 17a) have developed into an extensive separate branch of science—the theory of chemical reactors. Combustion in a reactor with ideal mixing is an example of the simplest thermal and gasdynamic situation, when the analysis requires only algebraic relations. This allows explicit demonstration of the basic features of exothermic chemical reactions in a flow which are also present in more complicated form in other combustion regimes—a laminar flame, diffusive combustion, detonation wave and others. Critical conditions of ignition and extinction and the existence of several regimes whose occurrence depends on the initial conditions—these are the most remarkable effects of combustion which attract the attention even of laymen. The relative ease of recording them makes them a convenient tool for physico-chemical research.

Experimental realization of a reactor with ideal mixing was first accomplished by J. Longwell and M. Weiss[1] who used it to study the kinetic characteristics of hydrocarbon–air mixtures. After this work there followed a series of studies by H. Hottel, G. Williams, M. Baker,[2] A. Clarke, A. Harrison, J. Otgers,[3] R. Schmitz, M. Grossball[4] and other authors in which various gaseous fuel mixtures were investigated, theoretical questions of the stability of regimes were considered, etc. P. Goldberg and R. Essenhigh applied the reactor to the study of combustion of coal gas suspensions.[5]

Since a reactor with ideal mixing ensures the maximum possible rate of exothermic chemical transformation, the results obtained in it on high-temperature disruption are an upper estimate of the permissible thermal intensity of furnaces and combustion chambers and serve as a standard for actual energy-producing devices which use combustion. In 1943 D. A. Frank-Kamenetskiĭ and I. E. Salnikov[6] analyzed the conditions of oscillation generation in chemical systems; together with the papers above, this article served as a beginning for research on non-steady regimes in reactors and the critical conditions of their occurrence.

The system of equations of non-steady combustion was analyzed on the "concentration–temperature" phase plane with the help of methods developed in the

[1] *Longwell J. P., Weiss M. A.*—Industr. and Eng. Chem. **47** 8, 1634–1643 (1955)

[2] *Hottel H. C., Williams G. C., Baker M. L.*—In: 6th Intern. Symp. on Combustion. NY: Reinhold, 398–411 (1957).

[3] *Clarke A. E., Harrison A. J., Otgers J.*—In: 7th Intern. Symp. on Combustion. London: Butterworths, 664–673 (1959).

[4] *Schmitz R. A., Grossball M. P.*—Combust. and Flame **9**, 339–342 (1965).

[5] *Goldberg P. M., Essenhigh R. H.*—In: 17th Intern. Symp. on Combustion. Pittsburgh, 1979, p. 315–324

[6] *Frank-Kamenetskiĭ D. A., Salnikov I. E.*—Zhurn. fiz. khimii **17**, 79–86 (1943).

qualitative theory of autonomous systems of second-order differential equations. On the phase plane, singular points of the equations correspond to steady states (of which there are, in general, an odd number), with stable states alternating with unstable ones. Under certain conditions periodic solutions appear which in the phase plane correspond to limiting cycles—closed trajectories encompassing one or several singular points. The existence of self-oscillating processes is caused by the presence in the system of positive or negative feedback. Positive feedback appears as the result of an increase in the reaction rate by self-heating and negative feedback—by its decrease because of consumption of the reacting material.[7-10]

Of much importance for chemico-technological process in circulating devices are the forced stabilization of unstable regimes and control of oscillation regimes using external effects (gas output, initial mixture temperature, etc.). Such regulation allows a chemical process to be run in a more intensive regime and output of the required materials to be optimized.[8,11,12] The maximum thermal intensity of a reactor may also be raised by preheating the initial reacting mixture by heat transfer from the combustion products through the walls in a circulating heat exchanger (thermal recycling). Such organization of the process allows the high reaction temperature to be combined with a high concentration of the reacting material. Several schemes for such reactors have been proposed and built;[9,13,14] having a combustion temperature which exceeds the adiabatic value allows, in particular, lean fuel mixtures to be burned in them which would otherwise not burn.

The principle of thermal recycling is also used in reactors with a boiling layer, in which the heat from the hot region of the reactor is transported to the cold region by circulating solid particles suspended in the gas flow.[15] Methods of the theory of chemical reactor regulation have been successfully used in other sciences as well. We note the model of Belousov–Zhabotinskiĭ, prcposed for the description of heart disease, of spasmatic contractions of the cardiac muscle.

[7] Volter B. V., Salnikov I. E. Ustoĭchivost rezhimov raboty khimicheskikh reaktorov [Stability of Chemical Reactor Regimes]. Moscow: Khimiĭa, 190 p. (1972).

[8] Aris R. Analiz protsessov v khimicheskikh reaktorakh [Analysis of Processes in Chemical Reactors]. Leningrad: Khimiĭa, 328 p. (1967).

[9] Boreskov G. K., Slinko M. G.—Khim. promyshlennost 1, 22–29 (1964).

[10] Zhabotinskiĭ A. M. Kontsentratsionnye avtokolebaniĭa [Concentration Self-Oscillations]. Moscow: Nauka, 178 p. (1974).

[11] Kafarov V. V. Metody kibernetiki v khimii i khimicheskoĭ tekhnologii [Cybernetic Methods in Chemistry and Chemical Technology]. Moscow: Khimiĭa, 379 p. (1968).

[12] Makhviladze G. M., Myshenkov V. I.—In: Gorenie kondensirovannykh sistem [Combustion of Condensed Systems]. Chernogolovka, 114–117 (1977).

[13] Weinberg F. J.—Phys. Bull. 33 3/4, 15 (1982).

[14] Hashimoto T., Yamasaki S., Takeno T.—In: Proc. of 9th Intern. Colloq. on Dynamics of Explosives and Reaction Systems. France, Poitiers, 218 (1983).

[15] Borodulya V. A., Gupalo Yu. P. Matematicheskie modeli khimicheskikh reaktorov s kipĭashchim sloem [Mathematical Models of Chemical Reactors with a Boiling Layer]. Minsk: Nauka i tekhnika, 211 p. (1976).

18

The Theory of Ignition
by a Heated Surface*

The problem of self-ignition of a gas confined in a vessel with two plane-parallel walls of identical temperature was solved by Frank-Kamenetskiĭ by investigating a steady regime which preserves equality between the released and evacuated heat at all points of the gas mixture, and by considering the limit of existence of this steady regime [1].

Generalizing the problem to the case of walls of different temperatures, we come to the question of ignition of a gas by a hot wall for given conditions of heat evacuation which are determined by the presence of a cold wall at a certain distance.

It is obvious here that it is the rate of the reaction and of heat release at temperatures close to the temperature of the hot wall that is the essential factor. Expanding the rate close to this temperature (T_i) into an exponential series using the method systematically developed by Frank-Kamenetskiĭ, we find

$$Q(T) = Ae^{-E/RT} = Q(T_i)e^{(T-T_i)E/RT_i^2}, \tag{1}$$

where $Q(T)$ is the amount of heat released in 1 cm^3 in 1 sec, which is equal to the reaction rate per unit volume times the reaction heat; E is the activation energy.

We transform the equation of heat conduction for the steady regime,

$$\frac{Kd^2T}{dx^2} = -Q(T) = -Q(T_i)e^{(T-T_i)E/RT_i^2}, \tag{2}$$

to dimensionless variables,

$$\theta = (T - T_i)\frac{E}{RT_i^2}, \tag{3}$$

so that $\theta = 0$ at the igniting wall;

$$\xi = x\sqrt{\frac{EQ(T_i)}{KRT_i^2}}, \tag{4}$$

after which it takes the form

$$\frac{d^2\theta}{d\xi^2} = -e^\theta. \tag{5}$$

*Zhurnal eksperimentalnoĭ i teoreticheskoĭ fiziki **9** 12, 1530–1534 (1939).

Below we give a numerical example which illustrates the order of magnitude of the temperature and length scale introduced by this transformation.

Let us consider the family of curves satisfying equation (5). As the two arbitrary constants we may take the two coordinates of the maximum of the curve.

Restricting ourselves to curves whose maximum is at $\xi = 0$ (all the other curves may be obtained from them by simple translation along the ξ-axis), we obtain the picture presented in Fig. 1. The higher the maximum of the curve, the more sharply it is bent in accord with the greater value of $e^{(\theta)}$, and the larger is the temperature gradient, since the amount of evacuated heat is larger.

These properties may be deduced analytically if we observe that all of the curves in Fig. 1 may be obtained from one another by the transformation

$$\theta' = \theta - m; \qquad \xi' = \xi e^{m/2}, \tag{6}$$

which does not change equation (5), as may be verified by substitution. The transformation shows that as the temperature maximum grows the integral curves narrow and become steeper. If the equation of the curve with a maximum $\theta = 0$ at $\xi = 0$ is

$$\theta = \varphi(\xi); \qquad \varphi(0) = 0; \qquad \varphi'(0) = 0, \tag{7}$$

then we obtain a curve with a maximum $\theta = m$ by transforming

$$\theta' = \varphi(\xi'); \qquad \theta - m = \varphi(\xi e^{m/2}); \qquad \theta = m + \varphi(\xi e^{m/2}). \tag{8}$$

We note that the envelopes of the family (see Fig. 1) of curves yield the limits of self-ignition, and thus serve as an illustration of Frank-Kamenetskiĭ's theory. Indeed, if we place two walls of temperature θ_0 at the points A and A_1, we find two steady regimes. No steady regime corresponds to a distance $C - C_1$ between the walls. Finally, $B - B_1$ is a limiting distance, the limiting diameter of a plane-parallel vessel at the given temperature θ_0, with B and B_1 located on the envelopes of the integral curves of the family under consideration.

Let us compare this result with Semenov's [2] interpretation of thermal explosion (Fig. 2), which operates with quantities averaged over the volume. When the coefficient of heat transfer per unit volume is decreased (which may be accomplished by increasing the dimensions of the vessel) we obtain consecutively: two steady solutions A_1 and A_2, the explosion limit B, and absence of steady solutions for still smaller heat transfer (line C).

By analogy we conclude that, of the two steady solutions under the limit (see Fig. 1), only the lower one corresponds to a stable regime (A_1 in Fig. 2); the upper curve corresponds to an unstable solution from which a small deviation leads either to explosion or to the lower stable regime (the point A_2 in Fig. 2).

It is curious that by considering one isolated integral curve (Fig. 3), it would be difficult to establish the fact that for a constant maximum tem-

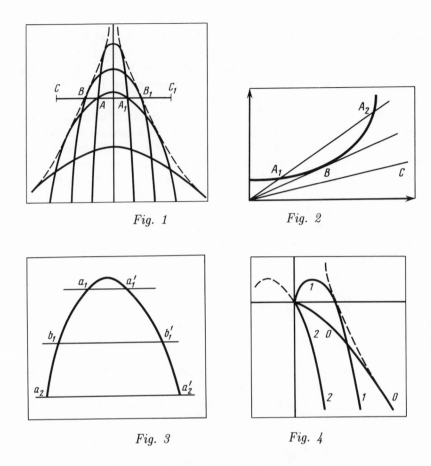

Fig. 1

Fig. 2

Fig. 3

Fig. 4

perature the steady solutions which it describes, depending on where the boundary conditions are applied ($a_1 - a_1'$ or $a_2 - a_2'$), correspond either to stable or unstable solutions. The boundary between them (b and b') is determined by the tangent points of our integral curve with the envelopes of the entire family (again see Fig. 1).

As we know from geometry, it follows from this that at these points the integral curve is intersected by infinitesimally close curves of the same family.

This last result must be considered a natural one from the point of view that for the solution of questions of stability one must always consider, together with the given distance, other infinitesimally close ones as well.

We cannot make direct use of Fig. 1 for our problem of ignition of a cold gas by one heated surface, since we restricted ourselves in this drawing to functions which have a maximum at $\theta = 0$, i.e., those which are symmetrical with respect to the ordinate axis, which is possible only for symmetric boundary conditions.

Let us consider a family of curves which correspond to the condition given at the hot wall $\theta = 0$ for $\xi = 0$ (Fig. 4). We may assert that it is precisely the curve for which the point $\theta = 0$, $\xi = 0$ is a maximum (curve *0* in Fig. 4), that is least steep and higher than all the other curves over a large distance. Indeed, curve *1* with a positive temperature gradient near the wall reaches a maximum at $\xi > 0$, $\theta_{\max} > 0$.

Curve *2*, which has a negative gradient at the coordinate origin, may be considered as a continuation of the curve which has a maximum to the left, at $\xi < 0$, $\theta_{\max} > 0$. Both curves *1* and *2* must therefore fall for large ξ faster than curve *0* in accordance with the higher maximum θ on curves *1*, *2*.

Curve *0* turns out for large ξ to be the asymptotic envelope of the family and, consequently, gives the limiting relation between distance and the temperature of the cold wall at the ignition limit.

It is not difficult to show that for large ξ we have a straight line:

$$\frac{d^2\theta}{d\xi^2} = \frac{1}{2}\frac{d}{d\theta}\left(\frac{d\theta}{d\xi}\right)^2 = -e^\theta, \tag{9}$$

$$\frac{d\theta}{d\xi} = \pm\left[2\int_0^\theta -e^\theta\,d\theta\right]^{1/2} = \pm\left[2(1 - e^\theta)\right]^{1/2}, \tag{10}$$

$$\frac{d\theta}{d\xi} \to \sqrt{2} \qquad \text{for} \qquad \theta \to -\infty, \quad \xi \to \infty.$$

The constant value of the temperature gradient for large absolute values of negative θ in the region where the gas is cold and the amount of heat released in it may be neglected corresponds to a particular value of the thermal flux:

$$q_{\lim} = K\frac{dT}{dx} = K\frac{RT_I^2}{E}\left(\frac{E}{RT_I^2}\frac{Q(T_I)}{K}\right)^{1/2}\left(\frac{d\theta}{d\xi}\right)_{\lim} = \left(\frac{2KRT_I^2Q(T_I)}{E}\right)^{1/2}. \tag{11}$$

For more intensive heat transfer, two regimes are possible (curves *1* and *2* in Fig. 4); for less intensive heat transfer no stationary regime is possible, the gas is ignited by the hot wall.

If the heat transfer is carried out, as we assumed above, by the cold wall, the limiting heat transfer found yields the relation between temperature and the distance d from the hot wall:

$$K\frac{T_I - T_{\text{cold}}}{d} = q_{\lim}. \tag{12}$$

But it is easy to see that the same results will be valid for any other means of heat transfer as well, for example, by a cold flow of explosive mixture blowing over the igniting surface. It is only necessary that within a thin boundary layer near the igniting surface—within several units' variation of θ and ξ—the influence of the gas motion not be felt, and a purely conductive regime take place (one corresponding to the plane case so that the curvature radius of the igniting surface must be large compared to the length corresponding to a change in ξ by one unit). As we shall see, this requirement is usually satisfied at the limit—for a thinner boundary layer, if the temperature of the mixture is significantly lower than the temperature of the igniting surface, the heat transfer is so intensive that ignition does not occur. In this way the dependence of the ignition temperature on the dimensions and form of the heated surface, the flow velocity, gas pressure, etc. may be reduced to the study of the dependence of the heat transfer on all of these factors under the known laws of chemical reaction kinetics in an igniting mixture.

A remarkable qualitative conclusion which may be experimentally verified is that at the ignition limit the amount of heat released by the heated surface to the gas being ignited by heat conduction vanishes since the temperature gradient is equal to zero.

$T,$ °K	RT_1^2/E (temperature scale)	$[RT_1^2 K/EQ(T)]^{1/2}$ (length scale, cm)	$q,$ cal/cm$^2 \cdot$ sec	$q,$ kcal/m$^2 \cdot$ hr	σ^* kcal/m$^2 \cdot$ hr $\cdot$ deg
500	6	10^3	$1 \cdot 10^{-8}$	$4 \cdot 10^{-4}$	$2 \cdot 10^{-6}$
1000	25	2	$2.5 \cdot 10^{-4}$	10	$1.4 \cdot 10^{-2}$
1500	56	$4 \cdot 10^{-3}$	0.25	10^4	10
2000	100	$1.5 \cdot 10^{-5}$	133	$5 \cdot 10^{-6}$	3000
*At $T_0 = 300$°K.					

We present a table which illustrates the various orders of magnitude at the ignition limit for a fictitious reaction whose rate is equal to

$$\frac{dn}{dt} = -n \cdot 10^{13} e^{-80\,000/RT} \tag{13}$$

at atmospheric pressure. The heat of reaction is $80\,000$ cal/mole and the thermal conductivity is $2 \cdot 10^{-5}$ cal/sec $\cdot$ cm $\cdot$ deg.

The last column gives the heat transfer coefficient corresponding to maximum heat flux in the case where the temperature of the surrounding (cold) gas is 300°K. Under industrial conditions this coefficient is usually of the order of several units or tens of units (in industrial units). This corresponds to an ignition temperature of 1500°K in the given example. As we see, at 1500°K the length scale is already very small—$4 \cdot 10^{-3}$ cm.

We note that we considered only ignition related to the non-steady character of the temperature field in a gas with a constant temperature of the igniting surface. Growth in the surface temperature, in particular, as a result of a catalytic reaction, may change these relations; it will then not be possible to verify the conclusion with respect to the absence of heat exchange between the gas and the surface at the limit; it is not easy to establish whether the heating of the surface is a cause or consequence of the explosion.

In the absence of catalysis on the surface, similarity of the concentration and temperature fields is achieved precisely at the ignition limit if the coefficients of diffusion and thermal diffusivity are equal, since in this case both the diffusion gradient and the temperature gradient at the igniting surface are equal to zero, and the equations of diffusion and thermal conductivity with the chemical reaction may be reduced to the form of an identity (see our work on flame propagation [3]).

The presence of similarity allows us to easily estimate the effect of consumption near the surface on ignition. Since the ignition temperature is significantly lower than the temperature developed during the combustion of our mixture, the effect of consumption is not large.

In contrast, if the ignition temperature, under the given conditions of heat transfer calculated according to the formulas above, is not lower than the combustion temperature, ignition does not occur at all, no matter how we heat the surface; on the contrary, a rapid reaction occurs at the surface, which becomes covered by a layer of the reaction products. The absence of ignition for an ignition temperature which is higher than the temperature of combustion is natural since flame propagation is nothing but ignition of a cold gas by reaction products at the combustion temperature.

Summary

A geometrical interpretation of the theory of thermal explosion in a conductive region is given.

The thermal regime of the reacting mixture between two walls of strongly differing temperature is studied, and the minimum (limiting) amount of evacuated heat found for which ignition at the hot wall still does not occur (a steady regime is possible).

The effect of various factors—the forms and dimensions of the igniting surface, motion of the fuel, etc.—are reduced to their effect on the conditions of heat transfer.

The temperature gradient and conductive thermal flux at the igniting surface disappear at the ignition limit.

A numerical example of calculation of the quantities entering into the theory is given.

Institute of Chemical Physics   *Received*
USSR Academy of Sciences, Leningrad   *October 4, 1939*

References

1. *Frank-Kamenetskiĭ D. A.*—Dokl. AN SSSR **18**, 411 (1938).
2. *Semenov N. N.* Tsepnye reaktsii [Chain Reactions]. Leningrad: Goskhimtekhizdat, 555 p. (1934).
3. *Zeldovich Ya. B., Frank-Kamenetskiĭ D. A.*—Zhurn. fiz. khimii **12**, 100 (1938); here art. 19.

Commentary

Consideration of the problem of gas ignition in a plane vessel with two walls of different temperatures was a natural generalization of D. A. Frank-Kamenetskiĭ's theory of thermal explosion.[1] At the same time it is physically obvious that the processes at the hot wall play the decisive role, and the solution of the problem introduces essentially new features.

After this paper was completed, a number of studies were carried out which analyzed the structure of the family of steady solutions and the conditions of ignition in spherical and cylindrical vessels. Regarding these studies, see the monograph by Ya.B. *et al.*[2] In this same monograph one may find literature on various applications of the concept of thermal explosion in other problems of physics, in particular, in the physics of polymers where, as was first shown by A. G. Merzhanov and his colleagues, as a result of viscous heating, steady flow of the polymer becomes impossible in the motion of polymer melts.

This paper is one of the first applications of the asymptotic method in world scientific literature, a method which twenty years later has received widespread use. Now it is called the method of matched asymptotic expansions. Without introducing the terminology which later appeared, the author essentially made use of the full arsenal of this method, which today makes the problem studied in this article a textbook example of its application. An exposition of the general technique of the method of matched asymptotic expansions and numerous examples of its use may be found in monographs.[3,4]

[1] *Frank-Kamenetskiĭ D. A.*—Dokl. AN SSSR **18**, 411–415 (1938); Diffuziĭa i teploperedacha v khimicheskoĭ kinetike [Diffusion and Heat Transfer in Chemical Kinetics]. 2nd Ed. Moscow: Nauka, 491 p. (1967).

[2] *Zeldovich Ya. B., Barenblatt G. I., Librovich V. B., Makhviladze G. M.* Matematicheskaĭa teoriĭa goreniĭa i vzryva [Mathematical Theory of Combustion and Explosion]. Moscow: Nauka, 478 p. (1980).

[3] *Van Dyke M.* Perturbation Methods in Fluid Mechanics]. Moscow: Mir, 310 p. (1967).

[4] *Cole J.* Metody vozmushcheniĭ v prikladnoĭ matematike [Perturbation Methods in Applied Mathematics]. Moscow: Mir, 274 p. (1970).

II

Flame Propagation

19
A Theory of Thermal Flame Propagation*

With D. A. Frank-Kamenetskiĭ

The conception of the thermal propagation of a flame as the most common mechanism of combustion, related to successive ignition of the explosive mixture by heat generated in the reaction, is very far from new [1].

Existing theories [2–4], however, are unsatisfactory [5] since they make use of the concept of the "ignition temperature" of the mixture. Meanwhile, as we know [6,7], the phenomenon of self-ignition is naturally explained by the assumption of smooth growth of the rate of exothermic reaction with increasing temperature if the heat balance of the system is considered. The ignition temperature itself proves here to be dependent not only on the properties of the explosive mixture, but also on the size and form of the vessel in which it is measured and on similar factors. A rational theory must give an expression for the velocity of flame propagation in a mixture in which the chemical reaction kinetics are given as a function of the temperature and the concentration of the reacting substances.

A simultaneous analysis of the equation of heat conduction, taking into account the release of heat resulting from the chemical reaction, and the diffusion equation which governs the concentrations of the reacting substances and reaction products, and also allowing for variation of these concentrations due to the chemical reaction, leads to conclusions which are very important for subsequent study.[1]

*Zhurnal fizicheskoĭ khimii 12 1, 100–105 (1938).

[1]The role of diffusion was recently pointed out qualitatively in a paper by Jost [5].

In particular, it turns out that the relative increase in temperature and the relative changes in the concentrations of all the reacting substances are identically equal:

$$\frac{T - T_0}{T_1 - T_0} = \frac{a - a_0}{a_1 - a_0} = \frac{b - b_0}{b_1 - b_0} = \dots \tag{1}$$

Here T_0, a_0, b_0, ... are the temperature, concentration of substance A, concentration of substance B, ... in the initial mixture; T_1, a_1, b_1 ... are the same for the final reaction products (T_1 is the theoretical temperature of combustion); finally, T, a, b, ... are the temperature and concentrations at an arbitrary point in the flame zone, and are functions of the space coordinates and time, while T_0, a_0, b_0 ... and T_1, a_1, b_1, ... are constant for a given mixture.

The content of formula (1) may be formulated as the similarity of the concentration fields to the temperature field in the flame, which means that the temperature and concentrations are interrelated just as in an adiabatic reaction, although the reaction develops in time quite differently.

In the derivation of formula (1) the following assumptions must be made:

1. Equality of the thermal diffusivity (the ratio of thermal conductivity to heat capacity per unit volume) and the diffusion coefficients of the substances.

2. The absence of heat transfer by radiation.

3. The absence of heat transfer to the walls of the vessel in which the flame propagates.

4. The formula relates only to the initial and final substances, only the gross reaction being present; it is inapplicable to the concentrations of the intermediate products, which must be small compared to those of the reacting substances. However, an active role by the intermediate products or any centers during the chemical reaction does not invalidate the formula as applied to the initial and final products. Assumptions (2) and (3) imply that formula (1) is actually valid only in the vicinity of the flame itself: only there does it correctly describe the relation between the increase in the temperature and the change in the mixture composition due to the chemical reaction and to heat and diffusion currents in the flame. Farther away, at distances which are large compared to the width of the flame, where the reaction products cool without changing the composition, formula (1) is no longer applicable.

For the derivation of formula (1), let us compare the equation of heat conduction,

$$\frac{\partial T}{\partial t} = \kappa \Delta T - (\mathbf{V}\nabla)T + \frac{Q}{c},$$

—where κ is the thermal diffusivity, $\mathbf{V}$ is the gas velocity, c is the heat capacity per unit volume, and the boundary conditions are $T = T_0$ in the initial substance, $T = T_1$ in the combustion products—with the equation of diffusion

$$\frac{\partial a}{\partial t} = D\Delta a - (\mathbf{V}\nabla)a + M,$$

—where D is the diffusion coefficient and M is the rate of formation of the substance A due to the reaction; the boundary conditions are $a = a_0$ in the initial substance, $a = a_1$ in the combustion products. Introducing

$$\theta = \frac{T - T_0}{T_1 - T_0} \qquad \text{and} \qquad \alpha = \frac{a - a_0}{a_1 - a_0},$$

we obtain

$$\frac{\partial \theta}{\partial t} = \kappa\Delta\theta - (\mathbf{V}\nabla)\theta + \frac{Q}{c(T_1 - T_0)}.$$

$\theta = 0$ in the initial substance, $\theta = 1$ in the combustion products, and

$$\frac{\partial \alpha}{\partial t} = D\Delta\alpha - (\mathbf{V}\nabla)\alpha + \frac{M}{a_1 - a_0}.$$

$\alpha = 0$ in the initial substance, $\alpha = 1$ in the combustion products. It is obvious that

$$\frac{M}{a_1 - a_0} = \frac{Q}{c(T_1 - T_0)}; \qquad \frac{Q}{M} = \frac{c(T_1 - T_0)}{a_1 - a_0}.$$

When only the gross reaction is present, the ratio of the rate of heat release to the rate of change of the concentration is the same as that of total amount of released heat, $c(T_1 - T_0)$, to the total change in concentration, $(a_1 - a_0)$, when the reaction runs completely. But then, assuming $D = \kappa$, the differential equations as well as the boundary conditions for θ and α coincide:

$$\theta = \alpha; \qquad \frac{T - T_0}{T_1 - T_0} = \frac{a - c_0}{c_1 - a_0}.$$

On the basis of this we can express the concentrations which enter into the equation of the chemical reaction rate in terms of the temperature, which will permit us to confine ourselves to consideration of the equation of heat conduction alone.

In the flame zone all temperatures lying in the interval between the initial temperature and the theoretical temperature of combustion, $T_0 \leq T \leq T_1$, are realized simultaneously. The rapid increase in the reaction rate with increasing temperature according to the exponential law of Arrhenius leads to the result that the reaction in effect proceeds basically at high temperatures near T_1.

Comparing this conclusion with formula (1), we see that the reaction proceeds in a mixture which contains a large amount of the final combustion products and comparatively small amounts of the initial substances. From this certain qualitative conclusions may be drawn as to the character of the dependence of the velocity of flame propagation on various parameters. Thus, as compared with a classical reaction which has an initial rate equal to the maximum rate of autocatalytic reaction, auto-catalysis of a combustion reaction by the final products, which strongly delays self-ignition of the mixture, should not affect the rate and character of the process of flame propagation since the reaction in the flame proceeds in the zone of high concentrations of the final reaction products. Various admixtures may also exert a noticeable influence on the velocity of flame propagation only insofar as they affect the rate of reaction at very high temperatures, near T_1. Finally, the initial temperature and the mixture composition should influence the velocity of propagation principally by changing the theoretical temperature of combustion[2] T_1.

The idea that the reaction proceeds principally at a temperature which is close to T_1 also permits us to find closed analytical expressions for the velocity of flame propagation. The heat released during the reaction is partly spent on heating the reacting gas itself and partly carried away by heat conduction to neighboring elements of the gas volume. If the temperature of the zone in which the reaction effectively occurs is already close to T_1, the amount of the heat spent on heating the reacting gas up to its final temperature beyond the flame front, T_1, is small compared to the total released heat of reaction. Approximately, we can consider that all the heat from the reaction zone is carried away by conduction. The Fourier equation of heat conduction is:[3]

$$c\frac{dT}{dt} = c\frac{\partial T}{\partial t} + c(\mathbf{V}\nabla)T = \operatorname{div}(K\nabla T) + Q. \tag{2}$$

Here c is the heat capacity, K is the thermal conductivity, $Q = Q(t)$ is the volume rate of heat release resulting from the chemical reaction (in $cal/cm^3 \cdot sec$), while the dependence of the reaction rate on the concentrations is reduced, with the aid of formula (1), to a dependence on temperature; thermal expansion is also accounted for.

We neglect the heat spent on heating the mixture in the reaction zone itself (cdT/dt) and further consider the thermal conductivity in this zone to be constant and equal, for the sake of definiteness, to the thermal conductivity K_1 of the reaction products at T_1. Turning to the one-dimensional problem, we obtain the equation

$$K_1\frac{d^2T}{dx^2} = -Q(T). \tag{3}$$

This equation is easily integrated in quadratures. Taking into account the boundary condition

$$\frac{dT}{dx} = 0 \quad \text{at} \quad T = T_1, \tag{4}$$

[2]Throughout the present paper we consistently neglect the heat losses by radiation, by heat transmission through the cold parts of the apparatus, etc.

[3]In the derivation of similarity above we assumed the heat capacity and conductivity to be constant, which is no longer necessary here.

we find

$$\frac{dT}{dx} = \sqrt{\frac{2}{K_1} \int_T^{T_1} Q(T)\,dT}. \tag{5}$$

To find dT/dx at the (somewhat arbitrary) boundary of the reaction zone, we must substitute the temperature at the boundary of the reaction zone for the lower limit of the integral in formula (5).

Due to the rapid drop in $Q(T)$ with decreasing temperature[4] we find the limiting value $(dT/dx)_z$, which differs little from the actual value, at the boundary of the reaction zone according to the formula

$$\left(\frac{dT}{dx}\right)_z = \sqrt{\frac{2}{K_1} \int_0^{T_1} Q(T)\,dT}. \tag{6}$$

Finally, we equate the thermal flux from the reaction zone to the total amount of heat released in the flame per unit time

$$K_1 \left(\frac{dT}{dx}\right)_z = uL, \tag{7}$$

where u is the mass rate of combustion (in grams of the mixture per unit time on 1 cm^2 of the flame surface), and L is the calorific value of the mixture (in cal/gr).

Comparing (6) and (7), we obtain the final expression for the mass rate of combustion,

$$u = \frac{1}{L} \sqrt{2K_1 \int_0^{T_1} Q(T)\,dT}. \tag{8}$$

In considering this expression it should not be forgotten that

$$Q(T) = Q(T, T_0, T_1, a_0, a_1, \dots) \tag{9}$$

by formula (1), and that the final temperature

$$T_1 = T_0 + \frac{L}{c} \tag{10}$$

itself depends on the initial temperature and calorific value of the mixture.

The mass rate of combustion is related to the linear velocity of flame propagation by the simple formula $u = v\rho$, where v is the linear velocity and ρ is the density.

Substituting the density of the initial mixture, we obtain the linear velocity of flame propagation relative to the initial mixture v_0, and substituting the density ρ_1 of the reaction products at T_1, we obtain the velocity v_1 relative to the burnt substance.

For several very simple cases—mono- and bi-molecular reactions and reactions which do not depend on the concentration (so-called zero-order reactions)—it is easy to find simple formulas for the combustion rate.

---

[4]The rate constant, which obeys the Arrhenius law, decreases by a factor of e in the temperature interval RT^2/E, where E is the activation energy. The condition that the reaction occur near T_1 reduces to the requirement that $RT^2/E \ll T_1 - T_0$.

Neglecting the change in volume due to the reaction, we obtain the linear velocity of flame propagation with respect to the original mixture.

For a zero-order reaction with a reaction rate equal to

$$\frac{dn}{dt} = Ce^{-E/RT},$$

where n is the concentration of the reacting substance (in g/cm^3),

$$v_0 = \sqrt{\frac{2K_1 C \exp(-E/RT_1)RT_1^2}{LEn_0\rho_0}}.$$

The fact that the calorific value of the mixture L has appeared in the denominator is not in disagreement with the usual conceptions; in fact, at a constant initial temperature T_0, the theoretical temperature of combustion T_1 entering into the formula increases with L, so that the velocity v increases as a whole. Conversely, an increase in L at constant T_1 corresponds to a drop in T_0, so that the decrease in the velocity here is natural.

For a first-order reaction,

$$\frac{dn}{dt} = C'ne^{-E/RT},$$

the propagation velocity is

$$v_0 = \sqrt{\frac{2K_1 C' \exp(-E/RT_1)(RT_1^2/E)^2}{\rho_0 L(T_1 - T_0)}}.$$

For a second-order reaction,

$$\frac{dn}{dt} = C''n^2e^{-E/RT},$$

$$v_0 = \sqrt{\frac{4K_1 C'' \exp(-E/RT_1)n_0(RT_1^2/E)^3}{\rho_0 L(T_1 - T_0)^2}}.$$

The condition of applicability of all three formulas is[5]

$$\frac{E(T_1 - T_0)}{RT_1^2} \gg 1.$$

Results similar to those presented here were obtained by Lewis and Elbe [8], who used, however, a number of completely unjustified assumptions.

---

[5]We shall repeat the meaning and dimensions of all the notations: v_0—linear velocity of flame propagation relative to the unburned mixture, cm/sec; K_1—thermal conductivity at the theoretical temperature of combustion, cal/cm · sec · deg; ρ_0—initial density of the mixture, g/cm^3; n—concentration of the reacting substance, g/cm^3; n_0—initial concentration of the reacting substance, g/cm^3; L—calorific value of the mixture, cal/g; T_0—initial temperature, deg; T_1—theoretical temperature of combustion, $T = T_0 + L/c$, deg; E—heat of activation, cal/mole; R—gas constant, cal/mole · deg; C—rate constant of a zero-order reaction, gr/cm^3 · sec; C'—rate constant of a first-order reaction, sec^{-1}; C''—rate constant of a second-order reaction, cm^3/g · sec.

In conclusion, we should note that calculation of the flame speed places extremely strict requirements on the study of the kinetics of combustion reactions. As was indicated above, the rate of reaction at extremely high temperatures, of order 1000–2000°K, at which the reaction takes 10^{-2}–10^{-6} sec, proves to be essential.

The presence of parallel reactions, the formation of considerable amounts of intermediate products, active participation (autocatalysis, chains) of intermediate products in the course of reaction—all these deviations from the simplest kinetic schemes greatly restrict the range of applicability of the quantitative results of the proposed theory.

Physico-Chemical Laboratory   *Received*
USSR Academy of Sciences, Leningrad   *March 7, 1938*

References

1. *Mallard E., Le Chatelier A.*—Ann. Mines **4**, 296 (1883).
2. *Nusselt W.*—Ztschr. Dt. inorg. Chem. **59**, 872 (1915).
3. *Jouguet E.*—C. r. Bull. chim. Soc. **168**, 820 (1919).
4. *Daniell L. J.*—Proc. Roy. Soc. London A **126**, 393 (1930).
5. *Jost W., Müffling B.*—Ztschr. phys. Chem. A **181**, 208 (1937).
6. *Want-Hoff J.* Etudes de dynamique chimique. Paris, (1900).
7. *Semenov N. N.* Tsepnye reaktsii [Chain reactions]. Leningrad: Goskhimtekhizdat, 555 p. (1934).
8. *Lewis B., Elbe G.*—J. Chem. Phys. **2**, 534 (1937).

Commentary

This article was a turning point in the theory of combustion; it was the beginning of a new stage in the development of this theory. A review of works preceding the paper, and of subsequent theoretical studies of combustion, may be found in the monograph.[1] Here it is appropriate to dwell on the following several points.

Apart from the theory of combustion, the concept of self-sustaining waves (now sometimes called autowaves), developed in works by A. N. Kolmogorov, I. G. Petrovskiĭ, N. S. Piskunov,[2] R. Fisher[3] and the present paper, has proved to be extraordinarily fruitful for many problems in chemistry and biophysics. Some of these applications may be found in the review by V. A. Vasiliev et al.[4] Let us note waves of a fundamentally new type: spiral, self-sustaining vortex formations,

[1] *Zeldovich Ya. B., Barenblatt G. I., Librovich V. B., Makhviladze G. M.* Matematicheskaĭa teoriĭa goreniĭa i vzryva [Mathematical Theory of Combustion and Explosion]. Moscow: Nauka, 478 p. (1980).

[2] *Kolmogorov An. N., Petrovskiĭ I. G., Piskunov N. S.*—Byul. MGU. Ser. A **6**, 1–16 (1937).

[3] *Fisher P.*—Ann. Engenics **7**, 355–367 (1937).

[4] *Vasiliev V. A., Romanovskiĭ Yu. M., Yakhno V. G.*—Uspekhi fiz. nauk **128**, 625–666 (1979).

discovered experimentally in the Zhabotinskii-Belousov chemical reaction, in the cardiac muscle during arrhythmia, and in other biological processes.[5] A theory of these waves which is closely related in its ideas to the conception developed in the present article was proposed at the recent Symposium on synergetics (Pushino-na-Oke, July, 1983) by A. S. Mikhailov.[6] It should be mentioned, in fact, that at this Symposium around 100 reports were presented which applied the concepts developed in the present paper to new problems.

The problem considered in this paper of a self-sustaining propagation wave of flame is closely related to self-similar solutions of the second kind, considered by Ya.B. in his paper, "Gas Motion Under the Action of Short-Duration Pressure (Impulse)" (see article **9** of the present volume) nearly twenty years later. Indeed, we are dealing here with wave-like solutions, for instance,

$$T = F(x - \lambda t + c); \tag{1}$$

here the propagation velocity of the wave λ is found as an eigenvalue from the condition of existence as a whole of a solution of the form (1). We now take in (1) $x = x_0 \ln \xi$; $t = t_0 \ln \tau$, $c = c_0 \ln A$ and obtain

$$T = F\left(x_0 \ln \frac{\xi}{A^{c_0/x_0} \tau^{\lambda t_0/x_0}}\right) = \varphi\left(c_0, \frac{\xi}{A^{c_0/x_0} \tau^{\lambda t_0/x_0}}\right), \tag{2}$$

i.e., an exponential-type self-similar solution in which the exponent $\lambda t_0/x_0$ is found as an eigenvalue from the condition of existence as a whole of a self-similar solution of the form (2).

In the aforementioned paper by A. N. Kolmogorov, I. G. Petrovskiĭ and N. S. Piskunov[2] the same equation was considered,

$$\partial_t v = \kappa \partial_{xx}^2 v + f(v), \tag{3}$$

as in the present paper: however, unlike the latter, it was assumed that $f(v) > 0$ for $0 < v < 1$, and $f'(v) < f'(0)$ for $v > 0$. (Using similar notations, the present article assumes that the function $f(v)$ is identically equal to zero in some interval $0 \le v \le \Delta$ near the point $v = 0$.) It turned out here that the velocities of propagation of stationary waves form a continuous spectrum, $\lambda \ge \lambda_0 = 2\sqrt{\kappa f'(0)}$, and only the wave corresponding to the lower end of the continuous spectrum $\lambda = \lambda_0$ has physical meaning, i.e., is an asymptote at $t \to \infty$ of the solution of the appropriately formulated non-invariant initial problem.

In connection with the investigation of the problem of flame propagation in a reacting mixture, a paper by Ya.B.[7] posed the question of how to describe flame waves in intermediate regimes for which the function $f(v)$ at small v, though small, is still non-zero and is not convex, i.e., the condition $f'(v) < f'(0)$ is not satisfied.

[5] *Agladze K. I., Krinsky V. I.*—Nature **296**, 424–426 (1982).

[6] *Mikhailov A. S.* Teoriĭa vrashchaĭushchikhsĭa voln i vedushchikh protsessov v aktivnykh sredakh: Doklad na Simposiume "Avtovolnovye protsessy v biologii, khimii i fizike" [Theory of Rotating Waves and Leading Processes in Active Media: Report at the Symposium "Autowave Processes in Biology, Chemistry and Physics"]. Pushino-na-Oke, July (1983).

[7] *Zeldovich Ya. B.* Rasprostranenie plameni po smesi, reagiruĭushcheĭ pri nachalnoĭ temperature [Propagation of a Flame in a Mixture Reacting at the Initial Temperature]. Chernogolovka: IKhF AN SSSR, preprint, 1–15 (1978); Comb. and Flame **39**, 219–224 (1980).

The difficulty lies in the fact that, formally, at $f(0) \neq 0$ a solution of the stationary wave type does not exist at all! Ya.B. carried out the investigation by consistently applying an asymptotic approach using the smallness of $f(0)$. First, using an elegant transformation of the variables, he reduced the problem to a problem of the same type as in the work of A. N. Kolmogorov, I. G. Petrovskiĭ and N. S. Piskunov, so that in equation (3) the effective $f(v)$ vanishes for $v = 0$, even though it remains non-convex. Ya.B. noted that this condition is not necessary to establish an approximate auto-wave regime moving with instantaneous velocity $\lambda = \lambda_0$ which, however, in this case is time-dependent. Subsequently this conclusion was confirmed by numerical calculations.[8] A mathematical investigation of the problem in the general case of a non-convex function was performed by V. E. Gluzberg and Yu. M. Lvovskiĭ.[9]

[8] *Aldushin A. P., Zeldovich Ya. B., Khudyaev S. I.* Rasprostranenie plameni po reagiruĭushcheĭ gazovoĭ smesi [Flame Propagation in a Reacting Gaseous Mixture]. Preprint OIKhF AN SSSR. Chernogolovka, 1–9 (1979).

[9] *Gluzberg V. E., Lvovskiĭ Yu. M.*—Khim. fizika **11**, 1546–1552 (1982).

20

The Theory of the
Limit of Propagation of a Slow Flame*

This paper is a continuation of a series of theoretical studies carried out at the Institute of Chemical Physics which seek to give a description of various phenomena of combustion and explosion under the simplest realistic assumptions about the kinetics of the chemical reaction. A characteristic feature of the specific rate (rate constant) of chemical combustion reactions is its strong Arrhenius-like dependence on the temperature with a large value of the activation heat, related to the large thermal effect of the combustion reaction.

Works published on this subject include Semenov's fundamental theory of thermal explosion [1], Todes' analysis of the kinetics of thermal explosion [2], Frank-Kamenetskiĭ's calculation of the absolute values of the limit of thermal explosion [3], the theory of ignition [4], and finally, most closely related to the first part of this paper, the theory of flame propagation by Frank-Kamenetskiĭ and the author [5].

§1. Elementary Theory

The basic results of the last work mentioned amount to the fact that in a flame the chemical reaction runs primarily in a narrow part of the zone, in the region where the temperature is close to the maximum (theoretical) combustion temperature.

In accordance with the general expression which follows from dimensional considerations

$$u \sim \sqrt{\kappa/\tau} \tag{1}$$

(accurate to within dimensionless numerical factors calculated earlier [5]), where κ is the thermal diffusivity, i.e., the ratio of the thermal conductivity to the specific heat per unit volume.

Substituting the reaction time

$$\tau \sim e^{E/RT}, \tag{2}$$

we obtain the dependence of the propagation velocity on the combustion temperature

$$u \sim e^{-E/2RT_C}, \tag{3}$$

*Zhurnal eksperimentalnoĭ i teoreticheskoĭ fiziki **11** 1, 159–169 (1941).

Since the reaction runs at close to the maximum temperature, the governing factor is the chemical reaction rate, the magnitude of the Arrhenius expression at maximum temperature.

Taking into account the effect of thermal losses on the maximum temperature, and hence on the flame propagation velocity, and the inverse effect of a change in the flame propagation velocity on the magnitude of the thermal losses, we easily find the limit of propagation of the flame.

Due to the narrowness of the zone in which the chemical reaction actually occurs, in the approximation that we are developing we may neglect thermal losses during the reaction itself.

Thus, the thermal losses which affect the maximum combustion temperature are reduced to

a. losses of heat in the zone of heating of the mixture from T_0 to (practically) T_c;

b. loss of heat from the combustion zone by heat transfer in the direction of flame propagation by cooling combustion products.

We shall show that both mechanisms identically lead to relative thermal losses which are inversely proportional to the square of the propagation velocity.

In studying mechanism a we first establish that the width of the heating zone is inversely proportional to the flame velocity, in accordance with the long-known law of temperature variation (Michelsohn)

$$T = (T_c - T_0)e^{-ux/\kappa} + T_0. \tag{4}$$

Also proportional to u^{-1} is the absolute amount of heat which is released by the heating zone in unit time; referring it to the amount of material burned in unit time, which is proportional to the propagation velocity, we obtain the required result.

In mechanism b, under given conditions of heat transfer, each elementary quantity of combustion products is cooled at a certain rate. For a given time derivative of the temperature, the derivative with respect to the coordinate is inversely proportional to the propagation velocity, and the same is true of the heat flux in the direction of the flame which carries heat out of the

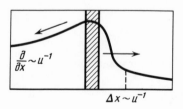

Fig. 1

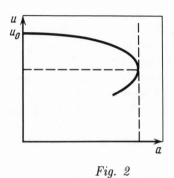

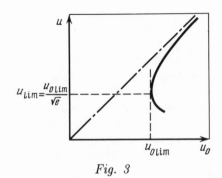

Fig. 2   Fig. 3

reaction zone (Fig. 1). Finally,

$$T_{\text{th}} - T_{\text{max}} = \frac{a}{u^2}. \tag{5}$$

Combining this with the dependence of the flame velocity on the temperature, we find

$$u \sim e^{-E/2RT_{\text{max}}}; \qquad u = u_0 e^{(T_{\text{max}}-T_{\text{th}})E/2RT_{\text{th}}^2}, \tag{6}$$

where u_0 is the velocity in the absence of heat transfer to the outside, i.e., when $T_{\text{max}} = T_{\text{th}}$. Denoting $b = aE/2RT_{\text{th}}^2$, we obtain the transcendental equation

$$u = u_0 e^{-b/u^2}, \tag{7}$$

which has a non-trivial solution only for

$$b < \frac{u_0^2}{2e}; \qquad T_{\text{th}} - T_{\text{max}} < \frac{RT_{\text{th}}^2}{E}, \qquad \text{with} \quad u > \frac{u_0}{\sqrt{e}}. \tag{8}$$

The limit of existence of the uniform propagation regime of the flame is reached at $b = u_0^2/2e$, and at the limit

$$T_{\text{th}} - T_{\text{max}} = \frac{RT_{\text{th}}^2}{E}; \qquad u_{\lim} = \frac{u_0}{\sqrt{e}}. \tag{9}$$

For given values of u_0 and T_{th} depending on the parameter a, i.e., for a given mixture depending on the conditions of heat transfer, we obtain the flame propagation variation curve shown in Fig. 2.

At the limit the derivative $\partial u/\partial a \to \infty$ and the curve is tangent to the vertical. The dashed continuation of the curve corresponds to a spurious root of the transcendental equation, an unstable, physically unrealizable regime.

By changing, for a given value of the parameter a, the quantities u_0, T_{th}, i.e., changing the composition of the mixture under given conditions of heat transfer, we obtain the picture depicted in Fig. 3 of the velocity variation near the concentration limit. We characterize the composition of the mixture by the quantity u_0 (the adiabatic flame propagation velocity). In the absence of thermal losses we would identically have $u = u_0$ (the bisecting line in Fig. 3).

In the case of a so-called zero-order reaction, i.e., one whose reaction rate is independent of the concentration of the reacting materials throughout the reaction, violation of the similarity of the concentration field and temperature field as a result of heat transfer to the outside, which does not have an analogue in diffusion, still does not over-complicate the problem. In this particular case we were able to follow in detail the effect of the heat transfer on the entire distribution of the temperature in the flame and on the propagation rate. In our approximation the activation heat is much larger than RT_{th}, and the results of this more detailed investigation confirm all of the formulas derived above (unpublished work by the author, 1938).

§2. Comparison with Other Theories

In the literature one encounters assertions that at the limit the flame velocity is exactly equal to zero. Bunte and Jahn [6, 7], taking this assumption as self-evident, use it for extrapolation of the dependence curves of the flame velocity on the composition of the mixture.

Direct experiments by Payman [8] show that at atmospheric pressure the velocity of uniform flame propagation in tubes of diameter 2.5–5 cm does not fall below 12–20 cm/sec. On the basis of work by Coward and Hartwell [9], who found the relation between the normal velocity (i.e., the velocity of the flame with respect to the gas which we used above) and the velocity of uniform flame propagation with respect to the walls of the tube in which the gas is enclosed, we know that the data observed by Payman correspond to a normal flame velocity of 4–8 cm/sec. The values of the normal flame velocity at the limit obtained by Coward and Hartwell themselves are of the same order.

In §1 we showed that, at the propagation limit due to heat transfer, the flame velocity must be non-zero.

However, this most important qualitative result is not in fact a new one since it has already been obtained by Daniel [10], albeit under completely different and unrealistic conceptions of the kinetics of chemical reaction in a flame.

According to (9) the limit corresponds to a specific decrease of the flame temperature: the stronger the heat transfer, the greater the velocity of the flame must be in order for it to propagate. This conclusion is also in agreement with experiment. As is known, the greater the flame velocity, the better the flame passes through narrow slits or capillaries [11]. The simplified one-dimensional investigation of the heat transfer of a flame propagating in a tube given in the previous section leads to the conclusion

$$a \sim \frac{T_c \kappa^2}{d^2}. \tag{10}$$

As the diameter is decreased, the heat transfer from a unit volume intensifies because of the increase in the ratio of surface to volume (d^{-1}); heat exchange per unit surface also intensifies. For a constant value of the Nusselt number the heat exchange coefficient is proportional to d^{-1}. Under the rough assumption that T_c and E change little from one case to the next, we come to the conclusion that at the limit the Peclet number (numerically equal for gases to the Reynolds number), based on the flame velocity (or adiabatic flame velocity u_0) has a specific value

$$\mathrm{Pe} = \frac{ud}{\kappa} \cong \frac{u_0 d}{\kappa \sqrt{e}} = \frac{\mathrm{Pe}}{\sqrt{e}} \cong \mathrm{Re} = \mathrm{const}. \tag{11}$$

Available experimental data are insufficient for an objective verification of relation (11).

In flame propagation in a tube the gas motion caused by the propagation of the combustion products significantly deforms the flame front and changes the velocity of uniform propagation. The effect of gas motion, determined by the Reynolds number, cannot be separated from the effect of heat transfer as the limit corresponding to a particular value of Re is approached. Therefore, experimentally, instead of the idealized picture in Fig. 2 for the velocity of uniform propagation one observes a continuous velocity increase up to very large values of the tube diameter, i.e., very small values of the heat transfer parameter a, which impedes verification of the relation

$$u_{\mathrm{lim}} = \frac{u_0}{\sqrt{e}}$$

We find an analogous expression, in which the value of the constant in (11) is expressed in terms of the combustion heat, specific heat, and the combustion, ignition and initial temperatures, in the work of Holm [12], who constructed the thermal balance of a hemispherical flame, with the diameter of the hemisphere taken equal to the tube diameter.

However, in our opinion, Holm's basic balance equation is incorrect. Heat losses are equated to the difference between the reaction heat and the enthalpy of the combustion products $\bar{c}_p(T_c - T_0)$; when the equation is written in this way it is incorrect to take T_c from someone else's experiments which were performed under different conditions. Even if T_c in fact changes little from one case to another, the small difference $Q - \bar{c}_p(T_c - T_0)$ is completely determined by the losses and depends on the conditions of the experiment.

It is also incorrect to take for the thermal losses of the entire flow the heat transport by conduction across the surface at which the gas reaches ignition temperature. The portion of the heat which goes to heat the mixture entering the flame is not lost at all, which is clear from consideration of the propagation of an infinite plane flame front. The portion of the heat in which we are interested which is truly lost for the flame differs significantly from the corresponding expression of Holm.

One thermal balance equation, even if it is correctly constructed, can give only the combustion temperature under given conditions. To find the limit, additional considerations are essential, for example, those establishing the minimum allowable combustion temperature or the dependence of the flame velocity on the combustion temperature, which Holm does not include.

Therefore Holm's calculation must be considered devoid of sense even within the framework of a theory which makes use of the temperature of self-ignition and which does not consider diffusion of the reacting materials in the flame.

§3. Application of Similarity Theory

Studying anew the differential equations of heat conduction, diffusion, gas motion and chemical kinetics under the conditions of a chemical reaction (flame) propagating in a tube, through a narrow slit or under similar conditions, using the methods of the theory of similarity we find the following dimensionless governing criteria:

1. The criterion of homochronity $\tau \kappa / d^2$, where d is the characteristic dimension of the system, κ is the thermal diffusivity and τ is the characteristic time of the chemical reaction. Due to the strong dependence of the chemical reaction rate on temperature (see the next two criteria), it is necessary to define the temperature to which the quantity τ relates. We will relate it to the theoretical temperature of combustion.

2. The dependence of the reaction rate on the temperature according to Arrhenius' law yields two criteria whose choice is in large measure arbitrary, for example,

$$\frac{E}{RT_{\rm th}} \quad \text{and} \quad \frac{T_{\rm th}}{T_0}.$$

In connection with the expansion of the exponent in a series studied by Frank-Kamenetskiĭ [3], a single criterion is sufficient to describe the dependence of the reaction rate on the temperature in the only temperature region of interest to us, that in which the reaction rate is high:

$$\frac{E(T_{\rm th} - T_0)}{RT_{\rm th}^2} = \theta. \tag{12}$$

The second criterion $T_{\rm th}/T_0$ enters, however, because of the necessity of describing the dependence of the material constants (viscosity, density and others) on the temperature.

3. As always, two criteria composed only of material constants enter: the Prandtl number $\eta c_p / \lambda = \nu / \kappa$, and the analogous criterion for diffusion, ν / D.

Here we use the following notations: η is the viscosity coefficient, λ is the heat conductivity, D is the diffusion coefficient, $\nu = \eta/\rho$ is the kinematic viscosity, $\kappa = \lambda/c_p\rho$ is the thermal diffusivity.

Thus, in the general case, all of the properties of the flame depend on all of the enumerated parameters. The condition of the propagation limit is written as

$$f\left(\frac{\tau\kappa}{d^2}, \frac{T_{\text{th}}}{T_0}, \frac{\nu}{\kappa}, \frac{\nu}{D}\right) = \text{const.} \tag{13}$$

The flame velocity should not enter into the governing criteria since it must itself be determined by the chemical reaction rate, the thermal conductivity and other properties of the mixture. From dimensional considerations it follows that the flame velocity in a tube is

$$u = \sqrt{\frac{\kappa}{\tau}} \, \varphi\left(\frac{\tau\kappa}{d^2}, \frac{T_{\text{th}}}{T_0}, \frac{\nu}{\kappa}, \frac{\nu}{D}\right). \tag{14}$$

The normal flame velocity in propagation of a plane front without heat transfer to the sides and without motion of the unburned gas does not depend on d and ν, whence follows:

$$u_{\text{N}} = \sqrt{\frac{\kappa}{\tau}} \psi\left(\theta, \frac{T_{\text{th}}}{T_0}, \frac{D}{\kappa}\right). \tag{15}$$

We express the chemical reaction time τ in terms of the normal flame propagation rate since the latter is fairly well studied experimentally as a function of various parameters (mixture composition, etc.).

We obtain the limit condition in the form

$$\chi\left(\frac{\kappa^2}{u_{\text{N}}^2 d^2}, \theta, \frac{T_{\text{th}}}{T_0}, \frac{\nu}{\kappa}, \frac{\nu}{D}\right) = \text{const} \tag{16}$$

or in a form solved for the Peclet number

$$\text{Pe} = \frac{u_{\text{N}} d}{\kappa} = \xi\left(\theta, \frac{T_{\text{th}}}{T_0}, \frac{\nu}{\kappa}, \frac{\nu}{D}\right). \tag{17}$$

Strictly speaking, except for the explicitly appearing reaction time τ or rate u_{N}, the form of the function ξ depends on the concrete kinetics of the chemical reaction, i.e., on whether we are dealing with a first- or second-order reaction or with an autocatalytic reaction, just as, except for the characteristic dimension d, the form of the functions depends on the concrete geometric properties of the system and will be different for a round capillary and a plane slit.

However, just as for different geometric forms, we are frequently able to introduce the concept of "hydraulic diameter" behind which differences in the geometric forms are hidden, we may assume that there exists an effective chemical reaction time which is identical in formulas (13)–(15), so that in this case the form of formulas (16) and (17) will not depend on the reaction kinetics. The basis for such an assumption is the fact that in a flame always, in both normal propagation and in propagation in a tube, the chemical reaction kinetics are significant in a particular temperature interval adjacent to the combustion temperature for small concentrations of the reacting materials compared with the initial concentrations. The only exceptions are cold flames which are not considered here.

In gas mixtures which are close in their physical properties (e.g., $CO-O_2-N_2$ or even $CH_4-O_2-N_2$), the dimensionless ratios of the material constants ν/κ, ν/D, D/κ are determined by the kinetic theory of gases, and are close to 1, changing little from one case to another. For sharply different gases, for example, in lean mixtures of hydrogen with air or oxygen, where this is not so, the ratio D/κ reaches 3–4 and we observe peculiar phenomena which deserve special consideration. Ignoring these cases as well, we find

$$\mathrm{Pe} \simeq \mathrm{Re} = \xi\left(\theta, \frac{T_{\mathrm{th}}}{T_0}\right). \tag{18}$$

Finally, the quantities θ and T_{th}/T_0 at the limit themselves change from one case to another much more weakly than the values of the diameter (0.1–50 mm) and the flame velocity (5–1000 cm/sec) entering into Re. Therefore, quite roughly as long as the experimental material is still very sparse, we may simply set at the limit

$$\mathrm{Re} \cong \mathrm{Pe} = \mathrm{const.} \tag{19}$$

The numerical value of the constant in formula (18), as experiments and calculations by Shchelkin and Shaulov in the combustion laboratory at the Institute of Chemical Physics show, is of order 100 (all quantities—flame velocity, thermal conductivity—are in the cold gas).

The general considerations developed above show that the result (10)–(18), found earlier from an elementary one-dimensional investigation of the problem, turns out to be approximately correct in considering realistic two- or three-dimensional flame propagation as well, with account taken of the gas motion. However, the analytic results of the elementary theory, for example, the decrease in the rate by a factor of $\sqrt{e}$ at the limit, no longer apply. In the non-one-dimensional case, for example in a capillary, the normal rate cannot be exactly established and is variable over the cross-section. Only the uniform propagation velocity may be rigorously determined. However, this latter depends on the tube diameter, via the Reynolds number or the homochronity criterion, not only due to the change in heat transfer, but also as a result of the change in the conditions of gas motion which unavoidably arises as the flame passes. Just from the impossibility of separating these two effects it is obvious that it is not possible to observe a curve for the uniform propagation velocity of the type in Fig. 2.

The last formula (19), just as the previous, more exact formulas (17) and (18), leads to compatible conclusions regarding the expected effect of pressure on the passage of the flame through capillaries and narrow slits.

This effect is dependent on the order of the reaction in the flame. For a reaction of order n with a rate proportional to p^n, the linear flame velocity $\sim p^{(n/2-1)}$; conditions (17) and (19) yield

$$dp^{n/2} = \text{const} \tag{20}$$

(the same condition as at the limit of thermal auto-ignition).

In particular, the generally observed dependence of the flame velocity on pressure between $p^{-1/2}$ and p^0 corresponds to a mono- or bimolecular reaction and leads to $d\sqrt{p} = \text{const}$ or $dp = \text{const}$. We are unaware of any systematic experimental investigations in this direction.

An important qualitative conclusion is that by increasing the diameter the heat transfer by conduction may be made arbitrarily small. In the absence of another mechanism of heat transfer, by increasing the diameter we may arbitrarily decrease the flame propagation velocity at the limit.

§4. Concentration Limits and Heat Losses by Radiation

Investigation of heat transfer in a flame by conduction to the vessel walls led us to the conclusion that the least allowable flame velocity (the velocity at the propagation limit) is inversely proportional to the diameter (dimension) of the vessel. Unlimited increase of the vessel diameter leads to a specific concentration limit only for a mixture composition which yields a flame velocity which is exactly equal to zero.

We pointed out above that the flame velocity at the limit is actually non-zero; experiment also shows that increasing the diameter to more than 5 cm practically does not expand the concentration limits and does not lower the flame velocity at the limit.

However, we did not account for the thermal radiation of gaseous combustion products (CO_2, H_2O), which is significant at high temperatures. Increasing the dimensions of the vessel leaves the amount of heat released in unit time by gases directly adjacent to the flame front practically unchanged, since we are dealing with a thin radiating layer whose thickness depends on the flame propagation velocity, not on the vessel dimensions, so that the decrease in heat radiation as a result of self-absorption is insignificant. One may imagine that the location of the concentration limits in sufficiently wide tubes (where independence from the diameter is achieved) is in fact determined by the heat radiation—at least in the case of ordinary explosive mixtures which yield H_2O and CO_2 in the combustion products. Thermal radiation of these two gases has been studied in detail over the last few years. The numerical data used below are borrowed from Hottel and Mangelsdorf [15].

The presence of the empirical formula [5], which describes the flame velocity in carbon mixtures in accordance with the modern theory of flame propagation based on chemical kinetics [16], allows reasonable extrapolation for these mixtures of the curves of the flame velocity as a function of the mixture composition. Let us pose a concrete problem: to calculate the concentration limits of flame propagation in mixtures of carbon monoxide with air.

In accordance with the elementary theory developed in §1, at the limit the combustion temperature decreases with respect to the theoretical temperature by a quantity RT^2/E and the flame velocity falls by a factor of $\sqrt{e}$. From the thermo-chemical data we find the theoretical temperature of combustion and the composition of the reaction products. According to the formulas of article [5], we find the flame velocity u_0 in the absence of losses as a function of the carbon monoxide content of the mixture. On the other hand, knowing the temperature $(T = T_{th} - RT_{th}^2/E)$ and composition of the combustion products, we find their thermal radiation and the flame velocity at which this thermal radiation causes a decrease in the temperature of combustion by the characteristic quantity RT^2/E. At the concentration limit $u_0 = u_{lim}\sqrt{e}$. We shall show here only the last part of the calculation which makes concrete the considerations of §1. We borrow from the graphs of Hottel[1] data on the radiation of CO_2 for the thinnest radiating layer—the product of the partial pressure p_{CO_2} (in atm) and the length of the ray S (in m): $pS = 0.003$m · atm. We note that we will have to calculate the radiation of even thinner layers, $pS \simeq 0.001$. In our extrapolation we disregard the self-absorption and represent Hottel's data with a formula which is linear in pS:

$$I = ApST^n = 0.01pST^{2.5}, \tag{21}$$

where I is the radiation (in kcal/hr · m^2) taken from the graph. In doing so we somewhat underestimate the radiation of layers for which $pS < 0.003$. The agreement between the formula with the values of A and n which we have selected and the experimental results of Hottel are clear from the data presented below:

T, °C	800	1000	1200	1400
T, K	1073	1273	1473	1673
I_{Hot} at $pS = 0.003$	1860	3000	4400	5800
$0.17pST^{2.5}$	1940	2960	4250	5840

Let us consider a plane layer of combustion products with thickness δ parallel to the flame front. In accordance with the adopted method of calculation for a plane-parallel layer, $S = 2\delta$ (see the chapter written by Hottel on radiation in the well-known reference by MacAdams [14]).

We substitute $p_{CO_2} = \%CO_2$ and take into account that the layer radiates in both directions. Finally the amount of heat given off by the layer in unit

[1]The graphs, smoothed by the addition of later data, were kindly made available for our use by Professor M. A. Styrikovich (Central Boiler and Turbine Institute).

time (an hour) is

$$2I = 2A\,\%\mathrm{CO_2}2\delta T^n = 6.8 \cdot 10^{-4} \cdot \%\mathrm{CO_2}T^{2.5}\delta\;\mathrm{kcal/hr \cdot m^2}. \tag{22}$$

Under the assumption adopted for the extrapolation of small self-absorption and a linear relation between I and pS, we obtain a definite volume rate of heat transfer

$$6.8 \cdot 10^{-4}\%\mathrm{CO_2}T^{2.5}\;\mathrm{kcal/hr \cdot m^3}. \tag{23}$$

Let us find the specific heat of the combustion products. The molecular specific heat is determined as a by-product of the calculation of the combustion temperature and is around 9 cal/g-mole · deg = 9 kcal/kg-mole · deg. The volume occupied by a kg-mole of the combustion products at the temperature T is equal to $V_N = 22.4\,\mathrm{m^3}$, and the specific heat of a unit of volume is

$$\frac{9.273\,\mathrm{kcal}}{22.4\,\mathrm{m^3 \cdot deg}} = \frac{110/T\,\mathrm{kcal}}{\mathrm{m^3 \cdot deg}}. \tag{24}$$

From here it is easy to find the cooling rate of the combustion products:

$$\frac{-dT}{dt} = \frac{6.8 \cdot 10^{-4}\%\mathrm{CO_2}T^{2.5}}{100/T} = 6 \cdot 10^{-6}\%\mathrm{CO_2}T^{3.5}\;\mathrm{deg/hr}. \tag{25}$$

For the sake of brevity we denote by u the limiting flame propagation velocity that we seek with respect to the original mixture at room temperature. The linear velocity with which the flame moves with respect to the heated and expanded combustion products, u_C, is larger in the ratio of specific volumes: the continuity equation yields

$$\frac{u}{v_N} = \frac{u_C}{v_C}; \qquad u_C = u\left(\frac{v_C}{v_N}\right) = u\frac{T}{T_0} = u\frac{T}{290}. \tag{26}$$

Let us find the temperature gradient which is obtained behind the flame front as a result of cooling of the combustion products as the flame front continues to propagate forward (see Fig. 1, §1)

$$\frac{\partial T}{\partial x} = -\frac{1}{u_C}\frac{dT}{dt} = \frac{6 \cdot 10^{-6}\%\mathrm{CO_2}T^{3.5}}{uT/290} = \frac{1.73 \cdot 10^{-3}\%\mathrm{CO_2}T^{2.5}}{u}\;\mathrm{deg/m}. \tag{27}$$

In this last formula u should be expressed in m/hr. We shall take the thermal conductivity of the combustion products to be equal to the thermal conductivity of air and find its value according to the Sutherland formula with the constants: $\lambda_0 = 0.0192\,\mathrm{kcal/hr \cdot m \cdot deg}$, $C = 125\,\mathrm{deg}$ (MacAdams). At high temperature

$$\lambda = \lambda_0\frac{1 + C/273}{1 + C/T}(T/273)^{1.2}$$
$$\simeq \lambda_0(1 + C/273)\sqrt{T/273} = 1.7 \cdot 10^{-3}\sqrt{T}. \tag{28}$$

The total amount of heat released from the flame front backward into the cooling combustion products is

$$\lambda\frac{\partial T}{\partial x} = \frac{3\cdot 10^{-6}\%CO_2 T^3}{u}\text{kcal/hr}\cdot\text{m}^2. \tag{29}$$

In agreement with the elementary theory of §1, the heat losses of the original mixture in the mixture heating zone in front of the flame (where a significant concentration of CO_2 is also attained through diffusion from the combustion zone) are numerically close to the losses through heat conduction into the products and also depend on the flame velocity, heat conduction and other parameters. We shall not consider them in detail, but instead shall double the last expression. Thus the total amount of heat lost in unit time by a unit of the flame surface is equal to

$$6\cdot 10^{-6}\%CO_2\frac{T^3}{u}. \tag{30}$$

The volume of the mixture which burns in unit time on $1\,\text{m}^2$ of the flame is equal to u by the definition of the normal velocity; its specific heat is equal to

$$u(9/22.4) = 0.4u\,\text{kcal/hr}\cdot\text{deg}, \tag{31}$$

where 9 is the molecular specific heat of the combustion products and 22.4 is the volume of 1 kg-mole.

The heat losses found above (30) lead to a decrease in the temperature of combustion by the quantity

$$\Delta T = \frac{6\cdot 10^{-6}\%CO_2 T^3/u}{0.4u} = \frac{1.5\cdot 10^{-5}\%CO_2 T^3}{u^2}. \tag{32}$$

At the limit we equate this quantity to RT^2/E and find the desired expression of the flame velocity (E is in cal/mole):

$$1.5\cdot 10^{-5}\%CO_2 T^3 u^2 = RT^2/E;$$

$$u = 2.74\cdot 10^{-3}\sqrt{\%CO_2 TE}\ \text{m/hr} = 7.6\cdot 10^{-5}\sqrt{\%CO_2 TE}\ \text{cm/sec.} \tag{33}$$

The table shows the values u_0 for various percentage compositions of mixtures of CO with air, calculated according to the formula of the theory of flame propagation [5]

$$u_0^2 = 1500(CO_{\text{Eff}}H_2O/CO_2^2)T\omega T_0^2/E, \tag{34}$$

where ω is the reaction probability taken with respect to a single collision of the molecules CO and H_2O. The table also gives the quantity $u\sqrt{e}$, where u is calculated according to formula (33). The entire calculation is carried out in two versions corresponding to the two expressions

$$\omega = e^{-25\,000/RT}\quad\text{(I)};\qquad \omega = 16e^{-35\,000/RT}\quad\text{(II)},$$

CO_{orig}, %	T_{th}, K	$CO_{2\,prod}$, %	$E = 25$ kcal/mole		$E = 35$ kcal/mole	
			u_0	$u\sqrt{e}$	u_0	$u\sqrt{e}$
8	1010	8	0.8	1.7	0.24	2.1
12	1330	12	3.6	2.4	1.9	2.9
16	1640	16	8.9	3.0	5.5	3.7
20	1930	20	16.2	3.6	15.5	4.4
24	2200	23.9	26.5	4.2	30	5.1
29.5	2355	25.9	39.1	4.4	47	5.5
37	2350	25.8	51.7	4.4	62	5.5
52	1935	20.1	50.1	3.6	48	4.5
64	1565	15.0	33.4	2.9	24	3.5
73	1275	11.3	18.1	2.3	8.9	2.8
82	970	7.5	5.4	1.64	1.45	2.0
88	760	5.0	1.2	1.3	0.15	1.5

where we substitute into both formula (33) and the preceding ones either $E = 25$ or $E = 35$ kcal/mole. The calculation is performed for a mixture saturated with water vapor at 15° C ($H_2O = 2\%$). We consistently disregard the change in the number of molecules in combustion, which does not exceed 12%. Flame velocities are given in cm/sec.

We find the concentration limits from the data in the table as those concentrations at which $u_0 = u\sqrt{e}$. For the lower limit we obtain 10%–13.5% CO and a velocity $u_0 = 2$–3.2 cm/sec, and for the upper limit, 81–87.5% CO, $u_0 = 2.1$–1.3 cm/sec. The figures given in Landolt and Bernstein's reference vary between 12.5–13.6–16.3% CO at the lower limit (the figure 16.3 is for a narrow tube), and 70–75–85% CO at the upper limit. Regarding the last figure for the upper limit it is indicated that it was obtained "for strong ignition." However, the methodology of the experiment does not allow us to judge whether flame propagation over the entire vessel occurred, or whether only a part of the mixture near the spark burned. I believe that the divergence between the normally cited figures for the upper limit 70–75% CO and those found in this paper, 81–87.5%, are most likely due to an error in extrapolation of the flame velocity u_0 in these mixtures, and does not cast doubt on the theory of the limit. For the flame velocity at the limit, Wheeler's empirical rule establishes a value ~ 15 cm/sec, however this is the velocity of *uniform propagation* of the flame in a tube which, as is well-known, exceeds by several times in wide tubes the normal flame velocity; our calculations relate precisely to this latter case so that here too the agreement must be considered satisfactory, at least by order of magnitude.

We find, finally, the diameter of the tube at which losses by radiation become equal to conductive heat transfer, so that further increase in the diameter has only a comparatively weak effect on the overall heat transfer and on the limit of propagation. Under typical conditions, for a 10% CO_2

content and $T = 1000°$K, the heat transfer by radiation comprises, according to the formulas developed,

$$6.8 \cdot 10^{-4} \% CO_2 T^{2.5} = 220\,000 \text{ kcal/hr} \cdot \text{m}^3.$$

Let us find the heat transfer by heat conduction. In a cylindrical vessel $Nu = 4$ and the heat transfer per unit volume is equal to

$$\alpha(T - T_0)S/V = \lambda d^{-1} Nu(T - T_0)S/V$$

$$= 16\lambda(T - T_0)/d^2 = 16 \cdot 0.054 \cdot 700/d^2 = 600/d^2 \text{ kcal/hr} \cdot \text{m}^3.$$

Equating the last expression to the heat transfer by radiation, we find

$$d^2 = 600/220\,000 = 27 \cdot 10^{-4}; \qquad d = 5 \cdot 10^{-2} \text{ m} = 5 \text{ cm},$$

also in satisfactory agreement with experiment. The effect of pressure on the propagation limits in very wide tubes is related to some dependence of the flame velocity on pressure. If u does not depend on the pressure (a bimolecular reaction in a flame), the concentration limits narrow as pressure is lowered; if $u \sim p^{-1/2}$ (a monomolecular reaction), the limits do not depend on the pressure. In all cases as the pressure decreases the vessel diameter increases as $p^{-1/2}$, after which further increase of the diameter no longer extends the limits.

We pause briefly on the question of the possible mechanism of the effect of admixtures which narrow the limits and flegmatize explosive mixtures. Above we related the limits to thermal properties of the mixture and to the flame velocity. The effect of small amounts of an admixture on the heat conduction, specific heat and thermal radiation cannot be significant. Apparently, as a rule, the role of extinguishing admixtures consists in decreasing the normal flame velocity. Preliminary experiments by Barskiĭ (CO–CCl_4, Cl_2) and Sadovnikov (H_2–$SnCl_4$) confirm this point of view.

I consider it my pleasant duty to express my gratitude to my colleagues in the gas combustion laboratory, to F. E. Yudin for his help with the calculations, to K. I. Shchelkin and Yu. M. Shaulov (Azerbaidzhan University) for permission to use their results, and to the Director of the Institute, N. N. Semenov, for his lively interest in this work.

Summary

1. When thermal losses are taken into account the combustion temperature depends on the flame propagation velocity. Consideration of this dependence, together with the Arrhenius dependence of the flame velocity on the combustion temperature, leads to a theory of the flame propagation limit; at the limit the decrease in the combustion temperature is RT^2/E and the normal flame velocity falls by a factor of $\sqrt{e}$ compared to the value in the absence of thermal losses.

2. Using the methods of the theory of similarity, we studied the problem of flame propagation in narrow tubes, in pores and similar conditions. At the limit a specific value of the Peclet number occurs, constructed from the flame velocity, the characteristic tube dimension and the thermal properties of the combustion products.

3. In very wide tubes thermal losses are due to radiation. On the basis of an earlier-developed theory of flame velocity in mixtures of carbon monoxide, the concentration limits of a CO–air mixture were calculated in satisfactory agreement with experiment.

Institute of Chemical Physics   *Received*
USSR Academy of Sciences, Leningrad   *October 2, 1940*

References

1. *Semenov N. N.*—Tsepnye Reaktsii [Chain Reactions]. Leningrad: Goskhim-tekhizdat, 555 p. (1934).
2. *Todes O. M.*—Acta physicochim. URSS **5**, 785 (1936); **9**, 153 (1939).
3. *Frank-Kamenetskiĭ D. A.*—Zhurn. fiz. khimii **13**, 738 (1939).
4. *Zeldovich Ya. B.*—ZhETF **9**, 1530 (1939); here art. 18.
5. *Zeldovich Ya. B., Frank-Kamenetskiĭ D. A.*—Zhurn. fiz. khimii **12**, 100 (1938); here art. 19.
6. *Bunte K., Jahn G.*—Gas. und Wassersch. **76**, 89 (1933).
7. *Jahn G.* Der Zundvorgang in Gasgemischen. Berlin: R. Oldenburg, 69 p. (1934).
8. *Payman L.*—Industr. and Eng. Chem. **20**, 1026 (1928).
9. *Coward H. F., Hartwell F. J.*—J. Chem. Soc. **9**, 2676 (1932).
10. *Daniel L. J.*—Proc. Roy. Soc. London A **126**, 393 (1930).
11. *Mallard E., Le Chatelier A.*—Ann. Mines **4**, 296 (1883).
12. *Holm I. M.*—Philos. Mag. **14**, 18 (1932).
13. *Michelsohn V. A.* Sobr. Soch. [Collected Works]. Moscow: Novyĭ Agronom **1**, 310 p. (1930).
14. *MacAdams W.* Teploperedacha [Heat Transfer]. Leningrad: ONTI (1936).
15. *Hottel V., Mangelsdorf K.*—Trans. Amer. Inst. Chem. Eng. **31**, 577 (1935).
16. *Zeldovich Ya. B., Semenov N. N.*—ZhETF **10**, 1427 (1940).

Commentary

The theory of the limits of propagation of slow combustion proposed in this paper is of fundamental value for accident prevention in working with explosives and for ecological questions in burning fuels in power installations: the emission of incomplete reaction products is due to quenching of the flame near the cold walls of the combustion chamber. The modeling principles have received widespread use: they are used for processing the extensive data on the limits of propagation which have been accumulated to date. The theory's assumption of a plane flame front which does not change its form when the heat transfer changes is better satisfied for condensed explosives, on whose laws of combustion the gasdynamics of the flow of combustion products has less effect, and in the propagation of an exothermic chemical reaction in a condensed medium. In these cases the theory has received good quantitative confirmation.[1,2]

In gas combustion complete quantitative verification is impeded by the complicated gasdynamics of the flow of reacting gas in a channel which leads to bending of the front and a change in the velocity of the flame with respect to the walls, as well as other effects. The natural convection has a strong effect on the combustion limits.[3,4] Near the flame propagation limit in a mixture of air with acetylene, spin and pulsating regimes have been discovered which are caused by instability of the hydrodynamic boundary layers in the combustion products.[5] Oscillations lead to periodic variation of the form of the flame front and increase the heat transfer to the walls. Theoretical analysis shows[6] that heat transfer facilitates the loss of thermodiffusion stability of the flame—the flame is transformed into a cellular one which has broader propagation limits, or takes on a pulsating character.[7,8] In practice, inert porous materials are used as flame barriers; flame propagation in them was studied by Yu. Kh. Shaulov.[9]

The flame velocity in porous media is determined by the effective longitudinal thermal conductivity, which strongly depends on the velocity of the gas. Quenching of the flame as the cold wall is approached and the resulting incompleteness of combustion of the fuel material have been the subject of investigation in many recent studies, both theoretical and experimental. In particular, the question of flame propagation in a mixture of methanol and air has been considered theoretically,[10] and the incomplete combustion of hydrocarbon mixtures was studied experimentally.[11]

Numerical methods have been used to study propagation of a bent laminar flame in a tube with varying radius.[12] Calculations showed that stabilization of the

[1] *Kondrikov B. N.*—Fizika goreniĭa i vzryva **5**, 51–60 (1969).

[2] *Maksimov E. I., Merzhanov A. G., Shkiro V. M.*—Fizika goreniĭa i vzryva **1**, 24–30 (1965).

[3] *Baratov A. N.*—In: Gorenie i vzryv [Combustion and Explosion]. Moscow: Nauka, 286–288 (1972).

[4] *Babkin V. S., Badalyan A. M., Nikulin V. V.*—In: Gorenie geterogennyhk i gasovykh sistemakh [Combustion in Heterogeneous and Gas Systems]. Chernogolovka, 39–41 (1977).

[5] *Golobov I. M., Granovskiĭ E. A., Gostintsev Yu. A.*—Fizika goreniĭa i vzryva **17**, 28–33 (1981).

[6] *Joulin G., Clavin R.*—Combust. and Flame **35**, 139–153 (1979); Combust. Sci. and Technol. **27**, 83–86 (1981).

[7] *Margolis S. B.*—Combust. Sci. and Technol. **22**, 143–169 (1980).

[8] *Sivashinsky G. I., Matkowsky B. J.*—SIAM J. Appl. Math. **40**, 255–260 (1981).

[9] *Shaulov Yu. Kh.* Issledovanie rasprostraneniĭa plameni cherez poristye sredy [The Study of Flame Propagation through Porous Media]. Baku: Izd-vo AN AzSSR, 215 p. (1954).

[10] *Brouning L. G., Perly R. C.*—Spring Mect. Centr. Sect. Combust. Inst. (1979).

[11] *Weiss P., Wrobel A. K., Keck J. C.*—Spring Mect. Centr. Sect. Combust. Inst. (1979).

[12] *Lee S. T., Tien J. S.*—Combust. and Flame **48**, 3–12 (1982).

flame is achieved by sectors located on the external boundary of the thermal and hydrodynamic layers; when the tube radius is decreased, the leading point of the flame shifts to the tube axis, after which combustion ceases. As was shown in a joint work by Ya.B. and G. I. Barenblatt,[13] the thermal limits of flame propagation may be established by solving the non-steady problem of ignition of a combustible gas and the transition to a steady propagation regime under heat transfer conditions: absence of realization of a steady solution indicates that the critical conditions have been reached.

The effect of natural convection in ignition by a limited thermal source was also analyzed:[14,15] the flame formed in different ways depending on the intensity of the source and its position in the closed vessel. At the propagation limit the kinetic peculiarities of the chemical transformation are strongly pronounced.[16] Yu. N. Shebeko, A. Ya. Korolchenko, and A. N. Baratov investigated[17] the effect of additions of HCl and HBr on the flame propagation limits in mixtures of CO with air; the strong change in the lower limit is related to the change of the chain mechanism of the reaction.

Besides the equilibrium radiation considered in this article, the concentration limits can also be affected by chemiluminescence which arises if, in the combustion, chemical compounds form with a non-equilibrium energy distribution along the degrees of freedom of molecules or an atom. The chemiluminescence itself cannot lead to the appearance of a flame propagation limit if only one specific energy fraction is emitted. However, forced luminescence in an optical resonator (in combustion lasers) can lead to quenching of the flame.

[13] *Zeldovich Ya. B., Barenblatt G. I.*—Combust. and Flame **3**, 61–74 (1959).

[14] *Zakharin E. A., Kashkarov V. P., Kramar V. F., Shtessel E. A.*—Fizika goreniĭa i vzryva **14**, 10–21 (1979).

[15] *Makhviladze G. M., Kopylov G. G.*—Fizika goreniĭa i vzryva **19**, 3–10 (1983).

[16] *Macek A.*—Combust. Sci. and Technol. **21**, 65–69 (1979).

[17] Shebeko Yu. N., Korolchenko A. Ya., Baratov A. N.—Khim. fizika **12**, 824–840 (1983).

21

Diffusion Phenomena
at the Limits of Flame Propagation.
An Experimental Study of Flegmatization
of Explosive Mixtures of Carbon Monoxide*

With N. P. Drozdov

In a hydrogen-lean explosive mixture one observes a significant difference between the concentration at which downward flame propagation ceases (for ignition in the upper part of the vessel), and the concentration at which upward flame propagation ceases (for ignition in the lower part of the vessel). In the concentration interval in which combustion is impossible when ignited from below, whereas when ignited from above the mixture does burn, the combustion process is characterized by a number of peculiarities. Observation shows that the flame propagates in the form of separate caps or pellets without covering the full cross-section of the vessel. In addition, chemical analysis establishes that combustion is far from complete.

We were able to explain all of these phenomena by proceeding from the fact that the diffusion coefficient of hydrogen in the mixture significantly exceeds the bulk thermal diffusivity of the mixture. This explanation, in agreement with experiment, explains the effect of substitution of hydrogen with deuterium, the absence of such phenomena in mixtures with an abundance of hydrogen, and the behavior of vapors of organic materials with air. In short the explanation may be summarized thus: when hydrogen is supplied by diffusion and heat removed by conduction in the chemical reaction zone a temperature may develop which significantly exceeds the theoretical combustion temperature. This results in conditions which are more favorable for combustion in the form of individual pellets or caps of flame carried by convective flows than for propagation of a flame front.

Similar considerations lead us to a prediction of the effect of the direction of flame propagation in the case of an explosive mixture in which combustion is suppressed by the addition of a largely immobile (i.e., with a large molecular weight and a small diffusion coefficient) flegmatizing material. This phenomenon was discovered and studied by us in mixtures of carbon

*Zhurnal fizicheskoĭ khimii **17** 3, 134–144 (1943).

monoxide flegmatized with carbon tetrachloride (the appearance of flame pellets in the ignition of hydrogen–air mixtures flegmatized with bromine was observed in 1932 by Zagulin and Chirkov at the Institute of Chemical Physics). The effect of carbon tetrachloride is examined in detail as a function of the mixture composition. The decline in the flame speed observed by Barskiĭ and Zeldovich when carbon tetrachloride is added is confirmed. Rational empirical formulas are given for the determination of the concentration of carbon tetrachloride necessary for combustion to stop. In accordance with the theory of flame propagation and the theory of the limit developed earlier, the extinguishing concentration of CCl_4 is related to the combustion temperature and with the concentrations in the reaction zone in a flame.

§1. White [1] was the first to systematically study the influence of the ignition site and the direction of flame propagation (with respect to the vertical) on the combustion of gas mixtures and in particular on the concentration limits. In all mixtures that have been studied the flame more willingly moves upward, not downward, which corresponds with the direction of the convective flows which lift the hot combustion products. When ignition occurs from above the limits are always somewhat narrower than when it occurs from below. In all cases a certain interval of mixture compositions is observed within which a mixture which has been ignited from above does not burn or burns only in the direct vicinity of the ignition source, while for ignition from below of the same mixture the flame passes to the top of the vessel. We cite certain data borrowed from the last supplementary volume of "Landolt's Physico-Chemical Tables."

Table 1

Mixture	Lower limit			Upper limit		
	($\uparrow$)	($\downarrow$)	B	($\uparrow$)	($\downarrow$)	B
H_2O–air	4.15	8.8	2.12	75.0	74.5	1.02
CO–air	12.8	15.3	1.20	72.0	70.5	1.05
Methanol–air	3.56	3.75	1.05	18	11.5	1.08
Ethyl ether–air	1.71	1.85	1.08	48	6.4	1.80

In Table 1 we give data for various mixtures on the percentages of fuel in mixtures of limiting concentration at the lower limit (i.e., for insufficient fuel) and at the upper limit (for an abundance of fuel, and insufficient oxygen). The direction of propagation is shown by the arrows. In column B we give the ratio of the limiting concentration for upward propagation ($\uparrow$) to the limiting concentration for downward propagation ($\downarrow$).

At the lower limit $B = (\downarrow)/(\uparrow)$; at the upper limit in reality it is the ratio of the oxygen concentrations, not that of the fuel concentrations, that is characteristic, and therefore we take $B = [100 - (\downarrow)]/[100 - (\uparrow)]$ here.

Analysis of all of the experimental material shows that the mixtures divide into two groups. In the first group the direction of propagation only weakly influences the conditions of propagation, which corresponds to a value of $B < 1.2$. This type of effect is not difficult to explain: the flame speed in a tube and the heat transfer conditions change noticeably when the direction of propagation is changed due to the effect of convection—for an unvarying combustion mechanism and almost constant flame propagation speed with respect to the gas. Formally this influence is not difficult to account for in the theory of the flame propagation limit [2] by introducing into the formulas an additional criterion, that of Grashoff, which characterizes the free convection.

In the present work, however, we will deal only with the second group of mixtures for which an extraordinarily strong difference between (↑) and (↓) and a value of B close to 2 are characteristic.[1] Such a very strong effect requires special explanation.

Lean hydrogen–air mixtures are a typical example.

Flame propagation in the interval between (↑) and (↓) is accompanied by incomplete combustion of the mixture. We cite several figures from the paper by Coward and Brindsley [3]; all experiments were performed in the same vessel with ignition from below:

H_2, % (init.)	4.35	4.7	5.1	7.2	8	9.1	10
Fraction burned	0.11	0.28	0.49	0.87	0.92	0.97	1.00

In a number of preliminary experiments (see methodology below) we observed flame propagation at the limits in hydrogen–air mixtures saturated with water vapor at $10°C$. For a 3.6% H_2 composition at ignition a pellet of flame appeared which traveled 15 cm up the tube. At 4.2% H_2 a cap-shaped flame front propagated up to the top of the tube. For ignition from above with an open bottom end of the tube the flame traveled the full length of the tube beginning at 8.5% H_2, and for an open upper end (near the ignition point) beginning at 9.8% H_2. At the upper limit for a mixture containing 73.5% H_2, 26.5% air combustion did not occur from either above or below; for a content of 72.6% H_2 and 27.4% air flame propagation occurred in both directions.

Thus these experiments confirm the nearness of B to unity at the upper limit and the magnitude $B > 2$ at the lower limit in hydrogen–air mixtures using our method.

Goldmann [4] observed quite curious phenomena when a spark is passed between platinum electrodes. In a mixture containing 3% H_2 no flame forms; when a charge is introduced tiny particles of platinum fly off the electrodes

---

[1] In subsequent work carried out at the combustion laboratory of the Institute of Chemical Physics, Kokochashvili showed that in the case of a mixture of bromine with hydrogen with an overabundance of hydrogen, B does not differ from unity, while with an overabundance of bromine B reaches five. *This note was added by the author in proof* (1943—*Editor's note*).

which luminesce due to heating by the catalytic reaction. In rich hydrogen mixtures near the upper limit with equal caloricity all of these phenomena are absent. In terms of caloricity per mole of the mixture the upper limit (75% H_2, 5% O_2, 20% N_2 for a hydrogen–air mixture) corresponds to 10% H_2: mixtures with smaller oxygen content, i.e., with lower caloricity, do not burn at all. Thus, phenomena in lean hydrogen mixtures differ significantly from the behavior of mixtures with insufficient oxygen, despite the common reaction chemistry. Goldmann points out the relation between the peculiarities of lean mixtures and the large diffusion coefficient of hydrogen. This explanation is confirmed by a comparison of hydrogen with deuterium carried out by Clusius and Cutschmidt: in accordance with the smaller diffusion coefficient of deuterium the quantity B for it is lower as a result of the increase in (↑). Klusius and Cutschmidt studied mixtures of hydrogen with pure oxygen and other gases, shown in Table 2.

Table 2

Mixture	Hydrogen			Deuterium		
	(↑)	(↓)	B	(↑)	(↓)	B
100% O_2	3.8	9.5	2.5	5.6	11.1	1.98
20% O_2, 80% N_2	3.9	9.6	2.46	5.6	11.0	1.96
20% O_2, 80% He	5.7	8.0	1.40	7.4	8.3	1.12
20% O_2, 80% Ne	3.5	7.0	2.00	4.2	7.7	1.83
0% O_2, 80% Ar	2.7	7.1	2.63	3.7	7.7	2.08

Their studies also confirmed incomplete combustion for ignition from below in the interval (↑)–(↓). However, neither by simply pointing out the large diffusion coefficient of hydrogen, nor by asserting that hydrogen has a higher diffusion rate than oxygen can we provide an explanation of the phenomenon.

In such binary mixtures as, for example, H_2–O_2, the mutual diffusion coefficients of hydrogen into oxygen and oxygen into hydrogen are equal— this is required by general considerations of the constancy of pressure in diffusion, and is in fact proved by the classical kinetic theory of gases. The coefficient of diffusion of H_2 into O_2 is also large in a lean mixture, as is the coefficient of diffusion of O_2 into H_2 in a rich mixture. Replacement of argon by helium increases the diffusion coefficient but does not decrease B.

§**2.** Analysis of Goldmann's observation leads to the correct answer; there is no doubt that in his experiment the temperature which was reached as a result of the catalytic reaction of a particle of platinum exceeds the theoretical combustion temperature of the mixture since at a temperature of 265–375° C it would not be possible to observe luminescence of the particles.

In the simplest case of a mixture of gases which are close in their properties the kinetic theory of gases asserts equality between the temperature of combustion and the temperature of the catalyst on whose surface the reaction runs to completion. Let us imagine a pellet of catalyst in an unbounded space filled with a mixture at rest. The amount of material of the ith type which is brought in unit time by diffusion to a unit of surface of the pellet is

$$n_i = D_i \frac{C_{i\infty} - C_{ir}}{r}, \tag{1}$$

where r is the pellet radius, D_i is the diffusion coefficient of the ith material, $C_{i\infty}$ is its concentration at infinity, and C_{ir} is the concentration at the pellet surface. The removed heat is written analogously thus:

$$q = K \frac{T_r - T_\infty}{r}, \tag{2}$$

where K is the thermal conductivity, and T_r and T_∞ are the temperatures at the surface and at an infinite distance from the pellet surface.

The reaction completeness condition means that on the surface of the catalyst the concentration of one of the components (for equality of the diffusion coefficients of the ith—of which there is not enough in the mixture) is equal to zero. Let this be the lth component. From this we find the flux of this component

$$n_l = \frac{D_l C_l}{r}. \tag{3}$$

From the stoichiometric equation of the reaction running on the catalyst surface we find the fluxes of the other materials. For example, if we have the reaction

$$2L + 3M = S, \tag{4}$$

then the relation between the fluxes is

$$n_m = \frac{3n_l}{2}, \qquad n_S = -\frac{n_l}{2}. \tag{5}$$

Hence with the aid of (1) we find the concentrations of the other components at the catalyst surface. In the example cited we obtain

$$C_{mr} = C_{m\infty} - \frac{3}{2}\frac{D_l}{D_S}C_{l\infty};$$

$$C_{Sr} = C_{S\infty} + \frac{1}{2}\frac{D_l}{D_S}C_{l\infty}. \tag{6}$$

All of the concentrations must be positive. If the calculation yields a negative value for one of the C_r, then the "deficient component" for which equation (3) is written was chosen incorrectly. Similarly we find the stationary temperature of the catalyst surface. Denoting the thermal capacity of the lth (deficient) component by Q_l we write the thermal balance

$$q = n_l Q_l; \qquad T_r - T_\infty + \left(\frac{D_l}{K}\right) C_{l\infty} Q_l, \tag{7}$$

where q is the amount of heat released in the chemical reaction in unit time per unit of catalyst surface.

We introduce the thermal diffusivity

$$\kappa = \frac{K}{c_p \rho}, \qquad (8)$$

where ρ is the density and c_p is the specific heat (not the concentration),

$$T_r = T_\infty + \left(\frac{D_l}{\kappa}\right)\left(\frac{C_{l\infty}Q_l}{\rho c_p}\right) = T_\infty + \left(\frac{D_l}{\kappa}\right)(T_C - T_\infty). \qquad (9)$$

In the previous formula the combustion temperature was introduced

$$T_C = T_\infty + \frac{C_{l\infty}Q_l}{\rho c_p}. \qquad (10)$$

In the above-mentioned case of a mixture of gases close in their properties the diffusion coefficient is equal to the thermal diffusivity and the temperature of the catalyst surface is equal to the theoretical combustion temperature of the mixture, defined by the elementary formula (10). But this equality is not a postulate, nor is it a direct consequence of the principles of thermodynamics. In the case of a mixture of strongly differing gases, for example hydrogen and oxygen, the equality is violated. In lean mixtures of hydrogen $D_{H_2-O_2}$ is much larger than the κ of a mixture which, for low hydrogen content, approaches $\kappa = D_{O_2-O_2}$. It is for precisely this reason that the high particle temperature is achieved in Goldmann's experiment. For an overabundance of hydrogen, when the deficient component is oxygen, $D_{H_2-O_2}$ is large as before, but κ is even larger, and tends to $\kappa_{H_2} = D_{H_2-O_2}$ in accordance with (9), $T_r < T_C$.

Thus, it is not the absolute value of D, but its ratio with κ that determines the character of the phenomena. Convection blurs the effect; in convective motion the particles of gas which carry quantities of material and heat are in the ratio of the concentration to the product of the specific heat and temperature, which corresponds to equality of the effective (related to the gas motion) coefficients of diffusion and thermal diffusivity. In all cases radiation from the surface of the catalyst lowers its temperature T_r.

The concepts which we develop above were used in the dissertation by Buben, written under the guidance of Frank-Kamenetskiĭ at the Institute of Chemical Physics in 1940. As Buben showed, together with the difference between D and κ, also significant is the role of thermal diffusion which acts, as a rule, in the same direction, i.e., increasing T_r compared to (9) in the case when the light component is deficient, $D_l > \kappa$ so that in accordance with (9) $T_r > T_C$. In contrast, when $T_r < T_C$ anyway, thermodiffusion decreases the flow of the heavy component to the heated surface and causes further decrease of T_r. Direct experiments by Buben, who studied the oxidation of hydrogen on platinum wires, showed that for a content of 2–5% H_2 in air,

$T_r - T_\infty$ is 2.5–2.65 times higher than $T_C - T_\infty$; on the other hand, for a content of 75–85% H_2 (3–5% O_2 in the mixture) $T_r - T_\infty$ is 3.3–3.6 times lower than $T_C - T_\infty$ in good agreement with his calculations.

All of what we have said above relates equally to the surface of a flame on which a rapid and full homogeneous chemical reaction takes place. In a lean hydrogen mixture on the surface of a flame at rest with respect to the gas to which hydrogen and oxygen are supplied by diffusion, a temperature significantly exceeding the theoretical temperature of combustion is established.

In contrast, in a flame which is moving with respect to the gas the temperature obviously cannot differ from the theoretical temperature. In this case, if we imagine a flame at rest in space, the reacting materials are supplied by the flow of gases to one side of the flame, and the heat is carried away by the combustion products leaving from the other side of the flame. The thermal balance yields precisely expression (10) for the combustion temperature. Diffusion processes inside the flame zone cannot affect this result [5, 6]. Thus, in hydrogen-lean mixtures combustion conditions for a flame that is not moving with respect to the gas turn out to be more favorable compared to conditions (temperature) in a flame that is propagating normally with a constant speed with respect to the gas. A flame at rest with respect to the gas can move around the vessel containing the explosive mixture only due to convective motion of the gas itself. Such a flame will be able to move only upward; its speed has no relation to the kinetics of the chemical reaction.

In the calculation which leads to (9) consideration of the diffusion flow from an infinite space was essential so that at a large distance we may assume that the concentration and temperature are constant in spite of the matter consumption and heat release at the surface of the catalyst or flame pellet (the spherical shape of the surface, chosen for simplicity, has no significance). In a vessel containing a limited amount of material, full combustion means the establishment of complete diffusional equilibrium which must be accompanied by the establishment of a thermal balance as well. Here the temperature will not differ from T_C in formula (10). It is possible to exceed this temperature only for incomplete combustion. In the case of hydrogen mixtures containing 4–5% H_2, there is no doubt of the impossibility of any sort of intensive chemical reaction at $T < T_C$ since T_C does not exceed 340–500°C. The possibility of diffusion of active centers from the combustion zone does not change this since below 550°C chain breaking prevails over branching, and the interaction of chains is not great.

Combustion at these concentrations is certainly possible (as experiment shows, all the way up to 8–9% H_2, i.e., to $T_C = 700$°C) only due to a mechanism which increases the flame temperature to above T_C. Consequently, combustion of mixtures containing from 4% to 8% H_2 must be incomplete.

If the temperature on the surface of a pellet is significantly higher than that which is minimally necessary for intensive reaction, the minimal temperature will be achieved at some distance from its surface and the reaction (flame surface) will move there, increasing the radius.

However, due to circumstances for which we did not account in the derivation of (9), as the dimensions of the pellet increase, the flame temperature will be lowered. The greater the radius, the less the intensity of the reaction per unit of flame surface and the greater the relative role of heat losses due to radiation of the combustion products (these latter grow approximately as $r^{3/4}$). In addition, as the dimensions increase, the role of convection increases, which moves the ratio of the effective values D/κ closer to unity.

Both factors (radiation and convection) lead to a reasonable value of the order of several millimeters of the radius at which their effect becomes noticeable and stops further growth of the pellets.

In the case of a flame cap moving convectively in the gas, the quantity r entering into formulas (1)–(3) should be understood as the thickness of the boundary diffusion and thermal layer determining the intensity of the corresponding processes, rather than the radius of the cap or hemisphere. It is significant that the flame then does not fill the entire cross-section of the tube and, together with the rise of the flame through the central part of the tube, descent of the cold mixture along the walls takes place; combustion is incomplete. With the help of the concepts developed above it is easy to explain the basic features of Table 2. With the replacement of hydrogen by deuterium the coefficient of diffusion decreases. Meanwhile the thermal diffusivity of the mixture, composed for the most part of oxygen and inert gas, changes very little. A decrease in D/κ leads to a decrease in $(\downarrow)\,/\,(\uparrow) = B$.

Replacing nitrogen with helium increases somewhat the diffusion coefficient of hydrogen, however, at the same time the thermal conductivity of the mixture grows significantly more strongly, D/κ decreases, and B decreases accordingly. The situation is somewhat more complex in the case of high-molecular organic materials mixed with air, where significant B is observed at the upper limit. An overabundance of fuel in the case of organic compounds has a significantly stronger effect than overabundance of an inert gas. We may convince ourselves of this by comparing the upper limit in mixtures with air (a) and the composition of mixtures critically diluted with nitrogen (b). For the comparison we represent each mixture as the sum of a stoichiometric methane–air mixture (9.5% $CH_4 + 90.5\%$ air) and the solvent. So, for methane we represent the limits

(a) 14.8% CH_4; 85.2% air and

(b) 6.4% CH_4; 93.6% mixture (13.7% $O_2 + 86.3\%$ N_2)

in the following form:

(a) 94.1% mixture (9.5% CH_4 + 90.5% air); 5.9% CH_4 and

(b) 67.4% mixture (9.5% CH_4 + 90.5% air); 32.6% N_2.

This effect of methane (5.5 times stronger than the effect of nitrogen) is caused by endothermic reactions into which the overabundant fuel enters with the products of complete combustion at the flame temperature (e.g., $CH_4 + CO_2 = 2CO + 2H_2 - 56$ kcal).[2]

At a diffusion coefficient of high-molecular organic vapors which is smaller than the diffusion coefficient of oxygen, conditions in the pellet will be more favorable than in a flame front moving with respect to the gas. We saw that the incompleteness of the combustion and the difference between ($\downarrow$) and ($\uparrow$) are related to this.

§**3.** There are substances known which have a specific flegmatizing effect on explosive mixtures, i.e., one which impedes their combustion. Whatever the mechanism of the flegmatizer's action might be, the intensity of the fleg-matizing effect depends on the relation between the concentrations of the fuel and oxygen, on one hand, and the flegmatizer on the other, as well as on the relation between the amount of burning material and the supply of flegmatizer to the flame front. In the case when the flegmatizer is a substance with a large molecular weight and a small coefficient of diffusion, combustion conditions in a pellet will be more favorable than at a flame front. Conditions in the pellet correspond to a mixture with a concentration of the flegmatizer which is less than the true one. Zagulin and Chirkov at the Institute of Chemical Physics in 1932 observed flame pellets in the ignition of hydrogen air mixtures, however, they did not continue to research this phenomenon and did not provide an explanation for it.

We expected analogous results for mixtures of carbon monoxide flegma-tized with carbon tetrachloride. The effect of CCl_4 on the limits of ignition were studied by Dutch scientists of the Iorissen school. However, for carbon monoxide their data reduce to the fact that for a 1% CCl_4 content mix-tures of CO with air burn with a CO content from 37.25 to 38.25%; at 2% CCl_4 content no CO mixtures with air burn. The effect of the direction of propagation was not noted at all [7].

We studied combustion of the mixture in a vertical tube with an inner diameter of 4.4 cm and length of 65 cm equipped with electrodes and closed with polished caps at both ends. We first introduced the carbon tetrachloride into the evacuated tube; the pressure was determined with an oil manome-ter. Next a mixture of CO, O_2, N_2 was introduced from a gasometer until atmospheric pressure was reached. After mixing, the mixture was ignited first with a charge (a high tension spark which was transformed into an arc

---

[2]From this point of view the sharp expansion of the upper limit of mixtures of a hydrocarbon with air and oxygen at high pressure is explained by the leftward shift in the equilibrium of the reduction reaction of the products of full combustion.

Table 3

Experiment #	CCl$_4$, %	Ignition from above ($\downarrow$)	Ignition from below ($\uparrow$)
108	2.75	Does not burn	Does not burn
109	1.5	Does not burn	Burns in pellets
110	2.6	Does not burn	Does not burn, tongue near spark
111	1.33	Does not burn	—
112	1.16	Burns for 5 cm	Burns as a continuous front
113	2.22	Does not burn	Does not burn
114	0.89	Burns completely	—
115	1.78	—	Burns in pellets
116	2.05	—	Two flame pellets formed, soon were extinguished

Table 4

CO, %	O$_2$, %	N$_2$, %	CCl$_2$, % ($\downarrow$)	CCl$_2$, % ($\uparrow$)	B	T_C, K
20	10	70	0.36	0.58	0.62	1970
23.5	11.7	64.8	0.58	1.03	0.57	2160
29	14.8	56.2	1.16	2.05	0.57	2390
40	20.4	39.6	2.14	3.75	0.58	2648
48	24	28	2.72	4.73	0.56	2740
66.6	33.4	—	4.55	—	—	2936

by an induction coil) between the upper pair of electrodes at the open upper cap. If the mixture did not ignite then, it was often used for ignition from below with the lower cap open and the upper one closed.

We show the results of a typical series of our experiments (Table 3). The initial mixture was 29% CO, 71% air, 1.8% H$_2$O (the percentages here and below are for the dry mixture).

Formation of a group of pellets occurred easily in CO–CCl$_4$ mixtures. The presence of a significant interval between ($\uparrow$) and ($\downarrow$) was distinctly pronounced. Due to good actinism the flame pellets could be photographed.

On the basis of the experiments in Table 3 we take the quenching concentrations of ($\downarrow$) to be 1.16 and ($\uparrow$) to be 2.05%. A similar series of experiments, presented in full in the dissertation of one of us (N. P. Drozdov), in the majority of cases (for details see §4) yield a constant ratio $B = (\downarrow)/(\uparrow) \approx 0.58$.

This ratio is less than unity because we are dealing with a concentration at the limit of a material which impedes flame propagation, whereas earlier $B > 1$ was calculated for a fuel or oxygen necessary for combustion. We present a table for a number of stoichiometric mixtures with various nitrogen dilutions for an H_2O content of 1.8% (Table 4).

The last column shows the theoretical combustion temperature calculated by us. The general interpretation of the difference between ($\downarrow$) and ($\uparrow$), relating it to the small coefficient of diffusion of CCl_4, may be considered reliable. Its details, however, remain not completely definite. A comparison of the diffusion coefficient D_{CCl_4} with the thermal diffusivity of the mixture or with the diffusion coefficient of the reacting materials would lead to the same conclusions, assuming the relation of the flow of CCl_4 and the flows of CO and O_2 in the flame and the diffusion coefficient and possible products of the dissociation of CCl_4 in the flame—HCl, Cl_2, CL—when determining the stationary concentration of these products in the flame.

§4. A detailed study of the molecular mechanism of the effect of CCl_4, possible in principle using the recently developed [8] mechanism of the combustion reaction of carbon monoxide, is outside the scope of our study.

From the standpoint of the theory of the propagation limit [2], the limit is determined by the relation between the flame speed and the heat transfer conditions; the probable mechanism for flegmatizing effects lies in decreasing the flame speed. In studying propagation in a horizontal tube, a decrease in the speed upon addition of CCl_4 was noted by Barskiĭ and Zeldovich (paper in preparation).* In the present work we observed the very strong influence of CCl_4 on the normal flame speed for ignition at the upper (open) end of a vertical tube in a mixture of $2CO + O_2 + 1.8\%$ H_2O; the normal speed without CCl_4 reaches 80–100 cm/sec, while at the limit, for a 4.5% CCl_4 content, it reaches 3.5–4 cm/sec (Fig. 1). This last value is close to the minimum normal speed observed at the limit in diluted lean mixtures of CO with air (6–7 cm/sec), and agrees with the cited [2] calculations. The effect of CCl_4 on the flame speed and the flegmatizing action of CCl_4 cannot be explained by its effect on the combustion temperature.

Direct calculation shows that for a mixture of 22% CO, 78% air, 1.8% H_2O, the introduction of 1% CCl_4 even raises the combustion temperature from 2079 to 2092°K (the difference, however, is close to the calculation error). In the combustion products at the theoretical temperature of 2080°K, once complete chemical equilibrium has been established, more than 90% of the hydrogen turns out to be bonded to chlorine in the form of HCl. Apparently it is precisely in the hydrogen bonding, necessary for combustion of carbon monoxide, that the flegmatizing role of CCl_4 lies.[3] The action of CCl_4

---

*Published in Zhurn. fiz. khimii **24**, 589–596 (1950).—*Editor's note.*

[3]Later experiments by Kokochashvili showed that HCL also has some, albeit weaker, flegmatizing effect in CO combustion. The experiments show that the action of chlorine-containing compounds is due not only to hydrogen bonding, but also to participation of atoms and radicals in combustion recombination processes. *Note made in proof (1943— Editor's note).*

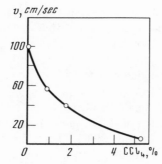

Fig. 1 Dependence of the normal combustion speed of CO on the concentration of CCl$_4$.

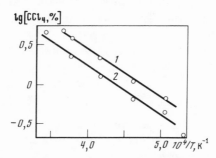

Fig. 2 Dependence of the logarithm of the concentration of CCl$_4$ on the theoretical temperature of combustion.

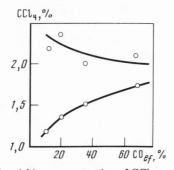

Fig. 3 Dependence of the extinguishing concentration of CCl$_4$ on effective CO concentration.

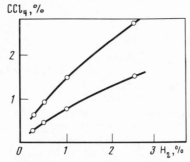

Fig. 4 Dependence of the extinguishing concentration of CCl$_4$ on the concentration of H$_2$ at a constant combustion temperature of 1990°K.

is approximately double that of molecular chlorine, studied by Lindeier [6].

In accord with our conceptions, combustion reactions run at a temperature close to the theoretical combustion temperature; it is natural to compare the concentration of the flegmatizer with the conditions in the reaction zone. In Fig. 2 we show the dependence of $\lg[CCl_4, \%]$ on $1/T_C$ for stoichiometric mixtures (see also Table 4).

In Fig. 3 we show the dependence of the quenching concentration of CCl_4 on the effective concentration of CO in the combustion zone [5, 6] for a constant combustion temperature of $2290°K$. For a large excess of CO, $B = (\uparrow)/(\downarrow)$ approaches unity; the reasons for this have not been established. $(\uparrow)$, $(\downarrow)$ and B are practically independent of the concentration of oxygen in mixtures with an excess of oxygen at a given combustion temperature, $B \approx 0.58$.

Finally, Fig. 4 shows the dependence of the quenching concentration of CCl_4 on the hydrogen content in a dry mixture of 20% CO, 80% air (measuring out different concentrations is methodologically easier for hydrogen, but not for water vapor). Comparing our data with the expression for the chemical reaction rate in a flame in the absence of CCl_4

$$W = Z\,H_2O \cdot CO_{eff}e^{-25\,000/RT}, \qquad (11)$$

we may establish that in a wide interval the extinguishing concentration is proportional to the reaction rate in the absence of CCl_4. In the last calculation, using data on stoichiometric mixtures, we took into account the change with temperature of the quantity CO_{eff} as a result of dissociation.

We consider it our pleasant duty to express our gratitude to G. A. Barskiĭ and V. I. Kokochashvili for their interest and aid in our work, and also the administration of Gorkiĭ University for providing one of us (N. P. Drozdov) the opportunity to carry out this work.

Combustion Laboratory　　　　　　　　　　　　　　　*Received*
Institute of Chemical Physics　　　　　　　　　　　*May 16, 1941*
Academy of Sciences USSR, Leningrad
Department of Physical Chemistry
Gorkiĭ University

References

1. *White A.*—J. Chem. Soc. **121**, 1268 (1922).
2. *Zeldovich Ya. B.*–ZhETF **11**, 159 (1941); here art. 20.
3. *Coward H. F., Brindsley M.*—J. Chem. Soc. **105**, 1859 (1914).
4. *Goldmann T.*—Ztschr. phys. Chem. **13**, 307 (1929).
5. *Langen J.*—Rec. trav. chim. **48**, 201 (1927).
6. *Lindeier H.*—Rec. trav. chim. **56**, 105 (1937).
7. *Zeldovich Ya. B., Frank-Kamenetskiĭ D. A.*—Zhurn. fiz. khimii **12**, 100 (1938); here art. 19.

8. *Zeldovich Ya. B., Semenov N. N.*—ZhETF **10**, 1116, 1427 (1940).

Commentary

Ya.B.'s paper with N. P. Drozdov on diffusion phenomena in a flame laid the foundation for a large number of papers on the thermodiffusional stability of various combustion regimes. We note that in the early papers by E. Jouguet, P. Danniel and others (see references in the article by B. Lewis and G. Elbe)[1] only thermal conduction was considered. The paper by B. Lewis and G. Elbe is remarkable not only because the authors reject the concept of ignition temperature. They also consider diffusion of active centers and reacting materials. For this reason the ratio of the diffusion coefficient to the thermal diffusivity was named the Lewis number (Le). However, a consistent and correct study of stationary combustion was given by Ya.B. and D. A. Frank-Kamenetskiĭ,[2] and the role of the Lewis number in perturbation theory is considered in the current article. This paper introduced for the first time a qualitative consideration of the governing role of the relation between the thermal diffusivity and the diffusion coefficient of a component which limits the rate of the chemical reaction. These considerations may also be found in Ya.B.'s monograph "The Theory of Combustion and Detonation of Gases" (see the present volume). Later they were developed into a quantitative theory of stability of a plane flame front with respect to spatial perturbations of its thermal and diffusional structure in Ya.B.'s article with G. I. Barenblatt and A. G. Istratov.[3] Under the assumption of a narrow zone of chemical transformation, compared to the zones of heating and diffusion, it was shown that at a Lewis number greater than unity, the flame front is unstable, while for Le > 1, it is stable. For Le = 1 there is a small margin of stability with respect to spatial perturbations.

For one dimensional perturbations the flame has neutral stability—see Ya.B.'s paper with G. I. Barenblatt.[4] Thermodiffusional instability of a plane flame front prevails over instability caused by the diffusion selectivity (a change in the composition of the reacting mixture in the zone compared to the original composition of the gas for different coefficients of diffusion of the components) considered in foreign literature[5] because the sensitivity of the reaction rate to changes in temperature is significantly higher than to changes in composition. Thermodiffusional instability has an aperiodic character, perturbations grow without oscillations. For Le ≪ 1 when there is a significant excess of the total (physical + chemical) enthalpy in the flame front, the flame also proves unstable—this instability was first hypothesized by B. Lewis and G. Elbe [See footnote 1].

The actual realization of such an instability was first accomplished in the theory

[1] *Lewis B., Elbe G.*—J. Chem. Phys. **2**, 537–551 (1934).

[2] *Zeldovich Ya. B., Frank-Kamenetskiĭ D. A.*—Zhurn. fiz. khimii **12**, 100–105 (1938); here art. 19.

[3] *Barenblatt G. I., Zeldovich Ya. B., Istratov A. G.*—Zhurn. prikl. mekhaniki i tekhn. fiziki **4**, 21–26 (1962).

[4] *Barenblatt G. I., Zeldovich Ya. B.*—PMM **2**, 856–859 (1957)

[5] Nestatsionarnoe rasprostranenie plameni [Non-Steady Flame Propagation]. Collection of articles. Ed. by J. Markstein. Moscow: Mir, 418 p. (1968).

of powders (see art. 25 in this volume). When in the solid phase the diffusion coefficient $D = 0$, we find the limit of one-dimensional stability of the combustion front. This form of instability is also realized in the propagation of the front of an exothermic reaction in a condensed medium.[6] Here the appearance of one-dimensional relaxation-type pulsations of the combustion speed near the stability boundary were discovered numerically, and an approximate analytical theory of pulsations was proposed by V. B. Librovich and G. M. Makhviladze,[7] and asymptotic methods were developed by B. J. Matkowsky and G. I. Sivashinsky.[8] With respect to two-dimensional perturbations the instability is realized more rapidly than to one-dimensional ones, and has an oscillatory character—perturbation waves "run" along the flame front.[9] An alternative to combustion in a plane flame front is combustion on the surface of a hot spherical volume, considered in the present article.[10]

Ya.B.'s more recent papers have been devoted to the study of nonlinear problems. In 1966 Ya.B. turned his attention to the stabilizing effect of accelerated motion through a hot mixture of a boundary of intersection of two flame fronts, convex in the direction of propagation, and proposed an approximate model of a steady cellular flame.[11] G. I. Sivashinsky, on the basis of this work, proposed a nonlinear model equation of thermodiffusional instability which describes the development of perturbations of a bent flame in time and, together with J. M. Michelson,[12] studied its solution near the stability boundary Le = Le$_{crit}$. It was shown numerically that the flat flame is transformed into a three-dimensional cellular one with a non-steady, chaotically pulsating structure. The formation of a two-dimensional cellular structure was also the subject of a numerical investigation by A. P. Aldushin, S. G. Kasparyan and K. G. Shkadinskiĭ,[13] who obtained steady flames in a wider parameter interval.

In a condensed medium thermodiffusional instability of the flame front in an exothermic reaction can lead in cylindrical charges to a self-oscillatory spin combustion regime (a reaction front which has a fracture with the leading igniting region still rotates about the axis of the charge as it moves pulsatingly along the charge); this phenomenon was discovered experimentally in the work of A. G. Merzhanov, I. P. Borovinskaĭa, Yu. E. Vologin and A. K. Filonenko.[14] To describe it Ya.B., together with A. P. Aldushin and B. A. Malomed, proposed a phenomenological approach[15] based on an analysis of the generalized equation of self-oscillations of

[6] *Shkadinskiĭ K. G., Khaikin B. I., Merzhanov A. G.*—Fizika goreniĭa i vzryva **7**, 19–24 (1971).

[7] *Librovich V. B, Makhviladze G. M.*—Zhurn. prikl. mekhaniki i tekhn. fiziki **6**, 136–141 (1974).

[8] *Matkowsky B. J., Sivashinsky G. I.*—SIAM J. Appl. Math. **35**, 465–469 (1978).

[9] *Makhviladze G. M., Novozhilov B. V.*—Zhurn. prikl. mekhaniki i tekhn. fiziki **5**, 51–59 (1971).

[10] *Selezneva I. K.*—In: Gorenie i vzryv [Combustion and Explosion]. Moscow: Nauka, 396 (1972).

[11] *Zeldovich Ya. B.*—PMM **1**, 102–104 (1966).

[12] *Sivashinsky G. I., Michelson J. M.*—Acta astronaut. **4**, 1177–1181, 1207–1213 (1977).

[13] *Aldushin A. P., Kasparyan S. G., Shkadinskiĭ K. G.*—Dokl. AN SSSR **247**, 1112–1115 (1979).

the distributed Van-der-Pohl oscillators, taking into account the nonlinear stabilizing effects at the curved reaction front. Numerical modelling of spin combustion was performed.[16]

Thermodiffusional instability in the complex field of gas flow exhibits itself in an unusual manner. Variation of the tangent component of the gas velocity along the flame front leads to a redistribution of the fluxes of heat and reacting components at the reaction zone and affects the reaction rate (stretch effect). As a result the flame may lose its stability or be extinguished altogether, while in other cases it may be more stable. The problem of a flame in a velocity field with a variable tangent component was first investigated by B. Karlovitz and his colleagues,[17] and later by A. M. Klimov[18] at Le = 1. In the more general case of Le $\neq$ 1 the stretch effect theory was developed by V. M. Gremyachkin and A. G. Istratov.[19] Ya.B., in a paper with A. G. Istratov, N. I. Kidin and V. B. Librovich,[20] showed that the stretch effect can lead to hydrodynamic stabilization of the curved front of a flame propagating in a tube with respect to the Landau instability.

The present article describes experiments on various limits of flame propagation in a vertical tube for upward and downward combustion. The influence of gravity on the flame and its propagation limits are given a theoretical explanation within the limits of the theory of hydrodynamic instability of a flame, considered by L. D. Landau:[21] if a heavier cold gas is located above the fuel, the hydrodynamic instability becomes stronger; in the opposite case the instability is suppressed. Recently P. Pelce and P. Clavin[22] analyzed the hydrodynamic instability of a flat flame front in a gravity field, taking into account the influence of the effects of molecular transport of heat and matter.

[14] *Merzhanov A. G., Borovinskaĭa I. P., Vologin Yu. E., Filonenko A. K.*—Dokl. AN SSSR **206**, 905–908 (1972); **208**, 892–894 (1973).

[15] *Zeldovich Ya. B., Aldushin A. P., Malomed B. A.*—Dokl. AN SSSR **251**, 1102–1106 (1980); In: Khimicheskaĭa fizika goreniya i vzryva. Gorenie geterogennykh i kondensirovannykh sistem [Chemical Physics of Combustion and Explosion. Combustion of Heterogeneous and Condensed Systems]. Chernogolovka, 104 (1980); Combust. and Flame **42**, 1–6 (1981).

[16] *Ivleva T. P., Merzhanov A. G., Shkadinskiĭ K. G.*—Dokl. AN SSSR **239**, 1086–1088 (1979); Fizika goreniĭa i vzryva **16**, 3–9 (1980).

[17] *Karlovitz B.* et al—In: 4th Intern. Symp. on Combustion. Baltimore, 613–618 (1953).

[18] *Klimov A. M.*—Zhurn. prikl. mekhaniki i tekhn. fiziki **3**, 49–54 (1963).

[19] *Gremyachkin V. M., Istratov A. G.*—In: Gorenie i vzryv [Combustion and Explosion]. Moscow: Nauka, 305–308 (1972).

[20] *Zeldovich Ya. B., Istratov A. G., Kidin N. I., Librovich V. B.*—Combust. Sci. and Technol. **24**, 1–13 (1980).

[21] *Landau L. D.*—ZhETF **14**, 240–245 (1944).

[22] *Pelce P., Clavin P.*—J. Fluid Mech. **124**, 219–226 (1982).

22

On the Theory of Combustion
of Non-Premixed Gases[*]

Let us consider a chemical reaction between two substances (a fuel and oxygen) accompanied by the formation of new substances—the combustion products—and release of heat.

We will consider the stationary process with continuous supply of the original substances and output of the products. A distinctive feature of this case is that the fuel and oxygen (or air) are supplied separately, i.e., are not premixed. Therefore, even in the case when the reaction rate constant for oxygen with the fuel is large, the intensity of combustion does not exceed some limit which depends on the rate of mixing of the fuel and oxygen.

We note that the very fact of combustion significantly changes the distribution of concentrations compared to the distribution of concentrations in mixing of the same gases without combustion.

For a very long time, since Faraday if not before, a qualitative conception has been established that the surface of a flame separates the region with oxygen and no fuel (the oxidation region) from the region without oxygen but with fuel (the recovery region).

Burke and Schumann [1] calculated the form of the flame surface for the very special case of combustion in parallel concentric laminar flows of fuel and oxygen or air. In doing so they did not consider in detail the phenomena taking place in the flame zone.

Most complete is the recently published work by Shwab [2] which was carried out in Leningrad as early as 1940. In it the turbulent plume of a flame is analyzed in detail for both the case when a pure combustible gas is supplied and for the case of a gas mixed with an insufficient amount of air, which we do not consider. Shwab finds the relations between the concentration fields of the gas, oxygen, and combustion products and of temperature, and the gas velocity field, also not considered by us. A number of the results obtained by Shwab (in particular, the constancy of the concentration of the combustion products and temperature) on the flame surface are reproduced in our article for the sake of completeness.

Essentially new is our detailed analysis of the zone and the chemical reaction kinetics (§5).

[*]Zhurnal tekhnicheskoĭ fiziki **19** 10, 1199–1210 (1949).

For laminar combustion on the basis of this investigation we are able to go further and determine the limits of intensification of combustion which result in the fact that for a large fuel and oxygen supply rate to the flame surface, the chemical reaction rate proves to be insufficient.

§1 General Equations

Let us consider a region in which the gas moves with a velocity $\mathbf{u}$, the density of the gas is ρ, the weight concentration of the component of interest is a, the diffusion coefficient is D, the thermal conductivity is λ, the thermal diffusivity is $\lambda/c\rho = \kappa$ and the temperature is T, with all of these quantities variable (dependent on the coordinates). The flux of the component a is given by the formula

$$\mathbf{q}_a = \rho a \mathbf{u} - \rho D \operatorname{grad} a. \tag{1}$$

The vector $\mathbf{q}_a$ gives the direction of flow and its magnitude in grams per cm^2 per second at a given point. The general equation of diffusion has the form

$$\operatorname{div} \mathbf{q}_a = L(a) = \frac{-\partial(\rho a)}{\partial t} + F_a. \tag{2}$$

The divergence of the flux is on the left-hand side, i.e., the difference between the amount of matter carried in and out by the flow per unit volume; on the right-hand side is the change in the amount of the substance a per unit volume $\partial(\rho a)/\partial t$ and the amount of a which forms per unit volume as a result of the chemical reaction.

$L(a)$ is an abbreviation for the differential operator

$$L(a) = \operatorname{div}(\rho a \mathbf{u}) = \operatorname{div}(\rho D \operatorname{grad} a). \tag{3}$$

In a steady flow

$$\frac{\partial(\rho a)}{\partial t} = 0; \qquad \frac{\partial \rho}{\partial t} = -\operatorname{div}(\rho \mathbf{u}) = 0, \tag{4}$$

$$L(a) = \rho \mathbf{u} \operatorname{grad} a - \operatorname{div}(\rho D \operatorname{grad} a). \tag{5}$$

For a stationary process

$$L(a) = F_a. \tag{6}$$

The equation of heat conduction has an analogous form.[1] The thermal flux is

$$\mathbf{q}_T = \rho T c \mathbf{u} - \lambda \operatorname{grad} T = \rho T c \mathbf{u} - \kappa c \rho \operatorname{grad} T, \tag{7}$$

where c is the specific heat per unit mass, which we take to be constant.

From this we obtain

$$\operatorname{div} \mathbf{q}_T = c L(T) = \frac{-c\partial(\rho T)}{\partial t} + Q, \tag{8}$$

where Q is the bulk rate of heat release.

---

[1] We disregard the heat transfer and radiation here (see below).

For a stationary process

$$L(T) = \frac{Q}{c}. \tag{9}$$

If the diffusion coefficients of the various substances—the original a and b and the combustion products g and h—and the thermal diffusivity are equal:

$$D_a = D_b = D_g = D_h = \kappa = D, \tag{10}$$

then the operators $L(a)$, $L(b)$, ..., $L(T)$ identically coincide in formulas of the form (6) written for the various substances, and in formula (9) for the temperature.

In a chemical combustion reaction the amount of the original materials entering into the reaction, the amounts of the combustion products that form and the quantity of heat generated are in definite, strictly constant relations to one another. Denoting the bulk reaction rate by F, we express all the other quantities in terms of it:

$$F_a = \frac{-F}{\alpha}; \quad F_b = \frac{-F}{\beta}; \quad F_g = \frac{F}{\gamma}; \quad F_h = \frac{F}{\eta}; \quad \frac{Q}{c} = \frac{F}{\tau} \tag{11}$$

via the constant positive coefficients $\alpha, \ldots, \tau$. The signs in (11) correspond to the fact that a and b are consumed and g, h, and the heat are produced in the reaction. The coefficients $\alpha, \ldots, \tau$ are placed in the denominator for convenience in subsequent calculations. As an example:

$$a = \mathrm{CH_4}; \quad b = \mathrm{O_2}; \quad g = \mathrm{CO_2}; \quad h = \mathrm{H_2O}; \quad c = 0.5\,\mathrm{cal/g \cdot deg};$$

the reaction heat is 192 kcal/mole of $\mathrm{CH_4}$.

Expressing F as the rate of consumption of all the materials entering into the reaction, in $\mathrm{g/cm^3 \cdot sec}$, we obtain

$$\alpha = 5; \quad \beta = 1.25; \quad \gamma = 1.82; \quad \mu = 2.22;$$
$$\tau = 1/4800° = 2.08 \cdot 10^{-4}\,\mathrm{deg^{-1}}.$$

With the help of these coefficients all the differential equations of the diffusion of the different substances and the equations of heat conduction for the chemical reaction take exactly the same form, with identical L and F in all the formulas:

$$L(\alpha a) = -F; \quad L(\beta b) = -F; \quad L(\gamma g) = F;$$
$$L(\mu h) = F; \quad L(\tau, T) = F. \tag{12}$$

From this, however, we may not conclude that the fields of all the quantities $a, \ldots, \tau$ are similar since the field of each quantity depends not only on the differential equation which this quantity satisfies, but also on the boundary conditions.

In one tube (I) a combustible gas is supplied; obviously inside the tube we have accordingly

$$\text{I.} \qquad a = a_0; \quad b = g = h = 0; \quad T = T_0, \tag{13}$$

where a_0 is the concentration of fuel in the gas entering combustion. This gas may be diluted by, e.g., nitrogen.

In another tube (II) air is supplied:

$$\text{II.} \qquad b = b_0; \quad a = g = h = 0; \quad T = T_0, \tag{14}$$

where b_0 is the concentration of oxygen in the air. When a flame burns in a free atmosphere, condition II relates not to the air tube but to the concentrations in the atmosphere at an infinite distance from the flame.

On the surface of the tubes, furnace, etc., the boundary conditions require that the flow of any substance across a material surface be equal to zero so that the normal surface of the component of the corresponding vector of the flow is zero. These conditions are identical for all the substances. The boundary conditions for temperature will be the same as the boundary conditions for $a, \ldots, h$, if we do not remove heat from the plume, i.e., if only thermally insulated, not thermally radiating surfaces are introduced; or if all walls with temperature T_0 are located where the gas temperature is equal to T_0.

We assume that these conditions are met so that the boundary conditions for all the substances and the temperature at the walls are the same.

§2. Analysis of the Equations. Equation of the Flame Surface

Let us turn now to an analysis of the equations. The chief difficulty in their direct solution consists in the fact that the reaction rate F is strongly dependent on the desired quantities a, b, T themselves.

Subtracting the first equation from the second we obtain

$$L(\alpha a - \beta b) = L(p) = 0, \qquad p = \alpha a - \beta b, \tag{15}$$

with the boundary conditions

$$\text{I.} \quad p = \alpha a_0; \qquad \text{II.} \quad p = -\beta b_0. \tag{16}$$

Thus the difference between the concentrations of fuel and oxygen, taken with the corresponding stoichiometric coefficients, obeys the diffusion equation into which the reaction rate F does not enter. This is the equation that Burke and Schumann considered to determine the form of the flame, basing this on the fact that the fuel may be considered to be negative oxygen. If we are dealing not with a slow reaction, but with combustion, then this means that the function F for a and b both non-zero is quite large. Since the total quantity of the substance which burns in unit time is limited by the amount of fuel supplied, the increase of $F(a, b, T)$ for given a, b, T implies that in fact in a flame the width of the zone in which a and b are simultaneously nonzero decreases, that the values of the quantities a and b themselves decrease in this zone.

At the limit in an infinitely rapid reaction a and b in the reaction zone tend to zero so that nowhere (except in an infinitely narrow zone) can a and b be simultaneously nonzero. Thus, in the case of combustion, by finding the distribution of p in space from the solution of the linear equation (15) with conditions (16), we easily find the fields for a and b as well:

$$\text{at } p > 0, \qquad a = p/\alpha, \quad b = 0$$
$$\text{at } p < 0, \qquad a = 0, \qquad b = -p/\beta \tag{17}$$

The condition $p = 0$ is in fact the equation of the flame surface. It is not difficult to verify that the fluxes of substance a, approaching the surface from one side, and substance b, approaching the surface from the other side, are in fact in a stoichiometric relation to one another. As proof we note that at the flame surface $p = a = b = 0$ so that the convective parts of the fluxes $\rho a \mathbf{u}$, $\rho b \mathbf{u}$, $\rho p \mathbf{u}$ are identically equal to zero. Therefore at the surface

$$\mathbf{q}_a = \rho D \operatorname{grad} a = \alpha^{-1} \rho D \operatorname{grad} p; \quad \mathbf{q}_b = \rho D \operatorname{grad} b = \beta^{-1} \rho D \operatorname{grad} p, \tag{17a}$$

and the quantity $\operatorname{grad} p$ has no singularity on the flame surface, where $p = 0$, since on this surface the F entering into the equations for a and b is large, but is not in the equation for p.

Burke and Schumann found the flame surface ($p = 0$) by integrating (15) for the simplest case of concentric flows of fuel and air moving with equal velocity. Their conclusions are in satisfactory agreement with experiment.

Let us turn to the equations for the temperature and combustion products and attempt to express these quantities in terms of a and b.

By analyzing the equation for one particular reaction product

$$L(\gamma g) = F$$

and comparing it to the equation for a and b

$$L(\alpha a) = -F; \qquad L(\beta b) = -F,$$

we see that we could exclude F from the equation by various means. However, if we also want to obtain the simplest boundary conditions, then we need to choose a new variable in a quite specific manner:[2]

$$r = \frac{a}{a_0} + \frac{b}{b_0} + \gamma g \left(\frac{1}{\alpha a_0} + \frac{1}{\beta b_0} \right). \tag{18}$$

It is easy to verify that then

$$L(r) = 0; \qquad \text{I.} \quad r = 1; \qquad \text{II.} \quad r = 1. \tag{19}$$

---

[2] Any combination $z = n(\gamma g + m\alpha a + (1 - m)\beta b)$, where n and m are any constants, yields $L = n[F - mF - (1 - m)F] \equiv 0$. However, in order for z to have the same values at $a = a_0$, $b = 0$ and at $a = 0$, $b = b_0$ it is necessary that $nm\alpha a_0 = n(1 - m)\beta b_0 = 1$, which in fact leads to the expression for r written above [equation (18)].

§3. The Distribution of the Reaction Products and Temperature

From equation (18) and the conditions (19) it is clear that over the entire region r satisfies the equation of diffusion and convective transport without sources or sinks (since the right-hand part of the equation is equal to zero). Here r is chosen such that $r = 1$ in all flows entering the region of interest, both in the gas flow (I) and in the flow of air or in the atmosphere (II). It is obvious that if in two mixing flows some substance (r) is contained in identical concentrations, then identically the same constant concentration of this substance will obtain throughout the mixing region. Mathematically, we may say that $r \equiv 1$ is a solution of (18) and (19). Hence we find the expression g in terms of a and b

$$g = \frac{\alpha\beta}{\gamma} \frac{a_0 b_0 - a b_0 - a_0 b}{\alpha a_0 + \beta b_0} \tag{20}$$

and completely analogous ones for h or T, obtained by substituting η or τ for γ. Thus the problem is solved.

In rapid combustion, when (17) applies, we obtain at the front ($p = 0$, $a = b = 0$)

$$g_{00} = \frac{\alpha\beta}{\gamma} \frac{a_0 b_0}{\alpha a_0 + \beta b_0}. \tag{21}$$

This result has a very simple and clear physical meaning: the quantities a_0 and b_0 characterize the concentrations of active substances in the burning gases. In order to obtain a stoichiometric mixture in which 1 gram would react, one must take $1/\alpha$ grams of substance a and $1/\beta$ grams of substance b, yielding $1/\gamma$ grams of substance g. Taking into account that the substances a and b in the original gases are diluted, we must take $1/\alpha a_0$ of one gas and $1/\beta b_0$ of the other. After combustion, an amount $1/\gamma$ of the substance g will be contained in a total amount of the combustion products equal to

$$\frac{1}{\alpha a_0} = \frac{1}{\beta b_0},$$

which in fact will yield the concentration g_{00} which satisfies formula (21).

Thus, for rapid combustion of non-premixed gases, in the reaction zone we obtain exactly the same concentration of combustion products as if we had mixed the burning gases in a stoichiometric ratio and carried out the chemical combustion reaction without any diffusion exchange.

In exactly the same way, in the absence of thermal losses by radiation or cooling surfaces in the flame, and for equal thermal diffusivity and the diffusion coefficient, it may be shown that the temperature in the combustion zone of a diffusion flame $T_{00} = (\alpha\beta/\gamma)a_0 b_0/(\alpha a_0 + \beta b_0)$ is identically equal to the temperature of combustion at a constant pressure of a stoichiometric mixture of the gases under consideration.

§4. Comparison of the Flame Temperature with Experiment

The conclusion at which we arrive above, that the temperature of a non-premixed flame is equal to the combustion temperature of the stoichiometric mixture, is in contradiction with experiment: it is well known from daily laboratory experience that in the combustion of a given luminescent gas in a Bunsen burner when the apertures for air suction are closed the temperature of the flame is lower than when the same gas burns with open apertures so that a ready mixture of gas with air enters the flame.[3]

However this disagreement with experiment is explained not by an error in the calculations, but by the fact that a condition of the applicability of the calculation assumed beforehand is not fulfilled: in the thermal balance of a laboratory burner we certainly cannot neglect the amount of heat given off by radiation.

For equal amounts of gas burned, i.e., for equal heat release, the size of the flame without supply of air is significantly larger than when air is sup-plied, and therefore the radiating surface is larger and the chemical energy released per unit surface is smaller. In addition, the luminescence of the flame without supply of air is greater as a result of the appearance within it of tiny particles of carbon which arise from the dissociation of the fuel's hydrocarbons; when air is supplied, the carbon disappears. The presence of soot in the flame of a non-premixed gas is also completely natural: let us consider a point A_0 in the recovery region (Fig. 1), i.e., inside the flame surface. Let this point be located near the flame surface: in this case the temperature is high at the point A_0, and the gas is already strongly di-luted by combustion products and nitrogen; however, it does not contain any oxygen. Heating in the absence of oxygen is in fact what causes the soot to form.

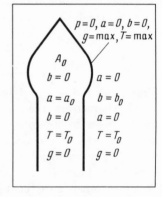

Fig. 1

In the flame of a mixture of gas and air, the gas is also heated before the flame front, but this heating occurs in the presence of oxygen and the pieces of hydrocarbon molecules that could become the initial centers of formation of soot are immediately oxidized. As a result the thermal radiation of the flame of the mixture turns out to be significantly lower and the temperature significantly higher than in the flame of a non-premixed gas, although the

[3]Here and below it is assumed that a stoichiometric amount of air is sucked in. One could consider—although we shall not do so—the case of an insufficient amount when two "cones" of flame form: an inner one (compressing the mixture) and an outer one (residual burning in the surrounding air).

initial "ideal" theoretical value of the combustion temperature is identical in both cases. As has already been indicated, this theoretical value is the temperature which should be realized in combustion in the absence of losses to radiation and side reactions, but with conductive and convective heat transfer between the flame and the gas and air taken fully into account.

Taking account of the conductive and convective heat transfer is a fundamental necessity since this heat transfer is inseparably linked to the processes which supply fuel and oxygen to the flame. Whatever the relation between the existing and supplied amounts of air and gas, when they are supplied separately the flame is always established in such a state as to have the fuel and oxygen supplied to the surface in stoichiometric relation to one another. For equivalence of the diffusion coefficient and the thermal diffusivity (particularly in turbulent motion, which provides such equality) the concentration of the combustion products and the temperature in the flame correspond precisely to combustion of a stoichiometric mixture (for equal losses to radiation); such is the conclusion of our calculations.

§5. Combustion Limit of Non-Premixed Gases

The method described above allows us to calculate the location of the flame surface for supply of any amount of gas and air for any low caloricity of the gas. This calculation is based on the assumption of a large chemical reaction rate at the flame surface (and at the combustion temperature), which leads to a narrow zone in which the chemical reaction runs and to the possibility of considering the flame a geometric surface.

It is obvious that with an insufficient chemical reaction rate we should expect deviations from this picture. By analogy with other combustion and explosion phenomena we may expect that a decrease in the chemical reaction rate with all other conditions equal first only causes some quantitative change—widening of the reaction zone, and then, after some critical value has been reached, the flame will be extinguished, combustion will become impossible and will stop and in its stead mixing of the cold gas and air without any reaction will occur. Below we shall attempt to analyze the critical conditions of quenching in the simplest, schematic case.

In 1940 we analyzed [3] the conditions of the possibility of combustion (flame propagation) in a ready mixture of gases. In this case the limit depended on lowering of the combustion temperature as a result of heat transfer to the side walls of the tube and heat release through radiation. Lowering of the combustion temperature in turn led to a decrease in the propagation rate of the flame, i.e., to a decrease in the heat release per second. When the speed of the flame is decreased, the relative heat losses increase, etc. Therefore we were able to write the critical condition of feasibility of combustion of a prepared mixture such that the decrease in the combustion temperature

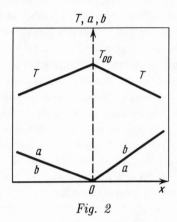

Fig. 2

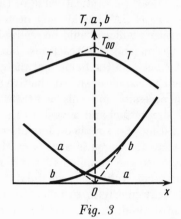

Fig. 3

from the action of thermal losses should not exceed a certain small limit (RT_C^2/E, where E is the activation heat of the combustion reaction and T_C is the combustion temperature).

However this theory of the effect of external thermal losses is inapplicable to the combustion of non-premixed gases. The reason is that lowering the combustion temperature in this case does not lead to a change in the amount of gas which burns per unit surface of the flame since the combustion rate is determined here exclusively by the rate of diffusion supply of oxygen and fuel to the flame surface and not by the propagation rate (which depends on the reaction rate) as for a ready mixture.

The combustion limit of non-premixed gases is determined by the temperature decrease, which depends on the final chemical reaction rate.

Let us consider the distribution of concentrations and temperature in the reaction zone. If the reaction rate were infinite, then the distribution would be given by Fig. 2. The dashed line shows the location of the zone in which $a = 0$, $b = 0$. We take it as the origin for the x-coordinate, with the x-axis directed perpendicularly to the flame surface. If the total amount of matter which reacts per unit surface in unit time is denoted by M, then the diffusion fluxes of a and b are accordingly equal to[4]

$$-\rho D\frac{\partial a}{\partial x} = \frac{M}{\alpha}\frac{1}{\alpha^{-1}+\beta^{-1}} = \beta M(\alpha+\beta);$$

$$\rho D\frac{\partial b}{\partial x} = \frac{\alpha M}{\alpha+\beta}, \qquad (22)$$

so that the distribution of the reacting components near the zone is given

---

[4]Since we are interested in the neighborhood of the reaction zone, in these formulas and below ρ and D should be taken at the temperature in the reaction zone, or approximately T_{00}.

by the formulas

$$x < 0; \qquad a = \frac{M}{\rho D} \frac{\beta(-x)}{\alpha + \beta}, \qquad b = 0;$$

$$x > 0; \qquad a = 0; \qquad b = \frac{M}{\rho D} \frac{\alpha x}{\alpha + \beta}. \qquad (23)$$

Using formulas (20) and (21) rewritten for T, we obtain the corresponding temperature distribution. In the general case[5]

$$T = T_{00} \left(1 - \frac{a}{a_0} - \frac{b}{b_0} \right), \qquad T_{00} = \frac{\alpha\beta}{\gamma} \frac{a_0 b_0}{\alpha a_0 + \beta b_0}. \qquad (24)$$

Substituting the expressions for a and b in (24) we obtain

$$T = T_{00} \left[1 - \left(\frac{M}{\rho D} \right) \frac{\beta(-x)}{a_0(\alpha + \beta)} \right], \qquad x < 0;$$

$$T = T_{00} \left[1 - \left(\frac{M}{\rho D} \right) \frac{\alpha x}{b_0(\alpha + \beta)} \right], \qquad x > 0. \qquad (25)$$

The corresponding line is also shown in Fig. 2.

How will the curves change if the reaction is not instantaneous?

It is obvious that for equal amounts of material M which burns on a unit of surface, the concentration gradient and the entire distribution of a and b far from the zone will not change. However, now the lines a and b can no longer have a sharp break at the coordinate origin as in Fig. 2. They will overlap as shown in Fig. 3, asymptotically approaching zero in the region occupied by the other component. The dashed line in this figure shows the distribution for an instantaneous reaction.

In order to find the distribution curves of the concentration in Fig. 3 we would have to integrate diffusion equations of the form (12), substituting a specific expression for the chemical reaction rate, e.g.,

$$F = abKe^{-E/RT}. \qquad (26)$$

Since F depends on a least three quantities, a, b and T, we shall have to consider a system of three second-order equations. However, thanks to the methods developed above we may first find p, which relates a and b, and then express T in terms of a and b and in this manner reduce the problem to one second order equation, e.g., for a, in which F will be expressed in terms of a and the known function $p(x)$.

However, for our purposes this method is too complicated, and we will obtain the desired results (albeit accurate only to within unknown numerical factors) using the methods of dimensional analysis and the theory of similarity.

---

[5]More exactly, $T_{00} = (\alpha\beta/\gamma)a_0 b_0/(\alpha a_0 + \beta b_0) + T_0$. In what follows we consider the initial temperature T_0 small compared to the combustion temperature and everywhere disregard T_0.

Let us introduce the effective quantities—the zone width x_1, the temperature in the reaction zone T_1, and the concentrations a_1 and b_1. The total amount of matter reacting throughout the zone is expressed in terms of these effective quantities

$$M = a_1 b_1 K x_1 e^{-E/RT_1}. \tag{27}$$

We shall express all of the quantities on the right-hand side in terms of x_1. In order to interrelate them, we note that we are given the dependence of a and b on x at distances from the zone which are large compared to the width of the chemical reaction zone x_1, and this dependence is linear, i.e., is characterized by specific, externally given, gradients $\partial a/\partial x$, $\partial b/\partial x$ or ratios a/x, b/x [see formula (23)].[6] From dimensional analysis we see that the relation between a_1, b_1 and x_1 should also be given by the same kind of formulas:

$$a_1 = \frac{M}{\rho D} \frac{\beta x_1}{\alpha + \beta}, \qquad b_1 = \frac{M}{\rho D} \frac{\alpha x_1}{\alpha + \beta}. \tag{28}$$

The relation of T_1 to a_1 and b_1 is given by formula (24) from which, substituting (28), we find

$$T_1 = T_{00}(1 - \varphi M x_1), \tag{29}$$

where $\varphi = [(\beta/a_0) + (\alpha/b_0)]/\rho D(\alpha + \beta) = [\alpha\beta/(\alpha + \beta)]/\rho D\tau T_{00}$.

Substituting into (27) we obtain finally the equation which relates M and x_1:

$$M = \frac{\alpha\beta}{(\alpha + \beta)^2} \left(\frac{M}{\rho D}\right)^2 K x_1^3 e^{-E/RT_0(1-\varphi M x_1)}$$

$$= \frac{\alpha\beta}{(\alpha + \beta)^2} \left(\frac{M}{\rho D}\right)^2 K e^{-E/RT_{00}} x_1^3 e^{-\varphi M x_1 E/RT_{00}}. \tag{30}$$

In the right-hand side of equation (30) we expanded the quantity in the exponent into a series using Frank-Kamenetskiĭ's method [4]. The quantity on the right-hand side which is dependent on x_1 has the form $x_1^3 \exp(-mx_1)$ and, consequently, passes through a maximum at some critical value

$$x_{\text{crit}} = \frac{3RT_{00}}{\varphi ME}, \tag{31}$$

$$M_{\text{crit}}^2 = \frac{\alpha\beta}{(\alpha + \beta)^2} \left(\frac{RT_{00}}{E}\right)^3 \rho D \left[\frac{\alpha + \beta}{\beta a_0^{-1} + \alpha b_0^{-1}}\right]^3 K e^{-E/RT_{00}}. \tag{32}$$

The meaning of the maximum of M as a function of x_1 is that for small x_1 the overlapped a and b are small, the reaction zone is narrow, and the concentrations of reacting substances in it are small.

For small x_1 and a given temperature T_{00}, the temperature T differs little from T_{00}, and $M \sim x_1^3$, $a \sim b \sim x_1 \sim \sqrt[3]{M}$, in accord with the known result

---

[6]The ratios have small variation only at distances larger than x_1 but smaller than the dimensions of the flame. At distances comparable to the flame dimensions, we may not disregard the convective terms in the diffusion equation.

for cold diffusion flames in a high vacuum [5]. For large x_1 the reagent concentrations are large, but the temperature decreases, which in fact leads to a decrease in the total amount of reacting material.

By intensifying combustion, intensifying the supply of the reacting substances a and b to the zone, we simultaneously increase the cooling of the reaction zone. As long as the reaction rate is sufficiently large, x_1 is less than the maximum and the temperature practically does not decrease. However at a certain critical value M_{crit} [formula (32)] a decrease in the temperature of the zone occurs, leading to further decrease in the reaction and further lowering of the temperature; disruption of combustion occurs and instead of combustion, mixing of the cold gases takes place. The maximum possible decrease (before disruption) of the temperature is equal to $3RT_{00}^2/E$. We note the curious similarity between expression (32) for M_{crit} and the expression for the amount of matter which burns in unit time in a flame which propagates through a stoichiometric premixed mixture of fuel and air, whose combustion in non-premixed form was considered above.

For propagation of a flame in a stoichiometric mixture, following the paper by Frank-Kamenetskiĭ and the author [6], we obtain in our notation here

$$M_{sm}^2 = (\alpha\beta)^2 \frac{1}{(\beta a_0^{-1} + \alpha b_0^{-1})^3} \left(\frac{RT_{00}}{E}\right)^3 6\rho D K e^{-E/RT_{00}}, \qquad (33)$$

which differs from (32) only by the factor $6\alpha\beta/(\alpha + \beta)$ which is not of importance since in the derivation of (32) we dropped numerical factors. This coincidence is very interesting from a fundamental point of view since it shows that the maximum intensity of combustion of a mixture and of non-premixed gases, if the mixing is sufficiently intensified, is of the same order.

In the theory of combustion of an explosive mixture we showed that the chemical reaction runs in a zone in which the concentration of the reacting gas (the one which is deficient in the mixture, or both in the case of a stoichiometric mixture) is quite small—of the order RT_{00}/E of the initial concentration. As the cited calculations show, in the combustion of non-premixed gases the concentrations of both reagents (fuel and oxygen) in the reaction zone are quite small. These concentrations depend on the intensity of the combustion: unlike an explosive mixture which has a characteristic quantity of combustion intensity (flame speed), the intensity of combustion of a flame of non-premixed gases M depends on external conditions. However, for maximum possible M, on the verge of disruption of the flame, the concentrations in the reaction zone do not exceed in order of magnitude a fraction RT_{00}/E of the concentrations in a stoichiometric mixture.

The limit of combustion intensity found above for non-premixed gases explains, at least qualitatively, the fact that in the flow of a fast jet out of a tube the flame is completely located at some distance from the mouth of the tube so that at the exit, where mixing of the reacting materials is most intensive, the flame is disrupted. Above we did not account for thermal losses by radiation. In accounting for them, it would seem that in formulas (29)–(32) instead of the theoretical value T_{00}, calculated from the specific heat, we will have to substitute the maximum possible temperature for instantaneous reaction; but when radiation is taken into account, we will have to substitute the temperature T'_{00}. This T'_{00} is lower than T_{00} even for an infinitely narrow zone, for $x_1 \to 0$, as a result of the radiation of the heated gases to the left and right of the reaction zone. The smaller M is, the lower is the temperature T'_{00}, since for small M less heat is released and the radiating zone is wider. When radiation is taken into account with a decrease in M because of the lowering of T'_{00}, M_{crit} will decrease as well according to (32) and for very small M there appears a second—lower—bound for M of flame disruption for too little combustion intensity. Finally, at low gas caloricity the upper and lower bounds of M may coincide and combustion will become completely impossible. Qualitatively, the picture is analogous to the simpler case of an exothermic reaction in a jet, accounting for heat transfer, considered by Zysin and the author [7]. We do not consider in the present paper the practically important, but more complex question of the limit of intensification of turbulent combustion of non-premixed gases; the complexity of this problem is related to the fact that in the presence of turbulence we are unable to relate the average reaction rate to the average temperature. Evidently here the limits of conditions of feasibility of combustion should be determined experimentally.

We note, finally, that with the help of equations (12) with a concrete form of the function F [e.g., (26)], it is also possible to solve the very interesting problem of the diffusion jump of fuel across the flame zone; as is shown in Fig. 3, the concentration of the mutually penetrating substances in the transition across the reaction zone falls sharply, but does not become zero. Since the temperature and reaction rate also fall on both sides of the reaction zone, the concentration of fuel which has already reached a certain distance from the flame in the oxidation zone no longer changes.

However, to solve this problem our approximate method is inappropriate; it is necessary to solve the equations as described in the text after formula (26).

Conclusions

We considered the distribution of the concentration of the combustion products and temperature in the combustion of non-premixed gases. It is shown that under the simplest assumptions these concentrations and temperature at the flame surface are the same as in the combustion of a premixed stoichiometric mixture of the gases considered.

The feasibility limit of intensification of combustion of non-premixed gases was found, and is dependent on the limiting rate of the chemical reaction. In order of magnitude this limit is close to the combustion rate of the stoichiometric mixture.

Received
October 10, 1948

References

1. *Burke H., Schumann V.*—Industr. and Eng. Chem. **20**, 998 (1928).
2. *Shwab V. A.*—In: Issledovanie protsessov goreniĭa naturalnogo topliva [Study of Combustion Process in Natural Fuels]. Moscow, Leningrad: Gosenergoizdat, 231 (1948).
3. *Zeldovich Ya. B.*—ZhETF **11**, 159 (1941); here art. 20.
4. *Frank-Kamenetskiĭ D. A.* Diffuziĭa i teploperedacha v khimicheskoĭ kinetike [Diffusion and Heat Conduction in Chemical Kinetics]. Moscow: Izd-vo AN SSSR, 266 p. (1947).
5. *Bentler R., Polanyi M.*—Ztschr. phys. Chem. **18**, 1 (1928).
6. *Zeldovich Ya. B., Frank-Kamenetskiĭ D. A.*—Zhurn. fiz. khimii **12**, 100 (1938); here art. 19.
7. *Zeldovich Ya. B., Zysin Yu. A.*—Zhurn. Tekhn. fiziki **11**, 501 (1941); here art. 17a.

Commentary

Combustion of non-premixed gases is determined by the diffusional and hydrodynamic supply of the material, and this has long been known. We note in this connection the fundamental paper by V. A. Shwab.[1] In the article above the kinetics have been introduced. Confirming the earlier-known limiting diffusion-hydrodynamic picture, Ya.B. also analyzed the disruption of the regime for excessive intensification and insufficient flow. An approach to the problem of incomplete combustion is also given. The simplest case of a one-stage irreversible exothermic reaction which is considered was generalized in subsequent papers to more complex chemical transformations which run through a large number of elementary events, with the participation of the initial substances, intermediate active centers, and final combustion products. It was considered that the rates of all the stages of the chemical transformation were sufficiently large and that all of these transformations occur in a single narrow zone near the maximum temperature. If the Lewis numbers of all the chemical components are equal to unity, then in a multi-stage transformation one may apply the Shwab-Zeldovich linear transformation which excludes the functions of the chemical reaction rates since they are related by stoichiometric re-

[1] *Shwab V. A.*—In: Issledovanie protsessov goreniĭa naturalnogo topliva [Study of Combustion Process in Natural Fuels]. Moscow, Leningrad: Gosenergoizdat, 231–248 (1948).

lations. It should be emphasized that the transformation is applicable to non-steady gasdynamic fields as well and therefore it is valid also for turbulent diffusion combustion.[2]

The most important result of this paper is the conclusion of the existence of a limit of diffusion combustion with intensification of the supply of the fuel and oxidizer to the reaction zone. It was later confirmed by D. B. Spalding,[3] who considered a simplified structure of the chemical reaction zone, and by A. Linan,[4] who studied a diffusion flame at the collision of two oppositely directed jets of fuel and oxidizer and used for a large reaction activation energy the method of matched asymptotic expansions. Three reaction regimes were isolated which were analogous to the combustion regime in an ideal mixing reactor. Two of these are stable and one—an intermediate one—is unstable, with transition between them determined by the critical values of the Damkohler number—the ratio of the characteristic times of the gasdynamics and the chemical reaction. The critical condition of quenching depends on stretching of the flame in a non-uniform gasdynamic field (stretch effect), and with strong stretching the flame is extinguished. This effect was used by N. Peters and F. A. Williams[5] to explain stabilization of diffusion combustion in turbulent jets. Extinguishing of a diffusion flame in the collision of oppositely directed jets was a convenient means of experimentally determining the kinetics of a chemical combustion reaction in the so-called Potter's burner.[6] Extensive experimental research using this method was carried out by G. Tsuji and I. Yamaoka[7] for gaseous reagents, and by L. Krishamurthy[8] for a flame located at the critical point of a solid fuel oxidizer gasifying in a flow.

Oppositely directed jets were also used to study the electrical properties of a flame.[9,10] Quenching of a diffusion flame when the supply of fuel and oxidizer is intensive is the reason why small particles of carbon and metals, and small drops of liquid fuel, do not burn and therefore it should be taken into account in calculating the conditions of ignition. On a similar basis a theory of ignition was proposed by V. B. Librovich[11] for hybrid solid rocket fuels (whose composition does not contain an oxidizer). In combustion along the boundary between a solid gasifying fuel and oxidizer the propagation speed of a diffusion flame is determined by the critical condition of quenching in the leading region;[12] a similar situation occurs in the

[2] *Vulis L. A., Yarin L. P.* Aerodinamika fakela [Aerodynamics of a Flame]. Leningrad: Energia, 216 p. (1977).

[3] *Spalding D. B.*—Fuel **33**, 255–273 (1954).

[4] *Linan A.*—Acta astronaut. **1**, 1007–1013 (1974).

[5] *Peters N., Williams F. A.*—Combust. Sci. and Technol. **30**, 1–17 (1983); AIAA Journal **21**, 423–429 (1983).

[6] *Potter A. E., Butler J. N.*—ARS Journal **29**, 54–59 (1959).

[7] *Tsuji H., Yamaoka I.*—In: 12th Intern. Symp. on Combustion. Pittsburgh, 97–105 (1969); 13th Intern. Symp. on Combustion. Pittsburgh, 723 (1971).

[8] *Krishamurthy L.*—Combust. Sci. and Technol. **10**, 21–28 (1975).

[9] *Jones F. L., Becker P. M., Heinsonn R. J.*—Combust. and Flame **19**, 351–362 (1972).

[10] *Kidin N. I., Makhviladze G. M.*—Inzh.-fiz. zhurn. **32**, 1034–1042 (1977).

[11] *Librovich V. B.*—Zhurn. prikl. mekhaniki i tekhn. fiziki **2**, 36–43 (1968).

propagation of a flame along the surface of a condensed fuel in a gas medium.[13]

In the paper above attention is drawn to the essential role of heat evacuation by radiation: in diffusion combustion hydrocarbon fuels heated in the absence of oxygen are subjected to pyrolysis and form a well-radiating soot. In particular, the radiation may change the condition of quenching of the flame for excessively intensive supply of reagents and determine the quenching of the flame for excessively weak supply of them. A study of the effect of radiation in diffusion combustion was conducted by A. D. Margolin and V. G. Krupkin.[14] A detailed review of research on flame quenching was published by F. A. Williams.[15]

[12] *Bahman N. N.*—Dokl. AN SSSR **129**, 1079–1082 (1959); *Librovich V. B.*— Zhurn. prikl. mekhaniki i tekhn. fiziki **4**, 33–36 (1962); *Bahman N. N., Librovich V. B.*—Combust. and Flame **15**, 143–147 (1970).

[13] *Rybanin S. S., Sobolev S. L., Stesik L. N.*—In: Khimicheskaĭa fizika goreniĭa i vzryva. Gorenie kondensirovannykh i heterogennykh sistem [Chemical physics of combustion and explosion. Combustion of condensed and heterogeneous systems]. Chernogolovka, 133 (1980).

[14] *Margolin A. D., Krupkin V. G.*—Dokl. AN SSSR **242**, 1326–1331 (1978).

[15] *Williams F. A.*—Fire Safety **3**, 163–175 (1981).

23

Numerical Study of Flame Propagation in a Mixture Reacting at the Initial Temperature*

With A. P. Aldushin and S. I. Khudyaev

A reaction at the initial temperature changes the characteristics of an explosive mixture before the flame front and introduces an element of non-steadiness into the process of propagation of the combustion wave. The method proposed in [1] to describe this effect consists in replacing the original non-steady problem by a quasi-steady one with adiabatically increasing initial temperature $T_a(t)$ and an effective source of heat release which takes this increase into account. We test this method below by comparing it directly with the results of a numerical solution of the original non-steady problem.

For simplicity we consider the case of similarity of the temperature and concentration fields, disregarding motion of the gas related to its expansion, and assume that the thermophysical characteristics of the mixture are constant. We assume that the mixture is ignited by the burning reaction products and that this also does not violate the similarity. Under these assumptions the propagation of an n-order exothermic reaction is described by the heat conduction equation

$$\frac{\partial \theta}{\partial t} = \frac{\partial^2 \theta}{\partial x^2} + \varphi(\theta), \qquad t > 0, \quad x > 0 \tag{1}$$

with the nonlinear source

$$\varphi = A(-\theta)^n e^{\theta/(1+\beta\theta)}. \tag{2}$$

Here t, x are the dimensionless time and coordinate along the propagation of the front; $\theta = (T - T_C)E/RT_C^2$ is the dimensionless temperature counted down from the combustion temperature T_C and measured in the characteristic intervals RT_C^2/E; E is the activation energy; $\beta = RT_C/E$; and A is the scale coefficient. In the numerical calculations we took $\beta = 0$.

The initial state, corresponding to catalytic ignition, was given in the form of an adiabatically burned layer of thickness x_0 before the initial explosive mixture with temperature θ_I. The boundary conditions assume an adiabatic wall or symmetric development of the process in both directions from the

*Fizika goreniĭa i vzryva **15** 6, 20–27 (1979).

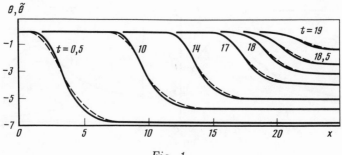

Fig. 1

coordinate origin:

$$\tau = 0 : \theta(x) = 0, \qquad 0 < x < x_0,$$
$$\theta(x) = \theta_{\mathrm{I}}, \qquad x > x_0;$$
$$x = 0 : \frac{d\theta}{dx} = 0; \tag{3}$$
$$x = \infty : \theta = \theta_{\mathrm{a}}, \qquad \frac{d\theta}{dx} = 0.$$

The gradient-free growth of the temperature far from the flame front is described by the ordinary differential equation

$$\frac{d\theta_{\mathrm{a}}}{dt} = \varphi(\theta_{\mathrm{a}}), \qquad \theta_{\mathrm{a}}(0) = \theta_{\mathrm{I}}. \tag{4}$$

Let us introduce a new variable,

$$\vartheta(x, t) = \frac{\theta(x, t)}{\theta_{\mathrm{a}}(t)}, \qquad 0 < \vartheta < 1, \tag{5}$$

which characterizes the relative magnitude of the temperature for a time-variable interval $\left(\theta_{\mathrm{C}}, \theta_{\mathrm{a}}(t)\right)$. Substituting (5) into (1) and taking (4) into account, we obtain the heat conduction equation with a modified law of heat release

$$\frac{\partial\vartheta}{\partial t} = \frac{\partial^2\vartheta}{\partial x^2} + \varphi_1, \qquad \varphi_1 = \frac{\varphi(\theta_{\mathrm{a}}, \vartheta) - \vartheta\varphi(\theta_{\mathrm{a}})}{\theta_{\mathrm{a}}}. \tag{6}$$

Unlike the source φ, the function φ_1 vanishes not only at the hot end ($\vartheta = 0$) but also at the cold end ($\vartheta = 1$) of the temperature interval. This last allows us to seek the solution of (6) in the form of a progressive wave

$$\vartheta = \vartheta(x - \int u\, dt).$$

Disregarding the non-steady term $\partial\vartheta/\partial t$ in the coordinate system $\xi = x - \int u\, dt$, related to the flame front, we have a boundary value problem for the determination of the wave velocity u:

$$-\infty < \xi < \infty, \qquad \vartheta''_{\xi\xi} + u\vartheta'_{\xi} + \varphi_1(\theta_{\mathrm{a}}, \vartheta) = 0,$$

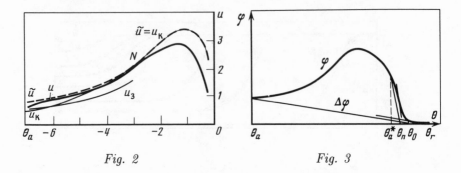

Fig. 2   Fig. 3

$$\vartheta(-\infty) = 0; \qquad \vartheta(\infty) = 1. \tag{7}$$

Returning to the temperature θ we obtain a quasi-steady problem of flame propagation with a time-dependent effective heat release source

$$\theta''_{\xi\xi} + u\theta'_{\xi} + \varphi_{\text{Eff}}(\theta, \theta_a) = 0, \qquad \theta(-\infty) = 0, \quad \theta(\infty) = \theta_a, \tag{8}$$

$$\varphi_{\text{Eff}} = \varphi(\theta) - \theta\theta_a^{-1}\varphi(\theta_a), \tag{9}$$

$$\frac{d\theta_a}{dt} = \varphi(\theta_a), \qquad \theta_a(0) = \theta_{\text{I}}. \tag{10}$$

The solution of the problem (8)–(10) is some approximation to the solution of the original problem (1)–(3) for sufficiently large times when the influence of the initial conditions disappears. To find the degree of correspondence between the solutions we performed numerical calculations on the determination of the non-steady velocity and temperature distribution in the combustion wave using direct and approximate methods.

In Fig. 1 the solid lines show the temperature profiles $\theta(x)$ for different times obtained by solution of the original non-steady problem ($n = 1$, $\theta_{\text{I}} = -7$, $\mathbf{x}_0 = 3$, $\beta = 0$). We note the increase in temperature in the original mixture as a result of slow adiabatic reaction. The heat conduction and diffusion in regions far from the front are insignificant since the profiles are practically gradient-free.

The dashed lines in Fig. 1 show the temperature profiles $\theta(t)$ for the quasi-steady problem superimposed on the actual values for the temperature $\theta_* = (\theta_C + \theta_a)/2$, where $\theta_C = 0$ is the combustion temperature and θ_a is the time-dependent temperature before the front. The close correspondence between the profiles (the difference is less than 10%) not only shows a good approximation to the exact solution, but also indicates the significant role of the half-sum of the temperatures as a reference coordinate of the flame, as was noted in [1].

The flame velocity as a function of the temperature before the front, calculated using various methods, is shown in Fig. 2. The curve u corresponds to the velocity of translation of the coordinate corresponding to the half-sum of the temperatures, $u = dx_*/dt$, $x_* = x(\theta_*)$, determined from the numerical solution of the non-steady problem (1)–(3) without any approximations. The line $\tilde{u}$ is the velocity, calculated using the quasi-steady method as the least eigenvalue of the boundary value problem (8)–(10) with an effectivized heat release rate. The curve $u_K = 2[\varphi'_{Eff}(\theta_a)]^{1/2}$ is the value of the flame velocity determined by the analytical formula of Kolmogorov–Petrovskiĭ–Piskunov [2], and the curve $u_Z = (2\theta_a^{-2} \int_{\theta_a}^0 \varphi \, d\theta)^{1/2}$ is the same according to the formula of Zeldovich–Frank-Kamenetskiĭ [3]. The line u_K coincides with the curve $\tilde{u}$ at the point N.

Extension of the approximate method of calculation of the instantaneous flame characteristics to the case of reaction of higher orders ($n > 1$) requires some modification of the procedure for constructing the effective source. Direct use of (9) leads to an incorrect boundary value problem (8), (9) as a result of the negative values of φ_{Eff} in the neighborhood of $\theta = 0$. Such a situation always arises when there is an inflection point on the heat release curve caused by braking of the reaction in the advanced stages of combustion (Fig. 3) for a reaction order higher than one or for the presence of a slow final reaction in a multi-stage process.[1]

The existence of a high-temperature region with a moderated heat release introduces a new element into the structure of the combustion wave [4]. Besides the normal heating and reaction zones with sharp gradients at the front, there appears a zone in which combustion and reaction run to completion with a slow increase in the temperature from θ_{infl} [the temperature at the inflection point $\varphi(0)$] to θ_C. As a result of the small gradients, this zone does not influence the process of propagation of the combustion wave or the structure of the preceding zones, which is also not affected by the later zones.

The autonomy of the process of completion of the combustion (burn-up) suggests a means of effectivizing the source in the presence of an inflection point on the curve $\varphi(\theta)$. To describe the phenomenon of propagation of the reaction, the correction to the heat release function should be carried out only in an interval of temperatures which excludes the burn-up zone. We shall further set

$$\varphi_{Eff} = \varphi - \varphi(\theta_a)\frac{\theta_0 - \theta}{\theta_0 - \theta_a}, \qquad \theta_a < \theta < \theta_0;$$

$$\varphi_{Eff} \equiv \varphi, \qquad \theta_0 < \theta < \theta_C. \qquad (11)$$

The temperature θ_0 is determined from a condition whose geometric meaning

---

[1]In combustion reactions of nitro-compounds and nitro-ethers it is possible that the deoxidation of NO, i.e., the reactions $2NO+2H_2 = N_2+H_2O$ and $2NO+2CO = N_2+2CO_2$, is such a stage.

is clear from Fig. 3:

$$-\varphi'(\theta_0) = \frac{\varphi(\theta_a)}{\theta_0 - \theta_a}. \tag{12}$$

The means chosen for effectivizing the source automatically excludes influence by the burn-up process on the correction to the source and is valid only for a description of propagation at a temperature before the front which is not too close to $\theta_C = 0$. This relates to a limitation of the quasi-steady method—at values of $\theta_a(t)$ larger than some critical θ_a^* (see Fig. 3), equation (12) loses its solution (at $n = 2$, $\theta_a^* = -1.7$). To describe the process of completion of the reaction, occurring in a quasi-adiabatic burn-up regime, we may again disregard (because of the small gradients) the heat conduction and consider the ordinary equation

$$\frac{\partial\theta}{\partial t} = \varphi(\theta), \tag{13}$$

which determines the temperature increase at each point in space. The initial condition for (13) is the temperature distribution at $\theta_a = \theta_a^*$.

It is not difficult to see that the coordinate x_*, corresponding to the half-sum of temperatures $(\theta_C + \theta_a)/2 = \theta_a/2$, moves backward as θ_a increases, which formally corresponds to a negative wave velocity in the process of burn-up. Indeed, the rate of change of the relative temperature $\vartheta = \theta/\theta_a$ in the burn-up regime is determined by the equation

$$\frac{\partial\vartheta}{\partial t} = \frac{1}{\theta_a}[\varphi(\theta) - \vartheta\varphi(\theta_a)]. \tag{14}$$

The velocity of the point x_* corresponding to a fixed value of ϑ changes sign together with the right-hand side of (14):

$$\left(\frac{dx_*}{dt}\right)_\vartheta = \frac{-(\partial\vartheta/\partial t)_*}{(\partial\vartheta/\partial x)_t}.$$

For a bimolecular reaction zero velocity is achieved at

$$\theta_a = -1.4 \, (\vartheta = 1/2).$$

The results of numerical calculations comparing the quasi-steady and non-steady approaches to the description of the process in the case of a second-order reaction are shown in Fig. 4 ($n = 2$, $\theta_I = -10$, $\beta = 0$, $x_0 = 3$). The notation and the meaning of the characteristics are the same as in Fig. 2. The velocity $\tilde{u}$, which is calculated as the smallest eigenvalue of the boundary value problem (8), (9) with effective source (11) at the propagation stage, is close to the velocity u of the point $x_*(\theta_a/2)$, found from the solution of the non-steady problem. The sign change of the propagation velocity of the wave in the burn-up process corresponds to the value $\theta_a = -1.4$ calculated earlier.

The broad region of applicability of the theory discovered in numerical calculations deserves independent analysis [2]. This theory, as is known, asserts equality of the flame velocity to the least value of the steady problem, determined by the slope of the heat release function in the neighborhood of the initial temperature,

$$u = 2\sqrt{\left(\frac{d\varphi}{d\theta}\right)_{\theta=\theta_a}}. \tag{15}$$

The authors of paper [2] considered the case of a convex heat release function $[\varphi'_\theta(\theta_a) \geq \varphi'_\theta(\theta)]$, e.g., $\varphi = k\theta(\theta - \theta_a)$, and proved that (15) is valid precisely for this case.

In the opposite case of a source which is concave in the initial region at $\varphi_{max} \gg (\theta_{max} - \theta_a) \cdot \varphi'(\theta_a)$, it is known that the magnitude of the velocity is determined by the region of the maximum of $\varphi(\theta)$, and heat release at low temperatures may be disregarded (the theory in [3]). There remained unclear the

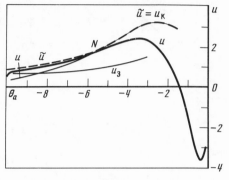

Fig. 4

question of whether the condition of convexity was *necessary* for relation (15) to hold. Just how does the transition from formula (15) [2] to the approximate solution [3] occur?

The numerical calculations shown in Figs. 2 and 4 show that the area of applicability of the theory of [2] is broader than might be assumed. The example given below demonstrates analytically its applicability in the case of a partially concave source as well.

Let us consider the steady problem of flame propagation

$$up = p\left(\frac{dp}{d\tau}\right) + \varphi(\tau), \qquad 0 < \tau < 1;$$

$$p(0) = p(1) = 0; \tag{16}$$

$$\tau = 1 - \theta\theta_a^{-1}; \qquad p = \frac{d\tau}{dx}$$

with a heat release source $\varphi(\tau)$ in the form (Fig. 5)

$$\varphi(\tau) = \begin{cases} \alpha\tau, & \tau < \tau_*; \\ \alpha N\tau_*, & \tau_* < \tau < 1. \end{cases} \tag{17}$$

We determine the magnitude of the combustion velocity u, i.e., the lower boundary of the spectrum of eigenvalues for problem (16), as a function of the source parameters N, τ_*.

Integrating (16) in the region $\tau_* < \tau < 1$, we find the position of the integral curve $p(\tau)$ at the point where the heat release function undergoes a

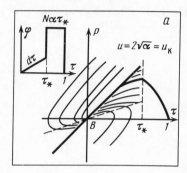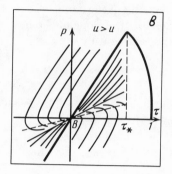

Fig. 5

jump:

$$p_* = p(\tau_*), \quad p_*\frac{u}{\varphi_0} + \ln\left(1 - p_*\frac{u}{\varphi_0}\right) = \left(\frac{u^2}{\varphi_0}\right)(\tau_* - 1),$$

$$\varphi_0 = \alpha N\tau_*. \tag{18}$$

The behavior of the integral curves in the region of linear variation of $\varphi(\tau)$ is shown in Fig. 5. The trajectories $p(\tau)$ entering the singular point B, as follows from the expression for the derivative,

$$p'_\tau = \frac{u(1 \pm \sqrt{1 - 4\alpha/u^2})}{2},$$

appear beginning from $u = u_* = 2\sqrt{\alpha}$. At $u = u_K$ all the integral curves are tangent to the line $p = \sqrt{\alpha}\tau$, which is one of the solutions of (16) in the interval $0 < \tau < \tau_*$; at $u > u_K$ the common [except for the trajectory $p_+ = (u/2)(1 + \sqrt{1 - 4\alpha/u^2})\tau$] tangent trajectory is the integral curve $p_- = (u/2)(1 - \sqrt{1 - 4\alpha/u^2})\tau$.

From Fig. 5a it is clear that at $p_* = p(\tau_*, N, u_K) < \sqrt{\alpha}\tau_*$ the minimum eigenvalue of problem (16), i.e., the desired flame velocity, is u_K. In the case of the inverse relation $p_* > \sqrt{\alpha}\tau$, the minimum u is determined by the equality (Fig. 5b) $p_* = p_+$.

Substituting into (18) $p_* = \sqrt{\alpha}\tau_*$, $u = 2\sqrt{\alpha}$ we find the limiting magnitude of the jump N_* for which the regime [2] still holds:

$$\frac{N_*}{2}\ln(1 - 2/N_*) = (1 - 2/\tau_*). \tag{19}$$

Solution (19) is shown in Fig. 6. It is obvious that the regime [2] is preserved for an arbitrarily large jump magnitude (the integral $\int \varphi\,d\tau$, however, remains bounded) if the latter is close to the combustion temperature.

Indeed, for $\tau_* \to 1$ from (19) we have

$$N_* \approx \frac{1}{2}(1 - \tau_*), \qquad u^2 = u_K^2 = 8\varphi_0(1 - \tau_*) = 2u_Z^2.$$

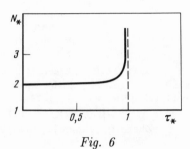

Fig. 6

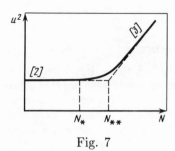

Fig. 7

For $N > N_*$ the flame velocity is determined by the equation obtained from
(19) when we substitute $p_* = p_+(\tau_*)$,

$$\frac{up_+}{\alpha N \tau_*} + \ln\left(1 - \frac{up_+}{\alpha N \tau_*}\right) = u^2 \frac{\tau_* - 1}{\alpha N \tau_*}.$$

Calculation of the derivative at the point $N = N_*$ yields $(du/dN)_{N_*+\varepsilon} = (du/dN)_{N_*-\varepsilon}$, i.e., the transition to the regime [2] occurs sufficiently smoothly (the continuity of the derivative is preserved).

Let us calculate the limit of the ratio $u^2/\alpha N = y$ for increasing N:

$$N \to \infty, \quad u \to \infty, \quad p_+ \to u\tau_*, \quad \ln(1 - y) + y\tau_*^{-1} = 0. \qquad (20)$$

Equation (20) always has solutions $y(\tau_*)$ which determine the asymptote
$u(N)$ for large values of the argument:

$$u \approx \sqrt{\alpha N y(\tau_*)}, \qquad 0 < y(\tau_*) < 1.$$

At large values of u the integral curve $p(\tau)$ in the region $0 < \tau < \tau_*$ takes
a slope $dp/d\varepsilon = u$, i.e., the value of p_*, and with it u, is determined by the
approximation [3] which excludes a slow reaction at low temperatures.

The character of the dependence $u^2(N)$ for fixed τ_* is presented in Fig. 7.
At $N < N_*$ we have the regime [2] ($u \equiv u_K$) which is smoothly replaced
by regime [3]. The arbitrary boundary of the regimes may be set to be
$N_{**} = 4/y(\tau_*)$. Figure 7 illustrates the means for estimating N_{**}.

In the case of values of τ_* which are close to unity, we have $N_{**} = 2\tau_*(1 - \tau_*)^{-1} = 4N_*$, $u(N_{**}) \approx 1.25u_Z(N_{**})$. The expression for u_Z takes the well-
known form, valid for a narrow reaction zone,

$$\tau_* \to 1, \qquad u_Z^2 = 2\varphi_0(1 - \tau_*).$$

The result of this investigation may be summarized in the conclusion that
flame propagation velocities are close to the larger of the values u_K and u_Z.
This fact is sufficiently clearly traced in Figs. 2 and 4 as well, which illustrate
the numerical calculations.

The authors thank K. V. Pribytkov and L. A. Zhukov for their help in the numerical calculations.

Received
December 15, 1978

References

1. *Zeldovich Ya. B.* Rasprostranenie plameni po smesi, reagiruĭushcheĭ pri nachalnoĭ temperature [Propagation of a Flame in a Mixture Reacting at the Initial Temperature]. Chernogolovka: In-t khim. fiziki AN SSSR, preprint (1978).
2. *Kolmogorov A. N., Petrovskiĭ I. G., Piskunov N. S.*—Byul. MGU. **A**, 1 (1937).
3. *Zeldovich Ya. B., Frank-Kamenetskiĭ D. A.*—Zhurn. fiz. khimii **12**, 100 (1938); here art. 19.
4. Aldushin A. P., Merzhanov A. G., Khaikin B. I.—Dokl. AN SSSR **204**, 1139 (1972).

Commentary

A short paper by Ya.B.[1] and his article above with A. P. Aldushin and S. I. Khudyaev mark an important final stage in the theory of the normal propagation velocity of a plane laminar flame as a fundamental concept in the theory of combustion. Following an initial paper with D. A. Frank-Kamenetskiĭ in 1938 (see article 19 in the present volume), in which the formula for the flame velocity was derived, in 1948 Ya.B. produced a ground-laying mathematical study of the problem[2] and rigorously proved that if one disregards the chemical reaction rate at the initial temperature of the combustible mixture, assumes similarity of the distributions of temperature and reagent concentrations and restricts oneself to a simple irreversible chemical reaction, then there exists a unique solution of the problem of a laminar flame propagating with constant velocity. The velocity itself is determined as an eigenvalue of the boundary value problem. Ya. I. Kanel[3] generalized the proof of existence and uniqueness to more complex chemical reactions.

If a chemical reaction runs at a noticeable rate at the initial gas temperature, the concept of normal flame velocity as a constant of the combustible mixture loses its meaning; the constant velocity may be understood only as an intermediate asymptote at times when the conditions of ignition have already ceased to have effect and changes in the properties of the initial combustible mixture may still be disregarded. This problem was the subject of discussion in a joint work by Ya.B. and G. I. Barenblatt.[4] If combustion occurs in a flow under conditions in a combustion chamber, one end of which is supplied with a reacting gas mixture with a high reaction rate, then combustion may take place in an induction regime (auto-ignition regime) when the molecular processes of heat and mass transfer become of secondary importance. This combustion regime was analyzed by Ya.B. together with R. M. Zaidel.[5]

[1] *Zeldovich Ya. B.*—Arch. Combust. **1**, 1–13 (1980).

[2] *Zeldovich Ya. B.*—Zhurn. fiz. khimii **22**, 27–48 (1948).

[3] *Kanel Ya. I.*—Dokl. AN SSSR **149**, 367–369 (1963).

[4] *Barenblatt G. I., Zeldovich Ya. B.*—Uspekhi mat. nauk **26**, 115–129 (1971); Ann. Rev. Fluid Mech. **4**, 285–312 (1972).

[5] *Zeldovich Ya. B., Zaidel R. M.*—Zhurn. prikl. mekhaniki i tekhn. fiziki **4**, 59–65 (1963).

In an unbounded space filled with combustible gas a finite reaction rate leads to the fact that the solution corresponding to constant velocity does not exist! In reality the flame does exist but propagates with a variable velocity, accelerating in the transition to regions of the gas more highly heated by reaction heat. The idea of such a regime and a method for calculating the velocity are given in the paper presented here. There is also a solution of another kind: if the initial state of the reacting gas is inhomogeneous—there exist spatial distributions of temperature and concentrations—then waves of the chemical transformation may arise as a result of the asynchronous occurrence of the reaction at different points in space. According to Ya.B.'s classification (see footnote 1), this is a regime of spontaneous propagation. Its velocity, determined by the initial conditions of the inhomogeneous state of the combustible gas, does not have any bounds; one or another combustion regime is realized from above depending on the relation between the velocity above, the normal flame velocity, the velocity of sound and the velocity of detonation.

Another branch of investigation of wave regimes of propagation of a chemical reaction was generated by the work of A. N. Kolmogorov, I. G. Petrovskiĭ, N. S. Piskunov (KPP),[6] and independently by R. A. Fisher,[7] in which wave propagation of an isothermic reaction is studied. Since the reaction rate vanishes only in the initial state of the substance and has here a maximum growth rate, it turns out that there exists a whole range of velocities of steady wave reaction propagation regimes, but only its lower bound, determined by the derivative of the reaction rate in the initial state, is stable.

Ya.B. studied the interrelation of two regimes of wave propagation—a thermal flame with dependence of the rate on the integral of the heat release function, and an autocatalytic reaction with a governing role for a local characteristic of the reaction rate—its derivative in the initial combustible mixture. This interrelation is most prominently exhibited in flame propagation in a combustible mixture which is reacting at the initial temperature. Restrictions on the form of the autocatalytic-type reaction rate function assumed in previous studies were removed and it was shown that intermediately asymptotic wave regimes of the type studied by KPP are realized for a wider class of reactions. It is necessary to compare the propagation velocity of the front calculated according to KPP with the velocity of the thermal flame theory of Zeldovich–Frank-Kamenetskiĭ and to select the larger one. This general theoretical consideration is supported by numerical calculations of the system of equations of combustion theory. The realizability of theoretical models of steadiness of wave regimes of flame propagation may be checked by solving the Cauchy problem of a chemical reaction entering a propagation regime when initiated by a local source. This technique is particularly useful when several steady propagation regimes are possible. Used in the work of KPP, it has been successfully applied both by Ya.B. in a joint paper with G. I. Barenblatt[8] and also in papers by Ya. I. Kanel,[9] and A. G. Merzhanov, V. V. Barzykin and V. V. Gontkovskaĭa.[10]

[6] *Kolmogorov A. N., Petrovskiĭ I. G., Piskunov N. S.*—Byul. MGU **A** 1, 16 p. (1937).

[7] *Fisher R. A.*—Ann. Engenics **7**, 355–364 (1937).

[8] *Zeldovich Ya. B., Barenblatt G. I.*—Combust. and Flame **3**, 61–74 (1959).

[9] *Kanel Ya. I.*—Mat. sb. **65**, 398–409 (1964).

[10] *Merzhanov A. G., Barzykin V. V., Gontkovskaĭa V. V.*—Dokl. AN SSSR **148**, 380–383 (1963).

III

Combustion of Powders

Oxidation of Nitrogen

24
On the Theory of Combustion
of Powders and Explosives*

§1. Introduction

The very interesting experimental work carried out in recent years by
Belyaev at the explosives laboratory of the Institute of Chemical Physics
of the USSR Academy of Sciences provides starting points for theoretical
investigation of a number of important problems, namely:

1. the spatial temperature distribution and state of matter in the
   combustion zone,
2. the combustion rate as a function of the conditions,
3. the conditions of the transition from combustion to detonation,
4. the ignition of an explosive material or powder and the conditions
   required for its combustion.

As is known, Belyaev [1] used convincing arguments to prove that in the
combustion of volatile secondary explosive materials the explosive material
is first heated to the boiling temperature, then evaporates, and the vapors
of the explosive material enter into chemical reaction after further heating.

*Zhurnal eksperimentalnoĭ i teoretichiskoĭ fiziki **12** 11/12, 498–524 (1942).

Using visual observation and photography Belyaev showed the presence of a dark zone between the surface of the liquid and the flame proper (the place where luminescence indicating intensive reaction occurs). In this way theories according to which the combustion reaction occurs in the condensed phase, or on the surface separating the liquid and gas under the action of direct impact by energy-rich molecules of the reaction products, were refuted. In significant measure the problem of combustion of condensed explosive materials[1] was reduced to the theory of flame propagation in gases, which was developed by Frank-Kamenetskiĭ and the author [2] simultaneously with the first papers by Belyaev on combustion. Since combustion in the gas phase of an evaporated substance determines the observed combustion rate, the question of the rate of steady combustion did not contain anything new compared to the question of the combustion rate of a gas. Specific features of the problem appeared in the analysis of non-steady phenomena and the conditions necessary for the realization of steady combustion of a condensed substance, i.e., in the study of the limits of existence of the steady regime. On the one hand, one could expect a chemical reaction in the condensed phase even at a temperature approaching the boiling temperature; the rate of such a reaction may be significant due to the high density of the condensed phase. The reaction rate increases also with an increase in the boiling temperature, e.g., with an increase in the pressure. An expression was found by the author for the boundary above which the reaction in the condensed phase makes steady combustion impossible. The physical meaning of the boundary, however, only became fully clear later, after Belyaev [3] showed that boiling of liquid EM immensely facilitates the occurrence of detonation. It became clear that a chemical reaction in a liquid leads to boiling and frothing, and consequently to detonation, and the expression found for the boundary of possibility of steady combustion is the condition for the transition of combustion to detonation.

Another aspect studied by us is related to the observation that the amount of heat stored in the heating zone in the condensed phase, from the initial temperature to the temperature at which transition to the gas phase occurs (i.e., the boiling temperature), is quite large. Changes in this store of heat, which are absent only in a strictly steady regime, have a very large effect on the energy balance of combustion. The physical consequences occur along two lines:

1. The ignition of the EM and the initiation of steady combustion require the creation of the indicated store of heat; in this way the principles of the theory of ignition are established.

2. It turns out that a steady combustion regime under certain conditions becomes unstable, and the store of heat in the narrow heating zone is either rapidly consumed by the accelerated (due to this heat) flame, or is dispersed

---

[1]In what follows we shall use the abbreviation EM for "explosive material."

in the condensed phase layer as the flame is simultaneously slowed. The final result of deviation from the steady regime in both cases is quenching of the EM. An additional condition is formulated which is necessary for steady, stable combustion of a condensed substance. Thus is given a theory of the limit beyond which quenching of combustion follows.

Belyaev's experimental studies related to liquid secondary explosive materials—methylnitrate, nitroglycol, nitroglycerine—and to secondary explosive materials which are solid at room temperature, but which melt when ignited—trotyl, picric acid, etc.

We assume—at present this is only a hypothesis—that our theoretical investigations relate also to the combustion of powders, and especially to smokeless powders. Classical internal ballistics does not consider the question of the intimate mechanism and stages of the chemical reactions which comprise the essence of powder combustion. However, many of Belyaev's arguments and our—Frank-Kamenetskiĭ and the author—work remain applicable. The presently proposed hypothetical scheme of powder combustion differs only in that, instead of reversible evaporation, we must speak of a primary (irreversible) reaction transforming the powder into a gas, but not into the final combustion products, followed by the reaction of "combustion proper," accompanied by the release of the primary part of the reaction heat.

The reader interested in new physical concepts related to the combustion of powders, the theory of ignition and the limit of combustion, may omit without any loss §3, which is devoted to more formal problems, and §4, in which the problem of transition of combustion to detonation could not be developed to the point of comparison with experiment because of the absence of much necessary data and information on the role of other factors.

§2. Basic Aspects of the Theory of Steady Combustion. A Scheme of Powder Combustion

We shall recall and sum up here the basic aspects of the theory of steady combustion, including those which are scattered in the literature referenced above.

The initial state of the powder or EM is given and quite exactly defined. Under certain assumptions, for example, that complete chemical equilibrium is achieved or that the reaction runs to certain chemical products (nitrogen oxide), we may also determine the state—composition and temperature—of the combustion products. The choice of initial assumptions is controlled by experiment—at least for now, in the absence of sufficient information on the kinetics of chemical reactions at high temperatures.

The temperature T_C of the combustion products is, as a rule, high—1200–2000°C, and the combustion products are a mixture of gases. In the combustion zone the entire interval of temperatures from T_0 to T_C is realized, and one may show that under conditions of a non-detonational combustion process which is slow compared to the velocity of sound, no temperature discontinuity can exist; all intermediate temperatures are realized. Concurrent with the temperature variation from T_0 to T_C, variation of the aggregate state also occurs.

For the case of sublimable secondary EM, as Belyaev indicates, variation of the aggregate state occurs at a boiling temperature which corresponds to the external pressure. Indeed, the evaporation rate from the free surface of a superheated fluid is extraordinarily high,[2] and superheating of the fluid is practically impossible. Dilution of the vapors by reaction products diffusing from the combustion zone even lowers the temperature of the liquid surface somewhat compared with the boiling temperature.

More complicated and less-well studied from a physico-chemical point of view is the process of powder combustion. The significant influence of pressure on the combustion rate undoubtedly indicates that the gas phase plays a role.

Some researchers [4, 5] presume that the impact of molecules of the gaseous products of the reaction, which possess sufficient energy, causes the decay of the surface layer of molecules of the substances comprising the powder. Such a conception leads to a combustion rate which is proportional to the pressure and depends on the temperature of the combustion products.

Theories of a direct impact effect of gas molecules may also be called temperature discontinuity theories: they make sense only when the cold powder and hot reaction products are adjacent to one another. If the temperature of the powder and gas at the interface are identical, then activation as a result of thermal motion in the solid body is much more probable, not activation by the impact of a gas molecule.

When a temperature discontinuity between the gas and solid powder exists, there should arise a heat flux of enormous intensity, exceeding by hundreds of times the heat release of combustion. It is obvious that such a heat flux cannot exist in combustion.

We are forced to abandon naive molecular conceptions of direct impact

---

[2]We easily find the dependence of the evaporation rate on superheating by applying the principle of detailed equilibrium. The accommodation coefficient, i.e., the probability that molecules of vapor falling onto the surface of the liquid will stick, we take equal to 1. The number of molecules which evaporate in unit time is equal to the number of molecules which fall in unit time onto the surface at equilibrium pressure which, in turn, is equal to the product of half the number n of molecules in a unit volume of vapor and the average velocity c_x of the molecules in the direction normal to the surface.

If superheating of the surface is such that the equilibrium vapor pressure is a factor $1 + \beta$ greater than the external pressure, the evaporation rate, expressed as the linear velocity of vapor outflow from the surface, comprises $\beta c_x / 2$. A combustion rate of 1 mm/sec of the liquid EM corresponds at atmospheric pressure to a vapor outflow rate of order 50 cm/sec, to which corresponds β from 0.001 to 0.002 and superheating of 0.02–0.04. For an accommodation coefficient $\alpha \neq 1$ and otherwise equal conditions, the superheating increases proportionally to $1/\alpha$. Finally, the superheating is proportional to the combustion rate.

and to seek another mechanism for the combustion of a powder. Smokeless powder, based on multiatomic nitroethers of cellulose, under ordinary conditions or in a vacuum can be sublimated. When heated, it decays with release of a number of products, including those substances which are contained in the combustion products of a powder (carbonic acid, carbon monoxide, water vapors, and in part nitrogen oxide). Can we not assume that such decay, accelerated by the increase in temperature, is in fact a combustion process, i.e., that combustion and transformation into a gas are a single, inseparable stage or, formulated differently, that chemical reactions do not occur in the gas phase and that chemical transformation occurs only in the solid phase and in the transition of the solid phase into a gas?

The temperature of the combustion products is predetermined by the value of the reaction heat. If the ultimate reaction products are immediately released from the surface of the solid powder, chemical reactions and heat release no longer take place in the gas phase. Consequently, the reaction products released from the powder which have their final composition must also have the final temperature T_C: the powder itself (solid phase) must, under the assumptions made, have this same temperature at the boundary where release of the gas occurs.

This is precisely what appears completely improbable if we keep in mind the very high combustion temperature of a powder. The difference in the conditions of combustion from normal thermal decomposition lies in the rapidity of the process: at a combustion rate of 1 mm/sec, the effective width of the heating zone is less than 0.2 mm and the heating time is less than 0.2 sec (these magnitudes depend little on the temperature of the surface, as we shall discuss below).

For such rapid heating and energetic supply of heat, the growth in the temperature of the powder is limited only by endothermic reactions which are accompanied by its transformation into the gas phase.[3] The rate of these reactions at low temperatures is less than the rate of normal decomposition, but as the temperature increases, the rate of endothermic reactions grows more rapidly[4] and at some temperature T_d reaches a quantity equal to the combustion rate. This temperature will in fact be achieved at the interface between the powder and gas.

The chemical energy of the final combustion products is less than the chemical energy of the powder. It is precisely in the freeing of chemical

[3]The reaction does not have to be endothermic in order for the temperature of the solid powder not to reach the combustion temperature; it is necessary only that not all of the combustion heat be released in the reaction of gas-formation, i.e., that the reaction be less exothermic than the combustion reaction. It is this which makes combustion possible, i.e., the subsequent reaction of gases with heat release. The stipulation made here relates to all cases below when we refer to endothermic reaction.

[4]In this connection we should mention the enormous magnitude of the pre-exponential factor in the expression of the probability of evaporation in the case of multi-atomic molecules (see Langmuir [6]).

energy and its transformation to thermal energy that the essence of combustion lies. The chemical energy of the gaseous products which are obtained in the endothermic decomposition of the powder, obviously, is greater than the chemical energy of the powder. The products of this decomposition react further, i.e., they burn then in the gas phase with release of the bulk of the heat.

Thus, in our opinion, the analogue of evaporation of low-molecular EM in powder combustion is endothermic decomposition with the formation of energy-rich intermediate products. Let us note some differences: evaporation occurs without breaking of chemical bonds; we may expect that the temperature of decomposition of the powder T_d will turn out to be higher than the boiling temperature T_b of such substances as methylnitrate (63°C), nitroglycol (200°C), nitroglycerin (250°C) and trotyl (300°C). Evaporation is a reversible process and T_b may be found by measuring the vapor pressure. T_b depends on the pressure at which combustion occurs, but it is practically free of any direct dependence on the combustion rate (see footnote 2). In contrast, the temperature of decomposition T_d is determined kinetically by the condition that the decomposition rate is equal to the combustion rate, otherwise the flame front will approach the surface, the heat flux will intensify, and the surface temperature T_d will rise until it corresponds to the stated condition. We may expect that the only factor determining T_d is precisely the combustion rate; other factors, such as the pressure and the initial temperature T_0, affect T_d only insofar as they change the combustion rate. However, the dependence of T_d on the combustion rate in fact must also be weak. The rate of decomposition depends exponentially on the temperature; the activation heat, including the reaction heat for an endothermic reaction, is large. Under these conditions a small change in T_d is sufficient for the decomposition rate to change significantly. We disregard the change in T_d. Therefore, in what follows, unless stated otherwise, we will discuss the combustion of EM and powder together, denoting by T_b the temperature at the interface between the condensed phase and the gas (boiling temperature or decomposition temperature, as appropriate), and by L the thermal effect of transition from the condensed into the gas phase (latent heat of evaporation or heat of decomposition, as appropriate).

§3. Mathematical Theory of a Steady Regime and the Combustion Velocity

Let us introduce a system of coordinates in which the flame is at rest. For the sake of definiteness we shall make the coordinate plane YOZ coincident with the interface between the condensed phase (briefly, c-phase) and the gas, with the c-phase located to the left at $x < 0$. In a system in which the flame is at rest, the material must move. The velocity of the material u

has a positive sign. The velocity of the c-phase far from the flame at $x < 0$ coincides, obviously, with the combustion velocity u.

As x increases, accompanied by heating, transformation into a gas, and chemical reaction of the material, the velocity varies.

We shall begin by establishing the conservation equations. To this end we place a control surface O far from the flame; we shall mark the quantities which pertain to this surface with the index "0" since they correspond to the initial state of the material. In the area where we have placed the surface O, at sufficient distance from the flame, all the gradients are also equal to zero.

Wherever we now choose a second control surface P (we shall write quantities pertaining to it without index) we may write expressions for the conservation laws; in this steady regime the state of the material between the control surfaces is invariable, the amount of matter, the energy and other quantities between the surfaces do not change. The conservation laws therefore reduce to equality of the fluxes on the control surfaces.

Conservation of mass yields

$$\rho_0 u_0 = \rho u, \tag{3.1}$$

where ρ is the density and u is the velocity.

The conservation law may also be applied to each individual sort of atom. Let us have g sorts of molecules (the first index), consisting of f sorts of atoms (the second index), then we characterize the composition of the molecules by the stoichiometric numbers ν_{il} (the number of atoms of sort l in a molecule of sort i). Conservation of the atoms of a given sort l is written thus:

$$u_0 \sum c_{i0} \nu_{il} = u \sum c_i \nu_{il} + \sum \delta_i \nu_{il}, \tag{3.2}$$

where c_i is the concentration of molecules of sort i; c_{i0} is the same on the plane O; δ_i is the diffusion flux of molecules of sort i; $\sum$ is the summation sign over the index i from $i = 1$ to $i = l$.

Finally, the law of conservation of energy is written for a process running at constant pressure in the form

$$\rho_0 u_0 H_0 = \rho u H - \eta \frac{dT}{dx} + \sum \delta_i h_i^0, \tag{3.3}$$

where H is the specific enthalpy (the heat content, $H = E + pv$); η is the thermal conductivity, so that $\eta(dT/dx)$ is the flow of heat transported by molecular heat conduction; h_i^0 are the values of the chemical energy of various molecules, i.e., their enthalpies h_i at absolute zero.[5] The quantities h_i^0 and h_i satisfy the following equations:

$$h_i = h_i^0 + q_i = h_i^0 + \int_0^T c_p \, dT, \tag{3.4}$$

[5]As the zero enthalpy for all g different sorts of molecules we should take the enthalpy of f particular sorts of molecules containing all f sorts of atoms in various proportions.

$$\rho H = \sum c_i h_i = \sum c_i h_i^0 + \sum c_i q_i = \sum c_i h_i^0 + \rho Q = \sum c_i h_i^0 + \rho \int c_p \, dT.$$
$$(3.5)$$

We have introduced here, together with the enthalpy H, h_i, the thermal energy Q, q_i, whose meaning is clear from the formulas. In the expression for the flux of energy transported by diffusion, h_i^0 is used rather than h_i because the transport of the physical heat has already been accounted for in the thermal conduction.

The expressions written above are the most general and, in particular, hold in the presence of a chemical reaction as well since in a reaction only transformation of certain forms of energy into others and certain kinds of molecules into other kinds occurs, without any change in the total amount of energy or number of atoms.

We will choose the second control surface such (position I, index 1) that between O and 1 the chemical reaction may be disregarded. In this case the fluxes of individual sorts of molecules are also conserved since in the interval of interest different kinds of molecules are not transformed into others. Together with the equations written above the following g equations also hold

$$u_0 c_{i0} = u_1 c_{i1} + \delta_{i1}. \tag{3.6}$$

In particular, over the entire zone in which the chemical reaction does not occur, the final reaction products (substances for which $c_{i0} = 0$) may be present, but their flux is equal to zero in the chosen system of coordinates in which the flame is at rest.

The thermal energy is conserved since no transformation of chemical energy into thermal energy occurs.

Using the notations introduced above, we write the conservation of the thermal energy in the following form:

$$\sum u_0 c_{i0} q_{i0} = \sum u_1 c_{i1} q_{i1} + \sum \delta_{i1} q_{i1} - \eta \frac{dT_1}{dx}. \tag{3.7}$$

Using the preceding equations (3.6), we represent the law of conservation of heat in the following form:

$$\sum u_0 c_{i0}(q_{i1} - q_{i0}) = \eta \frac{Dt}{dx}. \tag{3.8}$$

Introducing the specific heat of the original mixture,

$$Q(T) = \frac{c_{i0} q_i(T)}{\rho}, \qquad \frac{dQ(T)}{dT} = c_p, \tag{3.9}$$

where c_p is the specific heat of the original substance, we obtain the equation

$$\rho_0 u_0 [Q(T_1) - Q(T_0)] = \rho u [Q(T_1) - Q(T_0)] = \eta \frac{dT_1}{dx}. \tag{3.10}$$

By integrating this equation we easily find the temperature distribution in the heating zone, where the reaction has not yet begun,

$$x = (\rho u)^{-1} \int \eta [Q(T) - Q(T_0)]^{-1} \, dT + \text{const.} \qquad (3.11)$$

We determine the value of the constant from the condition

$$T = T_B, \quad x = 0, \qquad x_1 = (\rho u)^{-1} \int_{T_B}^{T_1} \eta [Q(T) - Q(T_0)]^{-1} \, dT. \qquad (3.12)$$

This is the first concrete conclusion relating to the temperature distribution.

In the condensed phase we may assume with satisfactory accuracy

$$\mu = \text{const}, \quad \rho = \text{const}, \quad u = \text{const}, \quad c_p = \text{const}, \quad Q = c_p T. \qquad (3.13)$$

In this case we find (the index 1 shows that the calculation relates to the heating zone, where the reaction has not yet begun)

$$x_1 = \frac{\eta}{c_p \rho u} \ln \left[\frac{T_1 - T_0}{T_B - T_0} \right],$$

$$T_1 = T_0 + (T_B - T_0) e^{c_p \rho u x_1 / \eta}. \qquad (3.14)$$

We shall find the temperature distribution in the gas in those layers adjacent to $x = 0$ in which the chemical reaction has not yet started. The evaporation heat or the heat of endothermic reaction of gas-formation, L, is equal to the jump in the thermal energy of the original substance. Thus, at $x = 0$ at the phase-boundary the magnitude of the thermal flux experiences a jump. Using one prime for the c-phase and two primes for the gas, we construct the equation

$$x = 0, \quad T_B' = T' = T'', \qquad \eta'' \frac{dT''}{dx} = \eta' \frac{dT'}{dx} + \rho u L. \qquad (3.15)$$

Equation (3.10) remains valid.

Taking μ'' and c_p'' in the gas phase as constant (by taking the mean values of these quantities, the product ρu is strictly constant everywhere), we find the temperature distribution law in the gas in the zone where the reaction is negligibly small.

Performing the calculation, we find a solution which satisfies (3.10) and the boundary condition (3.15) in the following form:

$$T = T_0'' + (T_B - T_0'') e^{c_p'' \rho u x / \eta''}, \qquad x > 0, \qquad (3.16)$$

where

$$T_0'' = T_B - \frac{c_p'(T_B - T_0) + L}{c_p''}. \qquad (3.16a)$$

T_0'' may be defined as the temperature at which the gaseous EM would have an enthalpy equal to the enthalpy of the liquid EM at T_0 if the gas did not condense in cooling and its specific heat c_p'' remained constant. The numerator of the fraction in expression (3.16a) is the full amount of heat necessary for heating from T_0 to T_B and evaporation of a unit mass of EM.

Using (3.10) we shall give an approximate estimate of the width l of the zone in the gas from the surface $x = 0$ of the c-phase to the point where the reaction is complete. For a rough estimate we extrapolate the law (3.10) to $T = T_C$ and find the corresponding value of x. We obtain

$$l = \frac{\eta''}{c_p'' \rho u} \ln \frac{T_C - T_0''}{T_B - T_0''} \simeq \frac{\eta''}{c_p'' \rho u} \ln \frac{W + L}{L + c_p'(T_B - T_0)}, \qquad (3.17)$$

where W is the combustion heat.

As we shall see below, almost the entire chemical reaction runs at a temperature close to the temperature of combustion.

From this it follows: (1) that our estimate of the magnitude of l is sufficiently close numerically and (2) that the quantity l in fact represents the width of the dark space between the surface of the c-phase and the thin zone of intensive chemical reaction.

The expression for the diffusion flux, which we need to study the concentration distribution, has a complex form since we are dealing with a multi-component system under a temperature gradient. The diffusion flux of a given component depends in general not only on the concentration gradient of this component, but also on the concentration gradients of other components (the entrainment of one component by the others)[6] and on the temperature gradient (thermodiffusion).

The simplest case is a mixture of gases with no thermodiffusion or entrainment. Under a temperature gradient, diffusion equilibrium in such a mixture is achieved for constant partial pressure of each component.

It is obvious that we must require that the partial pressure, rather than the concentration, be constant since under a temperature gradient the total pressure is conserved, whereas the sum of the concentrations varies in proportion to ρ, i.e., in proportion to $1/T$.

The diffusion flux is proportional to $\delta_i \sim \operatorname{grad} p_i$; in order to bring the resulting expression in line with the ordinary definition of the diffusion coefficient D relating to the isothermic case, we write for a planar flame

$$\delta_i = -D \frac{c_i}{p_i} \frac{dp_i}{dx} = -\frac{D}{T} \frac{d(c_i T)}{dx}. \qquad (3.18)$$

---

[6]Unlike thermal diffusion, there has been almost no investigation of entrainment in the sense of the term given in the text. Its existence is easily established by considering a three-component system A–B–C in which the properties A and B are quite similar, while C differs sharply. Let us bring into contact mixture I: z parts of A and $(1 - z)$ of B on the one hand, and mixture II: z of C and $(1-z)$ of B on the other. If C is many times heavier than A and B, then A from the first mixture will diffuse into the second, displacing B until we have I: z parts of A, $(1 - z)$ of B, and II: z of C, $(1 - z)z$ of A, $(1 - z^2)$ of B. The concentration gradient of A caused the flow of B, whose concentration was constant; only later will slow diffusion of C begin, as a result of which all the concentrations will become equal. The theory of this problem was recently studied by Hellund [7].

We further consider

$$D = \frac{\eta}{c_p \rho} = \kappa \qquad (3.19)$$

where κ is the so-called thermal diffusivity.

We will consider, finally, that the chemical reaction runs without a change in the number of molecules and without a change in the mean molecular weight of the mixture (in essence, this is necessary even to have equality of the diffusion coefficients to one another and to the thermal diffusivity of the mixture).

In this case the density of the gas mixture is inversely proportional to the absolute temperature. We may rewrite the expression for the diffusion flux as

$$\delta_i = -\frac{D}{T} d(c_i T) \, dx = -\frac{\kappa \rho d(c_i/\rho)}{dx}. \qquad (3.20)$$

Using these assumptions about the diffusion, we may transform for the gas phase the basic equation of the law of energy conservation written in such a form that it is not violated even when in the presence of the chemical reaction (3.3)

$$\rho_0 u_0 H_0 = \rho u H - \eta \frac{dT}{dx} + \sum \delta_i h_i^0 = \rho u H - c_p \rho \kappa \frac{dT}{dx} - \kappa \rho \sum h_i^0 \frac{d(c_i/\rho)}{dx}$$

$$= \rho u H - \rho \kappa \left(\frac{dQ}{dx} + \frac{dH^0}{dx} \right) = \rho u H - \rho \kappa \frac{dH}{dx}. \qquad (3.21)$$

Factoring out the quantity $\rho u = \rho_0 u_0$, we obtain the differential equation

$$\frac{\kappa}{u} \frac{dH}{dx} = H - H_0. \qquad (3.22)$$

Its general solution has one arbitrary constant

$$\int (H - H_0)^{-1} \, dH = \int \frac{u}{\kappa} \, dx + \text{const};$$

$$H = H_0 + C e^{\int (u/\kappa) \, dx}. \qquad (3.23)$$

The solution remains finite for unbounded increase of x if and only if $C = 0$, so that

$$H \equiv H_0. \qquad (3.24)$$

Thus we have established that the specific (of a unit mass) enthalpy is constant throughout the space occupied by the gas phase, from $x = 0$ to $x \to \infty$; near $x = 0$ the temperature, composition and density of the gas vary rapidly due to transport processes and the chemical reaction. Our result establishes a relation between changes in the temperature and composition in the gas. At the same time we have clarified the assumptions under which this relation holds.

We recall that (3.24) was postulated without derivation by Lewis and Elbe [8] for the combustion of a gas.

In the paper cited [2] the similarity of the concentration fields (relative concentrations or partial pressures) and the temperature field were also established for the combustion of a gas, from which also follows constancy of the enthalpy throughout the combustion zone.

This similarity was established in [2] by consideration of the second-order differential equations of diffusion and heat conduction. Under the assumptions made about the coefficient of diffusion and thermal diffusivity, similarity of the fields, and therefore constant enthalpy, in the case of gas combustion occur throughout the space; this is the case not only in the steady problem, but in any non-steady problem as well. It is only necessary that there not be any heat loss by radiation or heat transfer to the vessel walls and that there be no additional (other than the chemical reaction) sources of energy. These conditions relate to the combustion of powders and EM as well, and were tacitly accounted for by us when we wrote the equations where the corresponding terms were absent.

In the case of combustion of a condensed substance, conservation of enthalpy and similarity occur only in the gas phase and only in part of the space. In the c-phase the diffusion coefficient is much smaller than the thermal diffusivity, and we have heating of the c-phase by heat conduction without dilution by diffusion. The enthalpy of the c-phase at the boundary, for $x \to 0$ (from the side $x < 0$), is larger than the enthalpy of the c-phase far from the reaction zone and larger than the enthalpy of the combustion products. The advantage of the derivation given here is that the constancy of the enthalpy in the gas phase and its equality to H_0 (H_0 is the enthalpy of the c-phase far from the combustion zone, at $x \to -\infty$) are obtained without regard to the state of the intermediate layers of the c-phase. We should particularly emphasize that the constancy of the enthalpy in the combustion zone occurs only for a steady process. The presence of layers of the c-phase with increased enthalpy opens the possibility in a non-steady process of a temporary change in the enthalpy of the gas and the combustion temperature (on this see §5).

Let us find the concentrations at the phase boundary at $x = 0$, $T = T_B$. Under the assumptions which were made in the derivation of enthalpy constancy we find

$$p_{\text{init}} = p\frac{T_C - T_B}{T_C - T_0''} = p\frac{W - c_p'(T_B - T_0)}{W + L}, \tag{3.25}$$

$$p_{\text{ult}} = p\frac{T_B - T_0''}{T_C - T_0''} = p\frac{L + c_p'(T_B - T_0)}{W + L}, \tag{3.26}$$

where p_{init} and p_{ult} are the partial pressures of the initial and ultimate substances.

Thus, the evaporation of EM occurs into a medium of diluted vapors; the corresponding decrease in the temperature of the surface according to the

Clapeyron-Clausius law is

$$-\Delta T_{\mathrm{B}} = \frac{RT_{\mathrm{B}}^2}{L}\frac{\Delta p}{p} = \frac{RT_{\mathrm{B}}^2}{L}\frac{L + c_p'(T_{\mathrm{B}} - T_0)}{W + L}. \tag{3.27}$$

This quantity, in nitroglycol for example, reaches $9°$. In fact, a combustion reaction is accompanied, as a rule, by an increase in the number of molecules and a decrease in the mean molecular weight, and the diffusion coefficient of the vapors proves smaller than the thermal diffusivity of the mixture. We should consider that the order of magnitude of ΔT_{B} here does not change.

The combustion velocity enters into the equations as a parameter; in the most general case, given an arbitrary value of the combustion velocity, we may always perform numerical integration of the system of second order ordinary differential equations which the distributions of temperature and concentration obey (the equations of heat conduction and diffusion, all accounting for the chemical reaction; the coordinate is the independent variable). It is convenient to integrate from the side of the ultimate reaction products $(x \gg 0)$. For an arbitrary value of the velocity, in general, the boundary conditions will not be satisfied at $x \ll 0$. For us it is sufficient to perform numerical integration to the point where the reaction rate may be disregarded, and to establish whether the conditions written above are satisfied there.

Performing the integration with various values of the parameter u, we locate by trial and error the value at which the boundary conditions are satisfied, and this is in fact the true value of the combustion velocity.

We find the state of the reaction products at $x \gg 0$, the point from which we perform the integration, from the conservation laws. These same laws are contained in the differential equations. In the general case of differing D and κ there are no simple relations between the concentrations and the temperature. However, the fluxes of various types of molecules and the heat flux are interrelated by equations (3.2) and (3.3). Therefore, if one of the conditions (3.6), (3.8) is satisfied, then the others will prove to be identically satisfied. Varying only the quantity u, we must succeed in satisfying one condition, which is always possible.

The method developed in [2] for integrating the equations of the distribution of the temperature and concentration in a flame is based on the following idea: the chemical reaction rate grows without bound with increasing temperature; in the presence of some temperature distribution reaction at a temperature which is close to the maximum must always be accounted for. As Frank-Kamenetskiĭ showed in a paper on thermal explosion [9], the reaction rate constant changes by a factor e for a temperature change by the quantity $\theta \simeq RT^2/E$, where R is the gas constant and E is the activation heat. This temperature interval is in fact the governing one; considering the interval θ small compared to the interval from T_{C} to T_0 (for this it is necessary that $E \gg RT_{\mathrm{C}}$), in the reaction zone we may disregard the change in the material constants and temperature. Strongly varying, with the rate of change proportional to the reaction rate, are the fluxes of heat and of the reacting components (but not the temperature and concentrations). Under the assumption $\theta \ll T_C - T_0$ the equation states:

$$-\frac{d}{dx}\eta\frac{dT}{dx} = -\eta\frac{d^2T}{dx^2} = \Phi, \tag{3.28}$$

where Φ is the volume rate of heat release in the reaction. Noting that at $T = T_C$, $dT/dx = 0$ we find

$$\frac{dT}{dx} = \left(2\eta^{-1}\int_T^{T_C}\Phi\,dT\right)^{1/2}. \tag{3.29}$$

Let us write the condition of thermal balance, which states that the entire amount of heat release in unit time is generated by the chemical reaction,

$$\rho W = \int \Phi\,dx. \tag{3.30}$$

But using (3.29) we replace

$$\int \Phi\,dx = -\eta\frac{dT}{dx} = \left(2\eta\int^{T_C}\Phi\,dT\right)^{1/2}, \tag{3.31}$$

with the integral taken everywhere over the entire region in which $\Phi > 0$.

Finally,

$$\rho_0 u_0 = \rho u = W^{-1}\left(2\eta\int^{T_C}\Phi\,dT\right)^{1/2}. \tag{3.32}$$

The rate of heat release depends not only on the temperature, but also on the concentration of the reacting materials. In the case $\kappa = D$ the relation of the concentrations to the temperature is indicated above.

In the case $\kappa \neq D$, as Landau showed, the concentration of the original substance in the reaction turns out to be κ/D times larger than in the previous case.

We write an approximate expression for the combustion rate by substituting

$$\int^{T_C}\Phi\,dT = \Phi_{max}\theta = \Phi_{max}\frac{RT_C^2}{E}, \tag{3.33}$$

$$\rho_0 u_0 = \left(2\eta\Phi_{max}\frac{RT_C^2}{W^2 E}\right)^{1/2}, \tag{3.34}$$

where Φ_{max} is the maximum value of the heat release rate.

The integral (3.32) and approximate (3.34) formulas are more convenient for practical use than the closed-form expressions which are obtained after integration in the simplest cases.

For what follows the following basic conclusions are fundamental: the combustion rate is proportional to the square root of the reaction rate in the gas near T_C. The combustion rate $\rho_0 u_0$ depends[7] on the pressure as $p^{n/2}$ for a n-order chemical reaction, $\Phi \sim p^n$, the combustion rate depends on the temperature as $\exp(-E/2RT_C)$, if E is the activation heat of the reaction, $\Phi \sim \exp(-E/RT_C)$.

As is clear from the above[8] the combustion rate depends on the properties (T_C, η, Φ) of the combustion products and of the layers most closely adjacent to them; the combustion rate depends on the properties of the c-phase insofar as the composition and temperature of the combustion products depend on the composition, caloricity and temperature of the c-phase.

With accuracy essentially up to a numerical factor of order 2 the formula for the flame velocity may be obtained even without integration of the second order equation. We begin with the obvious equality (3.30), replace the integral with the product $\Phi_{max}\Delta x_{eff}$, and estimate the effective width of the reaction zone Δx_{eff} from the temperature interval and order of magnitude of the temperature gradient

$$\Delta x_{eff} = \frac{\theta}{dT/dx} = \frac{\eta\theta}{\eta dT/dx} = \frac{\eta\theta}{\rho_0 u_0 W}. \tag{3.35}$$

Combining these estimates we obtain an expression which differs from (3.34) only by the absence of a numerical factor.

Considerations of the flame velocity in gases which also relate to EM and powders may also be found in [10].

With respect to combustion of nitroglycol the exact exponential dependence of the flame velocity on the combustion temperature was shown by Belyaev [11]. For powders the literature has long had empirical formulas, proposed by Muraour [4] and Jamaga [5]. The content of these formulas essentially reduces to the fact that the flame velocity depends on the combustion temperature of the powder. The actual form of the dependence, according to Jamaga, is Arrhenian, as indeed follows from the theory developed here. Muraour proposes a dependence of the form $u = \exp(A+BT)$. As Frank-Kamenetskiĭ showed [9], such a dependence is a good approximation to the Arrhenian one. We note that both authors (Jamaga and Muraour) relate the velocity to the combustion temperature in a closed volume, calculated from the condition of equality of the energy of the powder at T_0 and the energy of the powder gases.

By analogy with gas systems we call the combustion temperature in a closed volume T_{expl}, retaining the notation T_C for combustion at constant pressure.

In fact the combustion products which have just formed and are located

---

[7]The density of the c-phase, ρ_0, is practically constant.

[8]See especially the description of the general method for finding the rate by numerical integration.

at the very surface of the powder, always have a combustion temperature at constant pressure, which is calculated from the condition of equality of the enthalpy of the powder at T_0 and the enthalpy of the products at T_C.

In the case of a process in a closed volume, combustion products which appeared earlier also had, at the moment of formation, the same temperature T_C, however, after this, due to adiabatic compression during the pressure increase in combustion, the temperature of the gases grew. Thus, T_{expl} actually occurs only as an average quantity—cf. Mache's theory of explosion of a gas mixture in a closed volume [12].

The combustion rate indeed depends on T_C. Fortunately, the relation between the two temperatures is elementary:

$$T_{expl} : T_C = c_p : c_v,$$

and the transition from one to another does not change the form of Jamaga's and Muraour's formulas. These formulas, which relate to the velocity of combustion at high pressures of order 1000 atm, agree poorly with one another. The most probable value of the activation heat of the reaction which follows from the formulas is 24 kcal/mole.

There are a large number of papers, referenced in every textbook on internal ballistics, on the dependence of the combustion rate of a powder on the pressure. Various authors give dependencies which range from $u \sim p^{0.5}$ to $u \sim p$, corresponding to reactions from first to second order. In the case of powders there is a complicating factor: at low pressures the combustion products contain a significant amount of nitrogen oxide which disappears at high pressure. Thus, it is possible that the pressure change also affects the flame velocity as a result of the change in the direction of the reaction and the combustion temperature. Presently it does not appear possible to distinguish these factors numerically.

§4. Reaction in the Liquid Phase and the Transition from Combustion to Detonation[9]

We shall preface our presentation of the problems posed in the title of this chapter with a subtler analysis of the foundations of the theory.

We stated above that the reaction basically runs at a temperature close to the combustion temperature.

Indeed, the reaction rate reaches a maximum near the combustion temperature. The fact that this region is the crucial one is best revealed in an attempt to calculate the combustion rate under the assumption of a reaction which runs, for example, at 80% capacity with accordingly decreased thermal effect and combustion temperature: the rate turns out to be lower than the actual one. We may describe the situation thus: combustion at

[9] The discussion below relates to the combustion of liquid secondary EM.

a temperature close to T_C propagates fastest of all and causes a greater flame velocity for which reaction at any lower temperature does not have time to occur, even though the absolute value of the velocity at this lower temperature may be significant.[10]

In the case of gas combustion this situation is the rule, from which exceptions are quite rare.

In the case of combustion of a liquid EM a change in the aggregate state causes a sharp jump in its properties for constant temperature. The volume rate of heat release in a gas monotonically falls with decreasing temperature, beginning from a temperature close to T_B. At T_B, in the transition from a gas to a liquid, the volume rate of heat release increases sharply, discontinuously, due to the sharp increase in the density. As the temperature is lowered further, the density of the liquid remains constant and the volume rate of heat release again falls according to Arrhenius' law of the dependence of temperature on the reaction rate.

The width of the reacting layer also experiences a discontinuous increase as a result of the increase in heat conductivity in the transformation of the vapor to liquid.

In a gas the combustion rate falls with the combustion temperature—if the lower combustion temperature occurs for a given initial state as a result of incomplete combustion. The combustion rate, however, again increases when the combustion temperature does not exceed T_B so that the reaction is limited to the liquid phase; the reasons for the increase are indicated above. However, it is better here to speak not of the combustion rate, but of the propagation rate of the heating wave of the liquid due to reaction in the liquid phase. The maximum temperature achievable in a liquid is limited by the quantity T_B, just as the combustion temperature, in the strict sense, is limited by the quantity T_C. Calculation of the velocity of the heating wave in a liquid is not difficult [see formulas (3.32), (3.34)] if the kinetics of the chemical reaction in the liquid phase are known.

If the velocity of the heating wave thus calculated is less than the combustion rate then reaction in the liquid phase does not violate the steady character of combustion, but only insignificantly changes the temperature distribution in the liquid.

On the other hand, if the velocity of the heating wave is greater than the combustion velocity, a steady regime is impossible: indeed, during combustion the surface of the liquid is heated to the temperature T_B, however, at this temperature in the liquid a chemical reaction begins to run which heats the adjacent layers of the liquid before they are able to evaporate and burn. The temperature distribution in the liquid will turn out then to be non-steady.

What actually happens here? The heating wave will move forward away

---

[10]This explanation was first proposed by the author in a discussion at the Scientific Council of the Institute of Chemical Physics in 1940.

from the surface at which the combustion of the gas evaporates the liquid. In the temperature distribution a widening region of liquid will form which has been heated to the temperature T_B but which has not yet evaporated.

At the temperature T_B reaction in the liquid phase certainly does not stop. The temperature growth of the liquid stops at T_B, and the reason for this is that further supply of heat is used for evaporation of the liquid. At the free surface of the liquid evaporation occurs without delay or superheating and removes heat from the nearest superheated layers. However, as the heating wave moves forward and the region of the liquid heated to T_B expands, in this region the liquid superheats as a result of the chemical reaction and boiling occurs.

It was shown experimentally by Belyaev [3] that a boiling EM detonates when ignited.[11] The explanation he offered is very convincing: on the one hand, boiling sharply increases the surface of the layer dividing the liquid and vapor at which combustion occurs and thus increases the amount of EM burned in unit time and the mechanical effects (increase in pressure) which accompany combustion. On the other hand, the presence in the liquid of bubbles also very strongly increases the compressibility of the system and its sensitivity to mechanical effects.[12]

Thus we may foresee that when the heating wave velocity exceeds the combustion velocity, this leads to transition from combustion to detonation.

The condition of transition from combustion to detonation may be calculated in advance on the basis of experimental data with minimal knowledge of the reaction mechanism. In doing so we use: (1) the dependence of the combustion velocity on the pressure, measured in experiments as $u = u_1 p^m$; (2) the rate of heat release Φ_L during the liquid-phase reaction; this quantity may be found calorimetrically or calculated from the reaction rate measured on the basis of the rate of gas release. We represent the dependence of the heat release rate on temperature in the form

$$\Phi'_L = Ae^{-E/RT}. \tag{4.1}$$

(3) the physical constants of the liquid: the thermal conductivity η', the specific heat c'_p, the density ρ' and the evaporation heat L; this last may be found from the dependence of the vapor pressure on the temperature

$$p = Be^{-L/RT}. \tag{4.2}$$

Let us construct the expression for the propagation rate of the heating wave: the reaction runs only as much as needed to heat the EM from T_0 to T_B so that we must substitute into formula (3.32) the thermal effect $W' = c'_p(T_B - T_0)$. This thermal effect comprises only a small part of the

[11]Unfortunately, for methodological reasons, the experiments were performed with a very small amount of liquid, part of which evaporated. A more detailed investigation would be very desirable.

[12]In an unpublished work in 1939 Khariton observed a sharp drop in the sensitivity of liquid EM to impact when freed of bubbles.

thermal effect of the reaction which has run to completion and therefore we may neglect the change in the concentration of the reacting molecules.[13]

The expression for the heating wave velocity u' has the form

$$u_1' = \frac{\sqrt{2\eta'\Phi_L'(T_B)\theta'}}{\rho'W}$$

$$= \frac{\sqrt{2\eta'(RT_B^2/E)Ae^{-E'/RT_B}}}{\rho'c_p'(T_B - T_0)}. \tag{4.3}$$

Let us find the dependence of the velocity u' on the pressure: the pressure does not directly enter into the preceding formula, however, the surface temperature T_B depends on the pressure.

Comparing (4.2) and (4.3) we find

$$u = u_1'p^{E'/2L}, \tag{4.4}$$

where u_1' is calculated from (4.3) for T_{B1}—the boiling temperature at a pressure of 1 atm. We neglect any dependence of u' on T_B other than the exponential one. The magnitude of the exponent $E'/2L$ is greater than unity: thus, for nitroglycol[14] $E' = 40\,000$, $L = 14\,500$ kcal/mole, $E/2L = 1.38$. Thus with an increase in the pressure the velocity of the heating wave grows more rapidly than the combustion velocity. The increase in pressure eases the transition from combustion to detonation—in agreement with experiment. We find the condition of transition in the form

$$u' = u, \qquad u_1p^m = u_1'p^{E'/2L}, \qquad p = (u_1/u_1')^{2L/(E'-2Lm)}. \tag{4.5}$$

It should be noted that the use of the simple dependence (4.2) with constant L is possible only at $p \ll p_{\text{crit}}$. If the pressure obtained from formula (4.5) becomes comparable to p_{crit}, it is necessary to use in the calculation a refined curve of vapor pressure. As we know, the evaporation heat decreases as the temperature increases; formulas (4.4), (4.5) with constant L measured at low temperature overestimate the limiting pressure. Finally, if the pressure found exceeds the critical pressure, above which the transition from liquid to gas occurs continuously, the theory developed here is inapplicable.

At present, with only scarce experimental material, it would be premature to consider questions of superheating of the liquid and the release of gas reaction products in the form of bubbles.

We note in conclusion one fundamental fact: Belyaev's original argument consists in comparing the reaction probability and the probability of evaporation of a molecule of the EM located in the liquid. From the fact that the activation heat of the reaction is much larger than the evaporation heat he concludes that the probability of evaporation sharply exceeds the probability of reaction: as heat is supplied, consequently, evaporation will

[13]Only if the products do not exert a drastic—positive or negative—catalytic effect on the reaction rate.

[14]Measured for reaction in the gas phase.

occur and the reaction will begin only in the gas phase. Does not accounting for reaction in the condensed phase contradict these basic assumptions? Indeed these arguments cannot be applied without reservations. The probability of evaporation of an individual molecule cannot be written simply as $\nu \exp(-L/kT)$, where ν is some frequency. In this form the formula is valid only for molecules which lie in a superficial single-molecule layer, and even then only if we ignore condensation.

For molecules located inside the liquid, at all temperatures lower than a boiling temperature which corresponds to the external pressure, the probability of evaporation is strictly equal to zero. Meanwhile, the probability of reaction under these conditions, though small, is finite.[15] Only later, at a temperature which exceeds the boiling temperature, does the probability of evaporation become noticeable, when it grows rapidly and soon sharply exceeds the probability of reaction. It is precisely this that Belyaev [13] observed in experiments on the ignition of nitroethers by wires in which rapid supply of heat at atmospheric pressure led to evaporation rather than ignition or detonation.

§5. On Non-Steady Combustion

It does not appear possible to consider in general form, discarding the assumptions of steady propagation of the regime with constant velocity, the differential equations of heat conduction and diffusion in a medium in which a chemical reaction is also running.

Below we (1) disregard the chemical reaction running in the c-phase, (2) consider the temperature T_B of the surface constant, and (3) consider the relaxation time of processes running in the gas to be quite small compared to the relaxation time (time of variation) of the distribution of heat in the c-phase. Restricting ourselves to consideration of intervals of time which are significant compared to the relaxation time of processes in the gas, we shall consider that the state of the nearest layer to the gas, in which the chemical reaction is concentrated, corresponds at each moment to the heat distribution in the c-phase. After establishing this correspondence the problem reduces to consideration of the comparatively slow variation of the heat distribution in the c-phase.

As justification for the second point we note that in the case of reversible evaporation (secondary EM) T_B is practically independent of the evaporation rate and has only a weak logarithmic dependence on the pressure. In the

---

[15]Belyaev considered this question in unpublished calculations and showed that at atmospheric pressure and for low-boiling EM the number of molecules which evaporate in the surface layer is larger than the number of molecules which react in a layer 1 cm thick, i.e., the ratio of the evaporation probability to the probability of reaction is larger than the ratio of the number of molecules in a 1 cm thick layer to the number of molecules in the surface layer. However, the situation changes as the point of transition from combustion to detonation at high pressure is approached.

case of irreversible gasification (powder) T_B does not directly depend on the pressure and has only a weak logarithmic dependence on the rate of gasification.

As justification for the third point we cite some typical figures: we define the relaxation time of a process in the gas as the ratio of the width of the zone of temperature change to the velocity of the gas. In other words, we define the relaxation time as the time during which a certain particle of gas travels from the interface to the place where the reaction ends.

We analogously define the relaxation time of the heat distribution in the c-phase. We give a numerical example relating to combustion of a powder at atmospheric pressure.

The density of the powder is equal to 1.6 g/cm^3, the combustion velocity is 0.04 cm/sec. The thermal conductivity of the powder is $5.1 \cdot 10^{-4}$ cal/cm $\cdot$ sec $\cdot$ deg, the specific heat is 0.36 cal/g $\cdot$ deg, $\kappa = 1 \cdot 10^{-3}$ cm^2/sec. If we arbitrarily say that the distance κ/u, at which the heating changes by a factor e, is the zone width, we obtain $\kappa/u = 0.025$ cm and the time of combustion of this layer $\tau' = 0.025/0.04 = 0.625$ sec.

Let us perform an analogous evaluation for a gas. In the gas the velocity u is significantly larger, in accordance with the drop in density by a factor 10 000. The order of magnitude in the gas of the relaxation time is equal to $\tau'' \simeq 9 \cdot 10^{-5}$ sec, $\tau'/\tau'' = 6\,800$ times smaller than the relaxation time of the temperature distribution in a solid powder. As the pressure increases, the ratio τ'/τ'' decreases slightly; at 100 atm it is still equal to 75. These figures justify the approach adopted in point 3.

Since the relaxation time of combustion in the gas is very small, we are justified in considering that combustion in the gas at each moment is determined by the thermal state of the nearest thin layer of the c-phase to the interface: the temperature distribution in deeper layers does not exert direct influence on the process occurring at the surface. Conditions in the gas must be completely determined by the instantaneous value of the surface temperature and the temperature gradient in the c-phase near the surface. For this latter we introduce the abbreviated notation

$$\left(\frac{dT}{dx}\right)_{x=0} = \varphi. \tag{5.1}$$

We consider the temperature of the surface T_B to be constant (see above).

The basic quantity which characterizes combustion in a gas is the combustion temperature T_C. We find it by writing the heat balance per unit time. The c-phase, heated to T_B, burns, producing gaseous combustion products which are heated to T_C; the difference in the enthalpies of flow rates per second represents the amount of heat carried away into the depth of the c-phase

$$\rho'u'[H'(T_B) - H''(T_C)] = \varphi\eta'. \tag{5.2}$$

The product $\rho'u'$ may be equivalently written with either two primes (the sign of the gas phase) or one prime (the sign of the c-phase) since it is conserved. We have chosen the c-phase for the sake of definiteness. We compare expression (5.2) with the elementary expression which determines the combustion temperature in a steady regime T_{CS}

$$H'(T_0) = H''(T_{CS}). \tag{5.3}$$

We use the relation between the enthalpy, the specific heat, and the temperature and obtain after simple transformations

$$T_C - T_{CS} = \frac{c'_p}{c''_p}\left[(T_B - T_0) - \frac{\eta'\varphi}{c'_p\rho'u'}\right]. \tag{5.4}$$

In a steady combustion regime the relation between the quantities is such that the expression inside the square brackets is equal to zero, and, as should have been expected, the temperature T_C does not differ from T_{CS}, calculated from (5.3).

In a steady regime the c-phase, heated from T_0 to T_B, actually burns in the same way, however, the corresponding energy gain is compensated for by removal of heat into the c-phase. Exact compensation is natural since the heat removed into the c-phase is in fact used to heat the c-phase from the initial temperature T_0 to T_B.

Using the relation

$$(T_B - T_0) - \left(\frac{\eta'}{c'_p\rho'u'_S}\right)\varphi_S = 0 \tag{5.5}$$

(the index "S" denotes a steady state), we rewrite (5.4) in the form

$$T_C - T_{CS} = \Delta\left(1 - \frac{\varphi u'_S}{\tau_S u'}\right), \qquad \Delta = \frac{c'_p}{c''_p}(T_B - T_0). \tag{5.6}$$

The difference between the combustion temperature and its steady value depends not only on the value φ (i.e., on the heating in the c-phase), but also on the combustion velocity u'. In turn the combustion velocity depends on the combustion temperature,

$$u' = Ae^{-E/RT_C}. \tag{5.7}$$

For a given value of the temperature gradient φ we have two equations for determination of the two quantities u' and T_C. We study this system of equations by constructing the curves in the coordinate system u', T_C.

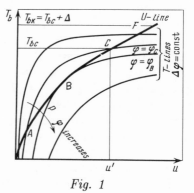

Fig. 1

In Fig. 1 the bold curve corresponds to formula (5.7). φ enters (5.6) as a parameter, and the family of thin lines corresponds to different values of φ. The direction of increasing φ is shown by the arrow. The thin lines represent a family of equilateral hyperbolas with common asymptotes: the ordinate axis, $u' = 0$, and the line parallel to the abscissa, $T_C = T_{CS} + \Delta$.

We shall call the bold line the U-curve since it shows the dependence of the combustion velocity on the temperature; we shall call the fine lines T-lines since formula (5.6) describes the dependence of the temperature on the combustion velocity. In the natural sciences the concepts "a depends on b" and "b depends on a" are not equivalent, whereas in mathematics $a = f_1(b)$, $b = f_2(a)$ and $f_3(a, b) = 0$ have completely identical meanings.

For large values of φ the curves do not intersect at all. A powder which is insufficiently heated from the surface does not burn. As it is heated and the heat propagates inside, the gradient φ decreases.

We reach, finally, the T-line which is tangent to the U-curve at the point C at the value $\varphi = \varphi_C$. Beginning from this value of φ the system of equations has a solution and combustion is possible. The critical value φ_C which corresponds to the tangent point determines the condition of ignition of the c-phase, and allows us to calculate the amount of heat and the rate of heating required. A calculation of φ_C will be performed below.

At still smaller values of φ the curves, as is easily verified, always have two points of intersection: this follows from the existence of a horizontal asymptote for the curve (5.6) and from the fact that at small T_C the exponent (5.7) decreases more rapidly than the algebraic curve (5.6) which has a vertical asymptote $u' = 0$ in common with the exponent.

In particular, at $\varphi = \varphi_S$ we certainly have a solution describing steady combustion; the equations, obviously, are satisfied by the solution $\varphi = \varphi_S$, $T_C = T_{CS}$, $u' = u'_S$ (the point S in Fig. 1).

We write out the value

$$\varphi_S = \frac{c'_p \rho'}{\eta'}(T_B - T_0)u'_S \tag{5.8a}$$

[see (5.8b) below].

With the change of T_0 from T_B and below the quantities φ_S grow due to the growth of the quantity $T_B - T_0$. Soon, however, the exponential decrease of u'_S overpowers the growth of $(T_B - T_0)$. The quantity φ_S passes through its maximum $\varphi_{S\,max} = \varphi_C$ as T_0 changes. Consequently, one value of φ_S corresponds to two different steady regimes with two different T_0 and two different u'_S. Thus we obtain confirmation of the fact that one value of the gradient corresponds to two values of u' and T_C. Each of them, with a corresponding value of T_0, describes a regime which satisfies the equations of steady combustion.

However, we have already established that the combustion regime in the gas phase must be determined by the value of φ regardless of the temperature distribution in the c-phase at great depth and, in particular, regardless of the value of T_0.

Let us compare the behavior of two possible regimes A and B corresponding to a single φ. Let us imagine that regime A is realized. A small drop in the combustion velocity u' will cause, in accordance with the T-curve, a drop in the temperature; the drop in velocity corresponding (U-curve) to the temperature drop will be even larger, and so on; continuing this will lead to unbounded decrease of the temperature, meaning quenching. If, however, starting from A we increase the temperature or the combustion velocity somewhat, then the same sequence of considerations will lead us to the state at point B. In contrast, the type of intersection of the T- and U-lines that we have at point B (and at all points above C) ensures stability: for example, as the velocity is increased the temperature grows weakly, and the corresponding increase in the velocity is small (less than the initial one) and so on; the sequence converges back to B.

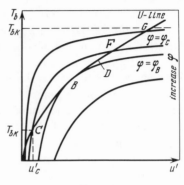

Fig. 2

Thus, for a given φ, of two possible combustion regimes of the gas the one which corresponds to the greater combustion velocity and higher combustion temperature is the one that always occurs.

We established above a method for finding the regime of combustion, i.e., u' and T_C, which is quickly established in the gas phase for a given state of the c-phase (i.e., given T_B and φ). In turn, depending on the combustion velocity, the temperature distribution in the c-phase itself changes slowly. The distribution and the gradient φ at each given moment depend on all the preceding values of the combustion velocity and on the entire previous thermal history of the powder. This dependence is determined by solution of the equation of heat conduction with a given value $T = T_B$ on the line $X(t)$, with $dX/dt = u'$.[16] It is not possible to write out this dependence analytically. In an elementary fashion from equation (5.8a) we find the limiting value of φ for prolonged maintenance of constant u':

$$\lim \varphi = \frac{c'_p \rho'}{\eta'} (T_B - T_0) u'. \tag{5.8b}$$

We easily verify that the point S (see Fig. 1) is stable with respect to slow changes in the temperature distribution in the c-phase. Let the point B be realized, $\varphi < \varphi_S$, $u' > u'_S$. To the greater value of u' corresponds in the established regime a greater value of φ. Consequently there will occur

[16]Here X is the coordinate perpendicular to the flame with respect to which the c-phase is at rest.

an increase in φ, accompanied by transition from one T-curve to another in the direction indicated by the arrow. The point of intersection of the T-curve and U-curve then shifts from B to S. At S the motion stops since here, not only does the combustion regime correspond to the thermal state of the c-phase, but the state of the c-phase corresponds to the combustion velocity as well. In the same way we may study the establishment of steady combustion after ignition; it is described in the graph as the rise from C to S.

We tacitly assumed that the point S, describing the steady regime, is higher than the tangent point C, as is shown in the drawing.

In fact the opposite case is also possible, where the point S lies below C so that the intersection of the U- and T-lines at the point S is of the type shown by point A (Fig. 2).[17]

What behavior do we expect from the system in this case? Let us imagine that the conditions for steady combustion are fulfilled, especially the corresponding gradient φ_S. As we saw, the regime S is quite rapidly—within the small relaxation time of the gas—replaced by another regime D with the same $\varphi = \varphi_S$, but with a larger combustion velocity. In accord with the increased combustion velocity, slow growth of φ occurs, accompanied by movement from D to C.

We shall prove for an arbitrary point E lying on the segment DC that in the case of Fig. 2 φ will grow.

φ_E is steady at u'_E and the corresponding T_{0E}, which is higher than the T_0 to which the point S corresponds; it is clear that $T_0 < T_{0E}$ from the fact that S lies below E for smaller T_C. If we have [cf. (5.8a) and (5.8b)] the identity

$$\varphi_E = \frac{c'_p \rho'}{\eta'}(T_B - T_{0D})u'_E,$$

then for the same velocity and T_0 the value to which φ tends is

$$\lim \varphi = \frac{c'_p \rho'}{\eta'}(T_B - T_0)u'_E > \frac{c'_p \rho'}{\eta'}(T_B - T_{0E})u'_E = \varphi_E,$$

$$\lim \varphi > \varphi_E.$$

The proof of the increase of φ remains valid at the endpoint of the segment at C as well.

Thus, growth of φ does not cease at C.

However, any further growth of φ leads us to T-lines which do not intersect the U-curve and combustion stops.

The physical processes which occur here are as follows: the externally heated c-phase begins to burn, but it burns at a velocity exceeding the steady velocity and with a temperature of the combustion products which exceeds the calculated theoretical temperature. Such intensive combustion occurs

---

[17]We recall that S describes the steady regime for given T_0.

due to irreversible discharge of the heat stored in ignition. After this supply has been exhausted the transport of heat into the c-phase reaches such a large value that the flame is extinguished. Thus, when ignited, instead of steady combustion we obtain a fairly energetic, but soon extinguished, flare-up.

Steady combustion, possible in the sense that it is possible to construct a regime satisfying the equations of steady combustion, proves to be fundamentally unrealizable as a result of its instability, described in detail above.

Let us find the conditions which determine what case occurs, whether the point S ends up above or below C.

We shall vary the initial temperature T_0 of the c-phase with constant chemical composition and constant pressure.

In its original form (5.2) we see that the family of T-lines (in which each curve corresponds to a specific value of φ) does not depend on T_0. We may rewrite (5.2) in the form

$$T_{\mathrm{C}} = T_{\mathrm{CB}} - \frac{\eta'}{\rho' c_p'} \frac{\varphi'}{u'}, \qquad \text{where} \quad H''(T_{\mathrm{CB}}) = H'(T_{\mathrm{B}}), \qquad (5.9)$$

so that T_{CB} is the theoretical combustion temperature of the c-phase heated to its highest temperature[18] T_{B}.

Thus, for all values of T_0 one and the same diagram with a single U-curve and family of T-lines is appropriate. Only the location of the point S, which always lies on the U-curve, depends on T_0. The relation between T_{CS} and T_0 is elementary. Steady combustion is possible only at T_0 such that $T_{\mathrm{CS}} > T_C$, i.e., is larger than the temperature corresponding to the tangent point. A powder or secondary EM cooled to a lower temperature is not capable of burning steadily and, when ignited, produces only flare-ups, as described above. Let us calculate the coordinates of the point C at which the T- and U-lines come into contact.

The condition of the contact is

$$\left(\frac{dT_{\mathrm{C}}}{du'} \right)_U = \left(\frac{dT_{\mathrm{C}}}{du'} \right)_{T,\varphi}. \qquad (5.10)$$

The derivative on the left is taken along the U-line, and on the right along the T-line at constant φ. Let us calculate these quantities. We apply (5.7)

---

[18]Comparing (5.9) with (5.6) we obtain the identities

$$T_{\mathrm{CS}} + \Delta = T_{\mathrm{CS}} + \frac{c_p'}{c_p''}(T_{\mathrm{B}} - T_0) = T_{\mathrm{CB}} = \text{const}, \qquad (5.9\mathrm{a})$$

$$\Delta u_{\mathrm{S}}' \varphi_{\mathrm{S}}^{-1} = \frac{c_p'}{c_p''}(T_{\mathrm{B}} - T_0)u_{\mathrm{S}}' \left[\frac{c_p'' \rho'}{\eta'}(T_{\mathrm{B}} - T_0)u_{\mathrm{S}}' \right]^{-1} = \frac{\eta'}{\rho' c_p'} = \kappa = \text{const}. \qquad (5.9\mathrm{b})$$

and (5.9):

$$\left(\frac{dT_C}{du'}\right)_U = \left(\frac{du'}{dT_C}\right)_U^{-1} = \left(\frac{Eu'}{2RT_C^2}\right)^{-1} = \frac{(2RT_C^2/E)}{u'}, \qquad (5.11)$$

$$\left(\frac{dT_C}{du'}\right)_{T,\varphi} = \frac{\eta'/\rho c_p'}{u'^2} = \frac{T_{CB} - T_C}{u'}. \qquad (5.12)$$

The condition (5.10) is written thus:

$$\frac{T_{CB} - T_C}{u'} = \frac{2RT_C^2/E}{u'},$$

$$T_{CB} - T_C = \frac{c_p'}{c_p''}(T_B - T_0) = \frac{2RT_C^2}{E} \simeq \frac{2RT_{CB}^2}{E}, \qquad (5.13)$$

$$T_B - T_0 \cong \frac{c_p''}{c_p'}\frac{2RT_{CB}^2}{E} \cong \frac{c_p''}{c_p'}\frac{2RT_{CB}T_C}{E}.$$

Substituting into (5.7) we find at the tangent point

$$u_C' = u'(T_C) = \frac{u'(T_{CB})}{e}. \qquad (5.14)$$

The condition of the limit of steady combustion is related to the temperature dependence of the combustion velocity. The last formula states that steady combustion is possible only at an initial temperature at which the combustion velocity is not less than $1/e$, i.e., is not less than 37% of the combustion velocity of the c-phase heated completely to the temperature T_B. Above the limit the dependence of the steady velocity on the initial temperature does not undergo any changes.

It is interesting to compare the result with the theory of the flame propagation limit in gases which was developed by the author [14]. In the case of gases the existence of a limit depends only on heat transfer to the outside; combustion is possible only until the heat transfer reduces the combustion velocity to not more than $1/\sqrt{e}$, i.e., to 61% of the adiabatic velocity. An important difference consists in the very sharp drop in the velocity at the limit in the case of a gas. The fundamental difference lies in the fact that below the limit in a gas steady solutions do not exist; in the case considered here of combustion of the c-phase, steady solutions exist—theoretically—but they are unstable and therefore unrealizable.

The result obtained here agrees with experiment. Thus, in Belyaev's experiments [11] at the boiling temperature of nitroglycol $u'(T_{CB}) = 0.65$ mm/ sec.[19] The lowest measured velocity at $0°$C was 0.26 mm/sec (Andreev), i.e., 40% of $u'(T_{CB})$. According to a private communication by Belyaev, at lower temperatures steady combustion was not observable.

---

[19]The velocity observed at $184°$C was 0.62 mm/sec. At higher temperatures the formation of bubbles inhibits measurement. The figure cited in the text was obtained by extrapolation to $T_B = 200°$C.

Much interesting material on this problem is contained in the article by Andreev [15]. Thus, according to his observations, at atmospheric pressure and room temperature nitroglycerin does not yield steady combustion. Steady combustion may be observed at a pressure lowered to 230–380 mm where the boiling temperature decreases to 210–225°C instead of 245°C at 760 mm.

This example is particularly convincing because as the pressure is lowered the combustion temperature can only decrease, and the combustion velocity drops noticeably. Thermal losses to the outside, which are usually used to explain the limit, only grow; it is difficult to find another explanation for the observed fact.

The explanations offered by Andreev are of a qualitative character and anticipate the ideas developed here in more detail: thus, Andreev writes of the necessity of heating for the ignition of a condensed substance, and of the possibility of quenching as a result of dissipation into the body of the material of heat which was concentrated in a thin layer near the surface, etc.

In the case of combustion of condensed substances, consideration of non-steady processes, accounting for the non-adiabatic nature of combustion, which we do not present here, leads to the following conclusions:

1. For a given T_0 and given properties of the substance, the presence of heat transfer to the outside leads—for weak heat transfer—to a decrease in the flame velocity compared to the adiabatic steady value; for sufficiently strong heat transfer, it leads to the impossibility of combustion.

2. Under given external conditions (tube diameter, etc.) the relative heat transfer, on which the velocity and possibility of combustion depend, increases as the combustion velocity decreases. The existence of a limit for the feasibility of combustion at low pressure, depending on the drop in the combustion velocity with decreased pressure, is therefore natural (together with the possibility of existence of an upper limit which depends on the increase in T_B as the pressure is raised—see the example of nitroglycerin above).

3. The combustion regime at the limit (including the limit which depends on the heat transfer) is located at the boundary of stability of the steady regime. The boundary of feasibility of the steady regime (see [11]) is not attained.

4. In the one-dimensional theory we may calculate the velocity at the limit as a function of a quantity characterizing the heat transfer.

In all cases

$$u'(T_{CB})e^{-1/2} > u'_{lim} > u_{ad}e^{-1/2} \qquad (5.15)$$

where u'_{lim} is the velocity at the limit and u_{ad} is the velocity of steady adiabatic combustion [calculated from (5.3) without losses to the outside].

In the light of certain theoretical predictions, a detailed experimental study of the problem is desirable.

The contact condition found (5.14) is essential for a theory of ignition. The necessary condition for ignition is that, at a surface temperature T_B, the gradient should not exceed the value which corresponds to the contact of the curves φ_C. At the point C we find from (5.8) and (5.14)

$$\varphi_C = \frac{c_p'' \rho'}{\eta'}(T_{CB} - T_C)u_C' = \frac{\rho' c_p''}{\eta'}\frac{2RT_{CB}^2}{E}u'(T_{CB})e^{-1}. \tag{5.16}$$

The gradient value necessary for ignition does not depend on the temperature T_0 of the c-phase. However, the amount of heat and duration of heating necessary to achieve $\varphi = \varphi_C$ at the surface do depend on T_0.

To accuracy within an order of magnitude we find

$$x_C \cong \frac{T_B - T_0}{\varphi_C} \simeq (T_B - T_0)\frac{E}{2RT_{CB}^2}\frac{c_p'}{c_p''}\frac{\kappa' e}{u'(T_{CB})}, \tag{5.17a}$$

$$t_C \cong \left[\frac{T_B - T_0}{(E/2RT_{CB}^2)(c_p'/c_p'')}\right]^2 \frac{\kappa' e^2}{[u'(T_{CB})]^2}, \tag{5.17b}$$

$$q_C \cong \rho' c_p'(T_B - T_0)x_C = (T_B - T_0)^2\frac{E}{2RT_{CB}^2}\frac{c_p'}{c_p''}\frac{\eta' e}{u'(T_{CB})}, \tag{5.17c}$$

where x_C is the width of the heated layer, t_C is the necessary heating time, and q_C is the amount of heat spent on heating.

We note that the feasibility condition of steady combustion (5.13) may be written

$$(T_B - T_0)\left(\frac{E}{2RT_{CB}}\right)^2\frac{c_p'}{c_p''} \leq 1. \tag{5.18}$$

Substituting into (5.17), we obtain the upper limit of the corresponding quantities (the heating time and amount of heat needed for ignition) for the lowest temperature at which combustion is still possible. As the temperature T_0 is raised, the amount of heat and time needed for ignition fall as $(T_B - T_0)^2$; experiment agrees qualitatively with this conclusion. We note that, besides heating of the c-phase, ignition also requires that the EM vapors which form be ignited. The amount of heat needed for this is quite small compared to the amount spent on heating of the c-phase (approximately in the ratio of the relaxation times—see the beginning of this section).

Above we presented the question of non-steady combustion applied to secondary explosive materials. With respect to the combustion of smokeless powder, a possible complicating factor is its multicomponent composition. Of interest in this connection is the fact cited by Andreev [15] of steady combustion of nitroglycerin gelated by 1% nitrocellulose. In the case of smoky powders and pipe mixtures the role of condensed combustion products which adhere to the burning surface and accumulate heat may be important.

Finally, the considerations developed for the conditions of ignition and the feasibility of combustion may also be applied to the combustion of coal, liquid fuel, etc., due to oxygen in the surrounding medium. In these cases the temperature gradient in the c-phase (in coal or oil) also plays a role in the thermal balance. A number of substantial differences, particularly a different form of the combustion velocity curve as a function of the parameters, makes a special study, inappropriate here, essential.

Together with the questions of ignition and the feasibility (limit) of combustion, the concepts developed here are important for the combustion of EM or powder under variable conditions, in particular, at non-constant pressure. Variable pressure is accompanied by a variable combustion velocity, and to each value of the combustion velocity corresponds a particular value of the gradient φ which is established in the steady regime. It is at precisely this value φ_S that the steady value of the combustion velocity occurs. Meanwhile, for rapid pressure variation the temperature distribution in the c-phase is not able to keep up with the change in pressure; for a non-steady value of φ we should accordingly expect that the combustion velocity will also prove different from the steady value. For rapid pressure variation the combustion velocity turns out to depend not only on the instantaneous pressure, but also on its variation curve, which distorts the classical law of combustion.

As Khariton noted, these considerations allow us to explain the peculiar phenomenon of powder quenching in a gun barrel after the shell has left it: at high pressure the combustion velocity is large, the temperature gradient is large, and the heated layer of powder is thin. Firing of the shell is accompanied by a sharp drop in pressure. The powder is fully capable of burning at atmospheric pressure, but this combustion occurs at a comparatively low velocity, at small φ, and requires significant thickness of the heated layer. Therefore, heating of a powder which burns at high pressure may prove insufficient for ignition at atmospheric pressure; when the rapid drop in pressure occurs the temperature distribution is unable to readjust in time and the abnormally steep—for combustion at low pressure—temperature gradient leads to a lowered combustion temperature and to quenching of the powder granules, in particular those left in the barrel. It would be quite interesting to analyze in detail the change in the heat distribution in a powder in the internal ballistics problem, as in a manometric bomb or a gun, and its influence on the combustion law.

In conclusion I wish to express my sincere gratitude to my colleagues at the Institute, A. F. Belyaev, O. I. Leipunskiĭ, D. A. Frank-Kamenetskiĭ and Yu. B. Khariton for their interest in this work and valuable comments, and for the opportunity to discuss and use their unpublished results.

Summary

1. We propose a scheme for the combustion of a smokeless powder which provides for its transformation into energy-rich gas materials through a heterogeneous reaction and the subsequent reaction of these materials with release of combustion heat in the gas phase at the surface of the powder. This scheme is analogous to Belyaev's scheme for the combustion of EM.

2. We analyze a theory of steady combustion of condensed materials (secondary EM and powders) which determines the distributions of the temperature and concentrations and the combustion velocity.

3. Taking account of an exothermic reaction in the liquid phase leads to the impossibility of steady combustion at a high boiling temperature. When the limit is attained, transition from combustion to detonation occurs. Formulas are given for calculating the conditions of transition.

4. A non-steady theory of combustion of condensed materials is constructed which is based on the fact that the relaxation time of the heat distribution in the condensed phase is much larger than the relaxation time in the gas phase.

5. On the basis of the non-steady theory we predict the combustion limit which is attained when the combustion velocity falls to 37% of the combustion velocity at the boiling temperature. The limit depends on the internal instability of combustion, not on external thermal losses.

6. A theory of ignition of condensed materials is constructed on the basis of the non-steady theory.

The time and amount of heat required for ignition are basically determined by heating in the condensed phase and are proportional to the square of the difference between the boiling temperature and the initial temperature.

Combustion Laboratory   *Received*
Institute of Chemical Physics   *September 25, 1942*
USSR Academy of Sciences, Leningrad

References

1. *Belyaev A. F.*—Zhurn. fiz. khimii **12**, 93 (1938).
2. *Zeldovich Ya. B., Frank-Kamenetskiĭ D. A.*—Zhurn. fiz. khimii **12**, 100 (1938); here art. 19.
3. *Belyaev A. F.*—Dokl. AN SSSR **24**, 253 (1939).
4. *Muraour H.*—C. r. Bull. chim. Soc. **187**, 289 (1928); **4**, 613 (1933).
5. *Jamaga N.*—Ztschr. ges. Schiess Sprengstwesen **11**, 113 (1930).
6. *Langmuir I.*—J. Amer. Chem. Soc. **54**, 2806 (1932).
7. *Hellund A.*—Phys. Rev. **57**, 319, 328, 737, 743 (1940).
8. *Lewis B., Elbe G.*—J. Chem. Phys. **21**, 537 (1934).

9. *Frank-Kamenetskiĭ D. A.*—Zhurn. fiz. khimii **13**, 738 (1939).
10. *Zeldovich Ya. B., Semenov N. N.*—ZhETF **10**, 1116 (1940).
11. *Belyaev A. F.*—Zhurn. fiz. khimii **14**, 1009 (1940).
12. *Flamm P., Mache H.*—Wien. Ber. **9**, 126 (1917); *Mache H.* Die Physik der Verbrennungserscheinungen. Leipzig: Velta Co (1918).
13. *Belyaev A. F.*—In: Sbornik stateĭ po teorii goreniĭa [Collection of Papers on the Theory of Combustion]. Ed. by K. K. Andreev, Yu. B. Khariton. Moscow, Leningrad: Oborongiz, 7 (1940).
14. *Zeldovich Ya. B.*—ZhETF **11**, 159 (1941); here art. 20.
15. *Andreev K. K.*—In: Sbornik stateĭ po teorii goreniĭa [Collection of Papers on the Theory of Combustion]. Ed. by K. K. Andreev, Yu. B. Khariton. Moscow, Leningrad: Oborongiz, 39 (1940).

Commentary

The value of this paper derives above all from its correct theory of powder combustion in which chemical transformations occur in both the condensed and gas phases (with the latter exerting a governing influence on the former). But the first phase and the heated layer in it play an important role as a heat accumulator, which is what causes the peculiar phenomena in non-steady combustion.

The central thesis of the theory of the non-steady combustion of powders and explosives developed by Ya.B. in this article is the assumption of rapid readjustability of the gas phase of combustion compared to thermal changes in the condensed phase, which allows us to consider the gas phase as quasi-steady. This fundamental property of burning condensed materials allows us not only to significantly simplify the solution of the problem by reducing it to an analysis of the non-steady temperature distribution in the surface layer of the condensed material, but also not to carry out a detailed analysis of the complex structure of the combustion zone above the material (the multi-stage character of the chemical transformation, thermal decomposition, and gasification of the dispersed particles of condensed material and other processes).

Following Ya.B.'s basic idea, for the description of non-steady combustion we may use the steady dependencies of the combustion velocity of a powder on the initial temperature which are obtained in basic ballistic experiments. They may be recalculated in the form of dependencies of the combustion velocity on the temperature gradient near the powder surface which are usable for non-steady combustion as well. In his paper of 1943 (it was published later, in 1964),[1] on the basis of such a phenomenological approach Ya.B. proposed a theory of non-steady powder combustion under rapid pressure variation, including such a strong decrease that the powder is extinguished. (This effect has received broad practical application—by decreasing the pressure the thrust is cut off in powder-driven jet engines.)

In another paper Ya.B. considered[2] the interaction between pressure pulsations in a powder-driven rocket chamber with supersonic flow of the combustion products and pulsations in the combustion velocity. It turned out that for small sizes of the combustion chamber self-generation of oscillations appears, leading to

[1] *Zeldovich Ya. B.*—Zhurn. prikl. mekhaniki i tekhn. fiziki **3**, 126–138 (1964).

[2] *Zeldovich Ya. B.*—Zhurn. prikl. mekhaniki i tekhn. fiziki **1**, 67–76 (1963).

a self-oscillatory combustion regime. Ya.B. wrote two monographs in collaboration with other authors on the combustion of powders and solid rocket fuels. In the monograph "The Momentum of the Reaction Force in Powder Rockets"[3] an effective method was proposed for the thermodynamical calculation of the combustion products of a powder and their flow through a supersonic nozzle to obtain the reaction force. The monograph "Theory of the Non-Steady Combustion of a Powder"[4] is devoted to non-steady phenomena: ignition, variable-velocity combustion, quenching, propagation limits, the criterion of stable combustion.

In each of the above directions, Ya.B.'s ideas presented in the present paper were developed and partially modified in numerous subsequent theoretical and experimental studies. Thus, experiments showed[5] that the region of stable combustion of a powder is larger than was predicted by the theory. Theoretical studies which developed Ya.B.'s ideas of 1943 to account for temperature variation at the powder surface gave an adequate explanation of this fact.[6,7] B. V. Novozhilov generalized Ya.B.'s phenomenological approach to non-steady combustion to the case of variable temperature in the zone of the gasification reaction of a powder and developed a theory of nonlinear oscillations of the combustion velocity at periodically changing pressure and self-oscillations beyond the boundary of stability.[8] A comparison of the basic results of the theory, now known as the Zeldovich–Novozhilov theory, with experimental data obtained at the AS USSR Institute of Chemical Physics was performed in a paper by A. A. Zenin.[9]

The ability of the burning surface of a powder to intensify reflecting acoustic waves was studied in detail with respect to rocket combustion chambers,[10] and a theoretical analysis of this question was performed by S. S. Novikov and Yu. S. Riazantsev.[11] The critical conditions of powder quenching by a decrease in the pressure became the subject of research,[12] and this question was also developed theoretically.[13] The ignition of powders was studied experimentally,[14] and different

[3] *Zeldovich Ya. B., Rivin M. A., Frank-Kamenetskiĭ* Impuls reaktivnoĭ sily porokhovykh raket [The Momentum of the Reaction Force in Powder Rockets]. Moscow: Oborongiz, 186 p. (1963).

[4] *Zeldovich Ya. B., Leipunskiĭ O. I., Librovich V. B.* Teoriia nestatsionarnogo goreniia porokha [Theory of the Non-Steady Combustion of a Powder]. Moscow: Nauka, 131 p. (1975).

[5] *Korotkov A. I., Leipunskiĭ O. I.*—In: Fizika vzryva [Physics of Explosion]. Moscow: Fizmatgiz, 3rd Printing, 213–215 (1953).

[6] *Istratov A. G., Librovich V. B.*—Zhurn. prikl. mekhaniki i tekhn. fiziki 5, 38–44 (1964).

[7] *Novikov S. S., Riazantsev Yu. S.*—Zhurn. prikl. mekhaniki i tekhn. fiziki 1, 65–69 (1965).

[8] *Novozhilov B. V.*—Zhurn. prikl. mekhaniki i tekhn. fiziki 4, 157–161 (1965); 4, 73–78 (1970).

[9] *Zenin A. A.* Fizicheskie protsessy pri gorenii i vzryve [Physical Processes in Combustion and Explosion]. Moscow: Atomizdat, 215 p. (1980).

[10] *Margolin A. D., Pokhil P. F.*—Fizika goreniia i vzryva 7, 188–194 (1971).

[11] *Novikov S. S., Riazantsev Yu. S.*—Zhurn. prikl. mekhaniki i tekhn. fiziki 2, 57–63 (1966).

[12] *Assovskiĭ I. G., Istratov A. G., Marshakov V. N., Leipunskiĭ O. I.*—Fizika goreniia i vzryva 3, 232–238 (1967); 5, 3–11 (1969); 13, 200–205 (1977).

regimes of ignition by convective and radiative thermal fluxes were analyzed.[15,16] Since it had a different initial foundation, the theory of the non-steady combustion of powders and solid rocket fuels in the Soviet Union developed along a different path than similar studies abroad, where the assumption of quasi-stability of the gas phase of combustion was not used and its non-steady structure was frequently analyzed using numerical methods. Comparison of the various approaches to non-steady combustion was performed in a joint Soviet–American paper.[17]

Let us note that certain questions posed by Ya.B. in the present article are still unsolved: for example, the problem of the two stages of reactions of nitrogen in a gas—to NO and then deoxidation to N_2, and of the concrete structure of the gasification zone.

[13] *Istratov A. G., Librovich V. B., Novozhilov B. V.*—Zhurn. prikl. mekhaniki i tekhn. fiziki **3**, 139–144 (1964); **4**, 34–41 (1971).

[14] *Mikheev V. F., Levashov Yu. V.*—Fizika goreniĭa i vzryva **9**, 506–511 (1973).

[15] *Librovich V. B.*—Zhurn. prikl. mekhaniki i tekhn. fiziki **6**, 74–79 (1963).

[16] *Assovskiĭ I. G.*—Candidate Dissertation in Physics/Math. Moscow: AS USSR Institute of Problems of Mechanics (1973).

[17] *Summerfield M., Caveny L. H., Battista R. A., Kubota N., Gostinzev Yu. A.*—AIAA Pap. **70**, 667–672 (1970).

25

The Oxidation of Nitrogen
in Combustion and Explosions[*]

Introduction

§1. Throughout the XIX century attention was repeatedly drawn to the fact that nitric oxide is formed during combustion and explosions. A theoretical approach to this problem, however, became possible only in the beginning of the XX century when the problem of the fixation of atmospheric nitrogen and the development of chemical thermodynamics led to a study of the equilibrium oxygen–nitrogen–nitric oxide [1]. Fink [2] investigated explosions of mixtures of $2H_2 + O_2$ with oxygen and nitrogen. Nernst [1] was of the opinion that a reversible bimolecular reaction takes place at the highest temperature of the explosion, the composition approaching the equilibrium value at this temperature in dependence on the reaction rate. Nernst's formulas do not agree with a number of experimental facts. However, as was shown in our work, these facts do not call for the rejection of the notion of thermal reactions, but rather for an extension of Nernst's conception to the more complicated temperature conditions of explosions and a more intricate reaction mechanism.

Haber [3] investigated the formation of nitric oxide during the combustion of carbon monoxide and came to the conclusion that charged particles—electrons and ions—have an important catalytic effect on the reaction in the flame.

Bone and co-workers [4] studied the formation of nitric oxide in the combustion of carbon monoxide and arrived at some very strange results, e.g., yields of nitric oxide exceeding equilibrium values, a sharply negative influence of water vapor and hydrogen, etc. Bone concluded that the reaction of nitrogen with oxygen is caused by activation of the nitrogen by radiation from the carbon monoxide flame.

Polyakov and co-workers [5] made an especially detailed study of the influence of the initial pressure and diameter of the vessel on the formation of nitric oxide in an explosion. They came to the conclusion that combustion takes place according to a branching chain mechanism and that the breaking off of the chains in the volume, i.e., the reaction of the active centers with

[*]Acta Physicochimica URSS 11 4, 577–628 (1946).

nitrogen, leads to the formation of nitric oxide; the oxidation of nitrogen is thus induced by combustion.

Bone's data were not confirmed by Sadovnikov [6] who repeated his experiments. As for the assumptions of Haber and Polyakov concerning the role of electrons or chemical induction, it is noteworthy that they follow from the impossibility of explaining the data on the basis of a thermal reaction. We shall see later [7] that if the specific temperature conditions of the explosion are taken into consideration, the conception of a thermal reaction completely covers all the observed facts.

As appears from the preceding review, at the time the present investigation was undertaken, there was no generally accepted theory of the mechanism of nitrogen oxidation in a flame. Nor did such a theory arise as a result of the investigations (see [5]) pursued simultaneously but independently of our own.

§2. From the viewpoint of modern kinetics the problem splits up into three: (1) Is the reaction induced or thermal, i.e., does the chemical reaction of combustion and explosion affect the oxidation of nitrogen (through the agency of electrons, radiation or active centers) or do combustion and explosion simply act as heat sources bringing nitrogen and oxygen to a high temperature, while the oxidation of nitrogen occurs exactly as if the reaction were proceeding in a corresponding mixture of oxygen and nitrogen brought to the same temperature by any other means? (2) If the reaction is thermal, what laws govern it, what are its formal kinetics, i.e., how does the reaction rate depend on the concentrations of the reacting substances and what is the heat of activation? (3) What is the actual mechanism of the reaction, i.e., what are the elementary acts which bring to the observed formal kinetic laws?

Corresponding to these three problems our paper is divided into three parts. In the first (Sections 3–6) is set forth the experimental material on the amount of nitric oxide formed in explosions of mixtures of various combustibles with oxygen and nitrogen. It is shown that heating the mixture is equivalent to adding a corresponding amount of combustible. Thermodynamic computations show that in all cases the observed amount of nitric oxide is less than the equilibrium value at the maximum temperature of the gas. Direct experiments on the combustion of a gas in a stream have proved that the nitric oxide is not formed during but only after combustion. Thus, in Part I the thermal nature of the reaction is proved.

In Part II (Sections 7–10) are described experiments on the decomposition of nitric oxide which was added to the explosive mixture in advance. These experiments establish a proportionality between the rate of decomposition and the square of the concentration of nitric oxide and give the heats of activation for the formation and decomposition of nitric oxide. The similarity theory and the exact mathematical theory of a reversible bimolecular reaction in an explosion are developed. These theories give a good account of the influence of various factors—pressure, dimensions of the vessel, composition of the mixture, etc.—on the amount of nitric oxide formed. The absolute values of the reaction rates are found.

In Part III (Sections 11–12) it is shown that the assumption of a bimolecular reaction $N_2 + O_2$ requires an effective diameter of molecular collisions leading to the reaction 30–50 $\overset{\circ}{A}$, which is obviously too high; a reasonable agreement with experiment is given by the chain mechanism:

$$O + N_2 = NO + N; \; N + O_2 = NO + O.$$

Experiments on mixtures with low oxygen content (determined by the dissociation of the combustion products in the presence of excess fuel) prove the correctness of the chain mechanism.

Careful consideration of the conditions of the reaction in an explosion has made the determination of the yield of nitric oxide in the explosive products an exact kinetic procedure. By this procedure we were enabled to investigate the kinetics of a reaction taking place in the hundredth and thousandth fraction of a second at temperatures up to 2000–3000° and to establish the laws and mechanism of the chemical reaction between two of the simplest and most widespread substances—oxygen and nitrogen—thus solving a problem which had for a long time been the object of numerous investigations.

Part I. Thermal Character of the Reaction

§3. *Amounts of nitric oxide formed in explosions of various mixtures at low initial pressure.*

Experimental procedure. A schematic diagram of the set-up is shown in Fig. 1. Mixtures of carefully purified and dried gases were made up by pressure in the flasks A and B. The changeable reaction vessels P were connected through a ground joint. The explosive mixture in P was made up by mixing definite amounts of the mixtures from A and B. The explosion was then effected by heating the nichrome wire H. The initial pressure did not exceed 250–300 mm to avoid the bursting of P. The volume of P was accurately measured and comprised 2.8–3.2 liters.

After the explosion P was removed from the joint, and air allowed to enter. Then 20 cm^3 of 5% hydrogen peroxide was added and the flask made airtight. Several hours later the nitric acid was titrated with 0.2 N NaOH and methyl orange. The acid content of the peroxide was determined by control experiments and the amount found subtracted from the result. When NO is oxidized by the oxygen from air with the adsorption of water, HNO_3 and HNO_2 are formed; the HNO_2 is oxidized to HNO_3 by peroxide, this being necessary since HNO_2 decomposes methyl orange. In determining small

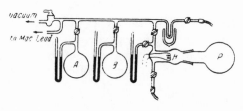

Fig. 1

quantities of NO we had recourse to adsorption by water, the nitrous acid in acid solution decomposing according to the reaction $3\,HNO_2 = HNO_3 + 2NO + H_2O$. The NO was again oxidized by oxygen from air, etc. In this case the titration was carried out after a lapse of 16–24 hours.

Results. As a result of the experiment we have a direct determination of the yield of nitric oxide. The volume of the flask being known, it is easy to express the yield in mole per cent with reference to one mole of the explosive mixture (i.e., in volume per cent with reference to the initial volume of the mixture). In Fig. 2 is plotted the yield of nitric acid expressed in mole per cent (on the ordinate axis) as a function of the per cent content of oxygen remaining after combustion $O_2' = O_2 - 0.5H_2$ for mixtures of H_2, O_2, N_2. The data for mixtures with different (48% H_2, 36% H_2, 32% H_2, 28% H_2) content of combustible are distinguished by different notations. The points to the right on the axis of abscissae give O_2' for mixtures containing no nitrogen, where nitric oxide cannot be formed. The points on the axis of ordinates give the amounts of NO formed in explosions of a mixture containing no excess of oxygen. Mixtures with a shortage of oxygen are conventionally plotted at negative O_2'. In these mixtures the nitric oxide is formed as a result of the dissociation of the combustion products and the amount produced is small. These mixtures are investigated in detail in Section 12 (Part III). As appears from the graph, in each series with a constant content of combustible, NO reaches a maximum about halfway between $O_2' = O$ and $N_2 = O$; this is natural since for NO to form there must be nitrogen and

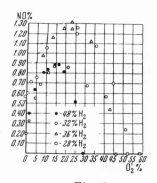

Fig. 2

free oxygen present.

At thermal equilibrium in the combustion products the relation

$$[NO] = C\sqrt{N_2 \cdot O_2'} \tag{3.1}$$

must hold. The brackets denote the equilibrium concentration of nitrogen oxide: the equilibrium constant C depends only on the temperature and is independent of the composition and pressure of the gas.

From the experimentally found content of NO in the explosion products we determined the behavior of a quantity C', defined by analogy with the equilibrium constant:

$$C' = \frac{NO}{\sqrt{N_2 \cdot O_2'}}; \quad NO = C'\sqrt{N_2 \cdot O_2'}. \tag{3.2}$$

In Fig. 3 this quantity is represented for the same experiments, the direct results of which are given in Fig. 2. As is evident from Fig. 3, C' actually varies little with the content of O_2 and N_2 in mixtures with a constant combustible content in which the explosion temperature may be considered in the first approximation to be constant. With an increase in the fuel content C' increases following the rise in the explosion temperature.[1]

A similar treatment of the data on the combustion of mixtures of ethylene with oxygen and nitrogen is shown in Fig. 4, where C' is plotted on the axis of ordinates and $O_2' = O_2 - 3C_2H_4$ on the axis of abscissae. These experiments bring out extremely clearly that C' is independent of O_2' and N_2 and grows with an increase in the combustible content.

We also investigated mixtures[2] of CO with O_2 and N_2; in this case the amounts of O_2' and N_2 varied within narrower limits which were insufficient for a complete check-up of the constancy of C'. However, in this case too, for a given fuel content the yield of nitric oxide was a maximum for the mixture in which $N_2 = O_2'$, just as was observed with the mixtures of hydrogen and ethylene, and as should be if C' is constant.

In the following graph the values of C' are compared for mixtures of different combustibles: hydrogen, ethylene, carbon monoxide[3] and methane;[4]

---

[1]We must point out here some deviations from this simple picture. C' increases sharply near the axis of ordinates; this can be explained by an increase in the actual amount of oxygen as compared with the value O_2' computed from the stoichiometry, as a result of the dissociation of the combustion products. The drop in C' at large oxygen content in the mixture containing 28% H_2 and in poorer explosive mixtures is due to the fact that an excess of oxygen lowers the combustion temperature (since its heat capacity is greater than that of nitrogen, and also because of the reaction $2H_2O + O_2 = 4OH$); it also depends on the reaction mechanism (see §11 and 12). The question has been analyzed quantitatively by Frank-Kamenetskiĭ [7].

[2]Dry gases were used; the carbon monoxide contained 1% H_2; the influence of small quantities of water vapor and hydrogen on the formation of NO was investigated in detail by Frank-Kamenetskiĭ [7].

[3]The carbon monoxide contained 1% H_2.

[4]The data on the formation of nitric oxide in explosions of mixtures of methane, oxygen and nitrogen were taken from an unpublished investigation by A. Kovalskiĭ and

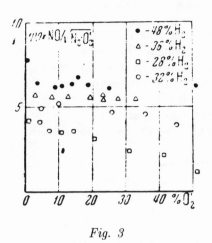

Fig. 3

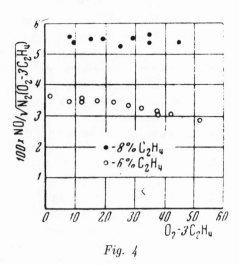

Fig. 4

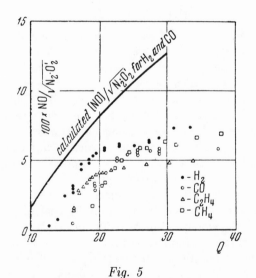

Fig. 5

for comparison we chose mixtures in which $N_2 = O_2'$; the amounts of oxygen and nitrogen in the combustion products are equal, which corresponds to the formation of a maximum amount of NO for a given fuel content. To compare the different fuels, the heat of combustion Q of the explosive mixture per mole of initial mixture[5] is plotted on the axis of abscissae; it is upon this

E. Mikhailova (Institute of Chemical Physics).

[5] The heats of combustion of the combustibles were calculated for combustion with the formation of water vapor.

quantity that the explosion temperature depends in the first approximation. On the axis of ordinates is plotted the quantity C'.

As appears from the diagram, the points for the different fuels fall fairly close together. The observed deviations can be explained in the first place by the fact that at equal heats of combustion of the initial mixture, the quantity and the heat capacity of the explosion products differ for different fuels. All the experiments represented in Fig. 5 were carried out at $p_0 = 200$ mm, except some experiments with hydrogen at $Q > 23$ kcal/mol, where $p_0 = 180$–150 mm. The highest value of C' equal to 0.089 was attained in the mixture 62% H_2, 34.5% O_2, 3.5% N_2.

The heavy line above the points represents the value of C, computed from the equilibrium concentration [NO] at the theoretical explosion temperature for H_2 and CO, and is discussed in §5.

The experimental material presented gives in the first approximation the general relations, governing the dependence of the amount of nitric oxide formed in an explosion on the composition of the mixture; this quantity is given by the expression:

$$NO = C'(Q)\sqrt{N_2 \cdot O_2'} \tag{3.3}$$

The graph of the function $C'(Q)$ should pass through the points of Fig. 5. The formula also describes the dependence of NO on the fuel content (on which C' and the quantity O_2' also depend) and its properties—heat value and composition. The above data and Fig. 5 refer to explosions at $p_0 = 200$ mm and to a volume of the vessel equaling 3 liters. Under other conditions the curve $C'(Q)$ is displaced but the general relationships remain valid. More exact and theoretically grounded formulas for the yield of nitric oxide in the explosion products will be derived in Part II, §8 and 9.

On the other hand, and this is the most important consideration here, it has been shown above that no stoichiometrical relations exist between the quantity of fuel and the nitric oxide formed, and that the specific chemical nature of the fuel is of no account but only its heat of combustion; it has been shown that the amount of nitric oxide formed is related to the oxygen content in the explosion products and not to the initial or average amount of oxygen during combustion. All this is rather difficult to reconcile with notions of chemical induction, the role of radiation, etc. and is a convincing argument in favor of the idea that the reaction of combustion is needed only to heat the mixture of O_2 and N_2 in which there then sets in a reversible thermal reaction (caused by the high temperature):

$$O_2 + N_2 \rightleftharpoons 2NO.$$

§4. Influence of the initial temperature on the yield of nitric oxide.

Experimental method. In the flask P (see Fig. 1) a mixture was made up at an initial pressure of 200 mm and a temperature of 20°C. Then an electric heater was fitted on to the flask which was heated together with the explosive mixture in it. We estimated the temperature of the mixture by the change in pressure. After a steady temperature was reached (varying in different experiments between 200–300°C) the mixture was exploded. The heater was then removed and the nitric oxides determined as in §3. It was shown by special experiments that even after the flask had been in the heater for fifteen minutes there was no loss of nitric oxide after the explosion. In some experiments the mixture was cooled before the explosion, the flask P being wrapped in a cloth and abundantly wetted with liquid nitrogen.

Results. Four series of experiments were carried out on hydrogen mixtures with $O_2' = N_2$, hydrogen–air mixtures, mixtures of 90% CO and 10% H_2 with $O_2^1 = N_2$, and mixtures of the same fuel with air. In each series parallel experiments were performed with and without heating. It can be seen from the description of the method that comparisons were drawn between mixtures with different initial pressures but the same density of the gas and, hence, with very close values of the explosion pressure.

In Fig. 6 is shown the dependence of the NO content on the H_2 content with and without heating in mixtures with $O_2' = N_2$; in Fig. 7 the same for the hydrogen–air mixture. Figures 8 and 9 represent the results of the same experiments as Figs. 6 and 7, the values $C' = NO/\sqrt{N_2 \cdot O_2'}$ (see §3) being plotted as ordinates against the heat of combustion of the mixture (see §3) as abscissae; in the experiments with heating the quantity of heat expended in raising the temperature of the mixture from 20°C to the heating temperature was added to the heat of combustion of the mixture. The heat of combustion of the mixture was expressed in percent of hydrogen; in the experiments with preheating to the true hydrogen content H_2 of the mixture was added a fictitious quantity H_2' equivalent to the heat supplied by the heater. As can be seen from Figs. 8 and 9 the results of the experiments carried out at different initial temperatures coincide.[6]

The results for carbon monoxide, which we do not quote for the sake of brevity, are exactly the same. Thus, preheating the mixture to 300° at a given content of N_2 and O_1' in the combustion products is equivalent to increasing the amount of H_2 by 2.7%. The increase in the amount of nitric oxide is especially large in the hydrogen–air mixture (almost double) since in these mixtures the quantity of H_2 is limited by the necessity for leaving a part of the oxygen to oxidize the nitrogen. The accuracy with which the

---

[6]With the exception of the mixtures with a high hydrogen content, 32% and higher, in which heating decreases and cooling increases the yield of nitric oxide. In these mixtures, as will be shown below, fundamental importance attaches to the decomposition of nitric oxide as the explosion products cool off from the temperature of the explosion to that of the vessel.

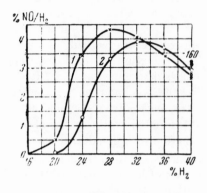

Fig. 6

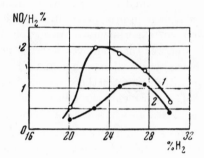

Fig. 7

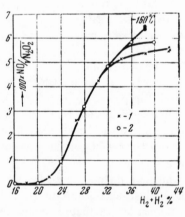

Fig. 8

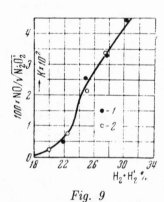

Fig. 9

dependence of C' on the heat value is borne out is greater in the experiments with preheating than when different fuels are compared (§3, Fig. 5), since the comparison is drawn between mixtures with very close compositions of the combustion products. For us special interest attaches to the equivalence between the heat expended in preheating and the chemical energy of the combustible with respect to the formation of nitric oxide as an argument against the assumption of chemical induction or the role of radiation, i.e., as an argument in support of the hypothesis of the thermal nature of the reaction.

§5. Comparison of the amounts of nitric oxide formed with the thermodynamical equilibrium quantities.

In the assumption of the thermal mechanism of reaction, the quantity of nitric oxide in the cooled products should not exceed the largest amount which can exist in chemical equilibrium with the oxygen and nitrogen. The equilibrium amount of nitric oxide increases with a rise in temperature. Hence the highest equilibrium quantity corresponds to the maximum temperature of the gas in the explosion and a comparison of the experimentally observed values of NO with the equilibrium values demands a knowledge of the maximum temperature of the explosion. The latter can be determined either from calculations based on the heat value and thermal capacity of the mixture, or experimentally from the maximum explosion pressure. The procedure worked out by us for accurately computing the dissociation, the theoretical combustion and explosion temperature and the equilibrium nitric oxide content [NO] at this temperature will be set forth elsewhere. The results of the computations for carbon monoxide and hydrogen at $p_0 = 200$ mm are shown in Fig. 5 for the same mixtures $2H_2 + O_2 + n(O_2 + N_2)$ and $2CO + O_2 + n(O_2 + N_2)$ and in the same coordinates (ratio $[C] = [NO]/\sqrt{N_2 \cdot O_2'}$ against heat of explosion per mole mixture) as the experimental data represented in the same figure. The ratio $[C]$ closely approaches the value of the equilibrium constant at maximum temperature, differing from it only insofar as the expression for the equilibrium constant contains the true concentrations, whereas the ratios plotted as ordinates contain the concentrations of nitrogen and oxygen computed from the stoichiometry without taking into account the change in the concentrations of oxygen and nitrogen in the formation of nitric oxide and in the reactions of dissociation. For a given mixture, i.e., for a given value of the abscissa, the observed and the equilibrium yields of NO for each fuel are in the same ratio as the values C' and $[C']$ plotted in Fig. 5. As appears from the figure, in all the cases the observed yield of NO does not exceed 60–70% of the equilibrium amount [NO]; at any rate NO/[NO] < 1 as was to be expected in a thermal reaction. The same holds both for our data and the data of other authors under different conditions of pressure, size of the vessel, etc. An exception is formed only by the data of Bone [4] on explosions of carbon monoxide at high initial pressure. The yields which he obtained are in some cases twice as great as the equilibrium values; however, Sadovnikov's [6] experiments did not confirm Bone's findings.

The temperature computed on the assumption that there are no heat losses is at any rate the upper limit of the true temperature. In rapidly burning mixtures (in particular, hydrogen mixtures) the temperature measured is close to the computed value. In slowly burning mixtures the temperature determined from the pressure is considerably lower than the computed value. For the latter case we shall discuss the relation between the computed temperature and the value determined from the pressure and the yield of nitric oxide on the basis of our own experimental data.

Experimental procedure. The experiments were carried out in a thick-walled cylindrical bomb of 0.5 liter volume with a spark plug and a membrane manometer fitted into the top. The deformation of the membrane by the pressure deflected a mirror which threw a reflected pencil of light on a revolving drum with photographic paper. The apparatus was kindly given us by N. Gussev to whom our thanks are due. We investigated mixtures of carbon monoxide and air.

The initial pressure was 5 kg per cm². The carbon monoxide was condensed in a trap cooled with liquid nitrogen, whence it was evaporated into a bomb at a pressure of 1–2 kg per cm². Air was then admitted from a container bringing the pressure up to 5 kg per cm². The explosion products were collected in a glass flask in which the nitric oxide content was determined as usual. Control experiments showed that a delay in transferring the explosion products to the flask reduced the yield of nitric oxide in a geometrical progression by 10% per minute. This was apparently due to corrosion and in succeeding experiments the time of transfer did not exceed one minute.

Results. In a typical experiment on the explosion of a mixture of 23.5% CO with air at $p_0 = 5$ kg/cm² we obtained 0.33% NO. The computed explosion temperature $T_m = 2560°$K and [NO] = 0.99%. However, experimentally we observed a maximum pressure of 2.5 kg/cm² which corresponds to an explosion temperature $T = 1670°$K; the equilibrium content of nitric oxide at 1670°K equals [NO] = 0.11%. It might appear that a triple yield of nitric oxide compared to the equilibrium value at the measured temperature even in one experiment would prove the existence of a non-thermal mechanism regardless of any other findings. Häusser [8] drew just such a conclusion from similar data. However, as we shall see later, it is incorrect.

Let us consider with more detail the explosion temperature in a closed volume and its determination from the maximum pressure. It must be taken into account that we deal here not with a chemical reaction of combustion proceeding simultaneously throughout the entire volume of the vessel, but with the propagation of a flame from the point of ignition.

Denote $T_0 + Q/c_p = T_p$. The quantity T_p is the combustion temperature at constant pressure. At the instant when a given element burns its temperature increases by Q/c_p, where A is the heat of combustion of the mixture, c_p—the heat capacity at constant pressure. In the absence of heat conduction, as a result of the adiabatic compression of a given element before and after its combustion, during the combustion of the other parts of the mixture, the final temperature of the given element is higher than T_p and differs for different elements.

An analysis and computation of the non-homogeneous distribution of the temperature were first carried out by Mache [9], after whom we shall call this phenomenon the Mache effect. Lewis and Elbe [10] computed the effect taking into consideration the variable heat capacity and dissociation.

The Mache effect has some influence on the quantitative relations governing the formation of nitric oxide in rapidly burning mixtures. In the absence of heat losses the maximum pressure remains practically unchanged if the Mache effect is taken into account.

In the above example of the combustion of carbon monoxide the time found experimentally in which the pressure reaches a maximum is 0.4 sec; the combustion time of a single element according to an estimate based on the theory of flame propagation [11, 12], is less than 0.001 sec. The loss of heat in 0.4 sec is considerable; the increase in pressure takes place so slowly that the state of the gas does not change adiabatically upon compression, and despite the compression each element cools after combustion. However, in 0.001 sec the loss of heat is negligibly small and each element in burning does attain the temperature T_p.

In order to judge whether or not the amount of nitric oxide formed in explosions of slowly burning mixtures exceeds the equilibrium value, it is necessary to compare the observed yield with the equilibrium quantity $[NO]_p$ at the temperature T_p which is unquestionably attained in each element, no matter how low may be the maximum average temperature T_{exp} determined from the pressure.[7] In the example considered (23.5% CO)

$$T_p = 2180°K \quad [NO]_p = 0.46\%, \quad \text{whereas} \quad NO = 0.33\%.$$

Thus, the observed yield is actually less than the equilibrium value and the example cited does not permit us to conclude that the nature of the reaction is not thermal.

The above example is typical. Throughout the series of experiments with CO–air mixtures at $p_0 = 5$ kg/cm^2, the time of combustion and the heat losses are important and the quantities in question can be arranged in order of increasing magnitude as follows:

$$[NO]_{exp} < NO < [NO]_p < [NO]. \tag{5.1}$$

---

[7]The different elements do not reach the temperature T_p simultaneously; hence when a given element is at T_p, there are others which have not yet burned and still others which reached T_p earlier and by this time have already partially cooled off. Therefore the average temperature throughout the mass which we estimate from the pressure can remain continually below T_p, so that if the temperature T_{exp} observed experimentally (from pressure measurements) is lower than T_p this fact does not invalidate our arguments or refute our assertion that the maximum temperature of each element of the gas is not lower than T_p.

If H_2 is substituted for 5 or 10% CO the propagation of the flame is accelerated, the pressure and measured temperature increase and—contrary to the data of Bone—the experimentally found amount of NO also slightly increases.

Example: 25.5% CO + air, $t = 0.26$ sec, $T_{exp} = 1960°$K, NO $= 0.30\%$; 23% CO + 2.5% H_2 + air, $t = 0.07$ sec, $T_{exp} = 2190°$K, NO $= 0.41\%$.

We see the reason for this increase in T_{exp} and NO in a more rapid rise in pressure and an approach to adiabatic compression. Interest attaches to a circumstance observed in the experiments at $p_0 = 200$ mm in the glass apparatus where the amount of nitric oxide changed when the point of ignition was transferred to the center of the flask P (Fig. 1): the conditions of the chemical reactions in the flame front do not change; it is only the conditions of the subsequent compression of the combustion products which are affected. The question has been investigated in detail by Frank-Kamenetskiĭ [7]. We shall merely observe here that the influence of the conditions of compression of the combustion products on the yield of nitric oxide proves the thermal nature of the reaction, since the compression is effective after combustion, when the reaction of the fuel with oxygen has ended. Such an influence would be impossible from the point of view of an induced reaction.

The conception of a chain reaction in which the breaking off of the chains on the walls depends on the dimensions of the vessel or on the location of the point of ignition is untenable, since the chemical reaction in the propagation of a flame takes place in a narrow zone, the thickness of which is thousands of times less than the dimensions of the vessel and the reaction time of an individual volume element and even of the entire explosive mixture in the vessel is many times less than the time of diffusion of active centers to the walls of the vessel.

§6. Experimental investigation of the oxidation of nitrogen upon combustion in a stream at constant pressure.[8]

Experimental procedure. The experiment consisted in burning coal gas for a prolonged period in a mixture of oxygen and nitrogen in a special burner. The combustion products moved with great speed: at different points in the flow the gases were collected for determinations of their composition and nitric oxide content. It was thus possible to determine the progress of the reaction of the coal gas with oxygen and of the oxidation of nitrogen in a mixture which had spent varying times in the apparatus from the entrance into the burner to the point of collection.

A schematic diagram of the apparatus is represented in Fig. 10. Air entered from the general compressed air pipe line. Oxygen was obtained

[8]It should be noted that the experiments described in the following were performed after those of Sadovnikov [6], who investigated the reaction kinetics in an explosion in a closed vessel.

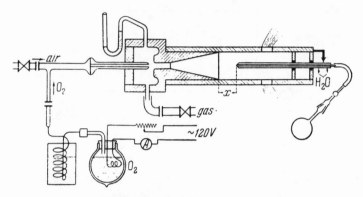

Fig. 10

by evaporation of liquid oxygen. The mixture of air and oxygen enters the central tube of the burner. Coal gas from the pipe line enters the annular space of the burner. A strong current of the oxygen–nitrogen mixture draws in the gas creating a reduced pressure in the annular space. The injector action is practically needed in order to guarantee sufficient gas consumption at the low pressure in the line (0.006 atm); at the same time strong injection insures good mixing of gas and air.

Combustion takes place in a widening cone. The reaction products enter further a stainless steel pipe. The products are collected for analysis through a capillary of 1.5 mm bore into a copper water-cooled tube. The gases were always collected at a point lying on the axis of the tube. The average composition of the coal gas was:

$$45\% \text{ CH}_4, \ 46\% \text{ CO}, \ 2.8\% \text{ N}_2, \ 0.2\% \text{ O}_2, \ 2\% \text{ CO}_2.$$

The nitric oxide was determined as usual—by absorption in water and titration. Control experiments on the combustion of the gas at low temperature and of analysis according to Kjeldahl (reduction of nitric and nitrous acids to NH_3 by the action of MgAl-alloy in alkaline solution, distillation and absorption of NH_3 by acid) convinced us that the gas contained no noticeable traces of sulphur and that the acidity as ordinarily determined depends on nitric and nitrous acids.

Results. In Fig. 11 the results of a typical experiment are represented by the points. The abscissae are the distances of the points of collection of the gas from the exit cross-section of the cone of the burner, the ordinates— the quantities of nitric oxide in the combustion products (in volume percent after condensation of the water vapor). The composition of the oxygen–air

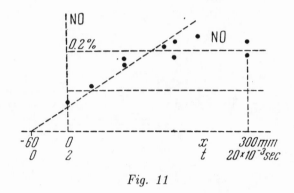

Fig. 11

mixture was 30.4% O_2, 69.6% N_2; the mixture was consumed at the rate of 1.78 normal liters per second. The combustion products contained 2.45% O_2, so that the quantity of oxygen fed to the burner was 1.075 times greater than that needed for complete combustion of the gas. The computed combustion temperature at constant pressure (without taking heat losses into account) was 2435°K. The rate of consumption of the oxygen–air mixture equaling 1.78 l/sec corresponds to a rate of 17.3 l/sec for the combustion products at 2435°K. Knowing the cross-section of the tube (11.2 cm^2), we find for the linear velocity of the combustion products about 15 m/sec. From this velocity we see the time spent in the tube by the gas collected at different points; the corresponding scale is shown in Fig. 11, bottom. This time includes the 0.002 sec spent in the cone. As appears from the figure, the formation of nitric oxide proceeds gradually for approximately 0.01 sec.

On the other hand, chemical analysis of the gases proves that already at the point—20 mm (i.e., when the collecting capillary reaches 20 mm into the cone the gas at this point having spent 0.001 sec in the cone before being collected) combustion of the gas has completely terminated, the oxygen content has fallen to 2.5%. Direct observation of the flame with the cone drawn out of the tube bears out the fact that combustion comes completely to an end in the cone, the rate of heat evolution being more than 500 000 000 cal/hr · m^3.

The experiment quite demonstratively shows that nitric oxide forms after combustion is over, thus disproving the theory of the induced oxidation of nitrogen.

Fig. 11 shows the theoretical curve for the thermal reaction of the formation of nitric oxide in the combustion products at 2435°K. The equilibrium amount of nitric oxide at this temperature is equal to 0.74%, the initial rate of formation of nitric oxide is 25 volume percent per second. As can be seen from the figure, the observed rate at first coincides closely with the computed value. Then a lag sets in which is explained naturally by the fall in the temperature of the combustion products in their motion through the tube. The theoretical curve is based on the expression for the reaction rate which we found in an investigation of the reaction kinetics in an explosion and which will be discussed in the following sections.

It has been pointed out in the literature that for nitrogen oxides to form during combustion in a burner at constant pressure increased pressure is necessary [3, 4, 13]. From this statement and the fact that the oxidation equilibrium of nitrogen is independent of the pressure far-reaching conclusions were drawn as to the non-thermal mechanism of the reaction. Actually in the papers mentioned the investigations were carried out on flames of small dimensions with an exceedingly small fuel consumption where the combustion products cooled off rapidly upon mixing with the surrounding gases. Increased pressure brought about an increase in the yield due to acceleration of the reaction and retardation of mixing. Our experiments, both those discussed above and earlier investigations, in which an even higher combustion temperature was reached (higher percent of oxygen and of hydrogen) have shown that with a greater expenditure of the fuel—1000 times greater than in the experiments cited [3, 4, 13]—the formation of nitric oxide during combustion can be observed at atmospheric pressure.

Part II. Reaction Kinetics in an Explosion

§7. Formation and decomposition of nitric oxide in an explosion.
Determination of the heat of activation of the reaction.

The hypothesis of the thermal nature of the oxidation of nitrogen leads to definite conclusions as to the relation between the formation and decomposition of nitric oxide in an explosion and the equilibrium concentration of nitric oxide at the explosion temperature. The equilibrium concentration may be computed independently from thermochemical data. The first task confronting us was to check up on these relations.

The degree to which the amount of nitric oxide obtained after the explosion approximates the equilibrium quantity characterizes the rate and time of the reaction in the explosion. By making the natural assumption that under similar conditions (volume of the vessel, initial pressure) the time of the reaction is the same, the influence of the explosion temperature on the reaction rate could be studied. The reaction $2NO + O_2 = 2NO_2$ in mixtures containing additional nitric oxide gave rise to some complications in the experimental technique and in the treatment of the results.

Experimental Procedure. The experiments were carried out in series with the same total concentration and the same initial density of the mixture at various initial contents of nitric oxide. Thus, in one series we investigated 190 mm of explosive mixture + 5 mm O_2 + 5 mm N_2, $p_0 = 200$ mm, and in another experiment 190 mm of explosive mixture + 10 mm NO, the initial pressure being $p_0 = 194.5$ mm due to the oxidation of NO to NO_2 and N_2O_2. Nitrogen dioxide reacts quickly with mercury, affecting the mercury manometers. The mixture was therefore prepared in the following order: first of all nitric oxide obtained by the action of nitric acid on copper and purified by fractional distillation was introduced into the vessel. The pressure of the nitric oxide was measured with a modified MacLeod manometer, which was then disconnected. The explosive mixture emerged from an intermediate vessel of known volume supplied with a gage; from the fall in the pressure of this vessel it was possible to determine the quantity of explosive mixture taken. The mixture was ignited ten minutes after it was made up. The initial pressure of the mixture containing NO_2 was not directly measured before the explosion. The yield of nitric oxide in the explosion products was determined as usual. The experiments were carried out with hydrogen mixtures, with side ignition (see Fig. 1).

As can be seen from the description of the method, immediately before the explosion the mixture contains no NO, but NO_2 and N_2O_2. Extrapolating to high temperatures Bodenstein's [14] data on the equilibrium and kinetics of the reaction $2NO + O_2$ it was possible to show that 99% of the NO_2 is decomposed to NO and O_2 in 10^{-3} sec at 2000°K and in $2 \cdot 10^{-5}$ sec at 3000°K.

The ease of decomposition of NO_2 is confirmed by observations on the discoloring of mixtures in explosions.

The decomposition of NO_2 to NO takes place considerably faster than the subsequent reactions of NO; we are thus justified in speaking of the reactions of NO, although actually the mixture contained NO_2 before the explosion. In computations of the theoretical explosion temperature of mixtures containing nitric oxide we have taken into consideration the energy spent on the reduction of NO_2 since this reaction precedes the reactions of NO which interest us.

Results. The results of a typical series of experiments are given below. The total composition of the mixture was 24% H_2, 38% O_2, 38% N_2, $p_0 = 200$ mm. NO_2 denotes the initial nitric oxide concentration (before the explosion); NO_k—the final concentration (after the explosion); NO_0 and NO_k are expressed in mm Hg at 290°K; the values in volume percent are obtained simply by dividing by 2.

Table 1

NO_0	0	1.29	2.26	3.50	7.16	10.08
	0.50	1.67	2.50	3.30	5.73	7.55

At small NO_0 the yield of nitric acid increases ($NO_k > NO_0$), at large NO_0 the decomposition of nitric acid predominates, $NO_k < NO_0$ and by interpolation we find the quantity—to be denoted {NO}—which remains unchanged during the explosion: $NO_0 = NO_k = \{NO\} = 2.90$ mm.

In the example quoted the theoretical explosion temperature T_m equals 2350°K and the equilibrium amount of nitric oxide at this temperature is $[NO] = 2.78$ mm. The quantities $[NO]$ and $\{NO\}$ are very close, as was to be expected on the assumption that formation and decomposition take place at a temperature close to T_m. At a higher explosion temperature, however, $\{NO\}$ is considerably less than $[NO]$. For example: with a mixture composition of 40% H_2, 40% O_2, 20% N_2, we obtain

Table 2

NO₀	0	2.63	7.15
NO_k	2.41	2.45	2.52

From the data in the table we find $\{NO\} = 2.447$ mm; $T_m = 2850$°K; $[NO] = 3.87$ mm. The lag between $\{NO\}$ and $[NO]$ indicates that con-

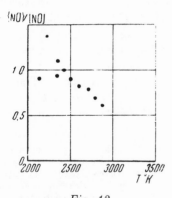

Fig. 12

siderable decomposition takes place after T_m during cooling. Fig. 12 represents the ratio $\{NO\}/[NO]$ as a function of T_m. As appears from the figure, at temperatures below 2500°K the ratio is on the average equal to unity— $\{NO\} \cong [NO]$, but at temperatures higher than 2500°K $\{NO\} < [NO]$.

In order to derive information on the reaction kinetics from results of the kind given in Tables 1 and 2 we write the equation of the reversible bimolecular reaction

$$\frac{dNO}{dt} = k' N_2 \cdot O_2' - k NO^2 \qquad (7.1)$$

where k' and k are the rate coefficients of formation and decomposition, respectively, and integrate (7.1) on the schematized assumption that the reaction goes on only in a definite time interval τ during which the temperature remains constant so that the factors k and k' are constant. From the experimental fact that the concentration $\{NO\}$ does not vary during the explosion it follows that

$$k' N_2 \cdot O_2' = k\{NO\}^2. \qquad (7.2)$$

Making this substitution and integrating the equation we find

$$k\tau = \frac{1}{2\{NO\}} \ln \frac{(\{NO\} - NO_0)(\{NO\} + NO_k)}{(\{NO\} + NO_0)(\{NO\} - NO_k)}. \qquad (7.3)$$

It can be assumed that in the series of experiments with the same initial pressure p_0 in vessels of the same dimensions the reaction time τ will be the same and will depend on the cooling rate. Experiments which are confined to an analysis of the final nitric oxide content in the explosion products can only yield in principle the product of the rate constant into the time $k\tau$, but not each of these quantities separately. By studying the dependence of $k\tau$ on the explosion temperature one can also find the temperature dependence of the activation heat of the reaction. In Fig. 13 $\log k\tau$ is plotted as a function of $1/T_m$. In each series with constant summary content of the mixture $k\tau$ was calculated by utilizing the data on the formation of nitric oxide, $NO_0 = 0$ (circles in Fig. 13) and on the decomposition of nitric oxide; $NO_0 = 7$ or 10 mm and is 2–5 times greater than $\{NO\}$ (triangles in Fig. 13).[9]

According to these data the heat of activation for the decomposition of nitric oxide, to which reaction the factor k refers, is $A = 82 \pm 10$ kcal/mole.[10] It should be especially noted that there is no systematic divergence between the data on the formation and on the decomposition of nitric oxide; this fact justifies the assumption that the rate of decomposition is directly proportional to the square of the nitric oxide concentration.[11] The investigation covered the temperature range from 2000°K to 2900°K in which the rate varies by a factor of 300. As appears from Fig. 13, except for the scattering due to the inevitable errors of the experiments and computations, the points actually do fall on a straight line in the coordinates $\lg k\tau$, $1/T_m$, i.e., the Arrhenius temperature dependence of the reaction rate holds. The thermodynamic relation between the rates of the direct and reverse reactions permits determining the heat of activation A' for the formation of nitric

[9]For mixtures with low hydrogen content and low explosion temperature in which it is difficult to determine $\{NO\}$ experimentally we utilized the equality $\{NO\} = [NO]$ established for temperatures below 2500°K (in the conditions of our experiments) and substituted in (7.3) the computed value of $[NO]$; then (7.3) coincides with Nernst formulas. The processing of the data obtained from mixtures with large hydrogen content and high explosion temperature in which $\{NO\} < [NO]$ contains assumptions which seem illogical or insufficiently grounded: (1) we refer the $k\tau$ found to the maximum explosion temperature T_m and substitute in (7.3) not the equilibrium quantity $[NO]$ at this temperature but the smaller quantity $\{NO\}$ which is in equilibrium with the oxygen and nitrogen at a lower temperature; (2) we assume that the reaction time τ of low temperature explosions is the same as in explosions at high temperatures where decomposition takes place during cooling. The exact theory of the kinetics of the reaction at varying temperature, set forth below (§9), justifies these assumptions.

[10]The absolute error ± 10 kcal/mole was determined as follows. We determined the values of A for which the mean square deviation of the experimental data from the relation $\ln k\tau = B - A/RT$ is twice as great as for the value of A which gives a minimum mean square deviation, the best value of B being chosen for each A. These values of A were found to be 72 and 92 kcal. The results of over 50 experiments were used in the calculations.

[11]The attempt to treat the data by proceeding from the assumption that the rate of decomposition of nitric oxide is proportional to its concentration leads to large systematic divergences.

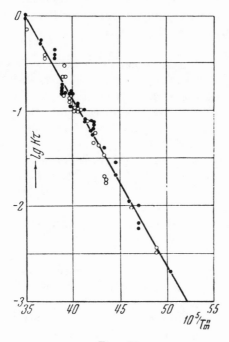

Fig. 13

oxide

$$A' = A + 2E = 82 \pm 10 + 2 \times 21.4 = 125 \pm 10 \, \text{kcal/mole}, \qquad (7.4)$$

where E is the heat of decomposition of one mole of nitric oxide.

Due to the unequal distribution of the temperature in the explosion products as a result of adiabatic compression during the explosion (Mache effect, see §5) the average values found for A and A' should be regarded as slightly too low. To take this into account and also to simplify the following calculations we shall further assume

$$A = 86 \, \text{kcal/mole} = 4E, \qquad (7.5)$$

$$A' = 129 \, \text{kcal/mole} = 6E. \qquad (7.6)$$

§8. The similarity theory of a chemical reaction in an explosion.

Let us consider the factors determining the yield of nitric oxide in the explosion products. They can readily be enumerated: the oxygen and nitrogen concentrations in the explosion products, the explosion temperature (which depends, in the first place, on the amount of combustible in the explosive mixture), the rate of cooling of the explosion products, the expression for the equilibrium constant $c = Be^{-E/RT}$ (heat of reaction E and factor B), the expression for the rate coefficient of the reaction of decomposition $k = De^{-A/RT}$ (heat of activation A and factor D). The rate coefficient for the formation of nitric oxide k' depends on k and c and hence need not be considered separately.[12]

We shall limit ourselves to practically instantaneous explosions. In this case it is possible to ignore the time in which the temperature reaches a maximum and the heat losses during that time. The maximum temperature may then be taken as the theoretical explosion temperature.

However, even after such simplifications the number of factors determining the yield of nitric oxide still remains large. The aim of the similarity theory consists in deducing the relationships between the quantities of interest in a simple form convenient for experimental investigation. The first step consists in computing the nitric oxide content [NO] at the theoretical temperature. This quantity which is expressed in terms of O_2', N_2, B, E, T_m provides a natural measurement unit for the yield of nitric oxide NO in the explosion products. We can thus define the dimensionless yield as the ratio NO/[NO].

The dimensionless yield of nitric oxide depends in a complicated way on the kinetics. Let us introduce the concept of the mobility of the reaction in an explosion, a quantity which characterizes the degree of approach to equilibrium; the mathematical definition of this quantity will be given below. If the mobility is low, the amount of NO formed during the explosion will be insufficient; the reaction will stop due to cooling and the yield will be small. On the contrary, at high mobility the dimensionless yield will be almost unity during the explosion, but the reaction will remain mobile during cooling when the equilibrium concentration falls off; the dimensionless yield will be less than unity due to decomposition in cooling.

The dimensionless yield depends on the mobility of the reaction; but we must decide on a measure of this mobility. We turn first to the rate coefficient of the reaction at the maximum temperature $k(T_m)$, for brevity denoted k_m. The dimensionless yield can, however, depend only on a dimensionless criterion. The rate coefficient of a bimolecular reaction has the dimensions $(\text{time})^{-1}$ $(\text{concentration})^{-1}$. Hence we obtain a dimensionless criterion of the mobility if we multiply the rate coefficient by the reaction time τ (which depends on the cooling rate) and the concentration. As concentration unit we have already chosen above the equilibrium concentration at the theoretical temperature [NO]. The dimensionless criterion of the mobility thus has the form

$$k_m \tau [\text{NO}] \qquad (8.1)$$

---

[12]It is possible, in principle, to consider k' instead of k, but as will be seen later, this would have complicated the formulas.

and the similarity theory leads to the relation

$$\frac{NO}{[NO]} = f(k_m \tau [NO]). \tag{8.2}$$

We adduce qualitative considerations to find the form of the function f in formula (8.2) for two limiting cases, when the mobility of the reaction is very low and when it is very high. The first case is realized in mixtures with a low fuel content. The dimensionless yield $NO/[NO]$ is then small throughout the process and decomposition of the nitric oxide may be neglected. The quantity of nitric oxide is directly proportional to the rate coefficient and to the reaction time. Neglecting decomposition of the nitric oxide we can write (cf. 7.1)

$$NO = k'_m \cdot N_2 \cdot O'_2 \tau = k_m [NO]^2 \tau \tag{8.3}$$

or, as a relation between dimensionless quantities,

$$\frac{NO}{[NO]} = k_m [NO] \tau. \tag{8.4}$$

Since the dimensionless yield is evidently less than unity and $NO < [NO]$, the last relation can only be satisfied if $k_m [NO] \tau < 1$.

The second limiting case when the mobility criterion is large $k_m [NO] \tau \gg 1$ is realized in mixtures rich in fuel and oxygen. To find the relation which interests us, we make use of the experimental fact that a preliminary addition of nitric oxide to these mixtures practically does not affect the yield in the cooled explosion products. Hence it can be concluded that in the given limiting case the absolute yield of nitric oxide does not depend on the maximum temperature attained during the explosion. Indeed let us compare two explosions under the same conditions; in one of them the temperature attains T_{m_1}, in the other T_{m_2}, where $T_{m_1} > T_{m_2}$ and both temperatures are so high that the mobility can be considered great.

In the second explosion in which the maximum temperature T_{m_2} is by condition higher, the temperature T_{m_1} is attained at some moment during the cooling. What is the difference between the condition of the products of the second explosion at this moment and the initial condition of the products of the first explosion when the temperatures of both are T_{m_1}?

The products of the second explosion at this moment have gone through a certain period of cooling and already contain a considerable amount of nitric oxide. The products of the first explosion contain no nitric oxide at first. But as was mentioned above, we know from special experiments on admixtures of nitric oxide to the initial mixture (see §7, Table 2) that the amount of nitric oxide in the mixture immediately after the explosion does not affect the yield of nitric oxide in the explosion products in the case of a high explosion temperature. Let us make use of the statement that the yield is independent of the maximum explosion temperature to determine the form of the function f in formula (8.2) at $k_m [NO] \tau \gg 1$. We write explicitly:

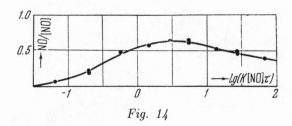

Fig. 14

$$NO = [NO]f(k_m[NO]\tau)$$
$$= c(T_m)\sqrt{N_2 \cdot O_2'}f(k_m(T_m)c(T_m)\sqrt{N_2 \cdot O_2'}\tau)$$
$$= Be^{-E/RT_m}\sqrt{N_2 \cdot O_2'}f(De^{-A/RT_m}Be^{-E/RT_m}\sqrt{N_2 \cdot O_2'}\tau). \quad (8.5)$$

In order that the quantity NO thus defined be independent of T_m it is necessary that the dependence on T_m of the equilibrium quantity [NO] and of the function f of the mobility criterion $k_m[NO]\tau$ cancel out. Hence f must be a power function with the exponent $-E/(A+E)$. Putting approximately $A = 4E = 86$ kcal/mole we obtain the limiting law for the yield of nitric oxide in mixtures with a high explosion temperature and high mobility of the reaction, which can be written in the form of a relation between dimensionless quantities

$$\frac{NO}{[NO]} = \alpha(k_m[NO]\tau)^{-1/3} \quad (8.6)$$

Substituting for [NO] and k_m their expressions (8.6) can be rewritten in a form showing that the yield is independent of the temperature T_m:

$$NO = \alpha(B\sqrt{N_2 \cdot O_2'})^{4/5}D^{-1/5}\tau^{-1/5}. \quad (8.7)$$

In these formulas α is a numerical constant which is to be found by integration (see §9), B and D are the pre-exponential factors in the expressions for the equilibrium constant and the rate coefficient.

Formulas (8.6) and (8.7) are an example of the power of the similarity theory which allows for deriving a by no means evident relation between the yield of nitric oxide and the reaction rate or the time τ from considerations of the temperature dependence of the yield of nitric oxide.

As appears from the formula one can find D from an experiment on an explosion at a very high temperature by measuring the nitric oxide yield in the products and the cooling rate without a knowledge of the temperature reached during the explosion. Due to the complicated temperature distribution in the reaction products (Mache effect, cf. §5) this method of computing D and the absolute value of the rate coefficient is extremely useful and has been used by us (see §10).

On the whole the similarity theory reduces the problem of the amount of nitric oxide formed in explosions to one series of experiments, which in principle is sufficient for determining the characteristic curve in dimensionless coordinates, NO/[NO] as a function of $k_m[NO]\tau$. Figure 14 represents such a curve obtained from experiments with hydrogen mixtures with equal oxygen and nitrogen content in the explosion products at $p_0 = 200$ mm and a volume of the vessel equal to 3 liters; the quantity $k_m[NO]\tau$ is plotted in the logarithmic scale. In plotting the curve we made use of the expression for the product $k\tau_1$ which we obtained in §7 from the data on the formation and decomposition of nitric oxide:[13]

$$k\tau_1 = 1.86 \times 10^6 e^{-82\,000/RT}. \tag{8.8}$$

Using this expression which refers to definite conditions and knowing how the cooling rate depends on these conditions (dimensions of the vessel, pressure, composition of the explosion products), we can compute the yield of nitric oxide in any case. T_m and [NO] are computed thermodynamically, $k_m\tau_1$ by formula (8.8); we take into consideration the difference between τ under the conditions of the given experiment and in the series of experiments on which the derivation of (8.8) was based. Then from the curve of Fig. 14 we find NO/[NO] corresponding to the new (because of the different τ) value of $k_m[NO]\tau$.

As appears from Fig. 14, NO/[NO] reaches a maximum of 0.65–0.68 at a given value of $k_m[NO]\tau$. A factor which makes $k_m[NO]\tau$ vary a given number of times will affect the yield of nitric oxide one way or another depending on the region in which the investigation is carried out, whether it lays before or after the maximum of the characteristic curve. Thus, it could have been foreseen that an increase in the pressure or in the dimensions of the vessel would increase the quantity of nitric oxide at small $k_m[NO]\tau$ (low explosion temperature) and decrease the yield at large $k_m[NO]\tau$. This conclusion was confirmed by Frank-Kamenetskiĭ [7] with respect to the dimensions of the vessel and by us (see §10) and later by Sadovnikov [6] over a much wider range with respect to the influence of the pressure.

§9. The mathematical theory of a reversible bimolecular reaction in an explosion.

We shall now bring to completion and numerically solve the problem of a bimolecular reaction under conditions of variable temperature, taking into consideration the different temperature dependence of the direct and the reverse reactions.

We shall limit ourselves to the case of an instantaneous explosion, i.e., an instantaneous rise in temperature to a maximum value, followed by cooling

---

[13]Expression (8.8) is the equation of the straight line drawn in Fig. 13, §7.

off of the explosion products. It is during cooling that the formation and decomposition of nitric oxide take place.

The differential equation of the reaction is

$$\frac{dNO}{dt} = k'N_2O_2' - kNO^2 \tag{9.1}$$

where k' and k depend on the temperature T and, hence, on the time t.

Let us introduce the instantaneous equilibrium amount of nitric oxide (NO), corresponding to the momentary value of the temperature; evidently

$$k'N_2O_2' - k(NO)^2 = 0 \tag{9.2}$$

so that

$$\frac{dNO}{dt} = k(NO)^2 - kNO^2. \tag{9.3}$$

We shall assume that the dependence of the rate coefficient k and the equilibrium concentration (NO) on the dimensionless time is the same for all the explosions:

$$k = k_m f_1\left(\frac{t}{\tau}\right); \qquad (NO) = [NO]f_2\left(\frac{t}{\tau}\right). \tag{9.4}$$

The dimensionless time t/τ is defined as the ratio of the time t to a characteristic constant τ have the dimensions of time to be defined later. Substituting, we obtain after simple mathematical transformations the equation

$$\frac{dNO/[NO]}{dt/\tau} = k_m[NO]\tau f_1\left(\frac{t}{\tau}\right)\left(f_2^2\left(\frac{t}{\tau}\right) - \frac{NO^2}{[NO]^2}\right). \tag{9.5}$$

Given the condition $NO/[NO] = 0$ at $t/\tau = 0$ we must find the limit of $NO/[NO]$ at $t/\tau \to \infty$. It follows from the form of the equation that at given f_1 and f_2 the quantity $NO/[NO]$ depends only on the product $k_m[NO]\tau$. This statement coincides with the content of the similarity theory introduced in the preceding section on dimensional considerations. We see that the validity of the theory depends on the existence of the functions f_1 and f_2, which must be the same for different explosions. But the rate coefficient and the equilibrium quantity depend on the temperature. Hence the form and the very existence of the functions f_1 and f_2 depend on the law of cooling. On the other hand, it is evident that the law of cooling must be formulated in such a manner that under the given conditions the cooling rate will depend only on the temperature of the gas, but not on the heat of activation of the reaction or the equilibrium quantity of nitric oxide. It can be shown that both conditions are satisfied only by the law:

$$\frac{dT}{dt} = -aT^2, \tag{9.6}$$

where T is the absolute temperature, a—a constant characterizing the rate of cooling. This law of cooling gives

$$\frac{1}{T} = \frac{1}{T_m} + at \tag{9.7}$$

$$k = De^{-A} = De^{-A/RT_m - aAt/R} = k_m e^{-aAt/R} \tag{9.8}$$

$$(NO) = \sqrt{N_2 \cdot O_2'} \cdot Be^{-E/RT} = \sqrt{n_2 \cdot O_2'} Be^{-E/RT_m - aEt/R} = [NO]e^{-aEt/R}. \tag{9.9}$$

We define the reaction time τ by the following formula:

$$\tau = \frac{R}{A'a} = \frac{R}{(A + 2E)a} = \frac{R}{6Ea}; \quad \tau = -\frac{RT^2}{6E}\left(\frac{dT}{dt}\right)^{-1}, \tag{9.10}$$

where A' is the heat of activation for the formation of nitric oxide, $A' = 6E$ [compare (7.6)].

Substituting the law of cooling and our definition of τ into Arrhenius' expression for the temperature dependence of the reaction rates and the equilibrium amount of nitric oxide, we find:

$$k' = k_m' e^{-t/\tau}; \quad k = k_m e^{-4t/6\tau}; \quad (NO) = [NO]e^{-t/6\tau}. \tag{9.11}$$

It should be recalled that k' is the rate coefficient for the formation and k for the decomposition of nitric oxide. Thus, the reaction time τ of formula (9.10) is the time in which the rate of the direct reaction of formation of nitric oxide decreases by a factor e.

The cooling law adopted actually does lead to universal laws for the variation of the constants and hence guarantees that the trend of the reaction will be similar in explosions with equal $k_m[NO]\tau$. Last but not least, the law of cooling (9.6) is in excellent agreement with the experimental data on the cooling of the explosion products over the wide range 1500–4000°K; it was proposed as an empirical law by Bone [4] and later confirmed by Sadovnikov [6].

The differential equation of the reversible reaction takes the form:

$$\frac{dNO/[NO]}{dt/\tau} = k_m[NO]\tau e^{-4t/6\tau}\left[e^{-2t/6\tau} - \frac{NO^2}{[NO]^2}\right]. \tag{9.12}$$

Given the initial condition

$$\frac{NO}{[NO]} = 0 \quad \text{at} \quad \frac{t}{\tau} = 0 \tag{9.12a}$$

we have to find by integration the quantity

$$\frac{NO}{[NO]} \quad \text{at} \quad \frac{t}{\tau} \to \infty \tag{9.12b}$$

as a function of the parameter $k_m[NO]\tau$.

The problem is easily solved for small $k_m[NO]\tau$. In this case NO/[NO] is also small, and neglecting the square of this quantity in the brackets, i.e., neglecting the decomposition of nitric oxide, we find

$$\text{at} \quad \frac{t}{\tau} \to \infty, \quad \frac{NO}{[NO]} = k_m[NO]\tau. \tag{9.13}$$

This formula coincides with the expression (8.4) of the preceding section and justifies our calling τ "the reaction time in the explosion."

To integrate the equation in the general case of arbitrary $k_m[NO]\tau$ we introduce new variables:

$$x = e^{-6/6\tau}(6k_m[NO]\tau)^{1/5}; \quad y = \frac{NO}{[NO]}(6k_m[NO]\tau)^{1/5} \qquad (9.14)$$

which permit rewriting equation (9.7) in the form

$$\frac{dy}{dx} = x^3(y^2 - x^2). \qquad (9.15)$$

For different values of the criterion of mobility we now have one differential equation with different initial conditions. Equation (9.15) is of the Riccatti type and cannot be integrated in elementary functions. It has, however, one remarkable property [15]. The complete integral of the equation with one arbitrary constant Γ can be written in the form:

$$y = \frac{\eta(\xi - \vartheta) - \Gamma\vartheta(\xi - \eta)}{\xi - \vartheta - \Gamma(\xi - \eta)}, \qquad (9.16)$$

if three particular integrals ξ, η, ϑ can be found. Such integrals were found by developing in a power series in x for $x \ll 1$, in $1/x$ for $x \gg 1$, and by numerical integration according to Runge's method in the intermediate region. Using (9.16) we plotted the entire curve

$$\frac{NO}{[NO]} = f(k_m[NO]\tau) \qquad (9.17)$$

as shown in Fig. 15. In the same figure are represented the limiting relationships obtained by integrating the equation. These are:

$$\frac{NO}{[NO]} = k_m[NO]\tau \qquad (9.18)$$

which coincides with the true curve at $k_m[NO]\tau \leq 0.3$ and

$$\frac{NO}{[NO]} = 0.745(k_m[NO]\tau)^{-1/5} \qquad (9.19)$$

which coincides with the true curve at $k_m[NO]\tau \geq 3$. In the same figure is also shown the curve of the maximum amount of nitric oxide $NO_{max}/[NO]$ produced in the course of each explosion at different $k_m[NO]\tau$ a quantity equal to NO_{max} could be obtained by instantaneously cooling the explosion products at the proper moment. The divergence between the curves $NO_{max}/[NO]$ and $NO/[NO]$ characterizes the decomposition of the nitric oxide during the natural uninterrupted cooling of the explosion products after the maximum concentration NO_{max} has been attained.

A comparison of the experimental data with the computations (the curves of Figs. 14 and 15, and the data on the decomposition of nitric oxide) shows a general qualitative agreement of experiment with the theory as regards both the form of the relationships and the numerical value of the maximum yield (experimental 0.65–0.68, computed 0.65). At the same time there are

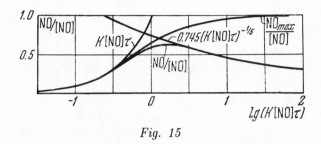

Fig. 15

definite quantitative discrepancies. For the concentration which remains constant during the explosion calculation gives at the low temperature limit

$$\frac{\{NO\}}{[NO]} \le \sqrt{\frac{2}{3}} = 0.82 \tag{9.20}$$

whereas experimentally (compare Fig. 12, §7)

$$\frac{\{NO\}}{[NO]} = 1. \tag{9.21}$$

At large $k_m[NO]\tau$ the calculated values of NO/[NO] are 12–13% lower than the experimental ones.

These deviations can be naturally explained by the non-homogeneous temperature distribution in the explosion products (Mache effect, §5). As a result of the large heat of activation for the formation of nitric oxide, the inequality of the temperature considerably increases the average reaction rate even at constant average temperature of the explosion products. The heat of activation for decomposition is smaller and the rate of this reaction increases to a lesser degree, which explains the increase in {NO}/[NO]. The true value of the rate coefficient is approximately half as great as was found in §7 and assumed in plotting Fig. 14, §8. Such a change in the coefficient corresponds to a displacement of the curve of Fig. 14 to the left, after which the experimental and computed curves coincide at large $k_m[NO]\tau$. At small $k_m[NO]\tau$ the experimental curve gives NO/[NO] twice as large as the theoretical value, which is to be explained by the influence of the Mache effect and agrees with the variation of {NO}/[NO] in these mixtures.

Without developing in detail the calculation of the reaction when taking into consideration the Mache effect, we shall draw a qualitative conclusion which will be of importance for the following: for reliable determination of the rate coefficient of the reaction one should use the data on explosions at high temperature comparing them with the limiting law with the exponent— $\frac{1}{5}$ [formula (9.19)] in which the maximum explosion temperature does not figure [compare §8, formulas (8.6) and (8.7)].

§10. Explosions at atmospheric pressure. Influence of the pressure and the absolute value of the coefficient D.

Experimental procedure. The explosions of the hydrogen mixtures at an initial pressure equal to atmospheric were carried out in a cylinder 330 mm long of 8 liters volume welded from a sheet of stainless chromium-nickel steel. The mixture fired by a spark plug screwed in the top. Before the experiment 100 cm^3 of water was poured into the cylinder so that the explosive mixture contained 2% water vapor.

After the explosion oxygen was introduced into the cylinder. Two hours later a probe of gas containing a small amount of unabsorbed oxides of nitrogen (about 5% of the nitric oxide formed) was taken in a glass flask and analyzed as usual. The water from the cylinder and the wash waters were poured into a measuring flask and the amount of nitric oxide absorbed in the cylinder was determined on a portion of the liquid. Due to corrosion of the metal the analytical procedure was modified: 50 cm^3 of solution was treated by 10 cm^3 of 5% H_2O_2 which converts HNO_2 into HNO_3 and the iron present in solution into the trivalent form. The hydrogen peroxide was decomposed by boiling and the boiling solution was titrated by 0.1 N NaOH in the presence of phenolphthalein. The indicator changed color only after coagulation of the ferric hydroxide.

The method was checked by analysis according to Kjeldahl (reduction of the fixed nitrogen to ammonia) and by analysis of measured amounts of nitric oxide let into the cylinder from a pipette.

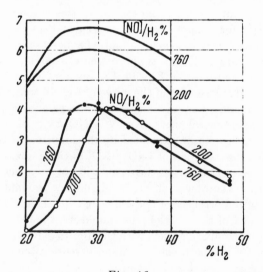

Fig. 16

Results. In Fig. 16 is plotted the ratio of the yield of nitric oxide to the combustible content in the mixture as a function of the latter at atmospheric pressure. The data refer to mixtures whose combustion products contain equal amounts of oxygen and nitrogen. For comparison we have plotted in the same figure the results of the experiments of §3 at a pressure of 200 mm. The maximum of NO/H_2 is displaced from 32% H_2 at 200 mm to 28% H_2 at 760 mm, without changing in height ($NO/H_2 = 4.1\%$). At a smaller content of H_2 the yield of nitric oxide increases with the pressure; at a greater H_2 content the yield decreases with a rise in pressure. This finding is in agreement with the theory.

The ratio NO/H_2 can be written in the form

$$\frac{NO}{H_2} = \frac{[NO]}{H_2} \cdot \frac{NO}{[NO]}. \tag{10.1}$$

According to the calculations of Frank-Kamenetskiĭ the first ratio $[NO]/H_2$ depends extremely feebly on the hydrogen content in the mixture. This quantity is also plotted in Fig. 16 for $p_0 = 760$ mm and $p_0 = 200$ mm. The position of the maximum of NO/H_2 is practically determined by the conditions under which $NO/[NO]$ reaches its maximum. At a given combustible content an increase in pressure brings about an increase in the criterion of mobility $k_m[NO]\tau$, due chiefly to an increase in the equilibrium concentration $[NO]$ which is proportional to the pressure, and to the greater time of cooling. It follows from the theory (§8 and 9) that at $k_m[NO]\tau$ less than the optimum value (giving a maximum of $NO/[NO] = 0.65$, an increase in the mobility increases the dimensionless yield; at $k_m[NO]\tau$ greater than the optimum value an increase in the mobility decreases the yield. Our data on the influence of the pressure are in complete agreement with these predictions of the theory.

Sadovnikov [6] measured the change in pressure in explosions of hydrogen mixtures at $p_0 = 760$ mm in a cylinder of 8 liters volume. His data on the cooling of the explosion products for different hydrogen mixtures all fit very well into an empirical formula containing the same constant for all mixtures:

$$\frac{1}{T} = \frac{1}{T_m} + 0.00163t, \tag{10.2}$$

where T is the actual temperature at a time t after the explosion; T_m is the maximum explosion temperature which is different for different mixtures. From formula (9.10) of the preceding section we find:

$$\tau = \frac{R}{6Ea} = \frac{1.98}{6 \cdot 21\,400 \cdot 0.00163} = 0.0095 \text{ sec.} \tag{10.3}$$

The reaction time for the formation of nitric oxide in an explosion is approximately equal to one hundredth of a second, which explains the difficulty of a direct experimental study of the trend of the process. Substituting in (8.7) the value of τ, the most probable experimental value of

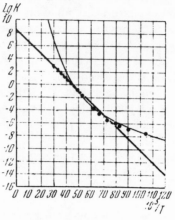

Fig. 17

$NO/(\sqrt{N_1 O_2'})^{4/5} = 0.133$, $B = 8/\sqrt{3} = 4.6$ which follows from the spectroscopic data, and the value of $\alpha = 0.745$ found in §9, we obtain:

$$D = 0.745^5 B^4 \tau^{-1} \left\{ \frac{NO}{(\sqrt{N_2 \cdot O_2'})^{4/5}} \right\}^{-5} = 3 \times 10^8 \; 1/\text{sec} \cdot \text{mm} \qquad (10.4)$$

$$k = D e^{-A/RT} = 3 \times 10^8 e^{-86\,000/RT} \; 1/\text{sec} \cdot \text{mm} \qquad (10.5)$$

The corresponding expression for the rate coefficient of the formation of nitric oxide has the form

$$k' = kc^2 = 6 \times 10^9 \, e^{-129\,000/RT} \; 1/\text{sec} \cdot \text{mm} \qquad (10.6)$$

where the time is measured in seconds, and all the concentrations in mm Hg at 17°C; 1 mm at 17°C equals 3.33×10^{16} molecules/cm^3 or 5.5×10^{-5} mole/liter.

The formation and decomposition of nitric oxide in heated vessels in a stream of a mixture of oxygen and nitrogen or nitric oxide were investigated by Jellineck [16]. In Fig. 17 Jellineck's experimental data are compared with our formula derived from an investigation of the reaction in an explosion. The logarithm of the constant of the bimolecular reaction of decomposition of nitric oxide is plotted as ordinate against the inverse temperature as abscissa. The circles represent Jellineck's data, the heavy straight line—our result (10.5). The dotted lines near the heavy line show approximately the temperature region in which lie our data on the reaction in explosions and the mean scattering of these data. The light curve represents the formula for the rate coefficient derived by Jellineck

$$k = e^{AT+B}. \qquad (10.7)$$

As appears from Fig. 17, our formula agrees well with Jellineck's data in the region of the highest temperatures attained by him. In the low temperature range investigated by Jellineck the reaction rate is considerably greater than follows from our formula; this can be explained by heterogeneous catalysis or by the small oxygen content in these experiments (see §11 and 12 on the role of oxygen).

Lastly, at explosion temperatures or in an electrical arc Jellineck's formula (10.7), which has no theoretical foundation, leads to a highly exaggerated value of the rate and is definitely inapplicable (see, e.g., Steinmetz [17]).

Part III. Mechanism of the Reaction of Oxygen with Nitrogen

§11. Mechanism of the oxidation of nitrogen.

The values found for the rate coefficients of the direct and reverse reactions are in good agreement both with the results of investigation of the reaction in an explosion and with other experimental data. However, as was first shown by P. Sadovnikov [6], though in a different form, the numerical value of the rate does not fit in with the concept of a bimolecular mechanism of the reaction. A simple calculation shows that the value of the rate coefficient for the formation of nitric oxide (10.6) corresponds to an effective cross-section of collision of an oxygen molecule with a nitrogen molecule of approximately 3×10^{-13} cm^2. This is 1000 times greater than the permissible value 3×10^{-16} cm^2. The divergence is many times in excess of the possible errors in the determination of the pre-exponential factor in the expression for the reaction rate.

On the suggestion of Prof. N. Semenov we investigated the assumption of a chain reaction of oxidation of nitrogen, which resolved all the difficulties.

Consider two reversible reactions:

$$O + N_2 = NO + N - 47 \, \text{kcal}, \tag{11.1}$$

$$N + O_2 = NO + O + 4 \, \text{kcal}. \tag{11.2}$$

Denote the rate coefficient of the first reaction by k_1, of the second reaction by k_2 and the coefficients of the corresponding reverse reactions by k_3 and k_4, respectively.

We now introduce notations for the equilibrium constants of the reactions (11.1) and (11.2), connect them with the rate coefficients and write down the expressions for the equilibrium constants at high temperature:[14]

$$c_1 = \frac{NO \cdot N}{N_2 \cdot O} = \frac{k_1}{k_3} = \frac{32}{9} e^{-47\,000/RT}, \tag{11.3}$$

---

[14]The expressions for c, c_1 and c_2 are derived from statistical mechanics on the approximate assumption that the masses of N and O are equal and the masses, moments of inertia and frequencies of N_2, NO and O_2 are equal; the differences in the symmetry and multiplicity of the main terms are taken into account. Comparison with the exact computations bears out the applicability of the simple formulas derived above.

$$c_2 = \frac{NO \cdot O}{O_2 \cdot N} = \frac{k_2}{k_4} = 6e^{-4\,000/RT}. \tag{11.4}$$

We have the identity

$$c_1 \cdot c_2 = \frac{k_1 \cdot k_2}{k_3 \cdot k_4} = \frac{NO \cdot N}{N_2 \cdot O} \cdot \frac{NO \cdot O}{O_2 \cdot N} = \frac{NO^2}{N_2 \cdot O_2} = c_2$$

where

$$c = \frac{NO}{\sqrt{N_2 \cdot O_2}} = \frac{8}{\sqrt{3}} e^{-21\,400/RT}. \tag{11.5}$$

The general kinetic equations are of the form:

$$\frac{dNO}{dt} = k_1 O \cdot N_2 + k_2 N \cdot O_2 - k_3 N \cdot NO - k_4 O \cdot NO, \tag{11.6}$$

$$\frac{dO}{dt} = -\frac{dN}{dt} = -k_1 O \cdot N_2 + k_2 N \cdot O_2 + k_3 N \cdot NO - k_4 O \cdot NO. \tag{11.7}$$

Applying the principle of stationary concentrations to the atomic gases O and N, we can put the last expression equal to zero and by means of it express the concentration of N in terms of O. Substituting into the preceding equation we obtain

$$\frac{dNO}{dt} = \frac{2.0}{k_2 O_2 + k_3 NO}(k_1 k_2 N_2 \cdot O_2 - k_3 k_4)NO^2). \tag{11.8}$$

We neglect the term K_3NO compared with K_2O_2 in the denominator[15] and use the relation between the rate coefficients and the equilibrium constants; finally we assume that the concentration of atomic oxygen is determined by the dissociation equilibrium of molecular oxygen

$$O = c_o \sqrt{O_2}. \tag{11.9}$$

On these assumptions we obtain for the reaction rate of nitric oxide

$$\frac{dNO}{dt} = 2c_0 k_1 N_2 \sqrt{O_2} - \frac{2c_0 k_1}{c^2 \sqrt{O_2}} NO^2 = \frac{2c_0 k_1}{c^2 \sqrt{O_2}}\{c^2 N_2 \cdot O_2 - NO^2\}$$

$$= \frac{2c_0 k_1}{c^2 \sqrt{O_2}}\{(NO)^2 - NO^2\}. \tag{11.10}$$

This expression coincides quite well with the equation for a reversible bimolecular reaction, which we used earlier. In particular, the rate of decomposition remains proportional to the square of the nitric oxide concentration, and the reaction rate is zero at the equilibrium concentration. The latter statement follows direction from the theory of non-branching chains, since for such chains the chemical energy of the centers cannot be utilized

---

[15] Both k_2 and k_3 are the rate coefficients of the exothermal reaction of an atom with a diatomic molecule, and are probably of the same order of magnitude. The concentration of oxygen is greater than that of nitric oxide. Finally, the proof of the permissibility of our approximation is furnished by the agreement of the conclusion with experiment (see §12).

for obtaining quantities of the main product (nitric oxide), exceeding thermodynamic equilibrium values.

As long as the oxygen concentration remains constant throughout the cooling of the explosion products, the previously developed similarity theory and the mathematical theory of a reversible bimolecular reaction in an explosion remain valid in their entirety.

The difference as compared with the former theory consists in that a more complex meaning is now attributed to k, the constant of the bimolecular reaction of decomposition of nitric oxide

$$k = \frac{2c_0 k_1}{c^2 \sqrt{O_2}}. \tag{11.11}$$

In the chain theory this "constant" depends on the oxygen concentration.

The experiments which we used to determine the rate coefficient of the reactions of formation and decomposition of nitric oxide referred to an oxygen concentration which varied within comparatively narrow limits near 150 mm O_2 at 17°C. Substituting in place of the old expression for the rate coefficient

$$k = 3 \times 10^8 e^{-86\,000/RT}; \quad k' = 6 \times 10^9 e^{-129\,000/RT} \tag{11.12}$$

the new expression:

$$k = \frac{2c_0 k_1}{c^2 \sqrt{O_2}} = \frac{3.6 \times 10^9}{\sqrt{O_2}} e^{-86\,000/RT}, \tag{11.13}$$

$$k' = \frac{2c_0 k_1}{\sqrt{O_2}} = \frac{7.5 \times 10^{10}}{\sqrt{O_2}} e^{-129\,000/RT}, \tag{11.14}$$

we practically do not influence the agreement between the calculated and experimental data. In particular, the theoretical curve for the formation of nitric oxide was computed from formula (11.13) for the experiment in a stream (§5, Fig. 11). Substituting the values of the equilibrium constant c and c_0, we obtain from a comparison of the experimental and theoretical (chain) expressions of the "constant" k the rate coefficient k_1 of the elementary reaction of atomic oxygen with molecular nitrogen (in seconds and millimeters Hg at 17°C):

$$k_1 = 1.1 \times 10^8 e^{-68\,000/RT} \; 1/\text{sec} \cdot \text{mm}. \tag{11.15}$$

The value of k_1 is quite reasonable; the heat of activation exceeds the heat of reaction for an endothermal reaction by 18 kcal:[16] the pre-exponential factor corresponds to a collision cross-section of less than $3 \times 10^{-16}\,\text{cm}^2$. The chain mechanism thus completely resolves the difficulties connected with the absolute value of the reaction rate.

The heat of activation for the oxidation of nitrogen, viz. (129 kcal), consists of the energy equal to 61 kcal which must be spent to form one atom

---

[16]These conclusions do not contradict the qualitative data of Spealman and Rodebush [18] on the reactions of atomic gases with the oxides of nitrogen.

of oxygen and which determines the temperature dependence of c_0 in the range 2000–3000°K and of the heat of activation for the reaction of an atom of oxygen with a molecule of nitrogen equal to 68 kcal.

It is highly interesting from the point of view of general kinetics that even for such simple and stable molecules as oxygen and nitrogen the chain mechanism presents kinetically an easier path for the reaction than the direction reaction following the collision of two molecules.

The mechanism of the dissociation of oxygen does not figure in our formulas. In the flame itself the content of atomic oxygen during the chemical reaction of the fuel with oxygen may exceed the equilibrium value. However, as appears from the formula (11.8) which does not contain the assumption of an equilibrium content of atomic oxygen, even if there is an excess of the latter, the yield of nitric oxide cannot exceed the amount required for equilibrium with oxygen and nitrogen; only the reaction rate will be increased. Due to the large heat of activation of the reaction of O with N_2 and the ease with which the reactions of O with the fuel take place, it should be expected that in the flame the time during which the concentration of atomic oxygen will exceed thermodynamical predictions will be insufficient for the formation of any noticeable amount of nitric oxides, so that the oxidation of the nitrogen will take place mainly after combustion, when the concentration of O is in thermodynamical equilibrium with the molecular oxygen present. Direct experiment fully bears out this point of view (§12, 6).

Thus the chain mechanism of the reaction of oxidation of nitrogen is in complete agreement with the thermal nature of the reaction, the rate and equilibrium of which depend solely on the temperature and the concentration of oxygen and nitrogen; the dependence, however, bears a more complicated character than was at first presumed.

§12. Influence of oxygen concentration on the reaction rate.

It has been shown in the preceding paragraph that the chain theory leads to an expression for the reaction rate which differs from that of the bimolecular mechanism only in that the rate coefficient depends on the oxygen content.

To check this conclusion it was necessary to study the reaction at extremely low oxygen content. At the same time it was desirable that the reaction of oxidation of nitrogen should not change the concentration of oxygen. We adopted a method entirely analogous to that of the buffer solutions used to keep constant the hydrogen ion concentration in solutions.

Mixtures with excess combustible (hydrogen) were exploded; the oxygen concentration in the explosion products of such mixtures was determined from the dissociation equilibrium of water vapor. All the experiments were carried out in glass flasks at an initial pressure of 200 mm; the nitric oxide was absorbed by water in the presence of air over a period of 24 hours. The composition of the mixture was determined by analysis before the explosion.

Consider an example: composition of the mixture 12.1% O_2; 28.5% H_2; 59.4% N_2; theoretical explosion temperature 2500°K; in the first approximation without considering dissociation reactions the composition of the combustion products is as follows: 24.2% H_2O, 4.3% H_2; 59.4% N_2 (the percentages are taken with reference to the initial mixture). The equilibrium amount of oxygen (at the theoretical temperature) is

$$[O_2] = c_5 \left(\frac{H_2O}{H_2}\right)^2 = 0.115 \text{ mm} = 0.06\%. \tag{12.1}$$

The equilibrium amount of nitric oxide at the equilibrium temperature is

$$[NO] = c\sqrt{[O_2] \cdot N_2} = c_6 \frac{H_2O}{H_2}\sqrt{N_2} = 0.25 \text{ mm} = 0.12\%, \tag{12.2}$$

$$c_6 = c\sqrt{c_5}. \tag{12.3}$$

The quantities c_5 and c_6 are the equilibrium constants of the reactions $2H_2O = 2H_2 + O_2$ and $H_2O + \frac{1}{2}N_2 = H_2 + NO$, respectively.[17]

The equilibrium amount of nitric oxide (0.12%) demands for its formation the entire available amount of oxygen (0.6%). However, in reality the formation of 0.12% NO merely causes a decrease in H_2O from 24.2% to 24.1% and an increase in H_2 from 4.3% to 4.4%; the equilibrium oxygen concentration changes by less than 5% of its magnitude.

The oxygen concentration in the above example is 300–1000 times less than in the previously investigated mixtures with excess oxygen. Hence the two expressions for the rate coefficient, (11.12) and (11.13) which practically coincide for all mixtures with excess oxygen, will give entirely different results in the example discussed: the chain mechanism (11.13) will give a reaction rate 30 times larger than the the bimolecular mechanism (11.12). Omitting the mathematical treatment we shall merely give the final results of the computations carried out on the two different assumptions of the bimolecular (index b) and chain (index c) mechanism and compare them with the experimental data.

Just as in §8 and 9 it can be shown that for mixtures with excess fuel there exist also two limiting regions:

1. A region of low mobility of the reaction in which the yield of nitric oxide is small compared with the equilibrium quantity and the decomposition of the nitric oxide may be neglected. In this region the yield of nitric oxide is proportional to the reaction rate. In mixtures with excess fuel the chain theory predicts a greater yield of nitric oxide than the bimolecular theory

$$NO = 1.1 \times 10^{15} e^{-245\,000/RT} \left(\frac{H_2O}{H_2}\right)^2 N_2, \tag{12.4a}$$

---

[17]The concentrations $[O_2] = 0.115$ mm and $[NO[= 0.25$ mm in formulas (12.1) and (12.2) are expressed not in partial pressures at the explosion temperature but in millimeters at 17°C, 1 mm at 17°C being equal to 5.5×10^{-8} mole/cm^3.

Table 3

Initial Mixture			Products		Computed Values*					Exper. Values*
$H_2\%$	$O_2\%$	$N_2\%$	$H_2O\%$	$H_2\%$	T_m	[NO]	$[O_2]$	NO (12.4a) **	NO (12.4b) ***	NO
27.8	11.2	61;	22.4	5.4	2300	0.08	0.02	0.00012	0.013	0.020
28.5	11.6	59.9	23.2	5.3	5435	0.12	0.04	0.0004	0.031	0.050
28.5	12.1	59.4	24.2	4.3	2500	0.24	0.11	0.0025	0.077	0.095
28.3	12.5	59.2	25.0	3.3	2540	0.42	0.30	0.0075	0.18	0.14
30.4	12.6	57.0	25.2	5.2	2550	0.28	0.14	0.0048	0.13	0.09

* $-$[NO], $[O_2]$ and the computed and measured concentrations of NO are given in millimeters of mercury at $17°C$.

** $-$values computed according to bimolecular mechanism

*** $-$values computed according to chain mechanism

$$NO = 2.5 \times 10^{12} e^{-187\,000/RT} \left(\frac{H_2O}{H_2}\right)^2 N_2. \qquad (12.4b)$$

The results of the computations and of the experiment are compared in Table 3. The experimental data support the chain mechanism.

2. A region of high mobility of the reaction in which the yield of nitric oxide in the explosion products does not depend on the maximum temperature attained but is determined by the conditions of decomposition during cooling. In this region the chain theory gives a greater (compared with the bimolecular theory) reaction rate for excess fuel and small oxygen concentration, predicts a small yield since here acceleration of the reaction furthers decomposition. We write down formulas analogous to (8.7) and (9.19)

$$NO = 0.11 \left(\frac{H_2O}{H_2}\right)^{0.5} N_2^{0.25} \qquad (12.5a)$$

$$NO = 0.11 \frac{H_2O}{H_2} N_2^{0.115} \qquad (12.5b)$$

The experimental and calculated data are compared in Table 4. Experiment quite unambiguously supports the proposed chain mechanism of the reaction with the participation of atomic O and N.

The present investigation was suggested by Prof. N. Semenov who also proposed the chain mechanism of the process. Our paper represents part of a large complex investigation carried out at the Institute of Chemical Physics in which took part G. Barskiĭ, N. Chirkov, D. Frank-Kamenetskiĭ, A. Kovalskiĭ, A. Nalbandian and P. Sadovnikov.

Table 4

Initial Mixture			Products		Computed Values*		Exp. Values*	
H₂%	O₂%	N₂%	H₂O%	H₂%	NO (12.5a) **	NO (12.5b) ***	NO	
39.5	17.6	42.9	35.2	4.3	1.0	0.15	0.17	
46.7	21.1	32.1	42.4	4.3	1.0	0.17	0.19	
53.9	24.8	21.3	49.6	4.3	1.0	0.19	0.20	
61.1	28.4	10.5	56.8	4.3	0.9	0.20	0.18	
43.2	16.6	40.2	33.2	10.0	0.62	0.058	0.062	0.066
50.0	20.0	30.0	40.0	10.0	0.64	0.063	0.062	0.070
56.6	23.3	20.1	46.6	10.0	0.62	0.075	0.062	0.074
66.8	28.4	4.8	56.8	10.0	0.47	0.080	0.060	0.062
43.2	16.6	40.2 + 2 mm NO			0.62	0.058	0.088	
50.0	20.0	30.0 + 2 mm NO			0.64	0.063	0.078	

   * –[NO], O_2 and the computed and measured concentrations of NO are given in millimeters of mercury at $17°C$.

  ** –values computed according to bimolecular mechanism

 *** –values computed according to chain mechanism

Summary

1. It is shown that the amount of nitric oxide formed in an explosion depends in the first approximation on the quantities of oxygen and nitrogen in the explosion products and on the heat of combustion of the explosive mixture, but not on the chemical nature of the combustible.

2. The heat communicated in pre-heating the mixture influences the yield of nitric oxide like the chemical energy of the explosion.

3. In all the experiments the yield of nitric oxide is less than the equilibrium amount at the maximum temperature attained.

4. When combustion takes place in a stream of gas, nitric oxide is formed after the fuel has completely burned.

5. The data set forth prove the thermal nature of the reaction: the formation of nitric oxide depends only on the high temperature of the explosion or flame but is not directly connected with combustion.

6. The decomposition of nitric oxide pre-mixed with the explosive mixture is investigated. The heats of activation for the formation and decomposition of nitric oxide are determined.

7. The similarity theory and an exact mathematical theory of the course of the reversible reaction during cooling of the explosion products are set forth. A universal relationship between the ratio of the yield to the equilibrium

value and the dimensionless criterion of mobility of the reaction describes the influence of all the various factors on the yield of nitric oxide.

8. The yield of nitric oxide in explosions of hydrogen mixtures at initial atmospheric pressure has been measured. The influence of the pressure in the range from 200 to 760 is in agreement with the theory.

9. The absolute values of the rate coefficients of the decomposition and formation of nitric oxide have been found.

10. The value found for the rate coefficients of the bimolecular reaction leads to an absurdly large collision cross-section of O_2 with N_2.

11. The chain mechanism of the reaction by way of oxygen and nitrogen atoms is discussed: $O + N_2 = NO + N$; $N + O_2 = NO + O$. It is found to be in agreement with the absolute value of the reaction rate and does not contradict the thermal nature of the reaction.

12. The dependence of the reaction rate on the oxygen content predicted on the basis of the chain mechanism is confirmed by experiments with small quantities of oxygen resulting from the dissociation of the combustion products in mixtures with excess fuel content.

13. The rate of the reaction between atomic oxygen and molecular nitrogen is determined from experimental observations on the reaction rate of nitrogen oxidation. The heat of activation of this reaction is 68 kcal per mole.

14. The final expression for the reaction rate with the most probable values of the constants can be written as follows:

$$\frac{dNO}{dt} = \frac{5 \cdot 10^{11}}{\sqrt{O_2}} e^{-86\,000/RT} \{O_2 \cdot N_2 \cdot 21 \cdot e^{-43\,000/RT} - NO^2\},$$

where the time is in seconds, the concentration in gram mole per liter. The expression in brackets is the difference between the squares of the equilibrium and actual concentrations of nitric oxide.

Institute of Chemical Physics
USSR Academy of Sciences, Moscow

*Received
September 5, 1945*

References

1. *Nernst W.*—Ztschr. anorg. und allg. Chem. **45**, 120 (1905); **49**, 213 (1906).
2. *Fink*—Ztschr. anorg. und allg. Chem. **45**, 116 (1906).
3. *Haber F., Coates M.*—Ztschr. phys. Chem. **69**, 338 (1909).
4. *Bone W.* Gaseous Combustion at High Pressures. London (1929).
5. *Elkenbard A. G., Genkina R. I., Polyakov M. V.*—Zh. Fiz. Khimii **13**, 464 (1939); *Genkina R. I., Poljakow M. V.*—Acta Physicochim. URSS **11**, 443 (1939).
6. *Sadovnikov P.*, to be published shortly.
7. *Frank-Kamenetskiĭ D. A.*—Acta Phys. Chim., in press.

8. *Häusser*—Stahl u. Eisen **41**, 956 (1921); Z. VDI **56**, 1157 (1912).

9. *Mache H.* Physik der Verbrennungserscheinungen, Leipzig (1918).

10. *Lewis, Elbe*—J. Chem. Phys. **3**, 63 (1935).

11. *Zeldovich Ya. B., Frank-Kamenetskiĭ D. A.*—Acta Phys. Chim. **9**, 341 (1938).

12. *Zeldovich Ya. B.* Theory of Combustion and Detonation of Gases (Russ.), Moscow-Leningrad (1944).

13. *Liveing, Dewar*—Proc. Roy. Soc. **48**, 217 (1830).

14. *Bodenstein M., Ramstetter*, Z. physik. Chem., **100**, (1922); *Bodenstein M., Lindner*, Z. physik Chem. **100**, 82 (1922).

15. *Weyr E.*—Abh. d. Böhm. Ges. d. Wiss. **6**, 8 (1875), cited from *Schlesinger* Lehrbuch der Differential Gleichungen (1912).

16. *Jellineck*—Z. anorg. Chem. **49**, 229 (1906).

17. *Steinmetz*—Chem. Met. Eng. **22**, 300, 500 (1920).

18. *Spealman, Rodebush*—J. Am. Chem. Soc. **57**, 1474 (1935).

26

Oxidation of Nitrogen
in Combustion and Explosions[*]

The formation of nitrogen oxides in combustion and explosions has been studied by many scientists, but to date there is no uniform opinion on the question of the nature and mechanism of this process. Nernst [1] believed that in an explosion, as a result of the high temperature, a direct bimolecular reaction of molecular oxygen with molecular nitrogen occurs. Haber [2] ascribed a catalytic role to the charged particles in the flame. Bone [3] considered that activation of nitrogen occurs in the flame. Polyakov [4, 5] considered nitrogen oxide to be a product of chain breaking in a chain combustion reaction.

Extensive research on the problem under a variety of conditions was carried out at the AS USSR Institute of Chemical Physics in 1935–1940. This research led to the quite definite conclusion of a thermal reaction, based on a chain mechanism, unrelated to combustion. The detailed results of this work will be published elsewhere in articles by the author, Frank-Kamenetskiĭ and Sadovnikov.[*2] Below we give a brief summary of the results obtained by the author.

From the standpoint of modern kinetics the problem falls into three parts:

1. Is the reaction induced or thermal, i.e., does the chemical reaction of combustion and explosion in some specific way—by electrons, radiation or active centers—influence the nitrogen oxidation, or do combustion and explosion only act as a heat source which maintains a high temperature, while the nitrogen oxidation reaction runs exactly as it would run in a corresponding mixture of oxygen and nitrogen at the same temperature using any other means of heating?

2. If the reaction is thermal, then what are its laws, i.e., what are the formal kinetics of the reaction and the dependence of the reaction rate on the concentrations of NO, NO_2, O_2 and the temperature? Finally,

3. The problem of the actual mechanism of the reaction, i.e., of the elementary events which lead to the observed formal kinetic laws.

[*]Doklady Akademii nauk SSSR **51** 3, 213–216 (1946).

[*2]These results were generalized in the book: *Zeldovich Ya. B., Sadovnikov P. Ya., Frank-Kamenetskiĭ D. A.* Okislenie azota pri gorenii [Oxidation of Nitrogen in Combustion].—*Editor's note.*

The following work was performed to solve the first question. The dependence of the amount of nitrogen oxide on the mixture composition in explosions of mixtures of various fuels (H_2, CO, C_2H_4, CH_4) with oxygen and nitrogen was studied. In all but the most diluted mixtures, for a given fuel percentage, the amount of NO was proportional to $\sqrt{N_2 \cdot O_2'}$, where O_2' is the concentration of oxygen in the combustion products. The ratio $NO/\sqrt{N_2 \cdot O_2'}$ uniformly grows as the fuel content is increased, i.e., as the temperature of combustion is increased. At identical temperatures it was approximately identical for different fuels. Such complete absence of any chemical specificity is an argument in favor of the thermal character of nitrogen oxidation.

The effect of the initial temperature of the explosion mixture on the amount of nitrogen oxide was studied using the mixtures $H_2 + O_2 + N_2$ and $CO + O_2 + N_2$. It was shown that the ratio $NO/\sqrt{N_2 \cdot O_2'}$ depends on the bulk energy (the sum of the thermal and chemical energy) of the mixture, so that heating is completely equivalent to increasing the fuel content, in accord with the thermal point of view. Exact calculations of the explosion temperature and the equilibrium chemical composition were carried out, in particular, of the equilibrium amount of nitrogen oxide at the explosion temperature. Using mixtures of carbon monoxide and air at an initial pressure of 5 atm, the case of slow combustion with large thermal losses during explosion was studied experimentally and theoretically and the relation was found between the average temperature (measured by the pressure) and the actual temperature of the gas. The amounts of nitrogen oxide in the explosion products, measured by chemical analysis, in all cases were less than the equilibrium amounts at maximum temperature, which is in accord with our understanding of thermal reaction.

After the work by Sadovnikov, who used special methods to study the time-development of nitrogen oxide formation and decomposition in explosion, we carried out an investigation of nitrogen oxide formation in a burner flame during the combustion of a luminescent gas mixed with oxygen and air. By taking gas from various places in the tube for analysis we were able to show that nitrogen oxide formation begins and continues after combustion is complete in a zone in which there are no fuel materials and no combustion reaction. These experiments directly prove the thermal nature of the reaction.

Solution of the problem of the magnitude of the rate and the dependence of the reaction rate on the temperature and, in a first approximation, on the composition of the reacting gas, required a number of experiments and calculations due to the complex conditions of an explosion. Explosions were performed of mixtures with known amounts of nitrogen oxide added beforehand. The nitrogen oxide content was determined after the explosion and, in particular, the amount of nitrogen oxide which remained unchanged in the explosion. The rate of the decomposition reaction of NO is proportional to the square of the concentration of NO. We determined the activation heat of nitrogen oxide decomposition, near 82 kcal/mole, and the activation heat of nitrogen oxide formation, near 125 kcal/mole. In mixtures with low fuel content (a low explosion temperature T_{max}) the amount of nitrogen oxide unchanged in the explosion was quite close to the equilibrium amount [NO] for T_{max}. The amount of nitrogen oxide which formed without preliminary addition of nitrogen oxide was small in this case due to the fact that the reaction was slow and cooling began before equilibrium was achieved. In fuel-rich mixtures (a high explosion temperature T_{max}), significant reverse decomposition of nitrogen oxide occurred during cooling.

Further development of the problem required that the specifics of the reversible reaction in explosion, i.e., under variable temperature conditions, be taken into account. A similarity theory was developed for a chemical reaction in an explosion: the dimensionless concentration of nitrogen oxide (which we defined as the ratio of the concentration to the equilibrium concentration at the highest temperature) represents a universal function of another dimensionless criterion which characterizes the mobility of the reaction in a given explosion. This criterion is constructed from the rate constant, the equilibrium concentration and the cooling time. Limiting expressions of the universal function were found both in the region in which the reaction does not have time to occur and the increase in reaction mobility raises the nitrogen concentration and in the region in which the reaction mobility is high and during cooling significant decomposition (increasing with the mobility) of the nitrogen oxide occurs.

For a cooling law which completely satisfies the requirements of the similarity theory we developed in full a mathematical theory of reversible bimolecular reaction. All of the numerical calculations for the formation of nitrogen oxide were performed, and the above universal function was found. The formation of nitrogen oxide in explosions of hydrogen mixtures at atmospheric initial pressure was studied. It was shown that the effect of a change in pressure between 200 mm and 1 atm is in agreement with the theory. Using data on the cooling rate of the explosion products and on the amount of nitrogen oxide in the explosion products, the absolute values of the reaction rates of formation and decomposition of nitrogen oxide were calculated using the theory. It was predicted that, just as with a pressure increase, increasing the size of the vessel increases the output of nitrogen oxide in lean fuel mixtures, and decreases the nitrogen oxide output in fuel-rich mixtures due to the change in the cooling time. This conclusion was confirmed by Frank-Kamenetskiĭ. Thus, a formal kinetics of the reaction was developed which also take into account the complicated conditions in which it runs. This allowed us to explain the dependencies observed in experiments and to find the absolute value of the reaction rate using experimental data.

It was shown that the absolute value of the reaction rate is not in accord with the assumption of a bimolecular mechanism since the pre-exponential factor far exceeds the gas-kinetic number of collisions between molecules of nitrogen and oxygen (this was first noted by Sadovnikov). Semenov's proposal of a chain mechanism for the following process was analyzed:

$$O + N_2 = NO + N,$$

$$N + O_2 = NO + O.$$

A chain mechanism leads to the expression for the reaction rate

$$\frac{dNO}{dt} = O_2^{-1/2}(K''N_2 \cdot O_2 - K'''NO^2),$$

in which K'' and K''' represent complex expressions comprised of the rate constants of the elementary events. The relation here between K'' and K''' is such that the reaction rate identically vanishes when the concentrations of nitrogen oxide, nitrogen and oxygen satisfy the condition of thermodynamic equilibrium. The absolute value of the reaction rate is in complete agreement with a chain mechanism. Since the decomposition rate is proportional to the square of the nitrogen oxide concentration, all of the conclusions from the formal kinetic calculations made under the assumption of a bimolecular reaction remain valid. A chain mechanism for the reaction does not contradict the idea of the thermal nature of the reaction since the concentration of active centers of the reaction—atoms of oxygen and nitrogen—is completely determined by the concentrations of molecular oxygen and nitrogen and the instantaneous temperature of the gas, while the fuel combustion reaction has no direct effect on the nitrogen oxidation reaction. The chain mechanism leads to a dependence of the rate constant, which would be invariable for a simple bimolecular reaction, on the oxygen content.

Experiments were performed with mixtures of very low oxygen content. By analogy with the method of creation and maintenance of small concentrations of hydrogen ions in buffer mixtures and in electrochemistry, we used mixtures with a fuel surplus in which the small concentration of oxygen was determined by a mobile dissociation equilibrium $2H_2O \rightleftharpoons 2H_2O + O_2$. The experiments showed with certainty the validity of the new expression for the reaction rate which follows from the chain mechanism.

Thanks to a thorough study of the conditions of reaction in explosion, the determination of the amount of a substance in the explosion products has become an exact kinetic method. This method has allowed us to study the kinetics of a reaction which runs within thousandths and hundredths of a second at temperatures reaching 2000–3000°K, and to completely clarify the laws and mechanism of the chemical reaction of the simplest and most abundant substances—oxygen and nitrogen.

Received
September 25, 1945

References

1. *Nernst W.*—Ztschr. anorg. und allg. Chem. **45**, 120 (1905).
2. *Haber F., Coates M.*—Ztschr. phys. Chem. **69**, 338 (1909).
3. *Bone W.* Gaseous combustion at high pressures. London (1929).
4. *Elkenbard A. G., Genkina R. I., Polyakov M. V.*—Zh. Fiz. Khimii **13**, 464 (1939).
5. *Genkina R. I., Poljakow M. V.*—Acta Physicochim. URSS **11**, 443 (1939).

Commentary*

The value of these two works for combustion science is great from various points of view. A detailed study of the kinetics of nitrogen oxidation at high temperatures is performed, and a mechanism proved and studied of chain non-branching reaction with active centers, atoms of oxygen and nitrogen, which plays a basic role in the formation and decomposition of nitrogen oxide. This mechanism is thoroughly verified experimentally—the results of the experiments and comparison with the theory are the subject of other sections of the monograph "Oxidation of Nitrogen in Combustion" which do not enter the present book and also of a detailed article by Ya.B.[1]

In addition, experiments following the theoretical scheme enabled Ya.B. to determine the activation energies of the elementary events which were essential for further study of nitrogen oxidation. The method proposed and theoretically justified of "hardening" the reaction products by rapidly cooling them (a full mathematical explanation of the "hardening" method is given) subsequently found broad application for the analysis of intermediate and final reaction products. Under conditions of varying temperature a description of the oxidation process on the basis of a number of criteria is introduced. As a similarity criterion the ratio of characteristic times is proposed: the relaxation time of the reaction and the cooling time of the combustible mixture. A universal function is found for nitrogen oxide output (with respect to the equilibrium value at maximum temperature) for a self-similar cooling law.

In chemical kinetics this kind of investigation is unique. The process occurs not only in combustion. In strong explosions in the atmosphere formation of nitrogen oxide plays an important role, changing the optical properties of the

---

*Article 25 in the original Soviet edition was an excerpt from the book: *Zeldovich Ya. B., Sadovnikov P. Ya., Frank-Kamenetskiĭ D. A.* Okislenie azota pri gorenii [Oxidation of Nitrogen in Combustion]. Moscow, Leningrad: Izd-vo AN SSSR, 1947, 147 p. In the current edition, this has been replaced with Ref. 1 below.

[1] *Zeldovich Ya. B.*—Acta Physicochim. URSS **21**, 577–628 (1946).

explosion products. Further oxidation of the nitrogen oxide leads to the formation of nitrogen dioxide which actively emits radiation in the visible portion of the spectrum.

The kinetics of nitrogen oxide formation and its further oxidation to the dioxide in a strong explosion was analyzed by Yu. P. Raizer[2] and presented in a monograph by Ya.B. and Yu. P. Raizer.[2a] Oxidation of nitrogen also occurs in impulse and continuous optical discharges in the air; the high plasma temperature in them ($\sim 15\,000°K$) is maintained by a laser beam. In optical discharge convection leads to rapid cooling of the plasma and to hardening of the nitrogen oxide.[3]

Nitrogen oxide forms as an intermediate product in the combustion of ballistic powders based on nitrocellulose. In the gasification of the powder the nitrogen compounds decompose with the formation of an NO_2 group which enters into reactions with other decomposition products, oxidizes, and yields NO. At low pressures the NO remains in the combustion products; at high pressures it is transformed into N_2. This slow concluding stage of a multi-stage process, as Ya.B. indicates, may pass through the reaction

$$2NO + 2H_2 = N_2 + 2H_2O, \qquad 2NO + 2CO = N_2 + 2CO_2.$$

The slow rate of transformation is confirmed by experiments on flame propagation in mixtures of NO with stable combustible gases (hydrogen, carbon monoxide, methane); some of them were performed by Ya.B. together with Yu. Kh. Shaulov.[4] NO and NO_2 have the property of rapidly reacting with unsaturated hydrocarbons and metal compounds introduced into the powder as catalysts—evidently, we should seek here one of the reasons for catalysis in the gas phase of the powder. The corresponding experiments were performed by A. I. Rozlovskiĭ.[5] Similar problems are considered in the monograph by A. P. Glazkova.[6]

Recently, work on nitrogen oxidation in the combustion of a fuel mixed with air has become extensive in connection with pollution of the environment by toxic combustion products (including NO). An good review of this research was done by A. N. Heiherst and I. M. Vincent,[7] and by A. Macek for coal combustion.[8] Detailed experimental and theoretical work on the formation of nitrogen oxides in turbulent gas flames was performed by P. Moreau and R. Borghi.[9]

In Ya.B.'s work it is assumed that atomic oxygen is in thermodynamic equilibrium with molecular oxygen. The combustion reaction may lead to a superequilibrium concentration of O atoms, which in turn accelerates formation of NO.

[2] *Raizer Yu. P.*—Zh. Fiz. Khimii **33**, 700–709 (1959).

[2a] *Zeldovich Ya. B., Raizer Yu. P.* Physics of Shock Waves and High-Temperature Hydrodynamic Phenomena. New York: Academic Press (1960).

[3] *Raizer Yu. P.* Lazernaĭa iskra i rasprostranenie razrĭadov [Laser Spark and Propagation of Discharges]. Moscow: Nauka, 406 p. (1974).

[4] *Zeldovich Ya. B., Shaulov Yu. Kh.*—Zh. Fiz. Khimii **20**, 1359–1362 (1946).

[5] *Rozlovskiĭ A. I.*—Dokl. AN SSSR **196**, 152–157 (1971).

[6] *Glazkova A. P.* Kataliz goreniĭa vzryvchatykh veshchestv [Catalysis of Combustion of Explosives]. Moscow: Nauka, 531 p. (1976).

[7] *Heiherst A. N., Vincent I. M.*—Progr. Energy Combust. Sci. **6**, 35–43 (1980).

[8] *Macek A.*—In: 17th Intern. Symp. on Combustion. Leeds, 65–71 (1978).

[9] *Moreau P., Borghi R.*—J. Energy **5**, 3–12 (1981).

The excess NO thus formed, compared to the amount calculated by the formulas, Ya.B. calls "prompt" NO. Its role in technical applications is, as a rule, insignificant.

A turbulent flame as a source of NO was also studied by V. Ya. Basevich, S. M. Kogarko, V. Yu. Pashkov and A. N. Tyurin[10] and by V. R. Kuznetsov.[11]

An important problem in research on NO formation in the combustion of coal and other nitrogen-containing fuels is to find the amounts of oxides supplied by atmospheric nitrogen and by nitrogen in the fuel. H. Semerjian and A. Vranos[12] consider that in the latter case the governing role belongs to the reactions

$$NH + OH = NO + H_2, \qquad HNO + O = NO + OH.$$

Various methods for lowering the concentrations of NO in combustion products have been proposed: lowering the temperature,[13] which decreases NO output from both sources; introduction of ammonia additives,[14,15] which reacts readily with NO; and use of a plasma jet of nitrogen atoms.[16] In coal combustion the nitrogen oxide is removed in a heterogeneous reaction on the surface of the coal particles.[17]

[10] Basevich V. Ya., Kogarko S. M., Pashkov V. Yu., Tyurin A. N.—Khim. Fiz. 11, 1557–1563 (1982).

[11] Kuznetsov V. R.—Izv. AN SSSR, MZhG 6, 3–9 (1982); Buriko Yu. Ya., Kuznetsov V. R.—Fiz. Goren. i Vzryva 19, 71–79 (1983).

[12] Semerjian H., Vranos A.—In: 16th Intern. Symp. on Combustion. Pittsburgh, 169–174 (1977).

[13] Kogarko S. M., Basevich V. Ya.—Fiz. Goren. i Vzryva 17, 3–11 (1981).

[14] Lucas D., Brown N.—Combust. and Flame 47, 219–224 (1982).

[15] Miller J. A., Branch M. K., Kee R. J.—Combust. and Flame 43, 81–88 (1981).

[16] Bechbahani H. F., Worris A. M., Weinberg F. J.—Combust. Sci. and Technol. 30, 289–294 (1983).

[17] Wendt J., Pershing D., Lee J., Glass J.—In: 17th Intern. Symp. on Combustion. Leeds, 77–85 (1978).

IV

Detonation

27

On the Theory of Detonation Propagation in Gaseous Systems*

In this article existing theories of detonation are critically examined. It is shown that the considerations which are used to select the steady value of the detonation velocity are unconvincing. In connection with the problem of a chemical reaction in a detonation wave, we refute objections against the idea of gas ignition by a shock wave which were expressed in the nineteenth century by Le Chatelier and Vieille. On the basis of this idea, we are able to rigorously justify the existing method of calculating the detonation velocity. We analyze the distribution of temperature, pressure and mass velocity at the front of a propagating detonation wave over the course of a chemical reaction. Assuming the absence of losses, pure compression of a gas in a shock wave before the beginning of the chemical reaction develops a temperature which is close to the combustion temperature of the given mixture at constant pressure. In addition, the specific volume and pressure are related by the equation of a line which passes through the point corresponding to the initial state of the gas (the Todes line):

$$p - p_0 = -D^2 \frac{v - v_0}{v_0^2}.$$

We study the influence of hydrodynamic drag and heat losses. Losses during the chemical reaction decrease the propagation velocity of detonation, which leads to a temperature decrease in the shock wave igniting the gas,

*Zhurnal eksperimentalnoĭ i teoreticheskoĭ fiziki **10** 5, 542–568 (1940).

a drop in the chemical reaction rate, and a further increase in losses. From these considerations we establish the existence of a limit of detonation propagation, where the maximum decrease in detonation velocity compared to the calculated theoretical value is small for reactions whose rate grows with the temperature. At the limit the chemical reaction is extended to a distance equal to several tube diameters. The theory developed allows us to calculate for a chemical reaction with known kinetics the detonation propagation velocity under real conditions with losses accounted for, the limits of steady detonation propagation the distribution of pressure, temperature, mass velocity, specific volume and concentration in a steadily propagating detonation wave. Conclusions of the theory relating to the detonation front structure and to the course of cooling and braking of the products are compared with experimental data.

Practical application of the one-dimensional theory developed to the calculation of the effects of losses on the detonation velocity is limited by the fact that even at the limit the reaction time is small and heat transfer and braking do not cover the entire cross-section of the tube. At the same time, in the vast majority of cases, long before the limit is reached one observes the so-called spin—a spiral-like or periodic propagation of detonation which is not described by our theory. Some thoughts are given concerning the dimensionless criteria on which the spin depends.

§1. *Propagation Velocity of Detonation. Classical Theory*

In accordance with the classical theory of propagation of detonation of Chapman [1], Schuster [2], Crussard [4], constructed by analogy with the theory of shock waves of Riemann [5], Hugoniot [6], Rayleigh [7], Rankine [8], assuming the total absence of any dissipative forces (transfer of heat or momentum to the outside, the effect of viscosity or heat conduction in the direction of propagation), the conservation equations may be written as follows:

$$\rho(D - u) = \text{const} = A_1, \tag{1a}$$

$$p + \rho(D - u)^2 = \text{const} = A_2, \tag{1b}$$

$$\rho(D - u)\left[E + \frac{(D - u)^2}{2}\right] + p(D - u) = \text{const} = A_3. \tag{1c}$$

Equations (1a), (1b), (1c) express the conservation laws of mass, momentum and energy, respectively, for a steady regime propagating with a velocity D, i.e., for the case when all the quantities p, ρ, u, E depend on the time and coordinate only in the combination $x - Dt$; thus, for example,

$$p = p(x, t) = p(x - Dt). \tag{2}$$

We recall the meaning of the notations: p is the pressure (in $g \cdot cm \cdot sec^{-2} \cdot cm^{-2}$), ρ is the density (in g/cm^3), E is the specific energy (in $g \cdot cm^2 \cdot$

sec$^{-2} \cdot$ g^{-1}), u is the velocity of the gas (in cm/sec), and E may include the chemical energy as well. All of the quantities just enumerated may vary from point to point in a propagating detonation wave (e.g., over the course of the chemical reaction), while the quantity D (propagation velocity of detonation, cm/sec) is constant for a given regime. The constants A_1, A_2, A_3 introduced by equations (1a)–(1c) have the simple meanings of mass flux (A_1), the sums of the momentum flux and the momentum of the pressure forces (A_2), and the sums of the fluxes of thermal, chemical and kinetic energy and the work of the pressure forces (A_3) across a surface moving with velocity D together with the detonation wave, so that the quantity $x - Dt$ on the surface is constant. Here the expressions for kinetic energy and other expressions are given in the coordinate system of an observer moving with velocity D together with the wave, e.g., the kinetic energy of a unit mass is $\frac{1}{2}(D - u)^2$, not $u^2/2$. Introducing the more convenient variable of inverse density—the specific volume v (cm^3/g)—and writing all the equations for the initial state of the explosive gas, we obtain the basic system of equations

$$\frac{D - u}{v} = A_1 = \frac{D}{v_0}, \tag{3a}$$

$$p + \frac{(D - u)^2}{v} = A_2 = p_0 + \frac{D^2}{v_0}, \tag{3b}$$

$$I + \frac{(D - u)^2}{2} = \frac{A_3}{A_1} = I_0 + \frac{D^2}{2}. \tag{3c}$$

We have introduced the enthalpy $I = E + pv$ (in cal/g, or g$\cdot$cm$^2\cdot$sec$^{-2}\cdot$g^{-1}).

All quantities for the initial state are denoted with the index "zero," and the velocity of the gas in the initial state is taken to be equal to zero (so that in this system of measurement all velocities are the velocities of the original unperturbed mixture).

Excluding D and u from the equations (3),

$$D^2 = v_0^2 \frac{p - p_0}{v_0 - v}, \tag{4a}$$

$$(D - u)^2 = v^2 \frac{D^2}{v_0^2} = v^2 \frac{p - p_0}{v_0 - v}, \tag{4b}$$

$$u = \frac{v_0 - v)D}{v_0} = \sqrt{(v_0 - v)(p - p_0)}, \tag{4c}$$

we obtain the basic equation—the so-called Hugionot dynamic adiabate [6]

$$I(p, v) - I_0(p_0, v_0) - \frac{1}{2}(v_0 - v)(p - p_0). \tag{5}$$

If in the relevant state the material has the same chemical composition as in the initial state so that the functions I and I_0 coincide (a shock wave without a change in the chemical composition), then the curve satisfying equation

(5) in the p, v-plane has the form $HCAM$ (Fig. 1). In particular, $p = p_0$, $v = v_0$ (the point A) is the obvious solution. In contrast, if the transition from the state p_0, v_0 to the state p, v is accompanied by an exothermic chemical reaction, so that for given p and v

$$I(p, v) \ll I_0(p, v),$$

since I_0 still contained the chemical energy which is gone in I, then in this case the relation between p and v, according to equation (5), is represented by a curve of the form $DFBGIJE$ (see Fig. 1).

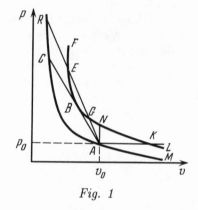

Fig. 1

According to formula (4a) relating the propagation velocity with the change in state, the curve $DFBGIJE$ is divided into three parts. The part JE corresponds to flame propagation with a velocity less than the velocity of sound in the original mixture—so-called deflagration. Normal velocities of flame propagation correspond to points which are quite close to the point J, at which $p = p_0$. In addition, by formula (4c), the velocity of the combustion products is negative, i.e., they move in the direction opposite the motion of the flame.

The part IJ of the curve corresponds to imaginary values of the propagation velocity and therefore does not correspond to any real process. Finally, the part $DFBGI$ of the curve corresponds to flame propagation with a velocity greater than the velocity of sound in the original gas, i.e., detonation, where the velocity of the gases has the same sign as the detonation velocity; the combustion products, compressed to high pressure and a density greater than the initial density, move in the direction of the original material.

In the diagram of Fig. 1 the detonation velocity may also take values from some minimum D to infinity, as the deflagration (slow combustion) velocity may vary from zero to some maximum D_1.

Experiment, however, shows a sharp difference between detonation and deflagration in this respect.

The velocity of deflagration is significantly less than the characteristic (maximum) value D_1 calculated from the gasdynamic considerations given above. The deflagration velocity is tens and hundreds of times less than the quantity D_1, and may vary strongly with small changes in the mixture composition (e.g., the addition of traces of hydrogen to mixtures of carbon monoxide) which do not change D_1 at all. In connection with this all theories of deflagration relate its velocity to the heat conduction and to the rate of chemical reaction in the mixture. Meanwhile, for detonation it is precisely the great stability of the value of the propagation velocity, depending little on external conditions, which is characteristic.

In a large number of cases, with all the accuracy that might reasonably be required, the measured detonation velocity coincides with the lowest velocity D possible on the branch $DFBGI$ of the curve in Fig. 1 [1].

The corresponding regime is also characterized by the remarkable property [3] that the velocity of sound in the combustion products is exactly equal to the detonation propagation velocity with respect to the combustion products:

$$c = D - u, \tag{6}$$

where c is the velocity of sound

$$c^2 = v^2 \left(\frac{\partial p}{\partial v} \right)_s. \tag{7}$$

The equivalence of these two conditions (Chapman [1] and Jouguet [3]) in the classical theory of propagation of detonation without losses was rigorously proved by Crussard [4].

For exact numerical calculations of the detonation velocity in the absence of losses, but accounting for dissociation, the dependence of the specific heat on the temperature, etc., Chapman's condition proves more convenient (see calculations by Ratner and the author [9]). However the physical meaning of Jouguet's [6] condition is much clearer, and it is to the latter that the ideas which we develop below on the mechanism of chemical reaction in a detonation wave lead.

§2. Selection of a Particular Value of the Detonation Velocity in Existing Theories

While the preceding conclusions, based on indisputable laws of mechanics, allow for the detonation velocity any value larger than or equal to D, selection of one particular value for the velocity and a corresponding particular state of the reaction products at a particular point on the segment $DFBGI$ of Fig. 1 requires the introduction of additional considerations.

As is obvious from the previous section, the Chapman–Jouguet conditions, in accord with experiment, select the tangent point B of the line ABC drawn from the point representing the initial state to the dynamic adiabate.

We are able to more or less convincingly exclude points which lie higher by noting that for the states DFB

$$c > D - u, \tag{8}$$

so that a perturbation (a rarefaction wave moving through the explosion products) can catch up with the front of the detonation wave and weaken it.[1]

The appearance of a rarefaction wave is related to the fact that at the detonation wave front there is an increased (compared to the initial value) density, from which it is clear that when the mixture is ignited in a closed tube at the ignition end there must be a region of decreased density since the overall amount of the substance, and therefore its average density, must be conserved. In analyzing steady propagation, the impossibility of maintaining a constant increased density [corresponding to constant positive, non-zero mass velocity in equation (1a) or (3a)], constant increased pressure and temperature is in fact the result of friction of the gas against the walls and heat transfer to the walls of the tube which contains the gas.

Anticipating a detailed investigation of heat transfer and braking in a detonation wave, it is not difficult to establish that the only possible final state of the combustion products following sufficient time after passage of the detonation wave is characterized by a temperature equal to the temperature of the walls due to the action of heat transfer, and by a gas velocity equal to zero due to the braking action of the walls. In accordance with equation (1a) [it is precisely this one equation of conservation of mass that remains valid despite the introduction of drag and heat transfer, which changes the form of equations (1b), (1c), (3b), (3c)], at zero gas velocity the density does not differ from the initial one.

Rarefaction (a drop in density and pressure compared to the conditions at the wave front) is present. But at the same time due to cooling a drop has also occurred in the velocity of sound to a value which is significantly less than the detonation velocity.

Consequently, if we were able to construct a regime in which, beyond the state represented by the point E on the segment DFB where (8) is satisfied (by the end of the chemical reaction), the material were to be subjected to braking and removal of heat, then at the same time there would appear a layer of material with a lowered (due to the lowered temperature) speed of sound, which would protect the detonation front from any additional rarefaction waves.

Thus the question of exclusion of the segment DFB in the rigorous theory of steady propagation takes on an outwardly completely different aspect. In fact, however, even in this more rigorous theory construction of a complete regime (with subsequent braking and cooling) in which relation (8) is

---

[1]Wendlandt [10] emphasizes the analogy between the overcompressed detonation wave on the branch BFD and a simple compression shock wave without chemical reaction which is also overtaken and weakened from behind by rarefaction waves. In contrast, a detonation wave at the tangent point, for which the Chapman–Jouguet condition is satisfied, is similar to a sound wave and is transformed into a sound wave when the thermal effect of the reaction goes to zero.

satisfied by the end of the chemical reaction turns out to be impossible. The conventional conception of a rarefaction wave which overtakes the detonation wave is quite close to the truth.

Completely unsatisfactory, however, are the arguments by which the lower branch BGE of the Hugoniot adiabate is excluded (see Fig. 1). Jouguet points out that the regimes described by the points on the segment BGE for which

$$c < D - u \qquad (9)$$

are supposedly unstable. Jouguet correctly notes [3] that due to a wave velocity (with respect to the combustion products) which is greater than the velocity of sound, no small perturbation of the state of the combustion products can overtake the wave front; on the contrary, the distance between the perturbed region and the front will increase. However, we cannot agree with the conclusion that this speaks to the "instability" of the wave, for the distance

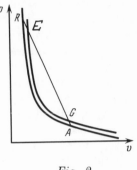

Fig. 2

between the perturbation and the wave increases, but the perturbation itself does not grow at all (in the presence of dissipative forces it even shrinks), i.e., it does not disrupt propagation of the wave.

Becker [11] points out that for a given detonation velocity (a particular slope of the straight line starting from the point A, e.g., $AGFH$—Fig. 1 or 2), the entropy on the lower branch (the point G) is smaller than at the point of intersection on the upper branch F [at which inequality (8) occurs]. Further, Becker writes: "It looks as if the detonation products at a given detonation velocity have the choice of transition either to the lower point (G) or to the upper one (H)"; and further, "If we imagine that the combustion products at the moment of their formation nonetheless take a state which, in the spirit of statistical physics, corresponds to a greater probability, then it is possible to conclude that they (the combustion products—*Ya. Z.*) decided for the point B (on the upper branch, the notations are ours—*Ya. Z.*) so that the lower part of the detonation branch will not correspond to any real process."

The unpersuasiveness of these arguments is obvious. The simplest example refuting such *a priori* statements is the propagation in a gas of compression shock waves: in a shock wave the entropy grows and the probability increases, but in order for this transition to a more probable state to occur, the gas' awareness of the growth in the entropy is not enough by itself—the motion of a piston compressing the gas is also needed.

The following thought experiment approaches the case of detonation even more closely: some small amount of energy is imparted by a light spot to a gas contained in a tube with a transparent wall. By turning the mirror we may easily cause any velocity of displacement of the place where the energy is supplied (even one greater than the speed of light). It is obvious that for supersonic speeds of displacement of the spot the state of the gas which is subjected to the action of the light will be described by precisely the point G (see Fig. 2), corresponding to a given velocity for which the increase in pressure and temperature is proportional to the energy of the light beam causing the change in state.

For the upper point F corresponding to the given velocity of the spot, in contrast, we obtain strong increase in the pressure, temperature and entropy which depends only on the velocity, but is practically independent of the power of the light ray itself.

It is obvious that the regime F (in the absence of a piston compressing the gas in addition to the action of the light spot) is not possible. The unfeasibility in detonation combustion of mixtures of regimes corresponding to the lower branch of the curve may only (and in fact, as we shall see below, does) follow from the specific mechanism of chemical energy release in detonation. Let us return to the above example of propagation of a perturbation for motion of a light spot: if we like, illumination of the gas can not only heat the gas due to absorption of the light, but can also cause a photochemical reaction with release of heat.[2]

What is the difference between true (conventional) detonation and such pseudo-propagation with externally imparted (by the motion of the mirror) velocity?

In our example heating of one volume element of gas followed (but did not result from) heating of the preceding element, following the motion of the light spot. But in the actual propagation of detonation *post hoc = propter hoc*, release of chemical energy in some volume is causally related to occurrence of the chemical reaction in the preceding (along the wave motion) volumes of gas. In this sense the qualitative arguments of Jost [12] are closest of all to the truth: on the lower branch (especially approaching the point I corresponding to constant volume) the detonation propagation velocity is larger than the velocity of sound and the chemical energy released at some distance cannot be transmitted to the wave front. Only by using such unconvincing arguments to exclude both the branches above and below the point B does the modern theory approach the only untouched point, that at which $c = D - u$ exactly. This is the tangent point B which yields, as was indicated above, the one definite, experimentally confirmed value of the detonation velocity. It is hardly worth mentioning here the attempts to obtain the tangent point as the only possible one, or from some other unprovable condition (the minimum of the detonation velocity or of the entropy, etc.) taken either in addition to or in place of the equations of hydrodynamics and chemical kinetics [12].

[2]See §4.

Finally, attempts to equate the detonation velocity with the velocity of certain molecules, atoms or radicals in the combustion products, after which the corresponding particles are declared to be active centers of a chain chemical reaction [13], are completely inadmissible today.

However good the numerical agreements, this is no more than doctoring and is a clear step backward compared with the thermodynamic theory presented; this is clear just from the fact that it is not at all obvious just which—the mean or the mean square, or some other—velocity of the molecules should enter into the calculation.

The author of the aforementioned theory himself points out the arbitrariness in the choice of the reaction carrier. In the case of detonation of a mixture of acetylene with nitrogen oxide, the velocities of atomic oxygen, atomic carbon and molecular nitrogen are calculated; the arithmetic mean (!) of these three quantities yields the measured detonation velocity with an accuracy of 0.6%.

Lewis points out [13] the desirability of merging his "chain theory of the detonation velocity" with the classical theory of Chapman–Jouguet. It is doubtful, however, that the latter will gain anything from such a merging.

§3. Modern Views on Chemical Reaction in a Detonation Wave

Thus, even in the classical theory of the dynamic adiabate and detonation velocity, which has had a number of indisputable achievements—the unquestionable equations (1)–(5) and a practically fully satisfactory means of calculating the detonation velocity—even in this theory we do not have complete clarity in its logical foundations. The situation for conditions of chemical reaction in a detonation wave is significantly worse.

The majority of researchers lean to the point of view, shared and developed further by the author, that the beginning of the chemical reaction, ignition, is related to heating of the gas above its auto-ignition temperature by adiabatic compression in a shock wave. It is precisely for this reason that we shall first consider other points of view and objections encountered in the literature against this conception.

Just for the sake of curiosity we may mention the "quantum-mechanical resonance" between states of the gas before and after passage of the detonation wave, which in some mystical way helps the chemical reaction to occur [14].

The impossibility of completion of a chemical reaction over a distance the length of the mean-free path, especially for reactions of any complexity requiring several collisions of a quite specific type, is sufficiently convincingly argued, for example, in the well-known book by Jost [12]. Thus, between two triple collisions a molecule undergoes about 1000 ordinary collisions at

ordinary densities. An activation heat of order 40 000 cal/mole, even at a
temperature of 3000°K, decreases the probability of reaction to 0.001, so that
again about 1000 collisions are required for an elementary reaction event.

As soon as we have established that the chemical reaction cannot occur
over a length of the order of the mean-free path, we eliminate all the theories
in which direct impact by the molecules of the reaction products against
molecules of the original materials plays a significant role. Indeed, between
the fresh, unreacted gas and the reaction products there is a more or less
wide zone in which the reaction is occurring and where the concentrations,
temperature, density, pressure and mass velocity undergo variation.

Since the width of this zone is significantly greater than the length of
the molecular mean free path, we should speak not of energy transfer by
direct impact, but of heat conduction and other dissipative processes in the
gas—diffusion and viscosity—related to the gradients of the temperature,
concentration and the velocity along the normal to the wave front.

It is precisely to the heat conduction along the wave (in the direction
of propagation of the detonation wave) that Becker [11] ascribes the basic
role in detonation propagation. Becker also, without studying the problem
in greater detail, points to the example of propagation of a shock wave in
which an increase in entropy occurs, in the final analysis, due to dissipative
processes, especially heat conduction in the wave front.

Analysis of the conditions of propagation of a normal flame allow us to
estimate the order of magnitude of the amounts of heat transported by heat
conduction.

Indeed, in normal (slow) combustion, which may propagate only due to
heat conduction,[3] the heat flux is a quantity of the same order as the combustion heat released in unit time. The width of the front should be of
the same order as the product of the chemical reaction time and the flame
propagation velocity.

Let us now compare the conditions in a slow flame with conditions in a
detonation wave. The order of magnitude of the temperatures and temperature differences is the same in both cases. For identical chemical reaction
times the zone width will turn out larger in the case of a detonation wave
than in that of slow combustion in the ratio of the velocities D/D''_{slow}; in this
case the gradients of temperature and concentration will fall in the inverse
ratio D''/D, as will the quantity of heat flux transported by heat conduction
(or diffusion flux). In contrast, for equal or close caloricity of the detonating
and deflagrating mixtures, the amount of heat released in unit time per unit
flame front surface is larger for detonation than for deflagration in the ratio
D/D''. Finally, assuming that in deflagration (normal combustion) the heat
flux transmitted by heat conduction is of the same order as the released heat

---

[3]Or due to another process of transport which is completely analogous in molecular
mechanism—diffusion of active centers. Replacement of heat conduction by diffusion
changes nothing in our conclusions.

of combustion, for detonation we obtain an estimate of the same ratio (the heat flux to the heat release)

$$\frac{K}{QD}\frac{\partial T}{\partial x} \cong \left(\frac{D''}{D}\right)^2 \cong 10^{-9} - 10^{-5}. \tag{10}$$

Here we use the following notations: K is the thermal conductivity (in $\text{cal} \cdot \text{deg}^{-1} \cdot \text{cm}^{-1} \cdot \text{sec}^{-1}$, or $\text{erg} \cdot \text{deg}^{-1} \cdot \text{cm}^{-1} \cdot \text{sec}^{-1} = \text{g} \cdot \text{cm} \cdot \text{deg}^{-1} \cdot \text{sec}^{-3}$); T is the temperature; x is the coordinate normal to the flame front (in cm); Q is the specific caloricity of the detonating material (in cal/cm, or $\text{erg/cm}^3 = \text{g/cm} \cdot \text{sec}^2$); D is the detonation velocity (in cm/sec); D'' is the deflagration velocity (in cm/sec).

The ratio (10) that we obtain is so small that there is no need to attempt to establish more exactly the relation between the heat transfer and heat of reaction in the various theories of normal combustion [3, 4, 15–18], or the accuracy of the temperature differences in the detonation wave, or to undertake other similar operations which can in no way change the basic results: the smallness of the heat flux in the direction of propagation of detonation; the adiabatic character (which holds with great accuracy as long as we do not consider heat losses to the walls of the tube) of the chemical reaction in the detonation wave; the impossibility of any noticeable role of heat transfer from the heated combustion products in ignition of the fresh, unreacted gas.

It remains to clarify the internal reasons for Becker's mistake, the deep difference between conditions in shock and detonation waves which causes the inconsistency of transferring the role of heat conduction from one case to the other.

In analyzing a steadily propagating compression shock wave we have no previously given characteristic quantity of time or length. Such quantities—the width of the shock wave front, the compression time in the shock wave—appear only when we introduce dissipative phenomena into the analysis: heat conduction and viscosity at the wave front. The width of the shock wave is determined (or, perhaps preferably, is calculated) in just such a way as to yield a sufficiently large temperature gradient for the heat flux to be sufficiently large and to ensure the required entropy growth, independent of the magnitude of the heat conduction and determined by the difference between the dynamic Hugoniot adiabate and the Poisson isentrope (static adiabate). In terms of similarity theory the shock wave as a whole, outside of a narrow band of the front, is self-similar with respect to dissipative quantities.

The smaller the heat conductivity, the greater must be the temperature gradient, and the smaller must be the front width and compression time in the shock wave. As Prandtl [19] and Becker [11] showed, any significant compression here makes the front width in the shock wave in the gas close to the length of the mean free path.

Things are quite different in a detonation wave. In this case we have a completely determined characteristic time—the duration of the chemical reaction: in combination with a particular linear velocity of detonation propagation we obtain the zone width of the chemical reaction, which can no longer (as in the case of a shock wave) vary with changes in the heat conduction. The chemical reaction cannot occur in the time of a single collision; many collisions of the molecules with one another will be required, and the zone width will be extended to a length many times the mean free path.

If, in accordance with Becker's calculations, it is only at a width of the order of the mean free path that the thermal flux by heat conduction becomes at all noticeable, then, naturally, for a zone stretched over many mean free paths, the thermal flux becomes negligibly small—in agreement with our result (10). It is appropriate only to again emphasize that in estimate (10) the use of the deflagration velocity was nothing more than a means of estimating the order of magnitude of the chemical reaction time at high temperature on the basis of known experimental data.

Of much interest is the attempt by Izmailov and Todes [20] to construct a theory of detonation. This work is not published anywhere and is known to the author from their report at the Scientific Council of the Institute of Chemical Physics in 1934.

Combining the first two basic equations of a steadily propagating regime, specifically the equations of conservation of mass and momentum, so as to obtain the equation

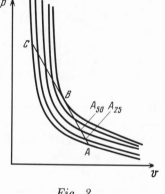

Fig. 3

$$D^2 = v_0^2 \frac{p - p_0}{v_0 - v}, \qquad (4a)$$

Izmailov and Todes give this equation a new interpretation, pointing out that in a regime of steady propagation with a single particular velocity, no matter how the chemical reaction proceeds, the specific volume and pressure must be linearly related according to formula (4a).

The motion in the p, v plane over the course of the chemical reaction should occur along the line ABC (see Fig. 1) which passes through the initial point A. This line, which Izmailov and Todes call the *isovel* line (i.e., corresponding to constant velocity), we shall call the Todes line.

The question then arises, in just what direction and how does the motion of a point indicating the state along the Todes line occur?

Izmailov and Todes assumed that the point ascends directly from the initial state A, which obviously lies on the dynamic adiabate for the fresh chemically-unchanged gas, to the point B, located on the Hugoniot adiabate corresponding to the fully completed chemical reaction. We may draw a row of intermediate adiabates which would correspond to passage of 25–50–75% of the reaction and thus easily establish the correspondence between the chemical reaction and motion of the point along the Todes line from A to $A_{25\%}$, $A_{50\%}$ and so on along the points of intersection with the corresponding adiabates (Fig. 3).

It is easy to see, however, the physical inadmissibility of the chosen direction of motion: of necessity the chemical reaction rate at low temperatures now enters into our considerations, temperatures at the initial value and near it, which should lead to enormous stretching of the zone.

At room temperature a mixture of hydrogen with oxygen reacts for years, whereas a detonation wave travels 2–3 km in a second.

The heat conduction, as we saw, in the case of detonation cannot be responsible for the initial rise in the temperature where the temperature is low and the chemical reaction rate is clearly insufficient.

Finally, all of the motion of the point on the path A–B occurs in the region in which the speed of sound is less than the propagation velocity of the regime, $c < D - u$. Just as in the analysis of pseudo-propagation in the example of a light ray sliding along the gas (see above), relation (9) contradicts the possibility of a causal relation between separate phases of combustion. It is as if the reaction at the initial temperature begins by itself, independently of the approach of the detonation wave, for it is impossible to transport ignition by a material agent in a continuous medium through a layer of mixture in which (9) holds true.

Leaving aside the physical conceptions of Izmailov and Todes concerning the path of motion of the system and the beginning of the reaction at the initial temperature, we fully utilize below the graphic and elegant concept of the Todes line (isovel).

The objections advanced by Becker [11] against the idea of ignition of an explosive mixture by a shock wave, where he cites Nernst [21] and Vant Hoff [22], are completely incorrect and are based on clear misunderstandings. Referring to the above authors, Becker states that adiabatic ignition requires much higher—up to 100 or 250 atm—pressures than the pressures available to us in a detonation wave.

In fact, Vant Hoff's 250 atm and Nernst's "pressure significantly higher than 100 atm" [21] are the pressures that explosive mixtures, adiabatically compressed to the temperature of auto-ignition, *may develop* in combustion.

In contrast, the pressures of adiabatic (isentropic, according to Poisson) compression which are *necessary* to bring an explosive mixture to auto-ignition are 19.5–23.9 atm according to Vant Hoff's calculations, and 25–40 according to Falk's [23] data quoted by Nernst.

In addition, in a shock wave (the Hugoniot adiabate) under strong compression the temperature increase significantly exceeds the heating in

Table 1

State of the Gas	Composition	p	v	T	Propagation Velocity, m/sec
Initial	A	1	1	$10°C = 283$ K	—
Detonation Wave	B	12.87	0.575	$2019°C = 2292$ K	1659
Shock Wave	C	12.13	0.241	$555°C = 828$ K	1210

A $H_2 + (1/2)O_2 + (3\,1/2)N_2$
B $H_2O + (3\,1/2)N_2$
C $H_2 + (1/2)O_2 + (3\,1/2)N_2$

isentropic compression (according to the Poisson adiabate) to the same pressure.

Thus, compression of a gas by a shock wave proves to be a necessary, possible, and indeed the only means of igniting a gas and causing a chemical reaction in a propagating detonation wave.

What are the characteristics of the shock wave propagating in front of the detonation wave and igniting the gas? We often encounter assertions (Jouguet [3], Sokolik [24], Crussard [4]) that the shock wave has the same pressure as the detonation wave (or a smaller pressure, Crussard [4]), and that the pressure in the shock wave propagating through the fresh unreacted gas is equal to the pressure corresponding to the tangent point of the dynamic adiabate, the pressure of the gas in the state in which the gas ends up at the moment the chemical reaction ends.

We show a comparison borrowed from Jouguet [3] between the initial state of the gas, the state of the gas at the moment the reaction is completed, and the state in a shock wave which develops precisely the ignition temperature (555°C; Table 1).

We see a wonderful numerical coincidence: for a mixture maximally diluted by nitrogen (at any greater dilution the mixture is no longer able to detonate), the detonation pressure is barely sufficient for auto-ignition of the mixture to occur in a shock wave of the same pressure. However, any smaller shock wave propagation velocity (compared to the detonation wave) for equal pressure makes it impossible to construct a steady regime in which an igniting shock wave propagates in front of the combustion products. In the example above it is unclear how the shock wave, whose propagation velocity is only 1210 m/sec, can ensure propagation of a chemical detonation reaction with a velocity of 1660 m/sec.

From an analysis (an inaccurate one at that, but we cannot dwell on that here) of the conditions for appearance of the detonation wave, in particular based on the experimentally established fact of the appearance, at the point of detonation generation, of a reverse, so-called retonation, compression wave which propagates through the combustion products, Crussard [4] concludes that an igniting shock wave must satisfy the conditions

$$p < p_{\text{det}} \qquad u > u_{\text{det}}. \tag{11}$$

Fulfillment of these two conditions ensures the "correct" regular appearance of detonation.

Practical application of the conditions (11) is complicated by the fact that when it arises, the detonation wave very often has a velocity and pressure which are larger than in the steady regime (see, e.g., data of Bone, Fraser, and Wheeler [25], according to which, in a mixture of CO and O_2, the detonation velocity at the moment when it appears reaches 3000 m/sec, whereas the steady velocity is equal to 1760 m/sec. References to super-pressures and breakage of the tube at the point where detonation begins are numerous).

Transfer of the conditions (11) to a shock wave which is continuously igniting a gas in steadily propagating detonation is absolutely impermissible: in this case there is no question of retonation waves. Any conditions more or less required for detonation *to be generated* have nothing at all to do with steady *propagation*.

The condition of equality of the pressures in a shock wave or detonation wave [i.e., condition (11)], which for Jouguet [3] "seems natural," in fact has no basis in the area of gasdynamics and velocities comparable with the speed of sound. More detailed experimental work on the limit of detonation refutes the coincidences found by Jouguet (see below).

If we seek a strictly steady detonation propagation regime in which the entire phenomenon moves in space with a single specific and constant veloc-ity, it is obvious that the shock wave igniting the gas must satisfy one and only one condition—it must propagate in the gas with the same velocity as that of the detonation.

In the early literature one may find isolated fragmentary accounts of shock waves with a propagation velocity equal to D. Thus, Vieille [26] in a note emphasizing the role of discontinuities (of shock waves) in the propagation of explosions for the detonating gas $2H_2 + O_2$ (speed of sound 510 m/sec, detonation velocity 2800 m/sec) finds the pressure in a shock wave of the same velocity equal to 40 atm and further discusses the possibility of attain-ing such a pressure in combustion without a change in the volume of a gas which has been previously compressed to several atmospheres.

Crussard [4] points out that for the mixture $2CO + O_2$, in which the mea-sured detonation velocity is variable and close to 1210 m/sec, in a shock wave propagating with a velocity of 1210 m/sec the compression tempera-ture is 720°C, so that at this temperature a noticeable delay in auto-ignition is still possible. In 1924–1925 Wendlandt [10], a student of Nernst, ener-getically defended the view that the gas is ignited by a shock wave with a velocity equal to that of the detonation wave. Wendlandt studied in detail the concentration limits of detonation propagation in explosive mixtures and measured the detonation velocity near the limit for ignition of the mixture under study by detonation of an explosive gas in a particular part of the tube. The steadiness of detonation propagation was established by comparison of the velocity on two segments of the path. Near the limit the velocity falls sharply, differing noticeably from the calculated value. The coincidences obtained by Jouguet prove to be completely illusory. On the contrary, the temperature in a shock wave of equal velocity (but not equal pressure) at the limit turns out in fact to be of the order of the temperature of auto-ignition with minimum delays, the temperature at the start of rapid reaction. We borrow from Wendlandt Fig. 4, which contains for hydrogen–air mixtures the results of all his experiments and calculations. The hydrogen content is plotted on the abscissa as a percentage, and the velocity is plotted on the ordinate axis in m/sec. Curve *1* gives the detonation velocities calculated according to the classical theory. Here too are plotted the results of calculations by Ratner and the author [9] (curve *2*) performed in 1940 using the current values of the specific heat and dissociation constants.

The dark dots mark the propagation velocities on the first section of the path, closest to the point of ignition, and the light dots mark those on the second, more distant, section.

Above the limit in the region of propagation the two velocities coincide; below the limit the detonation wave is extinguished, and the velocity on the second section is smaller than on the first.

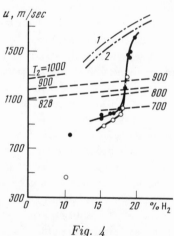

Fig. 4

Finally, the dotted lines indicate the velocities corresponding to shock waves in which the temperature indicated on these lines occurs (1000°K, 900°K, etc.). We see that at the limit it is in a shock wave of equal velocity (not of equal pressure, as Jouguet believed) that the temperature of auto-ignition with minimal delays is achieved.

In order to conclude this review of existing views of detonation propagation, we must still pause briefly on the phenomenon of spin in detonation and on several attempts to explain it.

Campbell and Finck [27] discovered a certain periodicity in photographs of detonation in certain mixtures.[4] A series of subsequent experiments showed that such periodicity may be related to propagation of the detonation front

---

[4]Periodicity of the distribution of luminescent particles in the cooling detonation products, which occurred so exactly that it cannot be considered accidental, was noted by Dixon [28].

along a spiral, i.e., by its essentially non-one-dimensional occurrence.

The only theoretical work [29] in which an attempt is made to describe such propagation, in a cylindrical tube in three dimensions, is not convincing since the "equations of helical motion" used as a basis are not justified in any way. Of great significance, however, is the relation calculated in this work between the spatial period and the diameter of the tube, which was excellently confirmed by experiment.

In contrast, in several very recent papers this relation is ignored, as are direct proofs of the helical nature of the propagation. The spin is described as a one-dimensional, but non-steady phenomenon.

Becker, in a 1936 paper [30], considers a mixture with an insufficient chemical reaction rate, for which steady detonation satisfying the Chapman–Jouguet condition is impossible since the temperature is not high enough. The supercompressed detonation wave, corresponding to the upper branch of the Hugoniot adiabate with a temperature higher than at the tangent point propagates, igniting the mixture. However, this wave is gradually weakened and transformed into a normal one. The detonation here is extinguished, but the shock wave which occurs from the still fairly lively reaction again causes detonation of the gas at a certain distance ahead: the detonation is again extinguished, and so on.

Jost [12, 31] considers a "discontinuous reaction," i.e., a reaction which begins only after a certain time τ has passed (induction period) after heating of the mixture. A shock wave entering the gas during the time τ propagates in the gas, gradually weakening, until the beginning of the reaction. When the induction period τ has passed, ignition of the compressed gas occurs and shock waves propagate from the point of ignition. The wave moving in front overtakes the initial, weakened wave, strengthens it, and everything begins over again.

Completely analogous views with respect to periodic change of the process were developed by Avanesov and Rukin [32], with the shock wave first moving forward, then being overtaken by the flame. In addition they also devoted particular attention to the chain character of the process, the role of active centers, and so on.

However, such considerations are still far from a theory of periodic propagation. Their unsubstantiated character is obvious. In order to prove the existence of a periodic regime, one must first of all consider possible steady regimes (complete absence of reaction, deflagration, detonation), and find their patterns and stability; then one must construct quantitatively the proposed periodic regime, and in particular, find its average propagation velocity, and show that this regime does not transform asymptotically into any steady regime.

Only by knowing the conditions of ignition in a steady detonation wave may we judge whether an explosive mixture which is unable to detonate with a normal velocity can be ignited by shock waves from a "still fairly lively reaction" (I quote Becker [30] verbatim—*Ya. Z.*).

Using the same conceptions of "discontinuous reaction" as Jost [31], it is quite possible to construct a strictly steady regime—a shock wave with subsequent additional compression to compensate for losses and a front of rapid chemical reaction at a constant distance (the product of the detonation velocity and the induction period) from the shock wave front. It is not clear whether the step-wise propagation described by Jost might not lead to just such a regime.

In this paper we shall restrict ourselves to consideration of one-dimensional steady propagation of detonation and the study of the conditions under which the reaction occurs, the influence of heat transfer and drag, the temperature and pressure distributions, etc., in a strictly steady regime.

In our opinion such calculations are necessary also as a starting point for any more complex theory of detonation propagation in three dimensions, theory of periodic regimes, etc.

§4. Ignition by Compression in a Shock Wave and the Selection of a Specific Value of the Velocity

In the following section we shall begin to construct a rigorous theory of steady propagation of detonation, accounting for losses, which are necessary in all cases for determination of the boundary conditions. Before this we shall give an elementary proof of the fact that the mechanism of ignition by a shock wave does indeed exclude the possibility of realization of the lower branch *BGI*.

The ideas of Le Chatelier [15], Berthelot [33], Vieille [26] on ignition by shock compression, supplemented by the definite indication of an igniting shock wave with velocity equal to D, and also the idea of Izmailov and Todes [20] that during the reaction there follows from the conservation laws a linear relation between the specific volume and the pressure, correspond to the following picture of the process in the p, v-plane (see Fig. 1): an instantaneous jump from point A to point C, and instantaneous compression without chemical reaction. Whereas in the initial state A the chemical reaction rate was negligibly small, the state C corresponds to a high temperature at which the chemical reaction runs at a significant rate. The motion of the point indicating the state along the Todes line ABC in the direction from C down to the point B corresponds to the occurrence of the chemical reaction.

At a larger detonation velocity, after the jump AH, motion occurs along the segment HF, and the impossibility in detonation of landing at the point G of the lower branch BGI thus follows directly from the mechanism of a chemical reaction which requires shock compression with a subsequent smooth slide along the Todes line in order to begin. Jump-wise motion along the Todes line (corresponding to shock waves) is possible only in the upward direction since only this direction corresponds to a rise in entropy.

It should be particularly emphasized that our considerations imply the possibility of propagation of detonation with a velocity exceeding that calculated from Jouguet's condition, with the state G on the lower branch of the Hugoniot adiabate realized in the reaction products. This occurs if the igniting agent, e.g., radiation of the reaction products, travels faster than the shock wave and causes chemical reaction in the material which is in state A or extremely close to it.[5] In this case, which corresponds to the original ideas of Todes and Izmailov on the motion along the segment AB, motion at another velocity is equally possible, for example, directly along the segment AG with realization of the point G on the lower branch despite the lower entropy. This last remark may be important for the theory of detonation of porous condensed explosives.

§5. Propagation of Detonation in a Tube Accounting for Braking and Heat Transfer

Let us consider the propagation of a detonation wave in a tube, taking account of heat transfer and braking against the side walls of the tube. We will restrict ourselves to the one-dimensional theory in which heat transfer and drag are uniformly distributed over the entire cross-section of the tube. We denote by x the coordinate measured from the detonation wave front toward the unreacted gas, in the direction of wave propagation. It is in fact on this coordinate alone in the steady and one-dimensional theory that all the following quantities depend:

$$\frac{d}{dx}\frac{D-u}{v} = 0; \qquad \frac{D-u}{v} = \frac{D}{v_0} = M, \tag{12}$$

$$\frac{d}{dx}\left[p + \frac{(D-u)^2}{v}\right] = -F, \tag{13}$$

$$\frac{d}{dx}\left[I + \frac{(D-u)^2}{2}\right]\frac{D-u}{v} = G - DF. \tag{14}$$

Here, in addition to the usual notations, D for the detonation velocity, u for the velocity of the material (both are measured with respect to the unreacted, unperturbed fresh mixture), p for the pressure, v for the specific volume (in cm^3/g), and I for the enthalpy of a unit mass, we have also introduced the following quantities: M for the mass flux per unit area of a surface moving at the detonation velocity[6] (in $g/cm^3 \cdot sec$; this quantity is equal to the mass velocity of combustion), F for the braking force per unit

---

[5] See §2.

[6] The quantity A_1 from formula (1a).

length of the tube with respect to its unit cross-section

$$F = \frac{-\xi u|u|}{2dv}, \tag{15}$$

where, from the definition of the dimensionless drag coefficient adopted in hydrodynamics,

$$\xi = \xi(\mathrm{Re}, l/d) > 0;$$

G is the amount of heat given off to the walls over a unit tube length in unit time with respect to a unit cross-section. If the heat exchange rate is determined by the heat transfer from the gas to the walls,

$$G = \frac{\alpha \xi |u|}{2dv}(I + \frac{u^2}{2} - I_{\mathrm{wall}}), \tag{16}$$

where I_{wall} is the enthalpy of the gas at the wall temperature and the coefficient $\alpha = 1$ if the Reynolds analogy between the heat exchange and friction is satisfied, $\alpha < 1$ in rough tubes (see any textbook of gasdynamics and heat transfer). In equation (14) the term $-DF$ represents the work by drag forces in a coordinate system moving with the wave.

The enthalpy I, besides a pair of variables determining the physical state of the material (e.g., p and S or p and v), also depends on the chemical variable n—the depth of occurrence of the irreversible chemical reaction which, for definiteness, we will equate with the concentration (dimensionless, g/g) of the final reaction product. It should be kept in mind that the reaction occurs, particularly at the beginning, irreversibly.

We write

$$I = I(p, v, n); \qquad dI = \left(\frac{\partial I}{\partial p}\right)_{v,n} dp + \left(\frac{\partial I}{\partial v}\right)_{p,n} dv + \left(\frac{\partial I}{\partial n}\right)_{p,v} dn; \quad (17)$$

$(\partial I/\partial n)_{p,v}$ is the chemical reaction heat, taken with the opposite sign, per unit mass of the reaction product. It is positive in an exothermic chemical reaction. It can be shown that reaction heat entering into the equation of the detonation wave is

$$-Q = \left(\frac{\partial I}{\partial n}\right)_{p,v} = -\frac{c_p Q_v - c_v Q_p}{c_p - c_v}, \tag{18}$$

where c_p and c_v are the specific heats, Q_p and Q_v are the reaction heats at constant pressure and volume. In the case of a reaction in which the number of molecules does not change,

$$Q = Q_p = Q_v. \tag{18a}$$

The derivatives, taken at a constant value of the chemical parameter, i.e., in the absence of an irreversible chemical reaction, may be transformed using thermodynamic relations. Comparing (17) with $I = I(p, S, n)$; $S =$

$S(p, v, n)$,

$$dI = \left(\frac{\partial I}{\partial p}\right)_{S,n} dp + \left(\frac{\partial I}{\partial S}\right)_{p,n} dS + \left(\frac{\partial I}{\partial n}\right)_{p,S} dn$$

$$= v\,dp + T\,dS + \left(\frac{\partial I}{\partial n}\right)_{p,S} dn = \left[v + T\left(\frac{\partial S}{\partial p}\right)_{v,n}\right] dp$$

$$+ T\left(\frac{\partial S}{\partial v}\right)_{p,n} dv + \left[T\left(\frac{\partial S}{\partial n}\right)_{p,v} + \left(\frac{\partial I}{\partial n}\right)_{p,S}\right] dn,$$

we find

$$\left(\frac{\partial I}{\partial p}\right)_{v,n} = v + T\left(\frac{\partial S}{\partial p}\right)_{v,n}, \tag{19}$$

$$\left(\frac{\partial I}{\partial v}\right)_{p,n} = T\left(\frac{\partial S}{\partial v}\right)_{p,n} = -T\left(\frac{\partial p}{\partial v}\right)_{S,n}\left(\frac{\partial S}{\partial p}\right)_{v,n}.$$

From now on we denote[7]

$$H^2 = -\left(\frac{\partial p}{\partial v}\right)_{S,n} = \frac{c^2}{v^2}, \qquad T\left(\frac{\partial S}{\partial p}\right)_{v,n} = v'. \tag{20}$$

The constant H introduced has the same dimensions as M; the quantity c is the speed of sound.

After simple algebraic calculations we reduce our equations (12)–(14) to the form

$$dp + M^2 dv = -F\,dx, \tag{21}$$

$$dp + H^2 dv = -\frac{v_0 - v}{v'} F\,dx + \frac{v_0}{v' D} G\,dx - \frac{Q}{v'} dn = -\xi F\,dx, \tag{22}$$

$$\xi = \frac{v_0 - v}{v'} - \frac{v_0}{v' D} \frac{G}{F} + \frac{Q}{v' F} \frac{dn}{dx} = \frac{v_0 - v}{v'} - \frac{v_0}{v D} \frac{G}{F} + \frac{Q}{v F(D - u)} \frac{dn}{dt}. \tag{22a}$$

For an energetic reaction $\xi > 1$; in contrast, after completion of the chemical reaction in the process of braking and cooling of the combustion products, $\xi < 1$.

For concrete calculations we must add the equation of chemical kinetics

$$\frac{dn}{dt} = f(n, T, p) = \varphi(n, p, v). \tag{23}$$

In what follows, however, the most general assertions regarding the reaction kinetics will suffice: the reaction rate is large in the region of small and medium burn-up, and goes to zero as the equilibrium state is approached.

The boundary conditions under which the equations are integrated are: for $x > 0$ the original mixture is in the initial state (A, see Fig. 1 or 5). At $x = 0$ we introduce a discontinuity (shock wave) which transforms the mixture to the state C, which is completely determined by the velocity of wave propagation D. After the wave has moved a large distance forward,

[7]For an ideal gas, $v' = v/(k - 1)$.

$x \to -\infty$, a state should be established in which, as a result of the action of braking and heat transfer,

$$F = 0; \qquad G = 0; \qquad u = 0; \qquad v = v_0;$$

$$T = T_{\text{wall}}; \qquad p = p_{\text{wall}} = \frac{QT_{\text{wall}}}{v_0}. \qquad (24)$$

Here we have already used equation (2) to derive $v = v_0$; T_{wall} is the temperature of the walls, p_{wall} is the corresponding pressure at the initial density and initial specific volume. The integral curves p, v, n as functions of x are fully determined by equations (12)–(14) or (21) and (22) and the initial conditions. The boundary conditions applied allow us to determine in addition the eigenvalue of the detonation velocity D which enters into the equations as a parameter directly through M and through the coordinates of the point C in the p, v-plane.

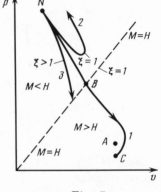

Fig. 5

Thus, our equations do not require the introduction of any externally imposed additional conditions, such as those introduced by Chapman [1] and Jouguet [3], in order to find a specific value for the detonation velocity. This is quite natural since the equations and boundary conditions (21)–(24) contain not only the conditions at the wave front but also the subsequent braking and cooling of the products.

Let us solve equations (15) and (16):

$$(M^2 - H^2)dv = (\xi - 1)F dx;$$

$$(M^2 - H^2)dp = (\xi M^2 - H^2)F dx. \qquad (25)$$

The line $M^2 = H^2$ is a singularity line of the equation (dotted line, Fig. 5). By defining detonation as a regime in which flame propagation occurs at a velocity greater than the speed of sound in the original gas, we find at the point A: $c < D$; $c^2/v_0^2 < D^2/v_0^2$, $H^2 < M^2$. After the shock compression, at the point C, as we know, $c > D$, $H^2 > M^2$; the shock compression is accompanied by a jump across the line $M = H$.

The final state of the combustion products (point S), in accordance with the boundary conditions, differs very little from the initial state—the difference exists only due to the change in the number of molecules. Once again at the point S, $H^2 < M^2$.

At the same time we recall that the quantity ξ of formula (22a) is significantly greater than unity in the zone of energetic chemical reaction; in the process of cooling and braking of the gas, in the absence or slowness of chemical reaction, $\xi < 1$.

The transition from C to S, requiring intersection of the integral curve and the line $H = M$, is only possible when we have simultaneously (curve *1*, Fig. 5)

$$H = M; \qquad \xi = 1. \tag{26}$$

It is not difficult to establish by analyzing equation (25) that as x decreases from 0 to $-\infty$, and if the value $\xi \le 1$ is achieved for $H > M$, then in the upper region the integral curve, without intersecting the line $M = H$, bends back and the boundary conditions cannot be satisfied (curve *2*, Fig. 5). This occurs if one attempts to construct a regime with the same detonation velocity in a mixture with lower caloricity than for the curve *1*.

If, in contrast, $\xi > 1$ at the time $M = H$ is achieved, (curve *3*, Fig. 5), then the integral curve, when it reaches the line $M = H$, has no extension in the real region. If we denote by γ the distance of the point from the line $M = H$ (positive in the upper region $M < H$), then for small γ we have the differential equation

$$\gamma d\gamma \sim (\xi - 1)Fdx; \qquad \gamma^2 \sim (\xi - 1)Fx + \text{const.} \tag{27}$$

If $\gamma = 0$ for $\xi > 1$ further decrease of x, ranging in the chosen coordinate system from 0 to $-\infty$, leads to imaginary values of γ. Case *3* is obtained if, without changing the velocity, we increase the caloricity of the mixture compared to the curve *1*.

After intersection of the line $M = H$ at $\xi = 1$ (curve *1*), fulfillment of the boundary conditions (24), i.e., the integral curve reaching the point S, is ensured. Substituting the values F and G in equations (22a) and (25), it is easy to verify that after completion of the chemical reaction the point S is a knot-like singular point through which all the integral curves in the lower region pass. In the lower region

$$H^2 < M^2, \qquad \frac{D - u}{v} > \frac{c}{v}, \qquad D > c + u; \tag{28}$$

it is natural that no matter how the cooling and braking of the reaction products take place, they will always lead to the final state (24), and transmission of a perturbation forward and a reverse effect on the regime are *impossible*.

Thus, analysis of equations (21) and (22) has led us to Jouguet's [3] condition [cf. (6)]:

$$H^2 = M^2, \quad \text{i.e.,} \quad \frac{c}{v} = \frac{D - u}{v}; \qquad c = D - u. \tag{29}$$

In addition, at the point where Jouguet's condition (29) or (6) is satisfied, the chemical reaction has not yet completed; $\xi = 1$ corresponds to a particular chemical reaction rate which balances the action of heat transfer and braking,

$$\frac{dn}{dt} = \frac{vG}{Q} + \frac{(v' + v - v_0)(D - u)}{Q}F, \tag{30}$$

i.e., it assumes the presence of some amount of fuel whose reaction is extended into the cooling zone, $M > H$, and does not affect the detonation velocity.

In the literature one encounters indications that in a slow chemical reaction not all the reaction heat is released in the detonation wave front, which in fact explains the lower velocity of detonation compared with that calculated according to the classical theory (Wendlandt [10], Lewis and Friauf [34], Rivin and Sokolik [35], Jost [12, 31]).

Formula (30) establishes that the losses to incomplete combustion in the wave front are related to the heat transfer and braking rate. The less the heat transfer and drag, the smaller the chemical reaction rate should be on the line $M = H$ for $\xi = 1$ (on the rear boundary of the wave front), and a smaller concentration of incompletely burned material is necessary to support the smaller reaction rate. Together with losses to incomplete combustion, which decrease the reaction heat at the front, it is necessary to take into account a second form of losses—braking and heat transfer in the front during energetic chemical reaction. In the zeroth approximation, considering the chemical reaction rate to be very large, $\xi \to \infty$ at the wave front, and $\xi = 1$ corresponds to an extremely small degree of incomplete combustion which tends to zero, and losses during the reaction also tend to zero. Our equations give in the limit the classical equations of a detonation wave together with Jouguet's condition, which is obtained here as a mathematical consequence of the equations and boundary conditions (21)–(24) and does not require special justification.

Integrating the equations to the point x_1 ($x_1 < 0$), at which $M = H$, $\xi = 1$, we find

$$\frac{D}{v_0} = \frac{D - u_1}{v_1}, \qquad p_1 + \frac{(D - u_1)^2}{v_1} = p_0 + \frac{D^2}{v_0} - \int_0^{x_1} F \, dx,$$

$$I_1 + \frac{(D - u_1)^2}{2} = I_0 + \frac{D^2}{2} + \frac{v_0}{D}\int_0^{x_1} g \, dx - v_0 \int_0^{x_1} F \, dx, \tag{31}$$

$$-\left(\frac{\partial p_1}{\partial v_1}\right)_{S,n} = \frac{D^2}{v_0^2}, \qquad I_0 = I_0' + Q_0(n_0), \qquad I_1 = I_1' + q_1.$$

In these last formulas we have split up the enthalpy by extracting the chemical energy from it: $Q_0(n_0)$ is the heat capacity of the original mixture and q_1 is the heat capacity of the incomplete combustion at the point x_1, where the amount of unreacted material at this point is determined from condition (30) in order to yield a reaction rate which balances the losses, $\xi = 1$.

Let us introduce for the sake of brevity the notations (the sign of D follows from formulas (9) and (10) and $x_1 < 0$ in the adopted system of measurement):

$$-\frac{v_0}{D} \int_0^{x_1} G\,dx = g > 0, \qquad -v_0 = \int_0^{x_1} F\,dx = f > 0. \qquad (32)$$

Taking the heat capacity of the fresh mixture to be large, both in comparison to the physical heat of the mixture at the initial temperature and in comparison to the losses:

$$f, g, q_1, I_0 \ll Q_0, \qquad (33)$$

we find, assuming constant specific heat, a deviation of the velocity from that calculated classically in the absence of losses

$$\frac{\Delta D}{D} = -\frac{1}{2(k_1^2 - 1)}\frac{f}{Q_0} - \frac{1}{2}\frac{g}{Q_0} - \frac{1}{2}\frac{q_1}{Q_0}, \qquad (34)$$

where k_1 is the exponent of the Poisson adiabate at the point x_1.

§6. The Conditions of the Chemical Reaction

The beginning of reaction in a detonation wave is related to compression and heating of the gas by a shock wave (the jump from A to C, Fig. 1 or 5). Let us consider the conditions of occurrence of the chemical reaction, accompanied by a change in the state which more or less closely follows the equation of the Todes line.

Let us compare the conditions in a detonation wave with conditions in slow, deflagration combustion. In the latter case the reaction rate in the initial state is negligible; the beginning of the reaction is related to heating of the mixture by heat conduction from the combustion products. Practically speaking, the reaction occurs entirely within a zone of temperatures which are quite close to the combustion temperature. The similarity between the molecular mechanism of the heat transfer and diffusion in gases corresponds to a composition of the reacting mixture approaching the composition of the combustion products (see the paper by Frank-Kamenetskiĭ and the author [18]). In contrast, in a detonation wave the substance is rapidly heated by compression to a temperature at which energetic chemical reaction takes place without a change in the composition.

Tentative calculations assuming constant specific heats, absence of dissociation and other similar simplifications show that in a shock wave propagating with a velocity equal to the detonation velocity (point C, Fig. 1 or 5), the gas density is 6 times greater than the initial density, and the pressure is twice as large as the pressure at the moment of completion of the chemical reaction (point B, Fig. 1 or 5) and 4 times larger than the pressure of explosion in a closed volume. The temperature is quite close (for a reaction

without change in the number of molecules) to the combustion temperature
of the given mixture at constant pressure. For numerical calculations of the
temperature see Wendlandt [10].

As has already been indicated, longitudinal heat transfer and diffusion in
a detonation wave may be disregarded so that the chemical reaction runs
almost adiabatically. In the case of
an autocatalytic reaction, the absence
of diffusion of the catalyzing products
may significantly delay the flow of the
reaction in the detonation wave. In the
course of reaction along the Todes line,
the heat release is tied to a significant
increase in temperature.

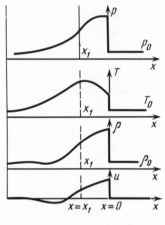

It is interesting that the temperature maximum on the Todes line is
shifted somewhat to the left (see Fig. 1
or 5) with respect to the tangent point
B so that between the maximum and
the point B there is a paradoxical region in which exothermic reaction and
heat release are accompanied by a decrease in the temperature as a result of

Fig. 6

the simultaneous expansion of the material. Heat release in this region is
accompanied by a growth in entropy. Tentative calculations show that the
maximum temperature exceeds the temperature at the point B at the end
of the reaction by about 50–100°.

Integration of our equations in the region of cooling and braking of the
reaction products leads to the approach of the integral curve to the end point
from the direction $v > v_0$, which corresponds in a steady regime, in accordance with the equation of conservation of matter, to a change in the sign
of the mass velocity $u < 0$, in qualitative agreement with the experimental
data of Dixon [28].

It may further be shown that the point x_1 itself (point B, Fig. 5), at
which intersection with the line $M = H$ takes place and Jouguet's condition
(29) is satisfied at the same time as $\xi = 1$ (30), does not correspond to any
singularities in the curves of pressure, density and other quantities. The
approximate form of the distribution of various quantities in space which
follows from the equation is shown in Fig. 6.

The conditions of the chemical reaction are extremely close to the conditions of adiabatic explosion.

As we know, the time of development of an explosion for any reaction
whose rate increases with the temperature, and—especially—for an autocatalytic reaction, is determined primarily by the smallest chemical reaction

rate at the lowest initial temperature [36, 37]:[8]

$$\tau \sim e^{E/RT_C}. \tag{35}$$

In expression (34) the first two terms are determined by the total reaction time in the interval 0–x_1 or in the p, v-plane from the point C to the intersection with the line $M = H$. The last term is determined by the reaction rate at the point x_1 itself at the intersection with the line $M = H$ in the p, v-plane, i.e., at a temperature which is significantly higher than the temperature at the moment of compression, $x = 0$, the point C.

In the absence of special reasons for a reaction whose rate increases with the temperature, the first two terms of expression (34)—the losses to braking and heat transfer during the reaction, which are determined primarily by conditions at the beginning of the reaction, at the point C—prove to be significantly larger than the last term, which describes incomplete consumption in the wave. The opposite may occur only in such systems where the chemical reaction is divided into two stages, with the second significantly slower, even at higher temperature, than the first. Examples of this kind, perhaps, are mixtures of dicyanogen with oxygen where formation of carbon monoxide occurs significantly more rapidly than its oxidation (Dixon [28]), or the decomposition of nitroethers where in the first stage nitrogen oxide easily forms (Berthelot [33], Appin, Belyaev [38]). In this case the second slow stage may occur entirely in the zone $x < 1$, $\xi < 1$ and not have an effect on the detonation velocity.

In the usual case of a single-stage reaction whose rate grows with the temperature this is not so and the basic losses are determined by the state of the matter, its velocity and the time of reaction near the point C (see Fig. 5). Using approximate relations for detonation and shock waves for a highly caloric mixture, we find

$$\frac{\Delta D}{D} = -\frac{2 + 2\alpha k_1^2 - 2\alpha}{(k_2 + 1)^2} \xi \frac{D\tau}{\alpha} \sim -e^{E/RT_C}, \tag{36}$$

where T_C is the temperature at the point C describing the shock compression of the gas by a wave propagating with the velocity D, E is the activation heat of the reaction, and k_1, k_2 are the exponents of the Poisson adiabates at the points B and C, respectively. For the other notations, see §5.

Taking into account the dependence of the temperature T_C on the detonation velocity [at the limit, for a reaction heat which significantly exceeds the initial heat content of the mixture (the last condition (33))]

$$T_C \sim D^2, \qquad \frac{\Delta T_C}{T_C} = \frac{2\Delta D}{D}, \tag{37}$$

---

[8]In slow flame propagation its velocity is determined by the maximum chemical reaction rate at a temperature close to the maximum temperature of combustion; the zone of low temperature and small reaction rate is overcome by the action of heat conduction.

we may easily write the transcendental equation which describes the propagation limit of detonation arising from the effects of losses,

$$\frac{\Delta D}{D_0} = -\Pi e^{2E\Delta D/RT_C^0 D} \tag{38}$$

(quantities with zeroes are calculated in the absence of losses),

$$\Delta D_{\lim} = \frac{-RT_C^0 D}{E}, \tag{39}$$

$$\Pi_{\lim} = -\frac{(2 + \alpha k_2^2)}{(k_1 + 1)^2}\xi\frac{D_0\tau_0}{d} = -\frac{2E}{ekT_C^0}. \tag{40}$$

Here τ_0 is the time of adiabatic reaction (period of induction of adiabatic explosion) at the initial temperature T_C^0; k_1 and k_2 are the specific heat ratios (adiabate exponents) at the points B and C in Fig. 1 or 5. For α, ξ, d—see formulas (15) and (16).

§7. Comparison with Experiment

The question of the deviation of the observed detonation velocity from the calculated value has been raised especially frequently in the last 10–15 years. The sharp drop in the velocity near the limit was explained by Wendlandt [10] (cf. details of his data below and in Fig. 4) by the fact that the reaction does not have time to occur within an "accessible" (Wendlandt's quotes) time and the heat release in the wave is less than the total reaction heat.

Lewis and Friauf [34] compared the detonation velocity in an explosive gas diluted with argon and helium. For equivalent dilution the calculated temperatures of detonation, the pressure and all dissociations are identical. The calculated detonation velocity is proportional to the square root of the density and is therefore twice as large in mixtures highly diluted with helium than in the same mixtures diluted with argon.

Experiments show that in mixtures diluted with argon the propagation velocity is closer to the calculated value and the product of the velocity and the square root of the density (molecular weight) is larger than in mixtures diluted with helium, which indicates smaller losses in mixtures with a smaller detonation velocity.

The authors write in this connection that at a larger wave velocity the chemical reaction does not have time to occur in the wave front and the incomplete combustion decreases the detonation velocity. Here the unsatisfactory nature of any such reasoning may be clearly seen: the chemical reaction rate has a completely different dimension from the detonation velocity. The assertion that the detonation velocity is greater than the chemical reaction rate, and that therefore the reaction does not have time to occur, is nonsense. Even from these trivial considerations it is clear that, without introducing any new time (or length—for a detonation velocity which has the dimensions length/time, this is equivalent) we cannot describe incomplete combustion or the deviation of the measured velocity from the calculated velocity.

In our theory such a time is introduced through losses—the time of braking and time of heat transfer. It is precisely the interrelation of these times with the chemical reaction time that determines the relative losses.

For a sufficient increase in the braking and heat transfer time we may imagine that the slowest reactions, e.g., combustion of dusts, will keep up with the detonation wave and can lead to detonation. Large losses in mixtures in which the detonation velocity is larger turn out to be related to intensification of the turbulent heat transfer and braking for an increase in the velocity proportional to D.

Completely lacking in the literature are indications of another form of losses—braking and heat transfer in the course of the chemical reaction (the quantities f and g in our formulas (32)–(34) in which incomplete combustion is denoted by q_1). Meanwhile, as has already been mentioned, it is precisely these losses, which depend on the smallest chemical reaction rate and its total duration, that are the most significant.

Table 2

H_2, %	D_{calc}, m/sec	D_{meas}, m/sec	τ, 10^{-4} sec	$T_{calc.\ shock}$ by D_{meas}, °K	E, cal/mole
19.6	1686	1620	0.55	1290	20 000
18.8	1645	1475	1.75	1120	9 000
18.5	1635	1300	3.94	930	5 000
18.3	1625	1207	5.30	840	

In Table 2 we show reaction times in a detonation wave propagating in a mixture of hydrogen and air using the data of Wendlandt [10]. We have taken in formula (36) $\xi = 0.02$, $\alpha = 0.25$, $k_1 = 1.3$, $k_2 = 1.4$ and ignored losses to underconsumption so that

$$\frac{\Delta D}{D} = 0.008 \frac{D\tau}{d}. \tag{41}$$

In the last column we give the activation heat of the reaction calculated from the dependence of the reaction time on the temperature (for each neighboring pair of points) [cf. (35)]. If we assume with Wendlandt that the last mixture (18.3%) lies at the limit, then the formula $\Delta D = kTD/2E$ yields $E = 5\,000$.

The values of the activation heat $\sim 5\,000$–$10\,000$ observed by Wendlandt resulting from the sharp drop in the velocity of detonation to 75% of the calculated value are completely unacceptable.

In later work by Breton [39] the velocity near the limit behaves very differently in different mixtures. In certain cases the velocity falls noticeably, in others it remains unchanged (hydrogen–oxygen mixtures). Finally, for ammonia with oxygen, according to Breton, the velocity near the limit increases, although in fact the scattering of experimental data is so large that one could equally argue that the velocity decreases.

Thus, the experimental picture of the limit of detonation is essentially unclear. Breton relates any decrease in the velocity as function of diameter to the spin and especially notes that in his experiments with mixtures spin *always* appeared near the limit so that our calculations cannot be directly applied.

Wendlandt's [10] chronoelectric technique does not allow us to observe the structure of the wave.

Let us note the interesting comment by Liveing and Dewar [40] who observed auto-inversion of the cadmium red line in the spectrum of detonation propagating toward the slit in the spectrograph. The conclusion of the authors concerning the temperature gradient at the wave front agrees fully with our conceptions. However, here too it is unclear whether the effect observed by Liveing and Dewar might not be related to reflection of the wave from the window with which the tube was closed [28].

According to our conceptions in a detonation wave the pressure, density and velocity of propagation fall as the chemical reaction progresses; the more slowly the chemical reaction runs, the slower this drop in pressure and the wider the zone of increased pressure and increased velocity.

A wave in which the reaction runs more slowly turns out to be more powerful and has a greater supply of energy, in accord with the fact that it is harder to generate. These arguments are evidently confirmed by the data of Rivin and Sokolik [35] according to which a mixture of carbon with oxygen (with a small addition of fire-damp) is no less able to cause detonation, if not more able, than the fire-damp and a mixture of ethane with oxygen.

§8. Immediate Tasks in the Development of the Theory

In the theory developed the pressure and velocity in the discontinuity without chemical reaction (a shock wave, point C, Fig. 1 or 5) which forms a detonation wave front are higher than are conventionally assumed for detonation, i.e., at the moment when (to accuracy to within the losses) the chemical reaction has completed (point B, Fig. 1 or 5, conditions (6), (29); cf. Fig. 6).

"Can a detonation wave (C) push before it a shock wave (B) of high pressure?" That this is possible is confirmed by the example of deflagration (slow combustion) where the combustion products, expanding, push before them the still-unburned mixture, and here the pressure of the combustion products is lower than the pressure of the unburned mixture.

Let us note that the relation between C and B is exactly the same as between the initial point A and the deflagration products on the branch JE; detonation is nothing more than deflagration of a shock-compressed gas which has been heated to a temperature higher than the temperature of auto-ignition.

The pressure momentum is conserved together with the momentum flux in a detonation wave of the structure calculated by us, as the sum $p + \xi(D - u)^2$ is conserved in a shock wave maintained by the motion of a piston where each of the terms is separately conserved. The stability of such a shock wave with respect to small perturbations is beyond doubt. It is clear that when the chemical reaction ceases, detonation will decay, but the shock wave will also change, entering a medium with different properties.

The first difference in a detonation wave which is favorable for stability is the fact that due to the decrease in the speed of sound (added to the mass velocity) to less than the propagation velocity, independence is achieved from the conditions behind the wave. On the other hand, application of small perturbations also leads to variation in the rate of the chemical reaction. It is impossible now to foresee the result of a calculation of stability with respect to small perturbations which might prove dependent on peculiarities of the chemical kinetics (autocatalysis, activation heat).

In any case, the fact that the entire chemical reaction determining the detonation velocity occurs in the region where the speed of sound is greater than the velocity of detonation is very satisfactory (the velocity becomes comparable only at the end of the reaction).[9]

In analyzing the influence of losses on the detonation velocity we have consistently restricted ourselves to the first approximation; in the zeroth approximation in the absence of losses, the state of the system changes as the reaction progresses in accordance with the equation of the Todes line. In the next, first, approximation, we found the losses and their influence on the velocity as the system varies along the zeroth approximation path. For a system with exactly known chemical reaction kinetics the influence of losses, the limit of propagation of detonation and so on should be found by direct integration of equations (21) and (22), and the eigenvalue of D may be found (since we are above the limit and the steady regime exists) by trial and error. The equations themselves, accounting for the dependence of the kinetics on the temperature, will ensure below the limit the absence of solutions which satisfy the boundary conditions.

It is very important that in the absence of a reaction, even if the Reynolds analogy is satisfied [$\alpha = 1$, see (15) and (16)], the relation between heat transfer and braking is such that in a steady regime of a certain velocity after compression in the shock wave additional heating of the gas occurs.

[9]Jost [12, 31] assumes that in normal detonation (i.e., the steady detonation which we are considering) the speed of sound is exactly equal to the velocity of detonation and hence deduces—verbally—the instability of normal detonation and its transformation into a periodic regime.

This effect becomes even stronger in a rough tube in which the braking increases faster than the heat transfer.

Taking account of such additional heating may extend somewhat the limits of detonation.

Continuing the refinement, we could take into account the increased heat transfer and braking at the beginning of the motion compared to stabilized motion with steady temperature and velocity profiles. But here we already overstep the boundary of applicability of the theory: from the moment that we begin to speak of a profile (distribution over the radius), a one-dimensional theory is no longer possible. We cannot distribute the heat transfer and braking in the short segments of interest to us over the entire cross-section.

A detailed analysis of the conditions of heat transfer and braking leads to the following conclusions: the Reynolds number of the flow is quite large so that in the stabilized flow we could consider the braking coefficient and the character of the flow to be practically (self-similarly) independent of the value of the Reynolds number.

However, the length of the stabilization sector still depends quite significantly on Re. Rapid chemical reactions evidently occur over a length which is insufficient by far for stabilization of the flow and for the braking and heat transfer to cover the entire cross-section. At the detonation limit the reaction runs over a length of the same order as the length l over which turbulization of the boundary layer occurs corresponding to a particular value of the Reynolds number constructed from this length

$$\mathrm{Re}_l = \frac{lu}{\nu} = \mathrm{const} \sim 5 \cdot 10^5. \tag{42}$$

Returning to the Reynolds number constructed from the tube diameter we find

$$l = 5 \cdot 10^5 \frac{d}{\mathrm{Re}}. \tag{43}$$

In view of this the boundaries of applicability of the one-dimensional theory developed above, particularly for establishing the influence of losses on the detonation velocity, are in need of additional study.

§9. On the Problem of Spin in Detonation

Experiment indicates the existence of completely idiosyncratic three-dimensional regimes of propagation of so-called spin of a detonation wave in which the instantaneous distribution also depends on the angle in the cylindrical system of coordinates, coaxial with the tube, in spite of the complete symmetry of the initial conditions.

At present the intrinsic reasons for this rotation are completely unclear. Moreover, the experimental conditions under which the spin arises have not been established with sufficient certainty, and particularly the relation between the spin and the detonation limit. On the basis of the analytical study of the simplest one-dimensional theory developed above we will attempt to use the methods of dimensional analysis to discover which quantities will enter into a future exact theory of the spin and the limit of detonation. First of all, by joining with the classical theory as a limiting case in the absence of losses and rapid reaction, we will obviously have the velocity of detonation without losses D and such dimensionless parameters as the ratio of this velocity to the speed of sound in the initial state, and the ratios of the pressures and volumes before and after the reaction. All of these dimensionless parameters vary little from one case to another—at least if the initial temperature of the mixture is not varied over a broad range and remains around room temperature.

Let us introduce a finite chemical reaction rate. One might think that the type of the chemical kinetics (autocatalysis or a classical reaction of some order) is in a certain sense not in itself essential; autocatalysis changes the absolute value of the induction period and places it in dependence on small admixtures in the original mixture, but the form of the kinetic curve itself hardly changes since, in a classical reaction as well, with any significant activation heat one observes significant self-acceleration related to the increase in temperature.

One way or another we have a characteristic time of the chemical reaction which, together with the detonation velocity, yields a characteristic length $D\tau$ of occurrence of the chemical reaction. Direct comparison of this length with the tube diameter does not make sense since all motions occur along the tube axis. Therefore we may expect that the characteristic distance of braking and cooling will enter—the quantity d/ξ where ξ is the dimensionless braking coefficient.[10]

At Reynolds numbers $\sim 10^5$ which are normally achieved in a detonation wave the braking coefficient may be considered practically constant so that, it would seem, in a turbulent, self-similar region at Re $\sim 10^5$ the dependence of the phenomenon on the Reynolds number disappears.

The results of our tentative calculations [formulas (41)–(43)] of the distance at which a stabilized regime is established and the braking and heat transfer cover the entire cross-section show the opposite: whereas in the stabilized flow the dependence on the Reynolds number disappears, the distance at which this stabilization occurs is very strongly dependent on the Reynolds number. At our large Reynolds numbers, long before stabilization, at a distance of $5 \cdot 10^5 d/\text{Re}$ turbulization of the boundary layer takes place.

The Reynolds number turns out to be very significant in particular for the two- and three-dimensional theories, which include analysis of the spin;

---

[10]It is obvious that our coefficient α characterizing the relation between heat transfer and braking will also enter here.

in the one-dimensional theory the stabilization region is marked only by a somewhat increased braking coefficient. Meanwhile, in reality (and in the two- and three-dimensional theory) in the stabilization region, which is in fact where the energetic chemical reaction occurs, the very nature of the phenomenon varies significantly in dependence on the Reynolds number, as do the portion of the cross-section affected by the braking and the character (laminar or turbulent) of the braked layer. Thus, from the condition at the point of turbulization of the sublayer $Re_l = 5 \cdot 10^5$ the length l also enters, independent of the diameter, where $l \simeq 5 \cdot 10^5 v/u$ or $5 \cdot 10^5 \nu/D$.

Thus we are dealing with three quantities of the length dimension which are significant for the process: the "chemical" length $D\tau$, the length of "stabilized braking" d/ξ, and the "turbulization" length $5 \cdot 10^5 \nu/D$. From these three lengths we may construct two dimensionless criteria, for example the Reynolds number $Re = dD/\nu$, and $D\tau\xi/d$ (the Rivin–Sokolik criterion).

The necessity of at least two criteria for description of the phenomena of spin and detonation limit may be established even by considering the comparatively meager experimental data currently available.

According to Rivin and Shchelkin, if one considers a wider interval of pressures and diameters (i.e., of values of the Reynolds number) than that with which Breton [39] worked, the relation between the spin and the limit ceases to be single-valued. Such a single-valued relation would necessarily follow from a theory with only the Rivin–Sokolik criterion $D\tau\xi/d$ in which we make only the wave front width concrete, replacing it with the tube diameter d or the braking distance $d/\xi = df$ (Re). One might think that when the Rivin-Sokolik criterion is increased we will move from the classical picture with a narrow plane front and a velocity which does not differ from the calculated value, to appearance of spin at one value of the criterion, and to the detonation limit at an even larger value of $D\tau\xi/d$. Such simple behavior is indeed observed in series of experiments performed in tubes of constant diameter at constant pressure (Breton), i.e., at a practically constant value of the Reynolds number.

Here we actually do become convinced that an increase in the criterion $D\tau\xi/d$ contributes to the appearance of the spin.

Meanwhile, if we take some mixture at atmospheric pressure in a tube of average diameter (15–25 mm) and decrease the pressure without changing the diameter or composition, it is apparently possible to achieve the limit in a number of cases without observing spin;[11] the decrease in pressure means an increase in the criterion $D\tau\xi/d$ and a decrease in the Reynolds number. In contrast, in the case of a mixture in which spin is observed under ordinary conditions, increasing the diameter does not lead to disappearance of the spin in spite of the decrease in $D\tau\xi/d$; instead of disappearance of the spin in wide tubes we observe the appearance of several spirals located like

[11]Private communication by M. A. Rivin and K. I. Shchelkin.

threads of a screw [41]. Thus we can evidently conclude that increasing the Reynolds number also contributes significantly to generation of the spin.

The material we used to compare our theory with experiment is incommensurately small in comparison with all the material which has been accumulated in the 60 years since detonation was discovered. In part the theory itself is at fault since it is extremely simplified and does not describe such peculiar phenomena as spin.

This also characterizes the very style of the experimental studies. Even until recently the hydrodynamic theory of the detonation velocity, which was excellently confirmed in experiments, created a sense of contentedness and did not inspire the search for the chemical reaction mechanism or investigation of the conditions at the detonation wave front. If our paper brings about new experimental studies which penetrate deeper into the essence of the phenomenon, then our task will have been accomplished.

Summary

1. The classical theory of the detonation velocity is presented. The ideas of Jouguet and Becker, which lead to the exclusion of a number of possible states of the reaction products corresponding to a larger detonation velocity, are unconvincing, which may be proved by thought experiment.

2. The views of various authors on the mechanism of chemical reaction in a detonation wave are considered. It is shown that diffusion of active particles and heating by thermal conduction cannot play a significant role in propagation of detonation. Ignition and the beginning of the chemical reaction are caused by instantaneous compression of the material and the related increase in temperature.

3. The course of change of the specific volume and pressure of the material in a detonation wave corresponding to these conceptions are studied; exclusion of the states indicated above (§1) and selection of a specific value of the velocity are consequences of the mechanism of the beginning of the chemical reaction described in §2 and of the equations of conservation which lead (Todes, Izmailov) to a linear relation between the pressure and volume in the absence of losses.

4. Equations are constructed for the chemical reaction, braking and heat transfer in a steadily propagating detonation wave. For rapid chemical reaction the equations yield a value of the detonation velocity which coincides with the classical value. In the next approximation the equations describe the influence of losses on the detonation velocity and the limit of detonation.

5. The one-dimensional theory developed is compared with available experimental data and peculiarities of the theory and limits of its applicability are indicated.

6. We express some ideas relating to the description of the experimen-

tally observed phenomenon of spin (spiral propagation) using dimensionless similarity criteria.

Gas Combustion Laboratory   *Received*
Institute of Chemical Physics   *March 16, 1940*
Academy of Sciences USSR, Leningrad

References

1. *Chapman D. L.*—Philos. Mag. **47**, 90 (1899).
2. *Schuster A.*—Philos. Trans. Roy. Soc. London A. **152**, 21 (1893).
3. *Jouguet E.* Mécanique des explosifs. Paris (1917).
4. *Crussard L.*—Bull. Soc. Ind. Minér **6**, 1 (1907).
5. *Riemann B.*—Ges. Werke. 2. Aufl., 156 (1870).
6. *Hugoniot H.*—J. école polytechn. **57**, 1 (1887).
7. *Rayleigh J. W. S.*—Proc. Roy. Soc. London A **84**, 247 (1910).
8. *Rankine M.*—Philos. Trans. **160**, 277 (1870).
9. *Zeldovich Ya. B., Ratner S. B.*—ZhETF **11**, 170 (1941), in press.
10. *Wendlandt R.*—Ztschr. phys. Chem. **110**, 637 (1924); **116**, 227 (1925).
11. *Becker R.*—Ztschr. Phys. **8**, 321 (1922).
12. *Jost W.* Explosions- und Verbrennungvorgänge in Gasen. Berlin (1939).
13. *Lewis B.*—J. Amer. Chem. Soc. **52**, 3120 (1930).
14. Sbornik po gasovym vzryvam [Collected Works on Gas Explosions]. Moscow: Aviaavtoizdat, 215 (1932).
15. *Mallard E., Le Chatelier A.*—Ann. Mines **4**, 296 (1883).
16. *Nusselt W.*—Ztschr. anorg. und alg. Chem. **45**, 120 (1905).
17. *Daniels P. J.*—Proc. Roy. Soc. London A **126**, 393 (1930).
18. *Zeldovich Ya. B., Frank-Kamenetskiĭ D. A.*—Zhurn. fiz. khimii **12**, 100 (1938); Acta physicochim. URSS **9**, 341 (1938); here art. 19.
19. *Prandtl R.*—Ztschr. Ges. Turbinenwes. **3**, 241 (1906).
20. *Izmailov S. V., Todes O. M.* Report at the Sci.-Tech. Council of the Leningrad Inst. of Chem. Physics (1934).
21. *Nernst W.* Theoretische Chemie. 8–10. Aufl. Berlin, 769 p. (1921).
22. *Vant Hoff J. H.* Vorlesungen über theoretische und physikalische Chemie. 2. Aufl. Berlin, 245 (1901).
23. *Falk H.*—J. Amer. Chem. Soc. **28**, 1517 (1906); **29**, 1536 (1907).
24. *Sokolik A. S.*—Zh. Fiz. Khimii **13**, 103 (1939).
25. *Bone W., Fraser R., Wheeler R.*—Philos. Trans. Roy. Soc. London A **235**, 29 (1936).
26. *Vieille P.*—C. r. Bull. chim. Soc. **131**, 413 (1900).
27. *Campbell C., Finck G.*—J. Chem. Soc., 1572 (1927).
28. *Dixon H.*—Trans. Roy. Soc. London A **200**, 315 (1903).
29. *Shchelkin K. I.*—Zh. Tekhn. Fiz. **4**, 730 (1934).
30. *Becker R.*—Ztschr. Elektrochem. **42**, 457 (1936).
31. *Jost W.*—Ztschr. phys. Chem. B **42**, 136 (1939).
32. *Avanesov D., Rukin S.*—Trudy Leningr. Khim.-Tekhn. In-ta **7**, (1939).

33. *Berthelot M.* Sur la force des matières explosives. Paris (1883).

34. *Lewis B., Friauf J.*—J. Amer. Chem. Soc. **7**, 3905 (1930).

35. *Rivin M. A., Sokolik A. S.*—Acta Physicochim. URSS **7**, 825 (1937).

36. *Semenov N. N.* Tsepnye reaktsii [Chain reactions]. Leningrad: Goskhimtekhizdat, 555 p. (1934).

37. *Todes O. M.*—Zh. Fiz. Khimii **4**, 71 (1933).

38. *Appin A. Ya., Belyaev A. F.*—Unpublished paper (Laboratory of Explosives, Institute of Chemical Physics).

39. *Breton M.* Thèses Univ. de Nancy, Paris (1936).

40. *Dewar J., Liveing G. O.*—Proc. Roy. Soc. London A **36**, 471 (1930).

41. *Bone W., Fraser R.*—Philos. Trans. **230**, 371 (1931).

Commentary

This paper practically completed the construction of the classical theory of detonation, i.e., the theory of a plane detonation wave as a steady complex propagating with a constant speed in the induced non-steady field of the gas flow. Analysis of an irreversible exothermic chemical transformation inside the wave permitted justification of the selection of the detonation speed, known earlier as the Chapman–Jouguet rule. The problem of selection of the detonation speed was also considered in a paper by A. A. Grib who, without going into the details of the internal wave structure, came to an analogous conclusion on the basis of analysis of the general laws of conservation in a detonation wave as a surface of gas-dynamic discontinuity within which accumulation of mass, energy and momentum does not occur. For selection of the speed an analogy between detonation and shock waves was used, together with the condition of continuous transition of one to the other as the heat of reaction is reduced. (A. A. Grib's paper was defended as a Ph.D. thesis in June, 1940 and published in 1944;[1] the value of this paper, which contains a detailed gas-dynamic analysis of the motion of the detonation products, was discussed in the introductory article).

The detailed analysis of the chemical and thermal structure of a detonation wave performed by Ya.B. revealed its important features, in particular the presence of high pressure behind the leading shock wave which is approximately twice as high as the pressure at the Chapman–Jouguet point. A number of Ya.B.'s papers are summarized in a monograph by Ya.B. and A. S. Kompaneets.[2]

A consistent theory of a normal (i.e., satisfying the Chapman–Jouguet rule) detonation wave allowed Ya.B. to produce experimentally a detonation wave with increased pressure[3] and also to theoretically examine the growth in pressure in a convergent cylindrical detonation wave.[4]

If an exothermic reaction follows the release of heat, then the final state of the detonation products corresponds to a pressure lower than normal and supersonic flow occurs. Ya.B. and S. B. Ratner apparently found such a situation in detonation

[1] *Grib A. A.*—PMM **8**, 148–160 (1944).

[2] *Zeldovich Ya. B., Kompaneets A. S.* Theory of Detonation. New York: Academic Press (1960).

[3] *Aivazov B. V., Zeldovich Ya. B.*—ZhETF **17**, 889–890 (1947).

[4] *Zeldovich Ya. B.*—ZhETF **36**, 782–792 (1959).

of a mixture of hydrogen and chlorine.[5] Undercompressed detonation, with lowered pressure, but increased speed is also realized in powder-form initiating substances with low density (see footnote 2).

Existence of a zone of increased pressure in the detonation of methane–air mixtures was demonstrated using a crude, but reliable method in a paper by Ya.B. and S. M. Kogarko, published in this book (article 28).

Using more subtle methods, by x-raying the detonating mixture, the reaction zone was studied by Ya.B. and M. A. Rivin.[6] Numerous foreign papers are also devoted to the detection and study of the reaction zone.

The article also investigates the influence of losses related to heat transfer and friction against the tube walls on the propagation of detonation, and finds the limit for the idealized case of a plane wave, i.e., the conditions for which propagation of detonation becomes impossible. This part of the paper is close to the well-known criterion of detonation capability of explosives formulated even earlier, in 1939, by Yu. B. Khariton. Further research on the role of losses may be found in the work of N. M. Kuznetsov.[7]

In a remarkable paper K. I. Shchelkin[8] discovered the influence of roughness of the walls on the transition from combustion of a gas to detonation. He explained this influence by turbulization of the gas. Following this Ya.B. noted[9] the role of the velocity profile, i.e., the non-uniform distribution of velocity over the tube cross-section. It has now become clear that both factors are very significant. Roughness of the walls has significant influence also on the losses, on the detonation speed, and on the limits.

A very important and fundamentally new factor found after the present paper by Ya.B. turned out to be the instability of the idealized one-dimensional model with a plane front and the appearance of a complex three-dimensional structure. Propagation of a bright spot at the detonation wave front was discovered as early as 1926.[10] For a long time, however, it appeared that this was an exotic phenomenon which occurred only in the case of slowly reacting mixtures near the limit. During this period K. I. Shchelkin[11] and then Ya.B.[12] constructed models of a separate bright region of ignition at the wave front. The works of N. Manson[13] and J. Fay[14] are closely related to this direction.

Ten years later K. I. Shchelkin's students Ya. K. Troshin and Yu. N. Denisov[15] discovered that a spin regime of propagation of detonation is a very frequent phe-

[5] *Zeldovich Ya. B., Ratner S. B*—ZhETF **11**, 170–183 (1941).

[6] *Zeldovich Ya. B., Rivin M. A.*—Dokl. AN SSSR **125**, 1288–1293 (1959).

[7] *Kuznetsov N. M.*—ZhETF **52**, 309–314 (1967); Zh. Prikl. Mekh. i Tekhn. Fiz. **1**, 45–49 (1968).

[8] *Shchelkin K. I.*—ZhETF **10**, 823–827 (1940).

[9] *Zeldovich Ya. B.*—Zh. Tekhn. Fiz. **17**, 3–26 (1947).

[10] *Campbell C., Woodhead D. W.*—J. Chem. Soc. **6**, 2010–3015 (1926).

[11] *Shchelkin K. I.*—Dokl. AN SSSR **47**, 501–509 (1945).

[12] *Zeldovich Ya. B.*—Dokl. AN SSSR **52**, 147–150 (1946).

[13] *Manson N.* Propagation des detonations. Paris, 194 p. (1947).

[14] *Fay J.*—J. Chem. Phys. **20**, 138–144 (1952).

[15] *Troshin Ya. K., Denisov Yu. N.*—Dokl. AN SSSR **125**, 110–113 (1959); Zh. Tekhn. Fiz. **30**, 4–9 (1960).

nomenon and is also realized in mixtures which are far from those near the limit. Besides the regular structure of spin, pulsating regimes of detonation are possible. A detailed presentation of this research with a theoretical interpretation is given in a monograph by K. I. Shchelkin and Ya. K. Troshin.[16] Research on spin was continued in Novosibirsk by B. V. Voitsekhovskiĭ, R. I. Soloukhin, M. E. Topchiyan, V. V. Mitrofanov and their colleagues who, by applying clever experimental techniques (self-compensation on film of a moving image of the pulsating detonation, stabilized detonation wave and others), decoded the subtle structure of complex detonation fronts and gave them a theoretical explanation. These results are presented in two important monographs[17,18] in which references are given to the original research of the authors performed in 1959–1963. Among works abroad of this period devoted to study of detonation spin we should mention the work of R. Duff.[19]

A. N. Dremin discovered and studied the phenomenon of detonation spin in condensed explosives as well.[20] In liquid explosives the detonation front is also sometimes non-ideally smooth, as was proved in work by Ya.B. and his colleagues[21] in which the smoothness was judged by reflecting light off of the wave front. One of the fundamental questions in the theory of detonation is this: why is a plane detonation wave unstable? K. I. Shchelkin proposed[22] a theoretical explanation of the instability based on the strong temperature dependence of the reaction rate in a gas which leads to intensification of the pressure perturbation caused by the bending of the leading shock wave.

Ya.B., together with R. M. Zaidel, showed[23] that nonlinear perturbations may also cause one-dimensional instability, i.e., lead to a pulsating detonation wave even without its being bent. Numerical calculations of the process of establishing the structure of a plane detonation wave also indicated one-dimensional instability.[24,25]

An investigation of stability of a plane detonation wave with a distributed chemical reaction with respect to spatial perturbations was performed by V. V. Pukhnachev[26] and J. Erpenbeck[27] who considered a supercompressed detonation

[16] *Shchelkin K. I., Troshin Ya. K.* Gazodinamika goreniĭa [Gas Dynamics of Combustion]. Moscow: Fizmatgiz, 256 p. (1963).

[17] *Soloukhin R. I.* Udarnye volny i detonatsiĭa v gazakh [Shock Waves and Detonation in Gases]. Moscow: Fizmatgiz, 175 p. (1963).

[18] *Voitsekhovskiĭ B. V., Mitrofanov V. V., Topchiyan M. E.* Struktura fronta detonatsii v gazakh [Structure of the Detonation Front in Gases]. Novosibirsk: Izd-vo SO AN SSSR, 315 p. (1963).

[19] *Duff R. E*—Phys. Fluids **4**, 1427–1440 (1961).

[20] *Dremin A. N., Savrov S. D., Trofimov V. S., Shvedov K. K.* Detonatsionnye volny v kondensirovannykh sredakh [Detonation Waves in Condensed Media]. Moscow: Nauka, 186 p. (1970).

[21] *Zeldovich Ya. B., Kormer S. B., Krishkevich G. V., Yushko K. B.*—Dokl. AN SSSR **158**, 1051–1053 (1964); **171**, 65–68 (1966).

[22] *Shchelkin K. I.*—ZhETF **36**, 600–606 (1959).

[23] *Zaidel R. M., Zeldovich Ya. B.*—Zh. Prikl. Mekh. i Tekhn. Fiz. **8**, 59–65 (1963).

[24] *Fickett W., Wood W. W.*—Phys. Fluids **9**, 5–12 (1966).

[25] *Zeldovich Ya. B., Librovich V. B., Makhviladze G. M., Sivashinsky G. I.*—Zh. Prikl. Mekh. i Tekhn. Fiz. **2**, 76–84 (1970); Acta Astronaut. **15**, 313–321 (1970).

[26] *Pukhnachev V. V.*—Dokl. AN SSSR **149**, 798–801 (1963); Zh. Prikl. Mekh. i Tekhn. Fiz. **6**, 66–70 (1963); **4**, 79–84 (1965).

[27] *Erpenbeck J. J.*—Phys. Fluids **7**, 684–693 (1964); **8**, 1293–1305 (1965).

wave (to analyze a normal detonation wave for stability is significantly more complicated due to the singular character of the Chapman–Jouguet point). It turned out that detonation waves are unstable up to high degrees of supercompression; the instability is manifested in the form of perturbations running along the wave front.

It should be emphasized, however, that proof of the linear instability of a plane wave by no means resolves the question of the very nonlinear and possibly even stochastic character of the wave structure which emerges when the perturbations are fully developed.

It is probable that ignition at bright spots occurs due to shock compression, however combustion of the remaining portion of the substance may be related to turbulent combustion. Satisfaction of the Chapman–Jouguet rule in spin detonation appears probable and is in agreement with experiment, however a rigorous proof of this assertion is lacking.

Close to the theory of gas detonation is the theory of transonic and supersonic flow in a tube and in nozzles developed by G. N. Abramovich and L. A. Vulis,[27a] and also the problem of supersonic motion of a body (a blunt or sharp bullet) in an explosive mixture.

Ya.B. and V. Ya. Shlyapintokh studied ignition of a gas with a bullet both experimentally and theoretically.[28]

Turning to the detonation of condensed EM we note that in this case the study of the equation of state of a dense gas in which repulsion of molecules is more important than their thermal motion turned out to be non-trivial (see the fundamental work by L. D. Landau and K. P. Stanyukovich).[29] Water-filled EM were studied by Yu. B. Khariton.[30] At present A. N. Dremin is developing ideas on the specific influence of a shock wave on the kinetics of reaction in an EM.[31] For gas-dispersion systems the structure of detonation waves has become the subject of numerous studies related to explosions of coal dust, husks in grain elevators, gas suspensions of dust in wood processing, etc.[32,33,34,35] Works on gas suspensions have also been published abroad.[36,37] In gas suspensions we may expect that the reaction rate is determined by diffusion and depends weakly on the temperature.

[27a] *Abramovich G. N., Vulis L. A.*—Dokl. AN SSSR **55**, 111–120 (1947).

[28] *Zeldovich Ya. B., Shlyapintokh V. Ya.*—Dokl. AN SSSR **65**, 871–879 (1949). Later this subject was studied by G. G. Chernyĭ, R.+I. Nigmatulin and their colleagues (see Abstracts of the 5th Congress on Mechanics, Alma-Ata, Nauka, 1981; and also the collection "Problems of Gas Dynamics and Mechanics of a Continuous Medium". Moscow: Nauka (1969), and references to previous work).

[29] *Landau L. D., Stanyukovich K. P.*—Dokl. AN SSSR **46**, 399–406 (1945).

[30] *Khariton Yu. B.*—In: Voprosy teorii vzryvchatykh veshchestv [Problems in the Theory of Explosives]. Moscow, Leningrad: Izd-vo AN SSSR, 7–28 (1947).

[31] *Dremin A. N.*—Fiz. Goren. i Vzryva **19**, 159–164 (1983).

[32] *Borisov A. A., Gelfand B. E., Gubin S. A., Gubanov A. V.*—In: Khimicheskaĭa fizika protsessov goreniĭa i vzryva. Detonatsiĭa [Chemical Physics of Combustion and Explosion Processes. Detonation]. Chernogolovka, 96–100 (1977).

[33] *Borisov A. A., Ermolaev B. S., Khasainov B. A.*—Khim. Fiz. **8**, 1129–1133 (1983).

[34] *Gubanov A. V., Timofeev E. I., Gelfand B. E., Tsyganov S. A.*—Khim. Fiz. **8**, 1133–1136 (1983).

[35] *Nigmatulin R. I., Vainshtein P. B. Akhatov I. Sh.*—In: Khimicheskaĭa fizika protsessov goreniĭa i vzryva. Detonatsiĭa [Chemical Physics of Combustion and Explosion Processes. Detonation]. Chernogolovka, 96–101 (1980).

In this case the instability of the plane front may disappear and the detonation will turn out to be one-dimensional, and the size and role of the reaction zone will increase.

In conclusion let us direct the reader's attention to the curious gasdynamic and kinetic features of detonation in systems of "liquid fuel film + gaseous oxidizer,"[38] and also in low-velocity regimes of propagation of detonation.[39] In these systems, which are characterized by a very non-one-dimensional structure of the detonation wave, a primary role is played by interactions between the shock waves and rarefaction waves propagating in different layers of the explosive system with different speeds.

[36] *Veyssiere B, Manson N.*—C. r. Acad. Sci. B **295**, 335–341 (1982).

[37] *Kauffman C. W., Wolanski P., Arisoy A.* et al.—In: Proc. of 9th Intern. Colloq. on Dynamics of Explosives and Reaction Systems. France, Poitiers, 113 (1983).

[38] *Lesnyak S. A., Nazarov M. A., Troshin Ya. K.*—In: Khimicheskaĭa fizika protsessov goreniĭa i vzryva. Detonatsiĭa [Chemical Physics of Combustion and Explosion Processes. Detonation]. Chernogolovka, 104–107 (1977); 99–103 (1980).

[39] *Dubovik A. V., Voskoboinikov I. M., Bobolev V. K.*—Fiz. Goren. i Vzryva **2**, 105–109 (1966).

28

On Detonation of Gas Mixtures*

With S. M. Kogarko

Rivin and Sokolik [1] were the first to notice that in a detonation wave the chemical reaction zone, in which the transformation of the original mixture into the reaction products occurs, must have a finite width which depends on the reaction rate.

Theoretical analysis of gas detonation leads to the conclusion that a shock wave propagates at the detonation front, compressing and heating the gas mixture. The chemical reaction runs in the already compressed gas, and it is only after completion of the reaction that the state of the explosion products calculated in the classical theory is attained (pressure p_C, velocity w_C, temperature T_C). In particular, in the wave front the velocity w_1 and the pressure p_1 of the compressed gas are approximately twice as large as in the reaction products: $w_1 \approx 2w_C$, $p_1 \approx 2p_C$. The amount of the compressed gas at the pressure p_1 and the thickness of this layer are proportional to the chemical reaction time, τ.

Only on the basis of this concept of the reaction mechanism was it possible to theoretically substantiate [2] the well-known fact that, of a number of possible values for the detonation speed which satisfy the equations of mechanics and thermodynamics, it is the smallest value, D_C, that is necessarily realized. Heat transfer from the gas to the tube walls and friction of the gas against the walls cause a decrease in the detonation speed and a corresponding decrease in the other quantities $(p_1, p_C, w_1, w_C, T_1, T_C)$. The greater the tube diameter d, the less the heat transfer and friction; the change in the detonation speed depends on (is proportional to) the ratio of the reaction time τ to the diameter d, τ/d.

Under the simplest assumptions, that the chemical reaction rate depends on the compression temperature T_1 (corresponding to a time τ_1), while the reaction itself runs uniformly over the entire cross-section of the tube, it may be concluded that propagation of detonation is possible only in the case when the ratio τ_1/d is less than some quantity. Starting from these assumptions, an investigation was performed at the Combustion Laboratory of the AS USSR Institute of Chemical Physics on gas detonation in a tube whose diameter was equal to 305 mm, i.e., 10–15 times greater than in ordinary laboratory conditions. It could be expected that detonation of mixtures with greater

*Doklady AN SSSR **63** 5, 553–556 (1948).

τ_1 than will detonate in narrow tubes would prove possible and that, due to the increase in τ_1 at the limit, it would be possible to experimentally detect the existence of a layer of compressed gas with higher (p_1) pressure at the wave front.

The experimental set-up consisted of a steel tube with internal diameter 305 mm and length 12.2 m equipped with viewing windows to record the propagation of detonation. Detonation was initiated by an explosive charge (30 g). To measure the pressure in the reflection of the detonation wave from the rigid wall at the end of the tube, crusher gauges were installed. Study of the propagation of detonation in hydrogen-air mixtures showed that the lower limit of detonation propagation of 19.6% H_2 in narrow tubes decreases to 15.0% in a tube of larger diameter, and the upper limit accordingly increases from 58.8 to 63.5% H_2.

We performed an estimate of the maximum pressures in the detonation wave by measuring the compacting of copper cones with diameter 5 mm and height 8.1 mm in specially constructed crusher gauges. Since the compacting time of a crusher cone ($\sim 10^{-4}$ sec) is significantly larger than the time during which the layer of compressed gas acts upon the pressure detector, in the general case we measure some intermediate value of the pressure (between p_1 and p_C). The wider the zone of increased pressure, the closer this pressure is to the maximum value, p_1.

The figure depicts the dependence we obtained of the pressure in reflection of the detonation wave from the end of the tube on the percentage content of hydrogen in the air. The broken line shows the corresponding dependence calculated from the classical detonation theory without consideration of the chemical reaction zone.[1]

From the figure it is clear that in hydrogen–air mixtures which are far from the composition at the limit, the experimentally measured pressure in reflection of the detonation wave from the wall differs little from the pressure calculated according to classical detonation theory. In mixtures near the limit the interrelation changes significantly. Thus, for example, at the lower limit in a mixture containing 15% H_2, the measured pressure is approximately three time higher than the

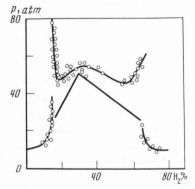

Pressure in reflection of a detonation wave from a rigid wall as a function of H_2 content in a hydrogen–air mixture.

[1]But taking into account the increase in pressure when the gas, moving with a speed w, stops at the end of the tube.

calculated value; at the upper limit in a mixture containing 63.5% H_2 the measured pressure is approximately twice as large as the calculated pressure.

The following explanation may be given for the sharp divergence between the experimentally determined values and those calculated according to the classical theory for the maximum pressures in the reflection of a detonation wave off the wall in mixtures near the limit, and the almost complete coincidence of these quantities in a mixture of stoichiometric composition. In stoichiometric mixtures and those close to them the chemical reaction rate in the wave is so large that the period of time from the compression of the gas by the shock wave to completion of the reaction is vanishingly small. As a result of this the duration of action by the reaction zone on the pressure-detecting column is also small compared to the tapering time of the column. In this case we measured the longer-acting pressure of the reaction products. As the mixture is made leaner or richer in hydrogen content the reaction rate in the mixture compressed by the shock wave will decrease, and at the limit the depth of the reaction zone will have been significantly extended. In this case the time of action by the pressure in the zone on the pressure-detecting column becomes comparable to the tapering time of the column and, consequently, here we measure a pressure which is closer to the pressure in the reaction zone than to the pressure in the reaction products.

Thus, our expectations with respect to the increase of τ at the limit and to extension of the limits when the diameter is increased have indeed been realized. However, a detailed photographic study of the wave showed that one of the assumptions, that of uniform reaction over the cross-section, is not justified. In accordance with the observations under laboratory conditions by Shchelkin and his colleagues [3], spin detonation is also found in a wide tube.

The results obtained must be interpreted from the point of view of the modern theory of spin detonation [4, 5]. In the works cited only the region in which ignition of the gas occurs (the so-called "head" of the spin or shock wave "kink") was considered. The conditions were found—the speed of the oblique shock wave and corresponding pressure p_2 and temperature T_2—at the spin head, i.e., the conditions under which ignition occurs in spin detonation. The size of the head b_2 and the reaction time τ_2 at the spin head under these conditions were considered quite small; the question of conditions for the existence/feasibility of spin detonation was not posed.

Further theoretical investigation of the problem led to the following conclusions: for ignition to occur at the spin head it is necessary that the ratio τ_2/b_2 be smaller than a certain quantity.[2] By studying conditions at the

―――――――――
[2]Since a rapid flow of gas occurs from the ignition zone at the spin head, rather than slow heat transfer, the critical value of τ_2/b_2 at the head is smaller than the critical value of τ_1/d for a normal wave in a tube.

shock wave kink from the standpoint of gasdynamics, we find that for a given distance between two neighboring heads[3] h, the wave amplitude at the kink decreases with the kink width b_2; also related to the wave amplitude is the temperature at which reaction begins in the kink. The value $T_2 = T_{20}$ calculated earlier [5] related to the case $b_2/h \to 0$; for $b_2/h \to 1$, obviously, $T_2 = T_1$ is the temperature in a plane wave. Thus as b_2/h increases, T_2 decreases, while the reaction time τ_2 increases. From this follows a physical conclusion: for small τ_2, small values of b_2 and h are possible; many spin heads may exist in a tube simultaneously, and the greater the tube diameter the more there may be. In a tube with given diameter d, as τ_2 increases the number of heads possible decreases. For certain τ_2 only one head is possible. Here $h = \pi d$; $b_2 \sim h$; $\tau_2 \sim b_2 \sim h \sim d$; the larger d is, the greater the corresponding τ_2 for one-headed spin. For even greater τ_2 spin detonation becomes impossible. Thus, as the limit is approached, the number of heads decreases. The limits of spin detonation should also expand as the tube diameter is increased.

The quantity τ_2 is related to the ignition time at the head. The total reaction time τ_3 of the entire gas mixture over the entire cross-section of the tube for spin detonation is determined by the time of turbulent propagation of the flame from the igniting heads over the entire cross-section. In a given tube this time decreases as the number of ignition points and the number of heads increases. As the limit is approached, when the number of heads decreases, the maximum magnitude of the reaction time $\tau_{3\,max}$ is attained. The speed of turbulent propagation is proportional to the detonation speed,[4] so that $\tau_{3\,max}$ (the reaction time of the entire mixture at the limit) is proportional to the tube diameter d.

Thus, the theoretical concepts developed here for the full picture of spin detonation lead to observed conclusions startlingly close to the conclusions made earlier for a normal detonation wave: expansion of the limit and increase of the maximum reaction time at the limit as the diameter is increased. These conclusions agree well with the results of experiments performed in very wide tubes.

Received
October 19, 1948

---

[3]When only one head is present, h should be understood to mean the distance from this head to itself, measured along the circumference, i.e., $h = \pi d$.

[4]Turbulent propagation occurs at the boundary of two gas layers moving with different speeds: the gas which has burned at the kink and the gas which has been compressed by the plane wave. From the calculation it follows that the difference in the velocities of these neighboring layers is proportional to the detonation velocity. The pulsation velocity at the boundary of the streams is proportional to the difference in velocities. The velocity of turbulent propagation, according to Shchelkin [6], is approximately proportional to the pulsation velocity.

References

1. *Rivin M. A., Sokolik A. S.*—Zh. Fiz. Khimii **10**, 692 (1937).
2. *Zeldovich Ya. B.*—ZhETF **10**, 542 (1940); here art. 27.
3. *Rakipova Kh. A., Troshin Ya. K., Shchelkin K. I.*—Zh. Tekhn. Fiz. **17**, 1409 (1947).
4. *Shchelkin K. I.*—Dokl. AN SSSR **47**, 501 (1945).
5. *Zeldovich Ya. B.*—Dokl. AN SSSR **52**, 147 (1946).
6. *Shchelkin K. I.*—Zh. Tekhn. Fiz. **13**, 520 (1943).

Commentary

This paper is an example of the first gas-dynamic–chemico-physical experiments with detonation waves in gas explosive mixtures which demonstrated the existence of a zone of increased pressure; the thickness of the zone is determined by the rate of chemical transformation. The paper was later continued.[1] Effects were measured for a large number of explosive gases and a theoretical interpretation of the results given more fully.

In combustible mixtures near the limit the wider chemical reaction zone intensifies the mechanical effect of the detonation wave on the crusher gauges, although the maximum pressure at the Jouguet point decreases: the crushing action is determined not only by the maximum magnitude of the pressure, but also by the duration of its action (impulse); this same factor plays an important role in other destructive effects of detonation waves as well. Subsequent development of experimental techniques led to the appearance of new methods for measuring the structure of detonation waves, even in the case of condensed explosives for which it is more complicated to perform measurements. Here the contribution of A. N. Dremin and his collaborators[2] is significant, as well as that of L. V. Altshuler and his colleagues.[3]

Papers studying the structure of a detonation wave are closely related methodologically to theoretical research on the reflection of detonation waves from barriers and the scattering of the detonation products, and Ya.B. was one of the initiators of such research.[4] These problems are presented in detail in a monograph by K. P. Stanyukovich.[5]

Let us return to an evaluation of the entire complex of Ya.B.'s work on the problem of detonation. Besides the present article by Ya.B. and S. M. Kogarko we may include Ya.B.'s fundamental work of 1940 (27), the second part of monograph (16) (pp. 196–229), and the articles indicated in the appropriate section of the

[1] *Kogarko S. M.*—Zh. Tekhn. Fiz. **29**, 128–134 (1959).

[2] *Dremin A. N., Shvedov K. K.*—Zh. Prikl. Mekh. i Tekhn Fiz. **2**, 154–159 (1964); *Vorobiev A. A., Dremin A. N., Savin L. I., Trofimov V. S.*—Fiz. Goren. i Vzryva **19**, 153–160 (1983).

[3] *Altshuler L. V*—UFN **85**, 197–258 (1965); *Altshuler L. V., Ashaev V. K., Balalaev V. V., Doronin G. S., Zhuchenko V. S.*—Fiz. Goren. i Vzryva **19**, 153–160 (1983).

[4] *Zeldovich Ya. B., Stanyukovich K. P.*—Dokl. AN SSSR **55**, 591–594 (1947); *Zeldovich Ya. B.*—ZhETF **36**, 782–792 (1959).

[5] *Stanyukovich K. P.* Neustanovivshiesĭa dvizheniĭa sploshnoĭ sredy [Non-steady flows of continuous media]. 2nd Ed., rev. and expanded. Moscow: Nauka, 854 p. (1971).

bibliography at the end of this volume. In the first part of (27) an idealized case without losses is considered, as was done by R. Doering, I. von Neumann and A. A. Grib (see the introductory article). However, when friction of the gas against the walls is taken into account [second part of (27)] the situation changes radically; selection of a particular reaction regime together with selection of the corresponding trajectory in the p,v-plane and of the detonation speed requires the solution of a differential equation. The detonation speed enters into the equations as a parameter. The existence of a solution depends on whether the integral curve passes through a singular point. At the singular point the derivative is determined by a fraction whose numerator and denominator simultaneously go to zero. The physical meaning of the singular point is related to the fact that at this point the velocity of the material is equal to that sound. Passage through this point is related to the transition from a subsonic to a supersonic regime. Such a principle for selecting the true solution was applied later, in 1942, by G. Guderley[6] in the problem of a convergent spherical shock wave and in 1954 by C. Weizsäcker[7] and later by Ya.B. (Art. 9) in the problem of a short-duration impact.

Of fundamental importance is the question of the limits of detonation, i.e., of the ability of a given explosive mixture to detonate. After the basic work by Yu. B. Khariton[8] it became clear that the limits of detonation are determined by the losses. For condensed EM, according to this work, the losses are related to expansion in the direction perpendicular to the direction of propagation of the detonation wave. In the case of detonation of a gas contained in a tube, the losses are related to heat transfer from the gas to the walls and to braking of the gas. It is obvious that only losses during the chemical reaction can be significant, can influence the detonation speed or cause detonation to cease, i.e., losses occurring before the Jouguet point is reached. Ya.B. obtained a very general result: at a chemical reaction rate which depends on the temperature according to the Arrhenius law, collapse and cessation of detonation occur at small losses of order RT/E.

This result is valid not only for idealized one-dimensional propagation, but also for spin propagation.[9] Laboratory experiments in smooth tubes confirm this thesis. Detonation either propagates with a velocity which is close to the theoretical value, or it is extinguished. Such chemical reaction regimes, which are essentially intermediate ones between combustion and detonation, are apparently realized in practice in catastrophic explosions in coal mines. To explain these phenomena with rough walls it is insufficient to point out the large hydraulic drag in a rough tube. One must explain why detonation does not cease! An answer is given in Ya.B.'s monograph (16) in the last section. There a propagation regime is described in which ignition occurs in the collision of the shock wave with protruding elements of the roughness. Combustion of the mixture as a whole is determined by the turbulent propagation of the flame from the points of ignition over the entire cross-section of the tube. Here the width of the reaction zone only weakly, not at

[6] *Guderley G.*—Luftfahrtvorschung **19**, 302–318 (1942).

[7] *Weizsäcker C.*—Ztschr. Naturforsch. **9A**, 269–282 (1954).

[8] *Rozing V., Khariton Yu. B_ذ*—Dokl. AN SSSR **26**, 360–361 (1940).

[9] *Zeldovich Ya. B., Kompaneets A. S.*—In: Sbornik, posvyashchennyĭ 70-letiĭu akademika A. F. Ioffe [Seventieth Birthday Festschrift for A. F. Ioffe]. Moscow: Izd-vo AN SSSR, 61–71 (1950).

all exponentially and not according to Arrhenius' law, depends on the shock wave parameters.

Let us add, finally, a reference to the freshest work investigating phenomena in the chemical reaction zone in solid explosives. S. Sheffield and D. Bloomquist[10] have performed sub-nanosecond measurements at the detonation wave front. They come to a very definite conclusion regarding the realization of a compressed, unreacted state of the material and state: "Computer calculations based on a model of the ignition and growth of a Zeldovich–von Neumann–Döring (ZND) detonation wave show good agreement with experiment."

[10] *Sheffield S. A., Bloomquist D. D.*—J. Chem. Phys. **80**, 3831–3844 (1984).

29

Flame Propagation in Tubes:
Hydrodynamics and Stability[*]

With A. G. Istratov, N. I. Kidin, and V. B. Librovich

INTRODUCTION

General Outline

The propagation of flames in channels has long attracted the attention of many investigators (Zeldovich, 1944; Tsien, 1951; Chernyǐ, 1954; Uberoi, 1959; Maxworthy, 1962; Borisov, 1978). The photographs show a curved shape of the flame front in channels up to considerable values of the Reynolds number, Re (Re is calculated from the flame propagation rate along the tube, the tube diameter, and the cold gas viscosity). The flame front has a stationary curved surface at Reynolds numbers equal to a few hundreds, although the hydrodynamic flame instability theory developed by Landau (1944), Darrieus (1944) and Markstein (1951) predicts stable propagation at Reynolds numbers just exceeding unity. The first consideration of this anomaly was that of Istratov and Librovich (1966).[1]

The photograph taken from the paper of Uberoi (1959) and presented in Fig. 1 shows the shape of the flame front propagating in a plane channel. It shows stagnation zones near the channel walls where the gas rests with respect to the flame front. The stagnation zones are caused by the refraction of stream lines at the extreme points of the flame front.

We will consider the cold-gas-convex surface of the flame front as a curved cell of the flame which had been formed after the plane flame lost its stability. The steady state of the convex flame is a result of the nonlinear hydrodynamic interaction with the gas flow field (see Zeldovich, 1966, 1979). In the linear approximation the flame perturbation amplitude grows in time in accordance with Landau theory, but this growth is restricted by nonlinear effects.

The flame propagating in channels has common features with freely propagating spherical flames which are known to be more stable to hydrodynamic disturbances (Shchelkin, Troshin, and Rakipova, 1947; Rozlovskiǐ and Zeldovich, 1954; Istratov and Librovich, 1969). Indeed, if starting from one

[*]Comb. Sci. and Technol. **24**, 1–13 (1980).

[1]It also has been under consideration in the recent paper by Zeldovich (1980).

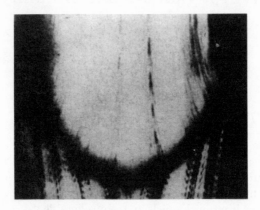

Fig. 1. Photograph of flow field and flame front shape from the work by Uberoi (1959). Flame is propagating downward in a vertical tube; small particles added to the gaseous mixture indicate streamlines before and behind flame front. One can see stagnation zone close to tube wall.

position of the front, we construct, by the Huygens' principle, its next position, the flame dimensions would increase, if the tube walls were not restricting. A peculiarity is that the expanding flame front runs into the cold wall and disappears and this condition provides the steady-state regime of flame propagation.

Theoretically, there always exists a stationary plane solution for the flame front propagating in a channel in the absence of the gravitational force, but this solution is unstable with respect to small hydrodynamic disturbances. In accordance with Landau theory the rate of growth of hydrodynamic disturbances is inversely proportional to the wavelength, so that the most rapidly growing are the short-wave disturbances (Landau, 1944; Zeldovich, 1966, 1979). But in this case the distortions in the flame front cause the changes in its structure and velocity, and the front cannot be treated as an infinitely thin hydrodynamic discontinuity. These changes in flame structure related to the transfer processes (diffusion and heat conductivity) may have a pronounced effect on the development of hydrodynamic instability. The disturbances with wavelengths of the order of a characteristic front thickness appear to be the most dangerous and grow at the highest rate.

In a channel the maximum size of the disturbance is restricted by the channel diameter. Small-scale disturbances with a wavelength considerably smaller than the diameter may also arise at any point of the curved flame front surface. Particularly, if the disturbance wavelength is comparable with the flame front thickness, the disturbance seems to be most dangerous. Nevertheless, because of the peculiar flow field in the vicinity of the curved flame front, the steady flame propagation appears to be stable up to Reynolds numbers of the order of hundreds. There exists a velocity gradient positioned along the flame (see Figs. 1,2), i.e., conditions arise for the appearance of a hydrodynamic stretching of the flame front and changing of its propagation rate—the so-called "stretch effect" (Karlovitz, 1953; Istratov and Gremyachkin, 1972). If the effective coefficient of thermal conductivity in the flame exceeds the diffusion coefficient of the main reactant or is equal to it (which is often the case), the effect of stretching reduces the flame speed, or improves the stability to hydrodynamic disturbances. However, of greater importance is the fact that flame stretching causes stretching of the wavelength for a given disturbance.

There is another effect which we wish to take into account. The time for the development of disturbances in a channel is limited. Along the front there exists a tangential gas flow, which transfers the developed disturbance to the channel walls, where it is quenched. In contrast to the freely expanding spherical flame with disturbances developing as standing waves, disturbances in channels propagate along the flame front as running waves. But, both for the flame in the channel and the spherical flame, one observes an increase in the wavelength of the initial disturbance with time (Istratov and Librovich, 1966; Zeldovich, 1979).

The present assumptions will be used below to interpret the steady propagation of a flame in a tube at rather large Reynolds numbers and to explain the stability of the curved flame front with respect to hydrodynamic disturbances of small amplitudes.

BASIC CONSIDERATION

Gasdynamics of Combustion in Tubes. Gas Flow before the Flame Front. Shape of the Flame Surface.
Consider the gasdynamic fields ahead of and behind the flame front separately. The gas flow ahead of the flame front can be assumed to be potential. The potential function ϕ satisfies the Laplace equation:

$$\Delta\phi = 0, \quad \mathbf{w} = \nabla\phi. \tag{1}$$

Consider the flame in a plane channel, and choose the cartesian coordinate system (x, y) moving with flame at a constant velocity equal to the flame propagation speed along the channel (Fig. 2), where y changes from 0 to R, x is read from the middle of the flame front into the combustion products side, so that $x = 0$, $y = 0$ is the leading point of the flame.

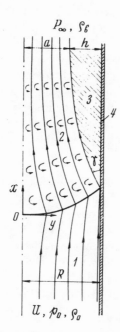

The solution of the equation in this coordinate system when the flow field is stationary may be presented in the form:

$$\phi = Ux + \sum_{n=1}^{\infty} A_n \exp(k_n x) \cos(k_n y)$$
(2)

where U is the velocity of flame propagation along the channel axis (in the moving coordinate system U is the gas velocity at $x = -\infty$), and A_n and k_n are constants. A bound of this solution far from the front (at $x = -\infty$) is taken into account in the Relation (2).

At first we shall account for the first term in the series of Eq. (2) only, namely, for the function $A_1 \exp(k_1 x) \cdot \cos(k_1 y)$. This function describes the general features of the cold gas flow qualitatively correctly. The subsequent

Fig. 2. Hydrodynamic model of flow for flame propagation in a channel (Region 1—potential flow of cold gas, region 2—vortex flow of hot products, region 3—stagnation zone, region 4—wall).

functions with more frequent zeroes specify the details of the flow, and do not change qualitatively its main properties.

In order to find the constants A_1 and k_1, we equate to zero the velocity component which is normal to the tube wall (the wall is gas-impenetrable). Due to the symmetry we also use the condition that the central streamline intersects the flame front at a right angle, and the velocity of the cold gas at that point is equal to the normal flame velocity u_n. Therefore

$$y = R, \qquad v = \partial_y \phi = 0,$$
$$y = 0, \qquad x = 0, \qquad u = \partial_x \phi = u_n$$
(3)

where R is the channel semiwidth, and u, v are components of velocity vector along the x, y axes, correspondingly.

From the first Condition (3) we obtain $k_1 = \pi/R$ and from the second, $A_1 = -\epsilon u_n/k_1$, where $\epsilon = U/u_n - 1$. The velocity vector components of the cold gas flow are written as

$$\frac{u}{u_n} = 1 + \epsilon[1 - e^{k_1 x} \cos(k_1 y)],$$

$$\frac{v}{u_n} = \epsilon e^{k_1 x} \sin(k_1 y) \tag{4}$$

As is seen, the maximum velocity of the gas flow is reached near the channel walls, where it is directed along the walls. At the channel axis the velocity is a minimum and has only the longitudinal component. At intermediate points of the channel cross-section the gas overflows from the axis to its periphery.

Knowing the flow field before the flame front, one can find the shape of its surface. For this, one should use the fact that the normal velocity component at the front must be equal to the normal flame velocity. In the simplest case, when the normal velocity of the flame is constant at all points of the surface, we obtain:

$$u \sin \theta - v \cos \theta = u_n \tag{5}$$

where θ is the angle between the tangent to the flame surface and the channel axis and $ctg\, \theta = dx'/dy$, where $x' = x'(y)$ is the equation of the surface.

After simple transformations of (5) we have:

$$\frac{dx}{dy} = \pm \frac{v u_n - u \sqrt{w^2 - u_n^2}}{u u_n + v \sqrt{w^2 - u_n^2}}, \quad w^2 = u^2 + v^2 \tag{6}$$

with the boundary condition $y = 0$, $x = 0$, where the velocity vector components u and v have the form as shown in Eq. (4). For a cold gas-convex front one should choose the minus sign in Eq. (6).

In extreme cases Eq. (6) can be simplified and integrated. At a U/u_n value that slightly differs from unity, $\epsilon \ll 1$ (the flame shape is close to plane), Eq. (6) yields:

$$x_f \simeq 4\sqrt{\epsilon} \frac{1 - \sqrt{0.5 \frac{\sin(k_1 y)}{\sqrt{1 - \cos(k_1 y)}}}}{k_1} \tag{7}$$

At $\epsilon \gg 1$ (a highly convex flame front), the integral of Eq. (6) is

$$x_f = \frac{\ln[k_1 y / \sin(k_1 y)]}{k_1} \tag{8}$$

For practically important intermediate values of ϵ, one can perform a numerical integration of Eq. (6). Results of numerical calculations of the flame shape in a channel at different values of the ratio of the flame propagation velocity along the channel to the normal flame velocity are given in Fig. 3.

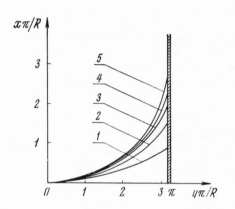

Fig. 3. Numerical calculations of the flame front shape at different values of parameter $\epsilon = U/u_n - 1$ (1: $\epsilon = 0.05$, 2: $\epsilon = 0.15$, 3: $\epsilon = 0.25$, 4: $\epsilon = 0.35$, 5: $\epsilon = 0.45$).

The distribution of pressure ahead of the flame front can be found from Bernoulli's integral. Since at $x = -\infty$ the flow is homogeneous ($u = U$, $v = 0$), and the pressure, $p(-\infty, y) = p_0$, is known then

$$p = p_0 + \rho_0 \frac{U^2 - u^2 - v^2}{2} \quad (9)$$

ρ_0 is cold gas density.

In conclusion of this section it should be noted that the main dimensionless parameters of the problem considered are: the ratio of the propagation flame velocity along the channel axis to the normal velocity of flame (parameter $\epsilon = U/u_n - 1$), and the degree of thermal gas expansion ($\alpha = T_0/T_b$; where T_0 and T_b are the initial and burned gas temperatures).

The Flow of Combustion Products. Conditions at the Flame Front. Determination of the Flame Propagation Velocity.

The flow of combustion products behind the flame front has a non-zero vorticity. When the gas crosses the flame front, which represents the gas-dynamic discontinuity surface where the velocity, pressure, density and gas temperature are step-like changing, the vorticity of combustion products is generated (Zeldovich, 1944, 1966, 1979, 1980; Borisov, 1978; Zeldovich et al. 1979; Tsien, 1951; and Chernyĭ, 1954).

Starting from the flame front the intensity of the vortices remains constant along each streamline, so that the region filled by combustion products is a rotational one. In some of the previous works mentioned, however, the existence of the stagnation zone behind the flame front has not been accounted for, so that the quantitative conclusions differ essentially from those of the hydrodynamic model presently under consideration. It should be noted that the boundary streamline of the stagnation zone is a tangential velocity component discontinuity surface or a vortex sheet. As a consequence of the rotational nature of the flow of combustion products, instead of Eq. (1) one should solve the following equations for the stream function

$$\rho_b \Delta \psi = -\omega(\psi) \tag{10}$$

and

$$\rho_b u = \partial_y \psi, \quad \rho_b v = -\partial_x \psi \tag{11}$$

where ω is the flow vorticity, which is constant along each streamline and can be calculated from the boundary conditions at the flame front and the cold gas flow field before it (discussed later), ρ_b is the hot gas density.

At the flame front the front-normal velocity component grows step-like due to variation of the gas density in accordance with the mass flow conservation law

$$\rho_0 u_n = \rho_b u_b \tag{12}$$

(u_b is the normal flame velocity with respect to the combustion products).

But the tangential velocity component u_τ is unchanged

$$u_\tau \mid_{-f} = u_\tau \mid_{+f} \tag{13}$$

where the signs "$-$" and "$+$" pertain to the cold gas and the hot combustion products respectively and the index $_f$ means that the values are taken at the flame front.

From the momentum conservation equation for the normal flow velocity component

$$p_{-f} + \rho_0 u_n^2 = p_{+f} + \rho_b u_b^2 \tag{14}$$

one can find the pressure difference at the flame front, which proves to be constant along the entire surface if the normal flame velocity is constant:

$$\Delta p = p_{-f} - p_{+f} = \rho_0 u_n^2 \frac{1-\alpha}{a}, \quad \alpha = \frac{\rho_b}{\rho_0} = \frac{T_0}{T_b} \tag{15}$$

The value of the stream function is the same on both sides of the hydrodynamic discontinuity:

$$\psi_{-f} = \psi_{+f} \tag{16}$$

From Eqs. (12)–(16) one can calculate on the hot side of the flame front

$$P(\psi) = p + \rho_b \frac{u^2 + v^2}{2} \tag{17}$$

and determine

$$\omega(\psi) = -\frac{dP(\psi)}{d\psi} \tag{18}$$

The last formulae can be obtained from the motion equation written in the Lamb form

$$\mathbf{w} \times \omega = \nabla(\frac{p}{\rho} + \frac{w^2}{2}), \quad \mathbf{w} = \mathbf{i}u + \mathbf{j}v \tag{19}$$

if a plane flow geometry and relations (11) are taken into account.

By making substitutions, the following expression for the vorticity distribution along the flame surface can be deduced:

$$\omega_{+f} = 0.5(\alpha - 1)\frac{d(u^2_{-f} + v^2_{-f})}{d\psi_{-f}} \tag{20}$$

Consider the condition, which determines the velocity of the curved flame front propagation in the channel. Inside the stagnation zone filled by combustion products the pressure is constant and is equal to the value at infinity (when $x = \infty$). Because of Bernoulli's integral along the streamline restricting the stagnation zone, the gas motion velocity remains unchanged. Since at $x = \infty$ the flow is plane-parallel ($p_\infty = $ const, $v = 0$), distributions of velocity u and of the stream function ψ_∞ are associated with the vorticity distribution:

$$\rho_b u_\infty(y) = \frac{d\psi}{dy} = -\int_0^y \omega(y)\,dy \tag{21}$$

Consider now the channel cross sections at $x = \infty$ and $x = -\infty$ (see Fig. 2), and write the mass flow conservation integrals as

$$\rho_b \int_0^a u^2_\infty(y)\,dy = \int_0^{\psi a} d\psi = \psi_a = \rho_0 U R \tag{22}$$

and the momentum integrals as:

$$p_\infty R + \rho_b \int_0^a u^2_\infty(y)\,dy = (p_0 + \rho_0 U^2)R \tag{23}$$

where a is a half-width of the cross section filled by the moving combustion products at infinity (in our coordinate system), and ψ_a is an extreme value of the stream function. The latter integral, with allowance for Eq. (22), can be expressed as

$$(p_0 - p_\infty + \rho_0 U^2)R = \int_0^{\psi a} u_\infty(\psi)\,d\psi \tag{24}$$

By writing Bernoulli's integrals along each streamline from the flame front to infinity, one can obtain:

$$u^2_\infty(\psi) = 2\frac{P_{+f}(\psi) - p_\infty}{\alpha \rho_0}, \tag{25}$$

so (24) can be rewritten as

$$(p_0 - p_\infty + \rho_0 U^2)R = \int_0^{\psi a} \sqrt{2\frac{P_{+f}(\psi) - p_\infty}{\alpha \rho_0}}\,d\psi \tag{26}$$

This expression, together with the boundary conditions at the flame front [Eqs. (12)–(16)] and Eq. (6) for the flame surface shape, determines the combustion front propagation velocity U as a function of the normal flame

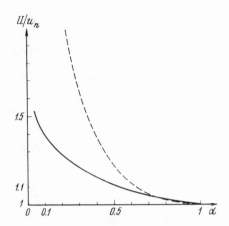

Fig. 4. Flame propagation velocity related to the normal flame velocity U/u_n along the channel axis as a function of the gas expansion in the flame front.
——————- numerical results of presented model.
- - - - - - - results of cellular model of Zeldovich (1966).

velocity u_n and the ratio $\alpha = \rho_b/\rho_0$:

$$U^2 + u_*^2 - 2u_n^2 \frac{1-\alpha}{\alpha} = \frac{2}{\alpha\rho_0} \int_0^{\rho_0 U R} u_n^2(1-\alpha^2) + u_*^2 - \alpha(1-\alpha)w_{-f}^2 \, d\psi \quad (27)$$

where u_* is the cold gas velocity at the contact point of the flame front with the channel wall. It is convenient to determine the solution of Eq. (27) in the following manner. Starting from some value of the parameter $\epsilon = U/u_n - 1$, define the surface shape and u_* by integrating Eq. (6). Then from Eq. (27) determine the corresponding value α.

An approximate analytical solution of the equations at $\epsilon \ll 1$, $1 - \alpha \ll 1$ results in the dependence:

$$\frac{U}{u_n} \simeq 1 + \frac{(1-\alpha)^2}{2} \quad (28)$$

[Note, also, that for any α there exists the trivial solution $u/n_n = 1$ (plane front), which is, however, unstable.] Numerical solutions of the equations are illustrated in Fig. 4.

The thickness of the jet of combustion products at $x = \infty$ is determined from the following integral:

$$a = \int_0^{\psi_0} \frac{1}{\rho_b u_\infty(\psi)} \, d\psi = \frac{1}{\rho_0} \int_0^{\rho_0 U R} \frac{1}{\sqrt{u_n^2(1-\alpha^2) + \alpha u_*^2 - \alpha(1-\alpha)w^2}} \, d\psi \quad (29)$$

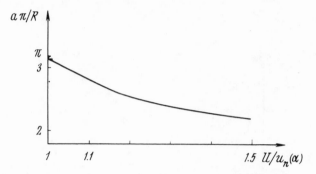

Fig. 5. Jet thickness a at $x = \infty$ versus dimensionless flame propagation velocity U/u_n.

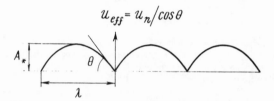

Fig. 6. Model of the cellular flame from the work of Zeldovich (1966).

An approximate solution at $\epsilon \ll 1$ yields α as

$$\alpha \simeq R(1-\epsilon) \simeq R\left[1 - \frac{(1-\alpha)^2}{2}\right] \tag{30}$$

and thickness of the stagnation zone h as

$$h \simeq R\epsilon \simeq R\frac{(1-\alpha)^2}{2} \tag{31}$$

Results of numerical integration of Eq. (29) are represented in Fig. 5. For ordinary flames $\alpha \sim 0.1$–0.2 the flame propagation velocity along the channel is greater than the normal flame velocity by a factor of 1.25–1.40. The stagnation zone in the combustion products occupies about a quarter of the channel cross section.

Comparison of Flame Propagation in a Channel and Cellular Flames
If one considers the disturbances of finite amplitude at the infinite thin flame surface from the point of view of Huygens' principle, then with time there appear angular points, which diminish the flame front surface, i.e.,

with time the disturbances become smoother. By allowing for this and the hydrodynamic instability effect, one can come to the conclusion that the growth of disturbances is restricted by a certain amplitude which is proportional to the wavelength and depends on the thermal expansion of the gas in the flame (Zeldovich, 1966).

By using these results one can derive the dependence of the flame propagation velocity on the degree of gas expansion. The disturbance amplitude (see Fig. 6) satisfies the following equation with negative nonlinearity:

$$\frac{dA}{dt} = \Omega A + \kappa A^2 \tag{32}$$

where $\Omega = k u_n f(\alpha)$ is linear characteristic frequency taken from the Landau dispersion equation, k is the disturbance wave number, $k = 2\pi/\lambda$, and λ is the wavelength, $\kappa = -2k^2 u_n^2/\pi^2$ (see Zeldovich, 1966). Equation (32) at $t \to \infty$ yields the stationary amplitude of the disturbance:

$$A(t \to \infty) = A_* = -\frac{\Omega}{\kappa} = f(\alpha)\frac{\pi^2}{2k} \tag{33}$$

The relation $A_* \sim k^{-1}$ means that the cell shape depends on α only, and not on the wavelength. In order to compare this result with the channel combustion problem, we assume that $k = 1$, and then obtain the amplitude of the curved flame corresponding to the channel width 2π:

$$A_* = f(\alpha)\frac{\pi^2}{2} \tag{34}$$

where $f(\alpha) = \{-1 + [(1-\alpha)^2/\alpha + 1]^{1/2}\}/(1+\alpha)$ is the positive root in the Landau dispersive equation.

In the Zeldovich work (1966) the cell shape was approximated by a parabola. So, to find the propagation velocity dependence on α, it is sufficient to calculate the length L of the parabola arc with the amplitude A_* and wavelength $\lambda = 2\pi/k$ (see Fig. 6):

$$L = 0.5\lambda \left\{ \sqrt{1 + \frac{4A_*}{\lambda}} + \frac{\lambda}{4A_*} \operatorname{arcsh}\frac{4A_*}{\lambda} \right\} \tag{35}$$

and to use the evident relation $L/\lambda = U/u_n$:

$$\frac{U}{u_n} = 1 + \epsilon = 0.5 \left\{ \sqrt{1 + (\pi f(\alpha))^2} + \frac{1}{\pi f(\alpha)} \operatorname{arcsh}(\pi f(\alpha)) \right\} \tag{36}$$

At $\epsilon \ll 1$ and $1 - \alpha \ll 1$, Eqs. (34) and (36) yield

$$A_* = \pi\sqrt{1.5\epsilon} \simeq 3.85\sqrt{\epsilon}$$

$$\epsilon = \pi^2\frac{(1-\alpha)^2}{24} \simeq 0.41(1-\alpha)^2 \tag{37}$$

(Compare these results with $A_* \simeq 4\epsilon^{1/2}$, $\epsilon \simeq (1-\alpha)^2/2$ obtained in the preceding section).

The dependence ϵ *vs* α derived from Eq. (36) is plotted on Fig. 4 as a dotted line. Solutions of Eq. (36) and of Eq. (27) show good agreement only at small values ($\epsilon < 0.05$) as would be expected, because the assumptions used in Eq. (32) are valid only at sufficiently small amplitudes $A_*/\lambda \ll 1$. Note that the changes of the gas motion before and after the curved front were considered indirectly through value $f(\alpha)$ taken from the Landau theory for a small disturbance.

Correction to the Flow Field Ahead of Flame Front Propagating in a Channel
The results of analytical and numerical calculations obtained in the previous section, and graphically depicted in Figs. 4 and 5 were derived, as noted above, when only one term was taken from the whole series of Eq. (2).

Now consider a contribution of the following terms to the series of Eq. (2). From the conditions at the channel wall and axis it follows that $k_n = nk_1$. Consequently, neither of the harmonics, excluding the first one, can be realized singly. For instance, for $n = 2$ the value of velocities $u_* = (u_*^2 + v_*^2)^{1/2}$ at the extreme point of the flame front at the wall, turns out to be less than u_n, that is inadmissible. However, to the first harmonic one can add the others separately with a small "weight" β, in order to fulfill the conditions at all frontal points.

$$\frac{u}{u_n} = 1 + \epsilon[1 - (1 - \beta)\exp(k_1 x)\cos(k_1 y) - \beta\exp(nk_1 x)\cos(nk_1 y)],$$

$$\frac{v}{u_n} = \epsilon[(1 - \beta)\exp(k_1 x)\sin(k_1 y) + \beta\exp(nk_1 x)\sin(nk_1 y)] \qquad (38)$$

Such a "mixing in" of separate terms of the series, as shown by the calculations, insignificantly affects the shape of the flame surface, with the exception of the region adjacent to the channel wall (see Fig. 7), but changes greatly the stagnation zone thickness and the flame propagation velocity. By calculating again the flame surface shape from Eq. (6) for each β and chosen harmonic number, and by solving the integral Eq. (27), one obtains for each β the dependence $\epsilon(\alpha)$. These dependencies are represented in Fig. 8. For each value β, for the harmonic number n and ratio α, there exists a value at which the solution vanishes. This value corresponds to the case when the flame front touches the channel wall at right angles, and the stagnation zone disappears. The numeric calculation shows that at the given value α, the critical value ϵ_{max} is independent of the number of the added harmonic. The dependence $\epsilon_{max}(\alpha)$ is represented in Fig. 8 by a dotted line. The dependence of the stagnation zone thickness on α at different values of β is shown in Fig. 9.

A similar procedure can be applied with the negative weight of harmonic added. In this case a critical value ϵ_{min} corresponds to the minimum of the propagation velocity when the jet thickness at $x = \infty$ tends to zero, and the stagnation zone occupies the whole channel cross section. One may suggest

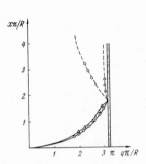

Fig. 7. Differences in the flame shape due to the added first and second harmonics with weight β. $\bigcirc \beta = 0$; $\triangle \beta = 0.1$, $n = 2$; $\epsilon = U/u_n - 1 = 0.2$.

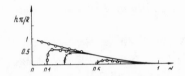

Fig. 9. Dependence of the stagnation zone thickness on α at different values of β ($\bigcirc \beta = 0$; $\square \beta = 0.03$; $\times \beta = 0.05$; $\triangle \beta = .01$).

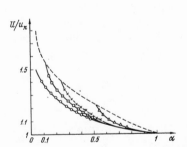

Fig. 8. Flame propagation velocity as a function of α at different values of β (which is the "weight" of second harmonic ($n = 2$)). ($\bigcirc \beta = 0$; $\square \beta = 0.03$; $\times \beta = 0.05$;; $\triangle \beta = 0.1$).

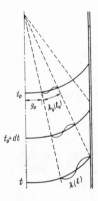

Fig. 10. Evolution of disturbances along the flame surface with time (t)—Wentzel-Kramers-Brillouin approximation.

that the real values of the propagation velocity at the given u_n and α lie between the extreme curves corresponding to the maximum and minimum of the propagation velocity.

The Stability of the Convex Flame to Small Disturbances

As noted above, the convex flame front in the channel can be considered as one cell arising due to the plane flame hydrodynamic instability up to the limit determined by the nonlinear stabilization. But for such a flame one may investigate the stability of the curved shape to a disturbance with a smaller wavelength, that is, with higher increment of growth. Suppose that at the flame surface one may locate many such waves. We will follow their evolution by using an approximate approach, which is an analog of the geometrical optics method or of the method by Wentzel-Kramers-Brillouin used in quantum mechanics (for details see the book by Landau and Lifshitz, 1954). In the inhomogeneous gas flow with the oblique curved flame surface one can consider the trajectories (beams) of the propagation of disturbances (see Fig. 10). They are moving along the channel together with the flame propagation and approach the channel walls due to the tangential component of velocity at the flame front. Because of the variation of the tangential velocity the disturbance wavelength does not remain constant; that is, there exists an effect of the stretching of the initial disturbance.

Thus, consider the disturbance of the wavelength λ_0 located not far from the channel axis at $y = y_0$ and determine how it would grow up until the moment of contact with the wall. Let A_0 be an initial amplitude of the disturbance, and A_w the finite amplitude near the wall. The shortwave perturbations are assumed to be small in order to use the linear approach. Since in accordance with Landau theory

$$\frac{dA}{dt} = \Omega A, \quad dl = u_\tau \, dt \tag{39}$$

(l is the coordinate along the flame surface), then the maximum A_w reached at the wall is expressed through the initial A_0 as

$$\frac{A_w}{A_0} = e^\Gamma, \quad \Gamma = \int_{l_0(y_0)}^{L(R)} \frac{\omega}{u_\tau} \, dl \tag{40}$$

In the approximation of the Wentzel-Kramers-Brillouin method, we take $\Omega = k(l)u_n f(\alpha)$ from the Landau theory on the hydrodynamic instability of the plane flame. The relative change in the disturbance because of the stretch-effect is equal to that in the tangential velocity component

$$\frac{d\lambda}{\lambda} = \frac{du_\tau}{u_\tau} \tag{41}$$

Taking into account Eq. (41) and $k(l) = 2\pi/\lambda$ (1) we obtain for Γ:

$$\Gamma = 2\pi f(\alpha) \frac{u_n u_{\tau 0}}{\lambda_0} \int_{l_0(y_0)}^{L(R)} \frac{1}{[u_\tau(l)]^2} \, dl \tag{42}$$

It should be noted here that $\Gamma \to \infty$ and, consequently, $A_w \to \infty$ if we take $l_0 = 0$, because the tangential velocity component $u_\tau = 0$ at the channel axis. The real disturbances, however, have finite dimensions and cannot be placed exactly at the leading point of the flame front. The disturbances with wavelength smaller than the thermal thickness of the flame decay in time and do not provoke the flame instability. The most rapidly growing disturbances, following Markstein (1951) theory, have a wavelength $\lambda_* = 2\kappa/[u_n(1 - \alpha)]$ (κ is the thermal diffusivity); and for moderately large Reynolds numbers of the flame front propagation along the channel (Re of the order of hundreds), they are much smaller than the channel width.

For small l_0 one may write

$$\psi(y_0) \simeq [(1 + \epsilon)y_0 - \epsilon \sin(k_1 y_0)]\rho_0 u_n \simeq \rho_0 u_n y_0 \tag{43}$$

In order to look for the most dangerous disturbance let us also take $y_0 = 4\pi\kappa/u_n(1 - \alpha)$. The integration of Eq. (42) at $\epsilon \ll 1$ yields

$$\Gamma = \epsilon^{1/2} \mathrm{Re} \frac{y_0}{4\pi} \left[\epsilon\pi \left(1 - \frac{y_0}{R} \right) + \mathrm{ctg} \frac{y_0 \pi}{2R} \right] \tag{44}$$

which, under the condition $y_0/R \ll 1$ approximately equals

$$\Gamma = (\epsilon)^{1/2} \frac{\mathrm{Re}}{2\pi} \tag{45}$$

where the Reynolds number $\mathrm{Re} = u_n R/\kappa$ is introduced. The results of the numerical integration of Eq. (42) at different ϵ, and at the corresponding values α (see previous sections) are represented in Fig. 11. The main part of the integral Γ is gained near the lower integration boundary. In other words, the disturbances grow significantly near the channel axis, but not at the channel wall.

Let us estimate the increase of the disturbance amplitude at the channel wall with respect to the initial value taken from the thermodynamic fluctuation theory. This theory predicts the initial disturbance of the order of $\lambda_0/(N)^{1/2}$, where N is the number of gas molecules in the volume λ_0^3. Therefore, the condition for reaching a noticeable value of the disturbance amplitude leads to the relation

$$\ln(N) = 2\Gamma \tag{46}$$

which yields, by using Eq. (45), the following critical Reynolds number Re_*

$$\mathrm{Re}_* = \pi(\epsilon)^{-1/2} \ln N \tag{47}$$

If we take $\lambda_0 = \kappa/u_n \sim 10^{-2}$ cm and $N \sim 5.3 \times 10^7$, $(\kappa/u_n(1 - \alpha))^3 \sim 5.3 \times 10^{16}(1 - \alpha)^{-3}$ of molecules, an estimate of the critical Reynolds number at $\epsilon \sim 0.1$ is equal to $\mathrm{Re}_* \sim 380$. The critical Reynolds number will be

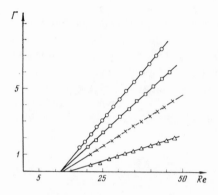

Fig. 11. Numerical calculations of the integral Γ as the function of Reynolds number Re at different values of the flame propagation velocity along the channel axis U/u_n ($\bigcirc U/u_n = 1.05$, $\alpha = 0.675$; $\square U/u_n = 1.15$, $\alpha = 0.4$; $\times U/n_n = 1.25$, $\alpha = 0.23$; $\triangle U/u_n = 1.35$, $\alpha = 0.12$).

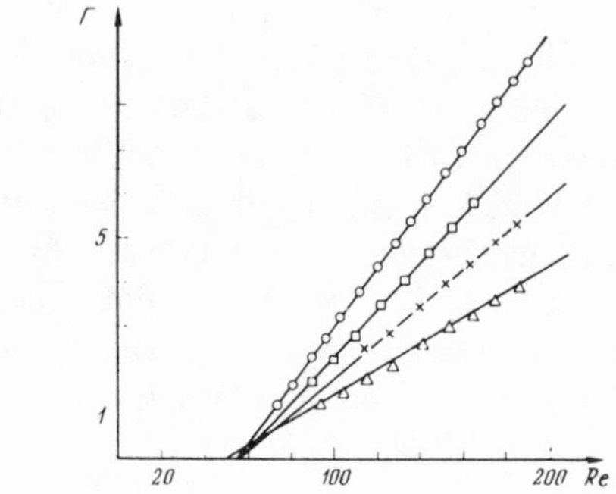

Fig. 12. Second approximation of calculations of the integral Γ ($\bigcirc U/u_n = 1.05$, $\alpha = 0.675$; $\square U/u_n = 1.15$, $\alpha = 0.4$; $\times U/n_n = 1.25$, $\alpha = 0.23$; $\triangle U/u_n = 1.35$, $\alpha = 0.12$).

greater if we take into account the dependence of the characteristic frequency ω on the flame front curvature following the Markstein theory (1951).[2] The solution of the plane flame instability problem considering the influence of the transfer processes (heat conduction, diffusion and viscosity) by means of the matched expansion series approach leads to the following expression for Ω (Istratov and Librovich, 1966).

$$\Omega = k(l)u_n f(\alpha)[1 - k(l)g(\alpha)\frac{\kappa}{u_n}] \tag{48}$$

The second correction term in brackets is important for disturbance wavelengths comparable with the thermal width of the flame, which now are under consideration. For Lewis and Peclet numbers equal unity the function $g(\alpha)$ has such a form

$$g(\alpha) = \frac{1 - \alpha - [1 + \alpha + 2\alpha f(\alpha)][1 + \ln \alpha/(1 - \alpha)]}{1 - \alpha - \alpha f(\alpha)} \tag{49}$$

Let us take again the most dangerous disturbances growing most rapidly, i.e. $-k_0 = 0.5u_n/[\kappa g(\alpha)]$, and by substituting Eqs. (48) and (49) into the

---

[2]Markstein's approach is the next approximation to the Landau theory accounting for the disturbance wavelengths comparable with the flame front thickness. It should also be noted that stretching of the flow field produces an additional effect of decreasing of the disturbance amplitude in the time that is related to the increase of the wavelength.

expression for Γ (42) we obtain:

$$\Gamma = 2f(\alpha)\frac{u_n u_{\tau 0}}{\lambda_0} \int_{\lambda_0}^{L} \frac{1}{u_\tau^2}\left[1 - \frac{2\pi u_{\tau 0}g(\alpha)\kappa}{u_n \lambda_0 u_\tau}\right] dl \qquad (50)$$

AT $\epsilon \ll 1$, Eq. (50) yields

$$\Gamma \simeq 3/[32\pi(2)^{1/2}][1 - (2\epsilon)^{1/2}/6]\mathrm{Re}. \qquad (51)$$

The results of the numerical integration of Eq. (50) are represented in Fig. 12.

Condition $A_w \sim \lambda_0$ yields the critical Reynolds number

$$\mathrm{Re}_* \simeq 16(2)^{1/2}(\pi/3)\ln N, \qquad (52)$$

which at the same typical parameters, as those used above, equals ~ 900.

DISCUSSION

An approximate analysis of the flame propagation in the channel performed in this paper includes the consideration of the inviscid gas flow fields before and after the curved flame front considered as a hydrodynamic discontinuity. The refraction of streamlines at the flame front demands the introduction of the stagnation zones in the combustion products where the gas is at rest with respect to the flame. The boundary between the combustion products flow and the stagnation zone, being the surface of the tangential velocity discontinuity, is unstable. The appearing small disturbances are increasing, and finally form a turbulent tail. Therefore, in reality, an ideal stagnation zone, taken into the analysis and shown in Fig. 2, does not exist. However, if the disturbances are not growing very rapidly and they have time to be carried away by the flow, then the considered scheme would be justified and would yield the correct results. Besides, the turbulent mixing is accompanied by momentum conservation, and therefore, it does not upset the equation used in the determination of the flame propagation velocity.

Processes that have not been taken into account in the scheme considered were the heat removal from the flame to the channel walls, and the influence of the gas viscosity. Not that one can provide the combustion conditions when these effects may be neglected (high Reynolds numbers). And, *vice versa*, there exist such regimes when these effects determine the whole process of the flame propagation.

In the scheme considered, as shown above, the convex flame front affects the hydrodynamics of the gas flow, and forms some velocity distribution ahead of it. This is associated with the pressure difference at the flame front. In other words, it is always necessary to solve a conjugate problem on the front propagation and the gas motion. Restricting the analysis by the first term in the series describing the flow field before the flame and taking into account the corresponding shape of the flame front, as was shown,

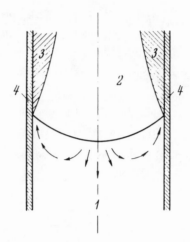

Fig. 13. Flow field before the flame front in the laboratory coordinate system. (Region 1—potential flow, region 2—vortex flow, region 3—stagnation zone, region 4—channel wall).

one can look for the flame propagation velocity form the integral mass and momentum conservation relations taken far from the flame in the cold and hot gases. More detailed description of the cold gas field demands, in its turn, the detailed consideration of the rotational flow of combustion products behind the flame, in order to find a relation for the flame propagation velocity.[3]

The velocity distribution ahead of the flame front in the room coordinate system is shown in Fig. 13. As seen from the figure, the gas motion is accompanied by the rotation and displacement of fluid elements, since the velocities near the axis and walls are directed in opposite directions. The rotation centers of each of the fluid elements are behind the flame front. But the flow is potential and $\omega = \text{curl } \mathbf{u} = 0$. In the coordinate system of the flame front, the velocities near the axis and walls have the same direction (see Fig. 2).

By describing the flame as a gasdynamic discontinuity, one neglects the thermodiffusional flame structure which determines the normal velocity of its propagation. Such an approach supposes that the normal flame velocity is an *a priori* given parameter or taken from experiment.[4] The stability

[3]It should be kept in mind that the direct application of numerical methods is complicated by the Landau instability effect.

[4]To take into account the stretch effect on the flame propagation rate, the consecutive theory needs the treatment of the flame structure. The phenomenological Markstein (1951) theory is the first in this direction.

to the shortwave disturbances also depends on the flame structure. Actually, in the analysis of the flame propagation stability up to the moderately high Reynolds numbers performed in the present paper, the structure of the flame has been taken into consideration by selecting the most dangerous wavelength of the small disturbances.

In conclusion we would like to mention that in order to verify the assumptions and conclusions made in this paper, it would be useful to perform a set of experiments on flame propagation at zero gravity with the visualization of flow in channels and tubes and with artificially created local disturbance. Similar experiments seem to be able to verify the basic theoretical conclusions of the present paper under appropriate conditions because the buoyancy usually affects the hydrodynamics of flow during the combustion and the flame front stability in the essential way.

APPENDIX

The Case of Cylindrical Symmetry

The conclusions drawn in previous sections and the method for obtaining the integral relations determining the flame propagation velocity along the plane channel are applicable to the case of the cylindric tubes also. The difference lies in the details of the formulas and in the quantitative results.

Equation (10) in cylindrical coordinates, allowing that the solution is bounded at the tube axis, can be solved as a series of Bessel's functions of the integer order

$$\phi = Uz + \sum_{t=0}^{\infty} A_i \exp(k_i z) J_t(k_i r) \qquad (53)$$

(r and z are the cylindrical coordinates).

By restricting ourselves to considering the first term in the series, and by satisfying the conditions at the tube walls and at the flame leading point

$$r = R, \quad v = 0,$$
$$r = 0, \quad z = 0, \quad u = u_n \qquad (54)$$

we obtain $k_0 = \lambda_1/R$, $A_0 = (u_n - U)/k_0$, λ_1 is the first zero of the Bessel's function $J_1(x)$. Introducing for simplicity the dimensionless values $u' = u/u_n$, $r' = k_0 r$, $z' = k_0 z$ and omitting the primes, we have

$$u = 1 + \epsilon[1 - \exp(z)J_0(r)],$$
$$v = \epsilon \exp(z)J_1(r),$$
$$\psi = (1 = \epsilon)r^2/2 - \epsilon \exp(z)r J_1(r) \qquad (55)$$

The equation for the flame surface takes the form

$$dz/dr = [-u + v(u^2 + v^2 - 1)^{1/2}]/[v + u(j^2 + v^2 - 1)], \qquad r = 0, \quad z = 0 \ (56)$$

The integrals of the mass flow and of the axial impulse component conservation yield:

$$2\alpha \int_0^a ru(\infty, r)\, dr = \lambda_1^2(1 + \epsilon) = 2\psi_a \tag{57}$$

$$\lambda_1^2[p_0 - p_\infty + (1 + \epsilon)^2] = 2\alpha \int_0^a ru^2(\infty, r)\, dr = 2 \int_0^{\psi a} \sqrt{2\frac{P_{+f}(\psi) - p_\infty}{\alpha}}\, d\psi \tag{58}$$

The latter, after considering Bernoulli's integral, is transformed into

$$\frac{(1 + \epsilon)^2}{2} + \frac{u_*^2}{2} - \frac{1 - \alpha}{\alpha} = \frac{2}{\alpha\lambda_1^2} \int_0^{\psi a} \{1 - \alpha^2 u_*^2 - \alpha(1 - \alpha)w_{?=f}^2\}^{1/2}\, d\psi \tag{59}$$

The integral Eq. (59) together with the surface shape Eq. (56) would determine the flame propagation velocity along the tube axis as a function of the normal flame velocity and the density ratio α. The width of the stagnation zone could be derived by integration

$$h = \lambda_1 - \{2 \int_0^{\psi a} \frac{1}{\alpha u_\infty(\psi)}\, d\psi\}^{1/2} \tag{60}$$

The performed numerical calculations for the case of cylindrical symmetry yielded the same qualitative dependence of the flame surface shape, the propagation velocity, and thickness of the stagnation zone, as in the case of the plane channel. The quantitative results for the cylindrical tube are somewhat different, e.g., the dimensionless propagation velocity along the tube axis at a real α proves to be 50 percent higher than in the case of plane symmetry.

USSR Academy of Sciences   *Received*
                                              April 17, 1980

References

1. *Aldushin A. P., et al.* (1980). Flame propagation in the reacting gaseous mixture. *Archivum combustionis*, **1**.
2. *Borisov V. I.* (1978). On the velocity of uniform flame propagation in a plane channel. *FGV* **14**, 26 (in Russian).
3. *Chernyĭ G. G.* (1954). Gas flow in a tube in the presence of the flame front. *Theoretical Hydromechanics, Moscow*, **4**, 31 (in Russian).
4. *Darrieus G.* (1944). Propagation d'un front de flamme. Essai de theorie de vitesses anomales de deflagration par development spontane de la turbulence. *Z. Phys. Chem.*, **88**, 641.
5. *Istratov A. G., Librovich V. B.* (1966). On the influence of transfer processes on the plane flame front stability. *PMM*, **30**, 451 (in Russian).

6. *Istratov A. G., Librovich V. B.* (1969). On the stability of gasdynamic discontinuities associated with chemical reactions. The case of a spherical flame. *Astr. Acta*, **14**, 453.

7. *Istratov A. G., Gremyachkin N. M.* (1972). On the stability of plane flame in the flow with the velocity gradient. *Combustion and Explosion, Nauka, Moscow*, pp. 305–308 (in Russian).

8. *Karlovitz B., et al.* (1953). A. Flame propagation across velocity gradients. B. Turbulence measurement in flames. *Fourth Symposium (International) on Combustion*, Williams and Wilkins, Baltimore, p. 613.

9. *Landau L. D.* (1944). On the theory of slow combustion. *Acta Phys.-Chem. URSS*, **19**, 77.

10. *Landau L. D., Lifshits E. M.* (1954). *Flow Mechanics*. Gostekhizdat, Moscow (in Russian).

11. *Markstein G.* (1951). Experimental and theoretical studies of flame front stability. *J. Aero. Sci.*, **18**, 199.

12. *Maxworthy T.* (1962). Flame propagation in tubes. *Phys. Fl.*, **5**, 407.

13. *Rozlovskiĭ A. I., Zeldovich Ya. B.* (1954). About the conditions for the spontaneous instability of normal burning. *DAN USSR*, **57**, 365 (in Russian).

14. *Shchelkin K. I., et al.* (1947). Measurements of normal velocities of acetylene-oxygen mixture flame. *ZhETF*, **17**, 1397 (in Russian).

15. *Tsien H. S.* (1951). Influence of flame front on the flow field. *J. Appl. Mech.*, **18**, 188.

16. *Uberoi M. S.* (1959). Flow field of a flame in a channel. *Phys. Fl.*, **2**, 72.

17. *Zeldovich Ya. B.* (1944). Comments on combustion of a rapid flow in a tube. *J. Tekhn. Phys.*, **14**, 162 (in Russian).

18. *Zeldovich Ya. B.* (1966). About some effects stabilizing the curved front of laminar flame. *PMTF*, **1**, 102 (in Russian).

19. *Zeldovich Ya. B.* (1979). Structure and stability of the stationary laminar flame at moderately large Reynolds numbers. Preprint of the Inst. of Chemical Physics, the USSR Acad. of Sci., Chernogolovka.

20. *Zeldovich Ya. B., et al.* (1979). Hydrodynamics of vortical flow behind the flame front propagating in tubes. *VIth International Symposium on Combustion Processes*. Abstracts. Karpacz, 83.

21. *Zeldovich Ya. B.* (1980). Structure and stability of steady laminar flame at moderately large Reynolds numbers. *Combustion and Flame* (in press).